ESSENTIALS OF PRECALCULUS MATHEMATICS

Dennis T. Christy

NASSAU COMMUNITY COLLEGE

HARPER & ROW PUBLISHERS
New York, Hagerstown, San Francisco, London

To Margaret Ellen,
who made my project
her project

Sponsoring Editor: George J. Telecki
Project Editor: Cynthia Hausdorff
Designer: Rita Naughton
Production Supervisor: Will C. Jomarrón
Compositor: Syntax International
Printer and Binder: Halliday Lithograph Corporation
Art Studio: J & R Technical Services Inc.

Essentials of Precalculus Mathematics

Library of Congress Cataloging in Publication Data
    Christy, Dennis T   1947–
        Essentials of precalculus mathematics.

    Includes index.
    1. Mathematics—1961–   I. Title.
QA39.2.C5   512′.1   76-10244
ISBN 0-06-041292-5

Contents

Preface   vii

PART ONE   ALGEBRA WITH AN INTRODUCTION TO TRIGONOMETRY   1

Chapter 1: Basic Concepts and Operations   3

1–1 Number Systems   3
1–2 Properties of Real Numbers   8
1–3 Operations with Real Numbers   12
1–4 Addition and Subtraction of Algebraic Expressions   9
1–5 Word Problems and Equations   21
1–6 Formulas   28
1–7 Ratio and Proportion   31
1–8 Inequalities   36
Sample Test: Chapter 1   45

Chapter 2: Functions and Graphs   47

2–1 Defining Functions   47
2–2 Variation   55
2–3 Functional Notation   60
2–4 Graphs of Functions   63
2–5 Distance Between Points   78
Sample Test: Chapter 2   83

Chapter 3: The Trigonometric Functions   85

3–1 Defining the Trigonometric Functions   85
3–2 Trigonometric Functions of Acute Angles   93
3–3 Right Triangles   101
3–4 Vectors   113
3–5 Special Angles and Cofunctions   121
3–6 Reducing Functions of Angles in Any Quadrant   127
3–7 Trigonometric Equations   134
3–8 Law of Sines   140
3–9 Law of Cosines   149
Sample Test: Chapter 3   156

Chapter 4: Products and Factoring   158

4–1 Laws of Exponents   158
4–2 Products of Algebraic Expressions   164
4–3 Binomial Theorem   171
4–4 Factoring   173
4–5 Solving Quadratic Equations by Factoring   182
Sample Test: Chapter 4

Chapter 5: Fractions   191

5–1 Algebraic Fractions   191
5–2 Multiplication and Division   198
5–3 Least Common Multiple   204
5–4 Addition and Subtraction   208
5–5 Complex Fractions   216
5–6 Equations That Contain Fractions   219
Sample Test: Chapter 5   227

Chapter 6: Exponents and Radicals   229

6–1 Zero and Negative Exponents   229
6–2 Scientific Notation   233
6–3 Fractional Exponents   235
6–4 Simplifying Radicals   240
6–5 Addition and Subtraction of Radicals   243
6–6 Multiplication of Radicals   244
6–7 Division of Radicals   247
6–8 Radical Equations   252
6–9 Complex Numbers   255
6–10 Trigonometric Form and De Moivre's Theorem   259
Sample Test: Chapter 6   264

PART TWO   ELEMENTARY FUNCTIONS   265

Chapter 7: Functions Revisited   267

7–1 Functions   267
7–2 Functional Notation   277
7–3 Operations with Functions   281
Sample Test: Chapter 7   286

Chapter 8: Topics Concerning Polynomials   288

8–1 Slope and Mathematical Change   288
8–2 Linear Functions   293

8–3 Parallel and Perpendicular Lines   299
8–4 Systems of Linear Functions   303
8–5 Determinants   311
8–6 Quadratic Functions   317
8–7 Quadratic Formula   327
8–8 Quadratic Inequalities   334
8–9 Polynomials   338
8–10 Rational Zeros of Polynomial Functions   343
8–11 Rational Functions   348
Sample Test: Chapter 8   355

Chapter 9: Exponential and Logarithmic Functions   357

9–1 Exponential Functions   357
9–2 More Applications and the Number *e*   361
9–3 Inverse Functions   369
9–4 Logarithmic Functions   373
9–5 Properties of Logarithms   376
9–6 Logarithms to the Base 10   379
9–7 Logarithms to Bases Other Than 10   387
9–8 Graphs on Logarithmic Paper   389
Sample Test: Chapter 9   395

Chapter 10: Trigonometric Functions of Real Numbers   397

10–1 Radians   397
10–2 Trigonometric Functions of Real Numbers   403
10–3 Evaluating Trigonometric Functions   413
10–4 Graphs of Sine and Cosine Functions   419
10–5 Graphs of the Other Trigonometric Functions   431
10–6 Trigonometric Equations   434
10–7 Inverse Trigonometric Functions   439
10–8 Trigonometric Identities   443
10–9 More on Trigonometric Identities   450
Sample Test: Chapter 10   455

Chapter 11: Conic Sections   457

11–1 Introduction   457
11–2 The Circle   460
11–3 The Ellipse   463
11–4 The Hyperbola   470
11–5 The Parabola   477
11–6 Conic Sections   483
Sample Test: Chapter 11   485

Appendixes   487

   A–1 Approximate Numbers   489
   A–2 Operations with Approximate Numbers   492
   A–3 Geometric Formulas   497
   A–4 Trigonometric Identities   499

Tables   501

   Table 1 Trigonometric Functions of Angles   503
   Table 2 Exponential Functions   508
   Table 3 Common Logarithms   509
   Table 4 Natural Logarithms (Base e)   511
   Table 5 Trigonometric Functions of Real Numbers   513
   Table 6 Squares, Square Roots, and Prime Factors   516
   Table 7 Pascal's Triangle   517

Answers to Odd-Numbered Problems   519

Index 565

Preface

This book is intended for students who need a concrete approach to mathematics. Set theory is omitted and notation is kept to a minimum. The text stresses problem solving and includes many applications that indicate the need for analyzing concepts with mathematical tools.

My experience has been that college students who need precalculus mathematics learn best by "doing." Examples and exercises are crucial. The text contains brief, precisely formulated paragraphs followed by many detailed examples. Each section has enough exercises for both in-class practice and homework. The problem sets are carefully graded and contain an unusual number of routine manipulative problems. There is nothing more frustrating than being stuck on the beginning exercises! So that each student may be challenged fairly difficult problems have been included. Above all, there is a wide variety of questions and discussions that show the student how mathematics helps us to master our world.

All the essential precalculus topics are covered, and there is adequate material for a two-semester course that meets five hours per week. As indicated in the table of contents, the first semester is an algebra course with an introduction to trigonometry. Students enrolled in engineering or technology programs require this early discussion of trigonometry. The student who starts at Chapter 1 should have a background that includes algebra and geometry. However, only a knowledge of arithmetic is taken for granted in the presentation.

The second semester is basically the "Math O" course recommended by CUPM. ("A Transfer Curriculum in Mathematics for Two Year Colleges," The Committee on the Undergraduate Program in Mathematics of the Mathematical Association of America, Berkeley, California, 1969.) The function concept plays a unifying role in the study of polynomial, rational, exponential, logarithmic, and circular functions. Analytical geometry is integrated throughout the text, and Chapter 11 discusses the conic sections. After this course the student can proceed directly to the study of calculus.

Necessarily, all the material needed for a course in college algebra and trigonometry may be found in the book. The topics appear in about the right order for such an option, and it should be easy to design a course that satisfies individual needs. Briefly, the text progresses as follows:

Chapter 1 reviews the essentials of elementary algebra.

Chapter 2 gives an early introduction to the function concept. Although the *ordered-pair* definition is discussed briefly, the *rule* or *machine* definition of a function is emphasized.

Chapter 3 covers the trigonometric functions defined on the degree measures of angles. The material is introduced through a general approach and proceeds to the traditional topics of triangle trigonometry.

Chapters 4, 5, and 6 deal with the standard topics from intermediate algebra. Trigonometric expressions are carried along simultaneously with the work in algebra to provide a smooth transition from work with algebraic expressions to work with trigonometric expressions.

Chapter 7 repeats some of the material from Chapter 2. The exercises are more difficult and a section on the arithmetic of functions is added. The function concept is so vital that it merits coverage in both semesters. Starting with Chapter 7 there is an effort to raise the level of the course and expect more of the student.

Chapter 8 covers topics concerning polynomials. Graphing of higher degree polynomials is deferred to calculus, and irrational zeros are left for a course in numerical analysis.

Chapter 9 discusses exponential and logarithmic functions. The idea of an inverse function is used and there is a special section on the number e.

Chapter 10 covers circular functions, although that term is not used. I prefer calling such functions "the trigonometric functions defined on real numbers." It is important to note that the connection between angles and real numbers is made through radians at the beginning of the chapter. Customarily, the relationship is not shown until after the discussion of circular functions, which often causes unnecessary confusion. Chapters 3 and 10 are written as independent units.

Chapter 11 discusses the conic sections as collections of points in the plane that satisfy particular geometric conditions.

The Appendix discusses computations with approximate numbers. For reference purposes, a list of geometric and trigonometric formulas is provided.

The preparation of this text has been a great learning experience for me and I am indebted to the many people who gave so generously of their time and effort. My editor, Professor Frank Kocher of Pennsylvania State University, deserves recognition for helping with the difficult job of making the mathematics precise without making the style and material too difficult for the intended audience. I am grateful also to all my colleagues at Nassau Community College and in particular to Professors James Baldwin, Bob Rosenfeld, John Schreiber, Michael Totoro, and Gene Zirkel for their support and constructive criticism during the extensive class testing of the manuscript. Special recognition is due Professor Abraham Weinstein, our chairman, for his unwavering support. Class testing with preliminary editions of a manuscript can lead to many administrative problems. In each difficulty Professor Weinstein provided the help and leadership that were necessary to overcome the problem. To my parents a special thank you. They have always enthusiastically supported my efforts and this undertaking was no exception. Finally, but most important, I thank my wife Margaret, who typed and proofread with me the entire manuscript. She embraced this endeavor in every respect, giving that special help only she could provide.

Dennis T. Christy

PART ONE

Algebra with an introduction to trigonometry

Chapter 1
Basic concepts and operations

1–1 NUMBER SYSTEMS

Mathematics is a basic tool in analyzing concepts in every field of human endeavor. In fact, the primary reason you have studied this subject for at least a decade is that mathematics is the most powerful instrument available to man in the search to understand the world and to control it. Mathematics is essential for full comprehension of technological and scientific advances, economic policies and business decisions, and the complexities of social and psychological issues. At the heart of this mathematics is algebra. Calculus, statistics, and computer science are but a few of the areas in which a knowledge of algebraic concepts and manipulations is necessary.

If your experience with algebra has been limited or unfavorable, you may hesitate to devote yourself to material that seems so abstract and technical. Nevertheless, if you fail to grasp at least the rudiments of this language, you will find the development and application of important concepts difficult to comprehend. Remember, you learn mathematics by doing mathematics, not by watching your teacher do it.

Algebra is a generalization of arithmetic. In arithmetic we work with specific numbers, such as 5. In algebra we study numerical relations in a more general way by using symbols, such as x, which may be replaced by a number from some collection of numbers. Since the symbols represent numbers, they behave according to the same rules that numbers must follow. Consequently, instead of studying specific numbers, we study symbolic representations of numbers and try to define the laws that govern them.

We begin our study of algebra by considering the types of numbers that are needed in a technological society.

1. Our most basic need is for the numbers that are used in counting: 1, 2, 3, ... ("..." means "and so on"). These numbers stand for whole quantities and are the first numbers that we learn.

EXAMPLE 1: When 500 students were sampled, 417 felt that the F grade should be abolished.

2. Our practical need to make precise measurements of quantities such as length, weight, and time makes the concept of fractions and decimals familiar to all of us.

EXAMPLE 2: The width of the room is $9\frac{1}{2}$ ft. The winning time in the race is 45.9 seconds.

3. Positive and negative numbers are needed to designate direction or to indicate whether a result is above or below some reference point.

EXAMPLE 3: In physics, the velocity of an object indicates both its speed and direction. A rocket traveling at a speed of 100 ft/second has a velocity of $+100$ ft/second when it is rising and -100 ft/second when it is falling.

EXAMPLE 4: In statistics, there is a rating system that assigns a positive rating to a score above the mean (average) and a negative rating to a score below the mean.

4. In technical work, we need special numbers, such as $\sqrt{2}$, π (pi), and $\sqrt{-1}$. We will discuss these numbers later.

We now formalize our work by giving specific names to various collections of numbers. *The collection of the counting numbers, zero, and the negatives of the counting numbers is called the integers (see Figure 1.1).*

$$\ldots, -5, -4, -3, -2, -1 \qquad 0 \qquad 1, 2, 3, 4, 5, \ldots$$

$$\underbrace{}_{\text{negative integers}} \qquad \underset{\text{zero}}{0} \qquad \underbrace{}_{\text{positive integers}}$$

Figure 1.1 Integers

The collection of fractions with an integer in the top of the fraction (numerator) and a nonzero integer in the bottom of the fraction (denominator) is called the rational numbers. Symbolically, a rational number is a number of the form a/b, where a and b are integers, with b not equal to ($\neq$) zero. The numbers $\sqrt{2}/3$ and $2/\pi$ are fractions.

They are not rational numbers because they are not the quotient of two integers. All integers are rational numbers because we can think of each integer as having a 1 in its denominator. (Example: 4 = 4/1.)

Our definition for rational numbers specified that the denominator cannot be zero. Let us see why.

$$\frac{8}{2} = 4 \text{ is equivalent to saying that } 8 = 4 \cdot 2$$

$$\frac{55}{11} = 5 \text{ is equivalent to saying that } 55 = 5 \cdot 11$$

If 8/0 = a, where a is some rational number, this would mean that 8 = $a \cdot 0$. But $a \cdot 0 = 0$ for any rational number. There is no rational number a such that $a \cdot 0 = 8$. Thus, we say that 8/0 is *undefined*.

Now consider 0/0 = a. This is equivalent to 0 = $a \cdot 0$. But $a \cdot 0 = 0$ for *any* rational number. Thus, not just one number a will solve the equation—any a will. Since 0/0 does not name a particular number, it is also undefined. Consequently, division by zero is undefined in every case. That is why the denominator in a rational number is not zero.

To define our next collection of numbers we now consider the decimal representation of numbers. We may convert rational numbers to decimals by long division. Consider the following examples of repeating decimals. A bar is placed above the portion of the decimal that repeats.

$$\frac{3}{4} = \begin{array}{c} 0.7500\ldots \\ \text{or} \\ 0.75\overline{0} \end{array} \qquad \frac{2}{3} = \begin{array}{c} 0.6666\ldots \\ \text{or} \\ 0.\overline{6} \end{array} \qquad \frac{8}{7} = 1.\overline{142857}$$

$$\begin{array}{r} .75\overline{0} \\ 4\overline{)3.00} \\ \underline{28} \\ 20 \\ \underline{20} \\ 0 \end{array} \qquad \begin{array}{r} .\overline{6} \\ 3\overline{)2.0} \\ \underline{18} \\ 2 \end{array} \qquad \begin{array}{r} 1.\overline{142857} \\ 7\overline{)8.000000} \\ \underline{7} \\ 1\,0 \\ \underline{7} \\ 30 \\ \underline{28} \\ 20 \\ \underline{14} \\ 60 \\ \underline{56} \\ 40 \\ \underline{35} \\ 50 \\ \underline{49} \\ 1 \end{array}$$

The decimals repeat because at some point we must perform the same division and start a cycle. For example, when converting $\frac{8}{7}$, the only possible remainders are 0, 1, 2, 3, 4, 5, and 6. In performing the division we had remainders of 1, 3, 2, 6, 4, and 5. In the next step we must obtain one of these remainders a second time and start a cycle, or obtain 0 as the remainder, which results in repeating zeros. Thus, if a/b is a rational number, it can be written as a repeating decimal. In summary, we may define a rational number either as the quotient of two integers or as a repeating decimal.

There are decimals that do not repeat, and the collection of these numbers is called the irrational numbers.

EXAMPLE 5: The numbers $\sqrt{2}, \sqrt{3}, \sqrt{5}, \sqrt{6}$, and $\sqrt{7}$ are irrational because they have nonrepeating decimal forms. A proof that $\sqrt{2}$ cannot be written as the quotient of two integers (and equivalently as a repeating decimal) is given in Exercise 41.

EXAMPLE 6: The number $\sqrt{4}$ is not irrational because $\sqrt{4} = 2$, which is a rational number. (*Note:* The symbol $\sqrt{\ }$ denotes the positive square root of a number. Thus, $\sqrt{4} \neq -2$. We discuss this concept in detail in Section 6-3.)

EXAMPLE 7: The number π, which represents the ratio between the circumference and the diameter of a circle, is a nonrepeating decimal (irrational number). The fraction $\frac{22}{7}$ is only an approximation for π ($\frac{355}{113}$ is a much better one).

Since an irrational number is a nonrepeating decimal and a rational number is a repeating decimal, there is no number that is both rational and irrational. *The collection of numbers that are either repeating decimals* (*rational numbers*) *or nonrepeating decimals* (*irrational numbers*) *constitutes the real numbers.* Real numbers are used extensively in this text as well as in calculus. All the numbers that we have mentioned (except the square roots of negative numbers, such as $\sqrt{-1}$) are real numbers. Unless it is stated otherwise, you may assume that the symbols in algebra (such as x) may be replaced by any real number. Consequently, the rules that govern real numbers determine our methods of computation in algebra. A graphical illustration of the various collections of numbers is given in Figure 1.2, where we think of a given collection of numbers as being contained in the appropriate rectangular region.

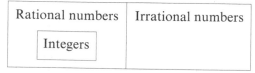

Real numbers

Rational numbers	Irrational numbers
Integers	

Figure 1.2

EXERCISES 1–1

In Exercises 1–20 classify each number by placing a check in the appropriate categories.

	Number	Real number	Rational number	Irrational number	Integer	Positive integer	None of these
Example:	14	✓	✓		✓	✓	
1.	-19	✓	✓		✓		
2.	5						
3.	1						
4.	0						
5.	π						
6.	45.9						
7.	$-25/3$						
8.	$-\pi$						
9.	$\sqrt{7}$						
10.	$\sqrt{9}$						
11.	$\sqrt{-4}$						
12.	$-\sqrt{4}$						
13.	$6/9$						
14.	$\pi/2$						
15.	$1/\sqrt{2}$						
16.	$\sqrt{-2}/3$						
17.	$0.\overline{3}$						
18.	$9.\overline{128}$						
19.	7%						
20.	200%						

In Exercises 21–30 express each rational number as a repeating decimal.

21. $\frac{4}{5}$ **22.** $\frac{5}{4}$

23. $\frac{1}{3}$ **24.** $\frac{2}{9}$

25. $\frac{5}{11}$ **26.** $\frac{5}{12}$

27. $\frac{37}{6}$ **28.** $\frac{26}{11}$

29. $\frac{10}{7}$ **30.** $\frac{100}{99}$

In Exercises 31–40 answer true or false. If false, give a specific counterexample.
31. All rational numbers are integers.
32. All rational numbers are real numbers.

33. All real numbers are irrational numbers.

34. All integers are irrational numbers.

35. The quotient of two integers is always an integer.

36. The quotient of two integers is always a rational number.

37. Every real number is either a rational number or an irrational number.

38. A number that can be written as a repeating decimal is called a rational number.

39. The nonnegative integers are the positive integers.

40. The nonpositive real numbers are the negative real numbers and zero.

41. The following is a proof that $\sqrt{2}$ is not a rational number. Read through it *slowly* and *carefully*. The method employed is that of *indirect proof*, one of the most powerful methods in mathematics. *We prove that $\sqrt{2}$ is not a rational number by showing that the assumption that it is a rational number leads to a contradiction.* Suppose that there is a rational number equal to $\sqrt{2}$. Then it can be represented by a fraction that has been reduced to lowest terms. Let a/b represent this fraction, where a and b are both integers. Since the fraction is in lowest terms, *a and b are not both even integers.* Since the square of this fraction is supposed to equal 2, we can write $(a/b) \cdot (a/b) = 2$. Multiplying both sides by $b \cdot b$, we find that $a \cdot a = 2 \cdot b \cdot b$. This equation tells us that $a \cdot a$ is double the integer $b \cdot b$, so $a \cdot a$ is an even integer. But if $a \cdot a$ is even, *then a must be even*, since the product of two odd numbers is always odd. Thus, a is double some other integer k (that is, $a = 2k$). If we substitute $2k$ for a in the equation $a \cdot a = 2 \cdot b \cdot b$, we get $(2k)(2k) = 2 \cdot b \cdot b$. Dividing both sides by 2, we see that $2 \cdot k \cdot k = b \cdot b$. In other words, $b \cdot b$ is double the integer $k \cdot k$, or $b \cdot b$ is an even integer. But if $b \cdot b$ is even, *then b must be even. We began by observing that a and b are not both even and end up by concluding that they are both even.* We were led to this contradiction by assuming that there is a rational number equal to $\sqrt{2}$. Therefore, we are compelled to reject this assumption.

42. Show that $1 + \sqrt{2}$ is irrational by an indirect proof. (*Hint:* The result of subtracting two rational numbers is always a rational number.)

1–2 PROPERTIES OF REAL NUMBERS

In order to understand the laws that govern the symbols in algebra, we must examine the properties of the numbers the symbols represent. Unless a problem states otherwise, algebraic symbols may be replaced by any real number. Thus, the rules for symbol manipulation are the same as those for the arithmetic of real numbers.

Here we list some of the properties that we assume from experience for addition and multiplication of real numbers. The addition and multiplication of negative numbers is not considered until the next section, so we shall illustrate each property in terms of positive numbers.

PROPERTY 1: Associative property of addition: If a, b, and c are any real numbers, then $(a + b) + c = a + (b + c)$. We obtain the same result if we change the grouping of the numbers in an addition problem.

EXAMPLE 1: $(2 + 3) + 4 = 2 + (3 + 4)$
$5 + 4 = 2 + 7$
$9 = 9$

PROPERTY 2: Associative property of multiplication: If a, b, and c are any real numbers, then $(a \cdot b) \cdot c = a \cdot (b \cdot c)$. We obtain the same result if we change the grouping of numbers in a multiplication problem.

EXAMPLE 2: $(0.05 \cdot 100) \cdot 2 = 0.05 \cdot (100 \cdot 2)$
$5 \cdot 2 = 0.05 \cdot 200$
$10 = 10$

PROPERTY 3: Commutative property of addition: If a and b are any real numbers, then $a + b = b + a$. The order in which we write the numbers in an addition problem does not affect their sum.

EXAMPLE 3: $2.05 + 0.61 = 0.61 + 2.05$
$2.66 = 2.66$

PROPERTY 4: Commutative property of multiplication: If a and b are any real numbers, then $a \cdot b = b \cdot a$. The order in which we write the numbers in a multiplication problem does not affect their product.

EXAMPLE 4: $4 \cdot 7 = 7 \cdot 4$
$28 = 28$

PROPERTY 5: Distributive property: If a, b, and c are any real numbers, then $a \cdot (b + c) = a \cdot b + a \cdot c$.

EXAMPLE 5: $2 \cdot (3 + 4) = 2 \cdot 3 + 2 \cdot 4$
$2 \cdot 7 = 6 + 8$
$14 = 14$

Remember, parentheses dictate which numbers are grouped together. If there are no parentheses, the multiplication is always done before the addition. For example, $2 + 3 \cdot 4 = 14$, not 20.

> **PROPERTY 6:** Fundamental property of 0: If a is any real number, then $a + 0 = a$.

EXAMPLE 6: $\sqrt{2} + 0 = \sqrt{2}$.

> **PROPERTY 7:** Fundamental property of 1: If a is any real number, then $a \cdot 1 = a$.

EXAMPLE 7: $\pi \cdot 1 = \pi$.

> **PROPERTY 8:** Existence of negatives: Every real number a has exactly one negative, denoted by $-a$, such that $a + (-a) = 0$.

EXAMPLE 8: $5 + (-5) = 0$, so -5 is the negative of 5 and 5 is the negative of -5.

Remember, there is a difference between a negative number and the negative of a number. Although -5 is a negative number, the negative of -5 is 5.

> **PROPERTY 9:** Existence of reciprocals: Every real number a (except zero) has exactly one reciprocal, denoted by $1/a$, such that $a(1/a) = 1$.

EXAMPLE 9: $5(\frac{1}{5}) = 1$, so $\frac{1}{5}$ is the reciprocal of 5 and 5 is the reciprocal of $\frac{1}{5}$.

EXAMPLE 10: $\frac{2}{7}\left(\frac{1}{2/7}\right) = 1$, so $\frac{1}{2/7}$ or $\frac{7}{2}$ is the reciprocal of $\frac{2}{7}$ and $\frac{2}{7}$ is the reciprocal of $\frac{7}{2}$.

These properties of the real numbers are the ones of most interest to us since they are the basis for the justification of many algebraic manipulations. As obvious as many of them must seem, it is essential that you understand them completely. For example, the commutative property states that the order of the numbers in an addition or multiplication problem is insignificant. However, if you stop to think, you will realize that in most situations order is very significant. Consider the consequences of taking a shower and then undressing!

EXERCISES 1–2

In Exercises 1–20 name the property illustrated in the statement.

1. $2 + 7 = 7 + 2$

2. $11 + 0 = 11$

3. $4(5 \cdot 11) = (4 \cdot 5)11$

4. $6(4 + 3) = 6 \cdot 4 + 6 \cdot 3$

5. $\sqrt{2} \cdot 1 = \sqrt{2}$

6. $-7.3 + 7.3 = 0$

7. $(2.5)3 = 3(2.5)$

8. $17(\frac{1}{17}) = 1$

9. $\pi \cdot 3 + \pi \cdot 8 = \pi(3 + 8)$

10. $(5 + 3) + (2 + 1)$
$= [(5 + 3) + 2] + 1$

11. $z(xy) = (zx)y$

12. $(xy)z = z(xy)$

13. $x + 0 = 0 + x$

14. $(-z) + z = 0$

15. $y + (x + z) = (x + z) + y$

16. $y \cdot 1 = 1 \cdot y$

17. $x(1/x) = 1$ if $x \neq 0$

18. $ax + ay = a(x + y)$

19. $y(z + x) = (z + x)y$

20. $(x + z) + y = x + (z + y)$

In Exercises 21–26 find the negative for each number.

21. 7

22. -11

23. $-\frac{2}{101}$

24. 4.23

25. 0

26. 1

In Exercises 27–32 find the reciprocal for each number.

27. 25

28. $\frac{1}{6}$

29. $\frac{5}{17}$

30. 1.76

31. 0

32. 1

33. Find, by two methods, the area of the rectangle shown and explain how the rectangle can be used to illustrate geometrically the distributive property.

a	Area 1	Area 2
	b	c

34. If x is a negative number, the negative of x is a ___ number.

In Exercises 35–37 show that $(b + c)a = ba + ca$ by filling in the property of the real numbers that justifies the step.

$(b + c)a = a(b + c)$ **35.** _____

$a(b + c) = ab + ac$ **36.** _____

$ab + ac = ba + ca$ **37.** _____

In Exercises 38–41 show that $a(bc) = b(ca)$ by filling in the property of the real numbers that justifies the step.

$a(bc) = (ab)c$ **38.** _____

$(ab)c = (ba)c$ **39.** _____

$(ba)c = b(ac)$ **40.** _____

$b(ac) = b(ca)$ **41.** _____

42. Show that $a + (b + c) = b + (a + c)$ by a procedure similar to the proofs above.

1–3 OPERATIONS WITH REAL NUMBERS

Real number line

The real numbers may be interpreted geometrically by considering a straight line. Every point on the line can be made to correspond to a real number and every real number can be made to correspond to a point. The first point that we designate is zero. It is the dividing point between positive and negative real numbers. Any number to the right of zero is called positive, and to the left of zero, negative. Figure 1.3 is the result of assigning a few positive and negative real numbers to points on the line.

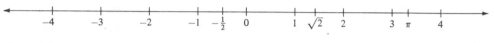

Figure 1.3

Absolute value

We are sometimes not interested in whether a number is to the right or left of zero but merely in the distance from the number to zero. We call this distance the *absolute value* or magnitude of the number. By convention this distance is given by a positive number or zero. The symbol $|a|$ means the absolute value of the number a. Examples 1–4 are illustrated in Figures 1.4–1.7.

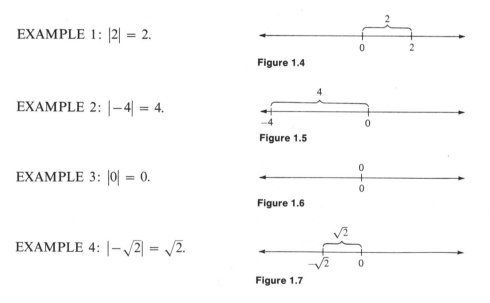

EXAMPLE 1: $|2| = 2.$

Figure 1.4

EXAMPLE 2: $|-4| = 4.$

Figure 1.5

EXAMPLE 3: $|0| = 0.$

Figure 1.6

EXAMPLE 4: $|-\sqrt{2}| = \sqrt{2}.$

Figure 1.7

Addition

Addition of two real numbers may be interpreted as movement on the number line. The initial starting point is zero. To add a positive number, we move to the right the desired distance. To add a negative number, we move to the left. Consider Figure 1.8, which illustrates the procedure. The addition of $2 + (-4)$ is shown on the real number line by drawing an arrow with a length of 2 units starting at zero and pointing to the right. From the tip of the arrow at 2 we draw an arrow pointing to the left with a length of 4 units. Since the tip of the arrow is at -2, we conclude that $2 + (-4) = -2$.

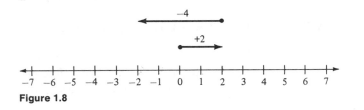

Figure 1.8

EXAMPLE 5: $2 + 4 = 6$ (Figure 1.9).

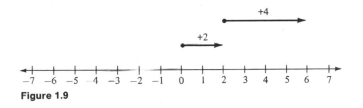

Figure 1.9

EXAMPLE 6: $(-2) + 4 = 2$ (Figure 1.10).

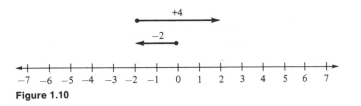

Figure 1.10

EXAMPLE 7: $(-2) + (-4) = -6$ (Figure 1.11).

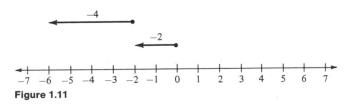

Figure 1.11

From these examples we can determine two basic rules for the addition of two real numbers.

> **RULE 1:** When adding two real numbers of the same sign, add their absolute values and use the sign they have in common.

EXAMPLE 8: $(-16) + (-4) = -20$ since $|-16| = 16, |-4| = 4, 16 + 4 = 20$, and the sign that -16 and -4 have in common is negative.

EXAMPLE 9: $5.4 + 0.9 = 6.3$. The sum of two positive numbers is positive.

> **RULE 2:** When adding two real numbers with different signs, subtract the smaller absolute value from the larger and use the sign of the number with the larger absolute value.

EXAMPLE 10: $14 + (-6) = 8$ since $|14| = 14, |-6| = 6, 14 - 6 = 8$, and the positive number 14 has the larger absolute value.

EXAMPLE 11: $(-\frac{5}{7}) + (\frac{2}{7}) = -\frac{3}{7}$ since $|-\frac{5}{7}| = \frac{5}{7}, |\frac{2}{7}| = \frac{2}{7}, \frac{5}{7} - \frac{2}{7} = \frac{3}{7}$, and the negative number $-\frac{5}{7}$ has the larger absolute value.

Subtraction

We define subtraction in terms of addition. If a and b are any real numbers, then $a - b = a + (-b)$. To subtract b from a we add to a the negative of b.

EXAMPLE 12: We know that $5 - 2 = 3$. If we use the definition above we obtain the same result, since $5 - 2 = 5 + (-2) = 3$.

EXAMPLE 13: $(-5) - 2 = (-5) + (-2) = -7$.

EXAMPLE 14: $5 - (-\frac{1}{2}) = 5 + \frac{1}{2} = \frac{10}{2} + \frac{1}{2} = \frac{11}{2}$.

EXAMPLE 15: $[(-7) - 2] - (-1) = [(-7) + (-2)] + 1 = (-9) + 1 = -8$.

Multiplication

It seems natural to define the product of two positive numbers as positive. Thus $4 \cdot 3 = 12$. To determine what sign to use for the product of a positive and a negative number we use the distributive property in the following problem.

$$5[2 + (-2)] = 5 \cdot 2 + 5(-2)$$
$$5 \cdot 0 = 10 + 5(-2)$$
$$0 = 10 + ?$$

We must define $5(-2)$ so that $10 + 5(-2) = 0$. Therefore, $5(-2)$ must equal -10. In every case the product of a positive number and a negative number is negative.

RULE 3: If the signs of two real numbers are different, the sign of the product is negative.

To determine the sign of the product of two negative numbers, consider the following problem:

$$-5[2 + (-2)] = (-5)2 + (-5)(-2)$$
$$(-5)0 = -10 + (-5)(-2)$$
$$0 = -10 + ?$$

We must define $(-5)(-2)$ so that $-10 + (-5)(-2) = 0$. Therefore, $(-5)(-2)$ must equal 10. In every case the product of two negative numbers is positive.

RULE 4: If the signs of two real numbers are the same, the sign of the product is positive.

EXAMPLE 16: $5(-6) = -30$.

EXAMPLE 17: $(-4)(-3.1) = 12.4$.

EXAMPLE 18: $9(-2)(-\pi) = 18\pi$.

EXAMPLE 19: $(-5)(-\frac{1}{5})(-2) = -2$.

EXAMPLE 20: $(-4)(-4)(-4) = -64$.

If two or more numbers are multiplied together, each number is a *factor* of the product. In Example 20 the factor -4 is used three times. An alternative way of writing $(-4)(-4)(-4)$ is $(-4)^3$. The number $(-4)^3$, or -64, is called the third *power* of -4. In general, by x^n, where n is a positive integer, we mean to use x as a factor n times.

$$x^n = \underbrace{x \cdot x \cdot x \cdots x}_{n \text{ factors}}$$

In the expression x^n, n is called the *exponent*.

EXAMPLE 21: 2^5 means to use 2 as a factor 5 times. Thus, $2^5 = 2 \cdot 2 \cdot 2 \cdot 2 \cdot 2 = 32$. The number 2^5, or 32, is the fifth power of 2. The exponent in 2^5 is 5.

EXAMPLE 22: $(-3)^4(-1) = (-3)(-3)(-3)(-3)(-1) = -81$.

Division

We define division in terms of multiplication. If a and b are any real numbers ($b \neq 0$), then $a \div b = a(1/b)$. To divide a by b we multiply a by the reciprocal of b. The sign in division is then determined in the same manner as multiplication.

EXAMPLE 23: $8 \div 2 = 8 \cdot \frac{1}{2} = 4$.

EXAMPLE 24: $(-10) \div 5 = (-10)\frac{1}{5} = -2$.

EXAMPLE 25: $(-9) \div (-3) = (-9)(-\frac{1}{3}) = 3$.

EXAMPLE 26: $6 \div (-\frac{12}{5}) = 6(-\frac{5}{12}) = -\frac{5}{2}$.

When working with fractions you should notice that the negative of $\frac{5}{2}$ equals $-5/2$ or $5/-2$. In general:

$$-\left(\frac{a}{b}\right) = \frac{-a}{b} = \frac{a}{-b}$$

Note: An amusing way to remember the rules that govern the sign in multiplication and division is called the Good Guy–Bad Guy theorem. This theorem equates positive ($+$) with good, and negative ($-$) with bad. Then:

If something good ($+$) happens to a good ($+$) guy, that's good ($+$); that is, $(+)(+) = (+)$

If something good ($+$) happens to a bad ($-$) guy, that's bad ($-$); that is, $(+)(-) = (-)$

If something bad ($-$) happens to a good ($+$) guy, that's bad ($-$); that is, $(-)(+) = (-)$

But if something bad ($-$) happens to a bad ($-$) guy, that's good ($+$); that is, $(-)(-) = (+)$

This scheme will never replace a proof based on the properties of real numbers, but it is easy to remember.

Evaluating algebraic expressions

An expression that combines numbers and algebraic symbols using the operations of addition, subtraction, multiplication, division, or extracting roots is called an *algebraic expression*. For example,

$$2x^3 - 5x + 3 \quad \text{and} \quad \frac{-b + \sqrt{b^2 - 4ac}}{2a} \quad \text{and} \quad \tfrac{1}{2}gt^2$$

are algebraic expressions.

If we are given numerical values for the symbols, we can evaluate the expression by substituting the given values and performing the indicated operations. Remember, do the operations in the following order: (1) find powers or roots; (2) perform all multiplications and divisions, and (3) perform all additions and subtractions.

EXAMPLE 27: Find the numerical value of $-x^2 + 5yz^2$ if $x = -2$, $y = 3$, and $z = -4$.

SOLUTION: $-x^2 + 5yz^2$, when $x = -2$, $y = 3$, and $z = -4$, equals

$$-(-2)^2 + 5(3)(-4)^2$$
$$= -(4) + 5(3)(16)$$
$$= -4 + 240$$
$$= 236 \quad \textbf{Answer}$$

EXAMPLE 28: If

$$m = \frac{y_2 - y_1}{x_2 - x_1}$$

evaluate m when $x_1 = 2$, $x_2 = -1$, $y_1 = -5$, and $y_2 = 3$. (*Note:* In the symbols x_1, x_2, y_1, and y_2 the numbers 1 and 2 are called *subscripts*. In this case x_1 and x_2 are used to denote different x values.)

SOLUTION:

$$m = \frac{y_2 - y_1}{x_2 - x_1}$$

so when $x_1 = 2$, $x_2 = -1$, $y_1 = -5$, and $y_2 = 3$, we have

$$m = \frac{(3) - (-5)}{(-1) - (2)} = \frac{8}{-3} \quad \textbf{Answer}$$

EXERCISES 1–3

In Exercises 1–6 find the absolute value of each real number.

1. 4
2. $-\sqrt{6}$
3. -3.148
4. 3π
5. 0
6. $-\frac{19}{5}$

In Exercises 7–40 evaluate each expression by performing the indicated operations.

7. $(-4) + (-3)$
8. $43 + (-21)$
9. $3 - 8$
10. $0 - 6$
11. $(-6) - (-4)$
12. $(-12) - 5$
13. $(-\frac{1}{5}) + 3$
14. $7 - (-\frac{3}{8})$
15. $(-0.87) + 0.33$
16. $0.17 - (-0.48)$
17. $(-0.1) - 4$
18. $(-3.3) + (-0.67)$
19. $(-11)6$
20. $(-8)(-3)$

21. $(-14) \div (-7)$

22. $89 \div (-89)$

23. $-(\frac{1}{8}) \div 4$

24. $(-\frac{2}{3}) \div (-\frac{4}{15})$

25. $(-5)(-\frac{4}{9})$

26. $(\frac{3}{16})(-2)$

27. $0 \div (-4)$

28. $(-4) \div 0$

29. $0.2(-0.2)$

30. $(-0.64) \div (-0.16)$

31. $(-32.4) \div (-3)$

32. $10.1(-11)$

33. $(-2) \div (6 - 5)$

34. $(5 - 3)(5 + 2)$

35. $-4 + 2(3 - 8)$

36. $(-4 + 2)(3 - 8)$

37. $(2 - 4)(3 - 6)$

38. $2 - 4(3 - 6)$

39. $-5(11 - 6) - [7 - (11 - 19)]$

40. $7 - 3[2(13 - 5) - (5 - 13)]$

In Exercises 41–60 evaluate each expression after setting $x = -2$, $y = 3$, and $z = -4$.

41. $x + y + z$

42. $4x - z$

43. $x^3 x^2$

44. y^3/y^2

45. $(x^2 + y^2)/z^2$

46. $20 - xyz^3$

47. $(-x)^2 + 2y$

48. $-x^2 + 2y$

49. $(x - y)^2 \div (y - x)^2$

50. $2x^3 + 3x^2 - x + 1$

51. $x - 2(3y - 4z)$

52. $(x - 2)(3y - 4z)$

53. $(z - x)(y - 3z)$

54. $z - x(y - 3z)$

55. $(2y - x)x - z$

56. $(2y - x)(x - z)$

57. $(x + y + z)(x + y - z)$

58. $x + y + z(x + y - z)$

59. $(x + 2)(y - 3) + (3 - z)(4 + x)$

60. $(5 - y)(z + 1) - (y - 2)(3 + x)$

In Exercises 61–70 $m = (y_2 - y_1)/(x_2 - x_1)$. Evaluate m for the following values of x_1, x_2, y_1, and y_2.

	x_1	x_2	y_1	y_2
61.	5	1	4	2
62.	1	4	6	8
63.	-3	-6	-1	-4
64.	-5	-2	-3	-6
65.	2	-6	-1	3
66.	0	-2	3	-4
67.	6	6	0	5
68.	-4	-4	7	-1
69.	-2	1	3	3
70.	-4	3	-2	-5

In Exercises 71–80 evaluate the expression $(x_2 - x_1)^2 + (y_2 - y_1)^2$ for the values of x_1, x_2, y_1, and y_2 given above.

81. Look back over the exercises and determine the sign of a product that contains:

  a. An even number of negative factors.

  b. An odd number of negative factors.

82. If $xy = xz$, does it follow that $y = z$?

83. If $xy = x$, what are the possibilities for x and y?

84. If $x^2 + y^2 = 0$, what are the possible values for x and y?

85. You own 100 shares of IBM stock. During a certain week you consult the financial page in the paper and write down the price that appears next to your stock. This number indicates the dollar gain or loss for each share on that particular day. At the end of the week your results are as follows: (Monday, $+5$), (Tuesday, -7), (Wednesday, -4), (Thursday, 0), (Friday, $+3$). How much money did you make or lose for the week?

86. A land speculator bought an acre of land for $1000, sold it for $2000, bought it back for $3000, and finally sold it for $4000. How much money did he make or lose in this series of transactions?

1–4 ADDITION AND SUBTRACTION OF ALGEBRAIC EXPRESSIONS

Variables and constants

In algebra, two types of symbols are used to represent numbers: variables and constants. A *variable* is a symbol that may be replaced by different numbers in a particular problem. Generally, letters near the end of the alphabet, such as x, y, or z, are variables. A *constant* is a symbol that represents the same number throughout a particular problem. Numbers, such as 2, $-\sqrt{5}$, and π, never change value and are called *absolute constants*. If we do not know the fixed number until we are given specific information about the problem, the symbol is called an *arbitrary constant*. Generally k or letters near the beginning of the alphabet, such as a, b, and c, are arbitrary constants.

> EXAMPLE 1: The sales tax T on a purchase is related to the price (p) of the item by the formula $T = kp$, where k is the sales tax rate. T and p are variables that may take on different values. The sales tax rate is fixed for any particular location so that k is a constant. For example, in New York City k is fixed at 8 percent. In another location k may be fixed at a different percentage.

Terms: Those parts of an algebraic expression separated by plus ($+$) or minus ($-$) signs are called *terms* of the expression.

> EXAMPLE 2: $x + (2x/5) - 3y$ is an algebraic expression with three terms, x, $2x/5$, and $3y$.

> EXAMPLE 3: $-4y + x - z + \pi xyz$ is an algebraic expression with four terms $-4y$, x, z, and πxyz.

> EXAMPLE 4: 6 is an algebraic expression with one term.

If the term is the product of some constants and variables, the constant factor is called the *coefficient* of the term. For example, the coefficient of the term $2x$ is 2, and the coefficient of ax^2 is a. Every term has a coefficient. If the term is x, the coefficient is 1 since $x = 1 \cdot x$. Similarly, if the term is $-x$, the coefficient is -1 since $-x = -1 \cdot x$. If two terms differ only in their coefficient (such as $4x$ and $-5x$), they are called *similar terms*. The distributive property indicates that we combine similar terms by combining their coefficients.

EXAMPLE 5: $3x + 7x = (3 + 7)x = 10x$.

EXAMPLE 6: $yz^2 - 10yz^2 = (1 - 10)yz^2 = -9yz^2$.

EXAMPLE 7: $2p - 3p + 4p = (2 - 3 + 4)p = 3p$.

EXAMPLE 8: $7x + 2y - 3x = (7 - 3)x + 2y = 4x + 2y$.

EXAMPLE 9: $-3x^2y + xy^2 - 7xy$. There are no similar terms, so we cannot simplify the expression.

To combine algebraic expressions it is sometimes necessary to remove the parentheses or brackets that group certain terms together. Parentheses are removed by applying the distributive property; that is, we multiply each term inside the parentheses by the factor in front of the parentheses. If the grouping is preceded by a minus sign, the factor is -1, so the sign of each term inside the parentheses must be changed. If the grouping is preceded by a plus sign, the factor is 1, so the sign of each term inside the parentheses remains the same. If there is more than one symbol of grouping, it is usually better to remove the innermost symbol of grouping first.

EXAMPLE 10: Remove parentheses and combine similar terms:

$$(4a + 3b - 12c) - (2a - b - 5c)$$

SOLUTION:

	$4a + 3b - 12c$		the sign of each term remains the same
add	$-2a + b + 5c$		the sign of each term is changed
	$2a + 4b - 7c$	**Answer**	

EXAMPLE 11: Remove parentheses and combine similar terms:

$$-2(3x - z) - (z - y) + 5(x + 2y)$$

SOLUTION:

	$-6x + 2z$		each term is multiplied by -2
	$ + y - z$		the sign of each term is changed
add	$5x + 10y$		each term is multiplied by 5
	$-x + 11y + z$	**Answer**	

EXAMPLE 12: Remove the symbols of grouping and combine similar terms:

$$7 - 2[4x - (1 - 3x)]$$

SOLUTION: $7 - 2[4x - (1 - 3x)]$

$= 7 - 2[4x - 1 + 3x]$ the sign of each term in the parentheses is changed

$= 7 - 2[7x - 1]$

$= 7 - 14x + 2$ each term within the brackets is multiplied by -2

$= 9 - 14x$ **Answer**

(*Note:* $9 - 14x \neq -5x$.)

EXERCISES 1–4

Simplify each expression and state how many terms are in the conclusion.

1. $6a + 4a + 9a$
2. $-3b - b + 7b$
3. $8a - 3b + 6a - 5b$
4. $5k - 6m - 7 + 3k - 4m + 2$
5. $5xy - 4cd - 2xy + 9cd$
6. $4p - 3q + 4p + q + 4p - 2q$
7. $2x^3 + 6y^2 - 5x^3 + 2y^2$
8. $2a^2b + 3ab - 4ab^2$
9. $x - 2(x + y) + 3(x - y)$
10. $k + (m - k) - (m - 2k)$
11. $-2(a - 2b) - 7(2a - b)$
12. $2a - (7a + 3) + (4a - 6)$
13. $-(x + y) + 4(x - y) + 7x - 2y$
14. $a(b - c) - b(a + c)$
15. $(2x^3 + 7x^2y^2 + 9xy^3) + (6x^2y^2 - 3x^3y)$
16. $(3c^2d + 2cd - 5d^3) - (9d^3 - 6c^2d - 2cd)$
17. $(3y^3 - 2y^2 + 7y - 1) + (4y^3 - y^2 - 3y + 7)$
18. $(2n - 4n^3 + 3n^2 + 4) - (6 - 2n^2 + 7n - n^3)$
19. $(b^3 - 2b^2 + 3b + 4) - (b^2 + 2b - 1)$
20. $(4x^3 + 7x^2y^2 - 2y^3) + (-7x^3 + 2x^2y^2 + 2y^3)$
21. $2[a + 5(a + 2)] - 6$
22. $10 - 4[3x - (1 - x)]$
23. $3a - [3(a - b) - 2(a + b)]$
24. $-[x - (x + y) - (x - y) - (y - x)]$
25. $2\{4x + 3[2 - (x + 1)] - x\}$

1–5 WORD PROBLEMS AND EQUATIONS

Word problems give us our first experience with one of the most challenging parts of mathematics: to obtain a mathematical description (model) of a practical situation. For as soon as we are able to describe the data mathematically, a solution can usually be obtained through algebraic methods. The first step in the process of translating the problem into mathematical language is to assign some variable to represent an unknown. If there is more than one unknown in the problem, try to represent them in terms of the original variable. Always use as few symbols as possible. Finally, use the information given in the problem to obtain a relationship between two algebraic expressions. In this section we discuss only the relationship of equality. A statement of equality between two algebraic expressions is called an *equation*.

EXAMPLE 1: The sum of a number and 5 is 21. Translate this statement into an equation.

SOLUTION: Let x represent the unknown number; then

$$x + 5 = 21$$

EXAMPLE 2: Taking one-third of a number gives the same result as subtracting 4 from the number. Translate this statement into an equation.

SOLUTION: Let y represent the unknown number; then

$$\tfrac{1}{3}y = y - 4$$

EXAMPLE 3: The sum of two consecutive integers is 67. Translate this statement into an equation.

SOLUTION: If z represents the first integer, $z + 1$ represents the next integer, so

$$z + (z + 1) = 67$$

EXAMPLE 4: The width of a rectangular lot is one-half the length and the perimeter of the lot is 5000 ft. Translate this statement into an equation.

SOLUTION: Let P represent the perimeter, L the length, and W the width; then

$$P = 2(L + W) \quad \text{and} \quad W = \tfrac{1}{2}L$$

Substituting $\tfrac{1}{2}L$ for W, we have

$$P = 2(L + \tfrac{1}{2}L)$$

Substituting 5000 for the perimeter, we have

$$5000 = 2(L + \tfrac{1}{2}L)$$

Once we represent the problem in mathematical symbols, we can attempt a solution. To solve an equation means to find *all* the values for the variable that make the equation a true statement. For example, consider the equation

$$x + 5 = 21$$

If $x = 16$, this equation becomes

$$16 + 5 = 21$$

Since the equation is a true statement when $x = 16$, we call 16 a *solution* of the equation. If we substitute any other number for x, the equation becomes a false statement. For example, if $x = 10$, the equation becomes

$$10 + 5 = 21$$

which is false. Thus, 16 is the only solution of the equation $x + 5 = 21$.

In this section we are concerned mainly with equations that have only one solution. However, two important exceptions are illustrated in Examples 5 and 6.

EXAMPLE 5: For what values(s) of x is the equation $2x + x = 3x$ a true statement?

SOLUTION: The equation $2x + x = 3x$ is a true statement no matter which number we substitute for x. An equation that is a true statement for all values of the variable is called an *identity*.

EXAMPLE 6: For what value(s) of x is the equation $x = x + 1$ a true statement?

SOLUTION: If we substitute any number for x in the equation $x = x + 1$, we will find that the number on the right side of the equation is always 1 greater than the number on the left side. Thus, this equation has no solution.

Algebraically, we solve an equation by isolating the variable on one side of the equation so that the equation has the form: x = number. Two equations are *equivalent* when they have the same solution. We change the original equation to an equivalent equation of the form x = number by performing any combination of the following steps:

1. Adding or subtracting the same expression to (from) both sides of the equation.
2. Multiplying or dividing both sides of the equation by the same (nonzero) number.

When performing steps 1 or 2 we always simplify the result.

EXAMPLE 7: For what value(s) of x is the equation $5x + 3 = 33$ a true statement?

SOLUTION: We first isolate the term involving the variable by subtracting 3 from both sides of the equation:

$$\begin{array}{rcr} 5x + 3 = & & 33 \\ -3 & & -3 \\ \hline 5x \quad = & & 30 \end{array}$$

Next we want the coefficient of x to be 1. Therefore, divide both sides of the equation by 5:

$$\frac{5x}{5} = \frac{30}{5}$$

$$x = 6$$

Thus, 6 is the solution of the equation. We can check the solution by replacing x by 6 in the original equation:

$$5x + 3 = 33$$

$$5(6) + 3 \stackrel{?}{=} 33$$

$$30 + 3 \stackrel{?}{=} 33$$

$$33 \stackrel{\checkmark}{=} 33$$

EXAMPLE 8: For what value(s) of x is the equation $\frac{10}{3}x + 7 = -x - 6$ a true statement?

SOLUTION:

$$\frac{10}{3}x + 7 = -x - 6$$

$$\underline{\phantom{\frac{10}{3}x}\; -7 \qquad\quad -7} \qquad \text{subtract 7 from both sides}$$

$$\frac{10}{3}x \qquad\; = -x - 13$$

$$\underline{+x \qquad\qquad +x} \qquad \text{add } x \text{ to both sides}$$

$$\frac{13}{3}x \qquad\; = \qquad -13$$

$$\left(\tfrac{3}{13}\right)\left(\tfrac{13}{3}x\right) = \left(\tfrac{3}{13}\right)(-13) \qquad \text{multiply both sides by } \tfrac{3}{13}$$

$$x = -3$$

Thus, -3 is the solution of the equation.

Check: Replace x by -3 in the original equation:

$$\tfrac{10}{3}(-3) + 7 \overset{?}{=} -(-3) - 6$$

$$-10 + 7 \overset{?}{=} 3 - 6$$

$$-3 \overset{\checkmark}{=} -3$$

EXAMPLE 9: Solve the equation

$$6(x + 1) = 14x + 2$$

SOLUTION:

$$6(x + 1) = 14x + 2$$

$$6x + 6 = 14x + 2$$

$$\underline{\; -6 \qquad\qquad -6} \qquad \text{subtract 6 from both sides}$$

$$6x \qquad = 14x - 4$$

$$\underline{-14x \qquad\quad -14x} \qquad \text{subtract } 14x \text{ from both sides}$$

$$-8x \qquad = \qquad\quad -4$$

$$\frac{-8x}{-8} \qquad = \frac{-4}{-8} \qquad \text{divide both sides by } -8$$

$$x \qquad = \tfrac{1}{2}$$

Thus, $\frac{1}{2}$ is the solution of the equation.

Check: Substitute $\frac{1}{2}$ for x in the original equation:

$$6\left(\frac{1}{2} + 1\right) \overset{?}{=} 14\left(\frac{1}{2}\right) + 2$$

$$6\left(\frac{3}{2}\right) \overset{?}{=} 7 + 2$$

$$9 \overset{\checkmark}{=} 9$$

Now that we can solve an equation, we can determine the final solutions for some word problems. The most difficult part is usually to describe the data mathematically. Use the guidelines given at the beginning of this section and remember that these problems gradually become easier with experience.

EXAMPLE 10: The total cost (including tax) of a new car is $4536. If the sales tax rate is 8 percent, how much did you pay in taxes?

SOLUTION: Let x represent the price of the car; then

Total cost = price of car + tax

$$4536 = x + 0.08x$$

$$4536 = 1.08x$$

$$\frac{4536}{1.08} = x$$

$$4200 = x$$

Tax = 8 percent of 4200 = 0.08(4200) = $336

EXAMPLE 11: Two trains whose rates differ by 10 mi/hour start at the same time from stations 375 mi apart. They meet in 2.5 hours. Find the rate of each train.

SOLUTION: Let x be the rate of the slow train.

	Rate (mi/hour)	Time (hours)	Distance (mi)
Slow train	x	2.5	$2.5x$
Fast train	$x + 10$	2.5	$2.5(x + 10)$

$$\text{Rate} \cdot \text{Time} = \text{Distance}$$

Distance traveled by slow train + Distance traveled by fast train = Distance between stations

$$2.5x + 2.5(x + 10) = 375$$

$$2.5x + 2.5x + 25 = 375$$

$$5x = 350$$

$$x = 70$$

Thus, the rate of the slow train is 70 mi/hour and the rate of the fast train is 80 mi/hour.

EXAMPLE 12: A chemist has two acid solutions, one 30 percent acid and the other 70 percent acid. How much of each solution must he use to obtain 100 pt of a solution that is 41 percent acid?

SOLUTION: Let x be the amount of the first solution in the mixture.

	Percent Acid	Amount of Solution (pt)	$=$	Quantity of Acid (pt)
First solution	30	x		$0.3x$
Second solution	70	$100 - x$		$0.7(100 - x)$
New solution	41	100		$0.41(100)$

$$\begin{array}{ll}\text{Quantity of acid} \\ \text{in first solution}\end{array} + \begin{array}{ll}\text{Quantity of acid} \\ \text{in second solution}\end{array} = \begin{array}{ll}\text{Quantity of acid} \\ \text{in new solution}\end{array}$$

$$0.3x + 0.7(100 - x) = 0.41(100)$$
$$0.3x + 70 - 0.7x = 41$$
$$-0.4x = -29$$
$$x = \frac{-29}{-0.4} = 72.5$$

Thus, he must mix 72.5 pt of the first solution with 27.5 pt of the second solution to obtain the desired mixture.

EXERCISES 1–5

In Exercises 1–30 solve each equation. If the solution is an integer, check your answer.

1. $6x = 18$

2. $-3x + 3 = 0$

3. $3x + 2 = 2x + 8$

4. $7y - 9 = 6y - 10$

5. $2z + 12 = 5z + 15$

6. $-6x + 7 = 5x - 4$

7. $3y - 14 = -4y + 7$

8. $3z - 8 = 13z - 9$

9. $8x - 10 = 3x$

10. $15 - 6y = 0$

11. $9 - 2y = 27 + y$

12. $15x + 5 = 2x + 14$

13. $4x - x = 3x$

14. $x - 1 = x + 3$

15. $2(x + 4) = 2(x - 1) + 3$

16. $2(x + 3) - 1 = 2x + 5$

17. $7x = 2x$

18. $3y - 7 = 2y - 7$

19. $4 - 3y = 3(4 - y)$

20. $7x - 2(x - 7) = 14 - 5x$

21. $5(x - 6) = -2(15 - 2x)$

22. $18(z + 1) = 9z + 10$

23. $3[2x - (x - 2)] = -3(3 - 2x)$

24. $2y - (3y - 4) = 4(y + \frac{3}{4})$

25. $\dfrac{x - 2}{7} = -1$

26. $\dfrac{2y - 7}{9} = 5$

27. $\dfrac{x - 9}{3} = 3x - 11$

28. $\dfrac{-14y - 17}{-5} = 2 + 3y$

29. $\dfrac{9x}{4} = \dfrac{18 + x}{2}$

30. $\dfrac{7x}{6} - 5 = x - \dfrac{44}{9}$

In Exercises 31–54 solve each problem by first setting up an appropriate equation.

31. Multiplying a number by 4 gives the same result as adding 6 to the number. What is the number?

32. Taking one-half of a number gives the same result as adding 3 to the number. What is the number?

33. If a number is decreased by 5, the result is twice the original number. What is the number?

34. Multiplying a number by 4 and adding 2 to the product gives the same result as dividing the number by −2. What is the number?

35. The sum of two consecutive integers is 89. Find the larger integer.

36. The sum of two integers is 35. One of the integers is four times the other. Find the integers.

37. During a test run machine A produced x cans; machine B turned out three times as many; machine C produced 10 more cans than A. The total production during the run was 5510. How many cans did each machine produce?

38. The sum of the three angles in a triangle is 180°. The second angle is 20° more than the first and the third angle is twice the first. What is the measure of each of the angles in the triangle?

39. The length of a rectangle is three times its width. If the perimeter is 160 in., determine the area of the rectangle.

40. A magazine and a newspaper cost $1.10. If the magazine cost $1.00 more than the newspaper, how much did each cost?

41. A stockbroker, after deducting his 10 percent commission, gives a client $315. How much was the customer's stock worth?

42. Suppose that 5 percent of your salary is deducted for a retirement fund. If your total deduction for 1 year is $725, what is your annual salary?

43. An aerospace worker whose salary is $10,000 accepts a 10 percent cut in wages because of his company's financial crisis. Later he is given a raise. What percent raise is needed if he is again to earn $10,000?

44. If the total cost (including tax) of a new car is $3317, how much did you pay in taxes if the sales tax rate is 7 percent?

45. Two trains whose rates differ by 8 mi/hour start at the same time from stations 427 mi apart. They meet in 3.5 hours. Find the rate of each train.

46. A passenger train and a freight train start from the same point at the same time and travel in opposite directions. The passenger train traveled twice as fast as the freight train. In 4 hours they were 396 mi apart. Find the rate of each train.

47. At 10 A.M. Mr. Sluggish started from his home, traveling by car at 40 mi/hour. At noon Mr. Swift started after him on the same road, traveling at 55 mi/hour. At what time did Mr. Swift overtake Mr. Sluggish?

48. A freight train averaging 45 mi/hour and an express train averaging 70 mi/hour leave the station at the same time. How long will it take the express train to get 80 mi ahead of the freight train?

49. A chemist has 5 qt of a 25 percent sulfuric acid solution. He wishes to obtain a solution that is 35 percent acid by adding a solution of 75 percent acid to his original solution. How much of the more concentrated acid must be added to achieve the desired concentration?

50. One bar of tin alloy is 35 percent pure tin and another bar is 10 percent pure tin. How many pounds of each must be used to make 95 lb of a new alloy that is 20 percent pure tin?

51. An alloy of copper and silver weighs 40 lb and is 20 percent silver. How much silver must be added to produce a metal that is 50 percent silver?

52. A 20 percent antifreeze solution is a solution that consists of 20 percent antifreeze and 80 percent water. If we have 20 qt of such a solution, how much water should be added to obtain a solution that contains 10 percent antifreeze?

53. Try each of the following number tricks with at least two numbers, then show algebraically why they work:

a. Choose a number.
　　Add three.
　　Multiply by two.

　　Add six.
　　Divide by two.
　　Subtract your original number.
　　Your result is six.

b. Choose a number.
　　Triple it.
　　Add the number one larger than your original number.
　　Add seven.
　　Divide by four.
　　Subtract two.
　　The result is your original number.

54. In the equation $ax = b$, a and b are constants. What combination of values for a and b results in an equation with exactly one solution? No solution? Infinitely many solutions?

1–6 FORMULAS

The translation of word problems into mathematical symbols often results in a general equation or formula that has applications in many fields. Consider the formula

$$C = \tfrac{5}{9}(F - 32)$$

This formula converts degrees Fahrenheit to degrees Celsius. If the Fahrenheit temperature is 41 degrees ($F = 41$), we could find the Celsius temperature (C) by substitution:

$$C = \tfrac{5}{9}(F - 32) = \tfrac{5}{9}(41 - 32) = \tfrac{5}{9}(9) = 5$$

Thus 41 Fahrenheit is equivalent to 5 Celsius.

If we wish to use this same formula to convert from degrees Celsius to degrees Fahrenheit, we would have to do more than just substitute. For example, if $C = 10$, upon substituting we get

$$10 = \tfrac{5}{9}(F - 32)$$

We have an equation that we must now solve for F:

$$
\begin{array}{ll}
10 = \tfrac{5}{9}(F - 32) & \\
\tfrac{9}{5}(10) = \tfrac{9}{5} \cdot \tfrac{5}{9}(F - 32) & \text{multiply both sides by } \tfrac{9}{5} \\
18 = F - 32 & \\
\underline{+\,32 = \quad\ \ +\,32} & \text{add 32 to both sides} \\
50 = F &
\end{array}
$$

This process is not difficult if we have only one conversion to make. But, if we have many conversions or if we wish to obtain the solution on a calculator, we need a formula for F. To obtain this we follow the same procedure as before.

$$C = \tfrac{5}{9}(F - 32)$$
$$\tfrac{9}{5}C = \tfrac{9}{5} \cdot \tfrac{5}{9}(F - 32) \qquad \text{multiply both sides by } \tfrac{9}{5}$$
$$\tfrac{9}{5}C = F - 32$$
$$\underline{+\ 32 \qquad\qquad +\ 32} \qquad \text{add 32 to both sides}$$
$$\tfrac{9}{5}C + 32 = F$$

Now if we want to change 10 Celsius to degrees Fahrenheit we could just substitute:

$$F = \tfrac{9}{5}C + 32$$
$$= \tfrac{9}{5}(10) + 32 = 18 + 32 = 50$$

The computation is now simpler with a formula for F.

EXAMPLE 1: The area of a triangle is equal to one-half the product of the altitude and the base ($A = \tfrac{1}{2}ab$). Find a formula for the altitude of a triangle in terms of its base and area.

SOLUTION:

$$A = \tfrac{1}{2}ab$$
$$(2)A = (2)\tfrac{1}{2}ab \qquad \text{multiply both sides by 2}$$
$$2A = ab$$

$$\frac{2A}{b} = \frac{ab}{b} \qquad \text{divide both sides by } b$$

$$\frac{2A}{b} = a$$

The altitude of any triangle equals twice the area divided by the base.

EXAMPLE 2: The perimeter of a rectangle equals twice the length plus twice the width ($p = 2l + 2w$). Find a formula for the length of the rectangle in terms of the perimeter and the width.

SOLUTION:

$$p = 2l + 2w$$
$$\underline{-\ 2w \qquad\qquad -\ 2w} \qquad \text{subtract } 2w \text{ from both sides}$$
$$p - 2w = 2l$$

$$\frac{p - 2w}{2} = \frac{2l}{2} \qquad \text{divide both sides by 2}$$

$$\frac{p - 2w}{2} = l \quad \text{or} \quad \tfrac{1}{2}p - w = l$$

The length is equal to one-half the perimeter minus the width.

EXAMPLE 3: The current in a circuit (I) is equal to the voltage (E) divided by the resistance (R). ($I = E/R$) Find a formula for the resistance in terms of the current and the voltage.

SOLUTION:

$$I = \frac{E}{R}$$

$$RI = R\frac{E}{R} \qquad \text{multiply both sides by } R$$

$$RI = E$$

$$\frac{RI}{I} = \frac{E}{I} \qquad \text{divide both sides by } I$$

$$R = \frac{E}{I}$$

The resistance in a circuit is equal to the voltage divided by the current.

EXAMPLE 4: The Z-score formula in statistics is $Z = (x - m)/s$, where x is a raw score, m is the mean, and s is the standard deviation. Solve the formula for the raw score (x).

SOLUTION:

$$Z = \frac{x - m}{s}$$

$$sZ = s\left(\frac{x - m}{s}\right) \qquad \text{multiply both sides by } s$$

$$sZ = x - m$$

$$\underline{+ m \qquad\quad + m} \qquad \text{add } m \text{ to both sides}$$

$$sZ + m = x$$

EXERCISES 1–6

In Exercises 1–10 find the value of the indicated variable in each formula.

1. $d = \frac{1}{2}at^2$; $a = 32$, $t = 3$. Find d.

2. $C = \frac{5}{9}(F - 32)$; $F = -13$. Find C.

3. $S = \frac{1}{2}gt^2 + vt$; $g = 32$, $t = 2$, $v = 40$. Find S.

4. $A = \pi(R^2 - r^2)$; $R = 13$, $r = 5$. Find A.

5. $R = \dfrac{kL}{d^2}$; $k = 8$, $L = 100$, $d = 5$. Find R.

6. $a = p(1 + rt)$; $a = 3000$, $p = 2000$, $r = 0.05$. Find t.
7. $P = 2(L + W)$; $P = 76$, $W = 27$. Find L.
8. $V = \frac{1}{3}\pi r^2 h$; $V = 51\pi$, $r = 3$. Find h.
9. $T = mg - mf$; $T = 80$, $m = 10$, $g = 14$. Find f.
10. $S = \frac{1}{2}n(a + L)$; $S = 85$, $n = 17$, $a = -7$. Find L.

In Exercises 11–30 solve each formula for the variable indicated.

11. $F = ma$ for m
12. $D = r \cdot t$ for r
13. $d = \frac{1}{2}at^2$ for a
14. $V = \frac{1}{3}\pi r^2 h$ for h
15. $P = mgh$ for m
16. $i = p \cdot r \cdot t$ for r

17. $A = \dfrac{a + b + c}{3}$ for b
18. $T = m \cdot g - m \cdot f$ for f

19. $S = \frac{1}{2}n(a + L)$ for L
20. $A = \frac{1}{2}h(b + c)$ for b
21. $a = p(1 + rt)$ for t
22. $a = p + prt$ for t

23. $A = \pi(R^2 - r^2)$ for R^2
24. $R = \dfrac{kL}{d^2}$ for d^2

25. $S = \dfrac{a}{1 - r}$ for r
26. $C = \dfrac{E}{R + S}$ for R

27. $I = \dfrac{2E}{R + 2r}$ for r
28. $\dfrac{I_1}{I_2} = \dfrac{r_1{}^2}{r_2{}^2}$ for I_2

29. $n = \dfrac{A + D}{A}$ for A
30. $xy' + y + y' = 2$ for y'
  (*read:* y prime)

1–7 RATIO AND PROPORTION

A *ratio* is a comparison by division of two quantities. For example, if the length of a room is 15 ft and the width of the room is 12 ft, the ratio of the length to the width is expressed in one of the following ways:

1. By the use of a colon: for example, 15:12 (*read:* 15 is to 12).
2. As a fraction: for example, $\frac{15}{12}$.

A ratio should always be expressed in the simplest form. Therefore, 15:12 would become 5:4 and $\frac{15}{12}$ would become $\frac{5}{4}$. It is important to remember that in expressing a ratio, the quantities must be measured in the same units whenever possible. Thus, the ratio of the length of a 1-ft ruler to the length of a 1-yd stick is not 1:1 but 1:3, because there are 3 ft in 1 yd.

EXAMPLE 1: Express in simplest form the ratio of a time interval measured at 30 seconds to another interval measured at 5 minutes.

SOLUTION: Since there are 300 seconds in 5 minutes, the ratio is 30:300, which simplifies to 1:10.

A *proportion* is a statement that two ratios are equal. For example, the ratios 4:5 and 8:10 are equal and form a proportion that may be written as 4:5 = 8:10 (*read*: 4 is to 5 as 8 is to 10) or as $\frac{4}{5} = \frac{8}{10}$. When working with proportions it is easier to write the ratios as fractions and then to use the techniques that we have developed for equations.

EXAMPLE 2: If 50 gal of water flows through a feeder pipe in 20 minutes, how many gallons of water will flow through the pipe in 32 minutes?

SOLUTION: Set up two equal ratios that compare like measurements.

$$\frac{32 \text{ minutes}}{20 \text{ minutes}} = \frac{x}{50 \text{ gal}}$$

(*Note*: It is easier if you set up the ratios so that the unknown is the numerator of a fraction.) Solve this equation.

$$\frac{8}{5} = \frac{x}{50}$$

$50(\frac{8}{5}) = x$ multiply both sides of the equation by 50

$80 = x$

Thus, 80 gal of water will flow through the pipe in 32 minutes.

EXAMPLE 3: A map has a scale of 1.5 in. = 4 mi. To the nearest tenth of a mile, what distance is represented by 14 in. on the map?

SOLUTION: Set up two equal ratios that compare like measurements.

$$\frac{14 \text{ in.}}{1.5 \text{ in.}} = \frac{x}{4 \text{ mi}}$$

Solve this equation.

$$\frac{14}{1.5} = \frac{x}{4}$$

$4\left(\dfrac{14}{1.5}\right) = x$ multiply both sides of the equation by 4

$37.3 = x$ (to the nearest tenth)

Thus, 14 in. on the map corresponds to approximately 37.3 mi.

EXAMPLE 4: A cylindrical tank holds 480 gal when it is filled to its full height of 8 ft. When the gauge shows that it contains oil at a height of 3 ft 1 in., how many gallons are in the tank?

SOLUTION: Set up two equal ratios that compare like measurements.

$$\frac{3 \text{ ft } 1 \text{ in.}}{8 \text{ ft}} = \frac{37 \text{ in.}}{96 \text{ in.}}$$

Therefore,

$$\frac{37 \text{ in.}}{96 \text{ in.}} = \frac{x}{480 \text{ gal}}$$

Solve this equation.

$$\frac{37}{96} = \frac{x}{480}$$

$$\overset{5}{\cancel{480}} \left(\frac{37}{\underset{1}{\cancel{96}}} \right) = x \qquad \text{multiply both sides of the equation by 480}$$

$$185 = x$$

Thus, there are 185 gal of oil in the tank when the oil is at a height of 3 ft 1 in.

EXERCISES 1–7

In Exercises 1–10 express each ratio in simplest form.

1. 21 ft to 14 ft
2. $48 to $120
3. 20 gal to 25 gal
4. 380 rpm to 75 rpm
   (rpm means revolutions per minute)
5. 60 in. to 2 ft
6. 12 ft to 8 yd
7. 150 seconds to 2 minutes
8. 2 dollars to 75 cents
9. 24 oz to 3 lb
10. 10 qt to 3 gal

In Exercises 11–30 answer the question by setting up and solving a proportion.

11. In a certain concrete mix, the ratio of sand to cement is 4:1. At this ratio, how many pounds of cement is needed to be mixed with 50 lb of sand?

12. A spring is stretched 6 in. by a weight of 2 lb. How much weight is needed to stretch the spring 15 in.?

13. A car travels 225 mi in 5 hours. At the same rate, how long will it take the car to travel 405 mi?

14. An object that weighs 48 lb on earth weighs 8 lb on the moon. How much will a man who weighs 174 lb on the earth weigh on the moon?

15. If 3 g of hydrogen is obtained by a certain process from 27 g of water, how much water would be required to produce 7 g of hydrogen by the same process?

16. If 20 ft of wire weighs 25 lb, what is the weight of 16 ft of the wire?

17. The tax on property assessed at $12,000 is $800. What is the assessed value of property taxed at $500?

18. The simple interest that an investor earns in one year is $55 if he deposits $1000. How much interest would he earn on $600?

P. 36

19. The commission of a real estate agent is $1800 if he sells a house for $30,000. How much commission does he earn for selling a house at $36,000?

20. The odds against the Mets winning the pennant are given at 17:6. If you bet $33 on the Mets and they win, how much did you win?

21. A stake 12 ft high casts a shadow 9 ft long at the same time that a tree casts a shadow 33 ft. What is the height of the tree?

22. The gable end of a building has the dimensions illustrated in the drawing on the left. If a carpenter wishes to keep the same proportions, what would be the vertical height of the structure illustrated in the drawing on the right?

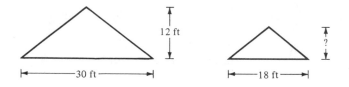

23. On a test you received a grade of 114 on a scale of 0–120. Convert your score to a scale of 0–100.

24. A cylindrical tank holds 370 gal when it is filled to its full height of 9 ft. When the gauge shows that it contains oil at a height of 6 ft 7 in., how many gallons (to the nearest gallon) are in the tank?

25. If 72 gal of water flows through a feeder pipe in 40 minutes, how many gallons of water will flow through the pipe in 3 hours?

26. A map has a scale of 1.5 in. = 7 mi. To the nearest tenth of a mile, what distance is represented by 20 in. on the map?

27. A car will travel 67 mi on 8 gal of gas. To the nearest mile, how far will the car travel on 11 gal of gas?

28. If 2 g of hydrogen unite with 16 g of oxygen to form water, how many grams of oxygen is needed to produce 162 g of water? (Assume that no loss takes place during the reaction.)

29. A certain antifreeze mixture calls for 5 qt of antifreeze to be mixed with 2 gal of water. How many quarts of antifreeze is needed for a total mixture of 65 qt?

30. A sociologist wishes to do a survey in a city where the census report indicates that the ratio of Protestants to Catholics is 5:2. He also determines that a suitable sample size is 420 people. How many of each religion should he interview if his sample is to have the same religious distribution as the city's population?

31. The power and beauty of mathematics is that one problem analyzed abstractly and mathematically can be applied to many different physical situations. For example, consider the following applications which refer to the drawing shown.

a. One of the most important phenomenon in our physical world is light. The simplest principle in the mathematical attack on the theory of this subject is the *law of reflection*. This law states that a ray of light which strikes a reflecting surface is reflected so that the angle of incidence (*i*) equals the angle of reflection (*r*). Many optical instruments, such as telescopes, work on this principle. This law also explains why you see yourself in a mirror. Light from your body strikes the mirror and is then

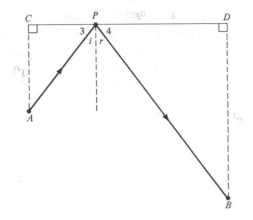

reflected back to your eyes. For a better lighting effect a photographer positioned at A aims his flash at position P on a reflecting surface to take a picture of a subject at B. If $AC = 4$ ft, $CP = 3$ ft, and $PD = 6$ ft, determine $AP + PB$ so that the photographer can set his camera at the correct f stop.

b. The same principle applies to a hard body that bounces off an elastic surface. That is, it rebounds at the same angle at which it strikes. A billiard player wishes to make a ball at A strike the cushion at P to hit another ball at B. If $AC = 2$ ft, $CD = 6$ ft, and $DB = 3$ ft, determine CP so that point P may be found.

c. Suppose that a company owns two plants located at A and B and that line segment CD represents a railroad track. The company wishes to build a station at P to ship its merchandise. Naturally they want the total trucking distance $AP + BP$ to be a minimum. Mathematically it can be shown that $AP + BP$ is a minimum when angle 3 equals angle 4. If $AC = 3$ mi, $CD = 10$ mi, and $DB = 5$ mi, determine DP so that the most desirable spot for the station may be found.

32. One of the classic problems from the history of mathematics is the ingenious method used by the Greek mathematician Eratosthenes to estimate the circumference of the earth in about 200 B.C. You might think the idea that the earth is a sphere was a new thought in Columbus' time. Actually the thought was new only to Europeans. In about A.D. 1200 Greek ideas about the physical world were introduced in Europe which had a profound effect on the society and were, in fact, a major cause of the Renaissance. One concept, held by most learned Greeks, was that the earth was round. Eratosthenes estimated the circumference of the earth by noting the following information, which is illustrated here. Alexandria (where Eratosthenes was librarian) was 500 mi due north of the city of Syene. At noon on June 21 he

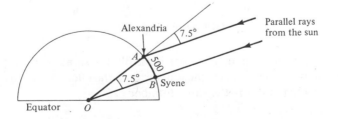

knew from records that the sun cast no shadow at Syene, which meant that the sun was directly overhead. At the same time in Alexandria he measured from a shadow that the sun was 7.5° south from the vertical. By assuming that the sun was sufficiently far away for the light rays to be parallel to the earth, he then determined that angle *AOB* was 7.5°. (Why?) Complete the line of reasoning and obtain Eratosthenes' estimate for the circumference. You will find the result remarkably close to the modern estimate of 24,900 mi, which is obtained with the same basic procedure but more accurate measurements.

1–8 INEQUALITIES

The solution to a problem is often not a single number but rather any number less than a certain number. For example, suppose that the average of your grades must be at least 60 if you are to pass this course. Then for any average that is *less than* 60 you will not satisfactorily complete the course. Similarly, the solution to a problem is often any number that is greater than a certain number. If the speed limit on an expressway is 65 mi/hour, you might get a ticket if you are traveling at any speed that is *greater than* 65 mi/hour. Statements involving these relationships are called *inequalities*. When we translate these problems into mathematical symbols the relationship "less than" is symbolized by $<$ (for example, $4 < 8$) and the relationship "greater than" is symbolized by $>$ (for example, $8 > 4$).

    Relations of "greater than" and "less than" can be seen very easily on the number line, as shown in Figures 1.12–1.14. The point representing the larger number will be to the right of the point representing the smaller number.

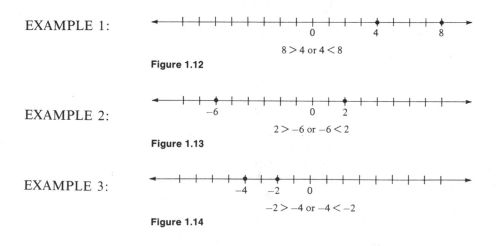

EXAMPLE 1:

$8 > 4$ or $4 < 8$

Figure 1.12

EXAMPLE 2:

$2 > -6$ or $-6 < 2$

Figure 1.13

EXAMPLE 3:

$-2 > -4$ or $-4 < -2$

Figure 1.14

    To solve an inequality, we use basically the same method that we developed to solve equations, but first we must determine the effect of adding or multiplying each side of an inequality by the same number.

EXAMPLE 4: $4 < 8$. If we add 3 to both sides of the inequality, the result, $7 < 11$, is a true statement.

EXAMPLE 5: $4 < 8$. If we subtract 3 from both sides of the inequality, the result, $1 < 5$, is a true statement.

EXAMPLE 6: $4 < 8$. If we multiply both sides of the inequality by 3, the result, $12 < 24$, is a true statement.

EXAMPLE 7: $4 < 8$. If we multiply both sides of the inequality by -3, the result, $-12 < -24$, is a false statement. We can make this statement true by changing the "less than" to "greater than" so that the result is $-12 > -24$. Thus, if we multiply an inequality by a negative number, we must change the sense of the inequality if the statement is to be true. The sense of the inequality means the direction in which the inequality symbol is pointing.

From the examples we can generalize to say that given any inequality, the result of each of the following is an equivalent (same solution) inequality:

1. Add or subtract the same expression to (from) both sides of the inequality.
2. Multiply or divide both sides of the inequality by the same positive number.
3. Multiply or divide both sides of the inequality by the same negative number and change the sense of the inequality.

EXAMPLE 8: For what values of x is the inequality $5x + 2 < 22$ true?

SOLUTION:
$$5x + 2 < 22$$
$$\underline{ - 2 \quad - 2} \qquad \text{subtract 2 from both sides}$$
$$5x \quad < \quad 20$$

$$\frac{5x}{5} \quad < \quad \frac{20}{5} \qquad \text{divide both sides by 5}$$

$$x \quad < \quad 4$$

Thus all the real numbers less than 4 will make the above inequality true. The graph of this solution (Figure 1.15) is the collection of points on the number line representing numbers less than 4. We put a blank circle at the point representing the number 4 since 4 is not a solution to the problem.

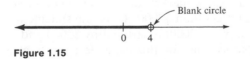

Figure 1.15

EXAMPLE 9: For what values of x is the inequality $-2x - 7 > 13$ true?

SOLUTION:

$$-2x - 7 > 13$$
$$\underline{+7 \quad +7} \qquad \text{add 7 to both sides}$$
$$-2x \quad > \quad 20$$

$$\frac{-2x}{-2} \quad < \quad \frac{20}{-2} \qquad \begin{array}{l}\text{divide both sides by } -2 \text{ and} \\ \text{change the sense of the inequality}\end{array}$$

$$x \quad < -10 \qquad \text{(see Figure 1.16)}$$

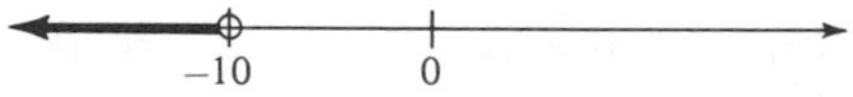

Figure 1.16

EXAMPLE 10: Solve the inequality $2x + 1 < 5x - 6$.

SOLUTION:

$$2x + 1 < 5x - 6$$
$$\underline{-1 \qquad \quad -1} \qquad \text{subtract 1 from both sides}$$
$$2x \quad < \quad 5x - 7$$
$$\underline{-5x \qquad \quad -5x} \qquad \text{subtract } 5x \text{ from both sides}$$
$$-3x \quad < \qquad -7$$

$$\frac{-3x}{-3} \quad > \quad \frac{-7}{-3} \qquad \begin{array}{l}\text{divide both sides by } -3 \text{ and} \\ \text{change the sense of the inequality}\end{array}$$

$$x \quad > \quad \frac{7}{3} \qquad \text{(see Figure 1.17)}$$

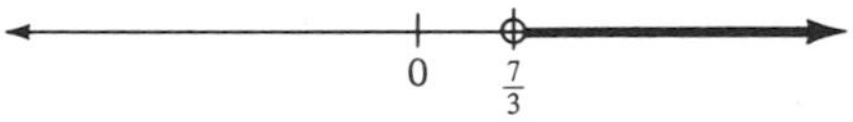

Figure 1.17

EXAMPLE 11: Solve the inequality $x > x + 1$.

SOLUTION:

$$x > x + 1$$
$$\underline{-x \quad -x} \qquad \text{subtract } x \text{ from both sides}$$
$$0 > \quad 1$$

Note that the resulting inequality is false. Since the original inequality is *equivalent* to a false statement, no replacements of x will make it true. There is no solution.

Sometimes the data from a problem allow for solutions that are not only less than a certain number but are also equal to that number. For example, if you have $2000 to invest in the stock market, the amount (x) that you invest can be exactly $2000 or any amount less than $2000. Thus, the statement "x is less than *or* equal to 2000" (written as $x \leq \$2000$) is satisfied by $2000 and all the amounts that are less than $2000. (Notice that for this example there is also the restriction that x cannot be negative.) Similarly, the statement "x is greater than *or* equal to 1000"

(written as $x \geq \$1000$) is satisfied by $\$1000$ and all the amounts greater than $\$1000$. Solutions to problems that use these relationships are obtained by use of the same procedure as outlined above.

EXAMPLE 12: For what value of x is the inequality $7x + 2 \leq 5x - 3$ true?

SOLUTION:
$$
\begin{array}{lll}
7x + 2 \leq & 5x - 3 & \\
\quad\, - 2 & \quad\, - 2 & \text{subtract 2 from both sides} \\
\hline
7x \quad\; \leq & 5x - 5 & \\
- 5x & \quad - 5x & \text{subtract } 5x \text{ from both sides} \\
\hline
2x \quad\; \leq & \quad -5 & \\
\end{array}
$$

$$
\begin{array}{lll}
\dfrac{2x}{2} \leq \dfrac{-5}{2} & \quad \text{divide both sides by 2} \\[2mm]
x \;\;\leq \dfrac{-5}{2} & \\
\end{array}
$$

The graph of this solution is the collection of points on the number line representing $-5/2$ or numbers less than $-5/2$ (Figure 1.18). We put a solid circle at the point representing $-5/2$ since $-5/2$ is a solution to the problem.

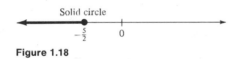

Figure 1.18

EXAMPLE 13: For what values of x is the inequality $2x + 1 \leq 7x + 4$ true?

SOLUTION:
$$
\begin{array}{lll}
2x + 1 \leq & 7x + 4 & \\
\quad\, - 1 & \quad\, - 1 & \text{subtract 1 from both sides} \\
\hline
2x \quad\; \leq & 7x + 3 & \\
- 7x & \quad - 7x & \text{subtract } 7x \text{ from both sides} \\
\hline
- 5x \quad\; \leq & \quad 3 & \\
\end{array}
$$

$$
\begin{array}{lll}
\dfrac{-5x}{-5} \geq \dfrac{3}{-5} & \quad \text{divide both sides by } -5 \text{ and} \\
 & \quad \text{change the sense of the inequality} \\[2mm]
x \;\;\geq \dfrac{-3}{5} & \quad \text{(see Figure 1.19)} \\
\end{array}
$$

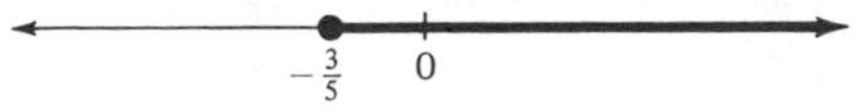

Figure 1.19

Inequalities can be used to describe an interval between two numbers. For example, suppose that the average (a) of your grades must be between 80 and 90 if you are to receive a grade of B in the course. We may describe this interval by using a pair of inequalities.

$$80 < a \quad \text{and} \quad a < 90$$

The inequality $80 < a$ specifies that the average must be above 80, while $a < 90$ specifies that the average must be below 90. The two inequalities are both true only for averages between 80 and 90 (see Figure 1.20). The statement "$80 < a$ and $a < 90$" is usually expressed in the compact form: $80 < a < 90$.

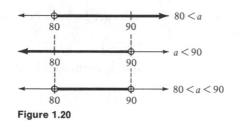

Figure 1.20

Usually an 80 average means a B grade and a 90 average means an A grade. We indicate whether one or both end points are included in the interval by using either the "less than" symbol ($<$) or the "less than or equal to" symbol ($\leq$), as shown in Figure 1.21.

Inequality	Interval	Comment
$80 < a < 90$	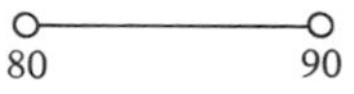 ○———○ 80 90	This interval excludes both 80 and 90.
$80 \leq a \leq 90$	 ●———● 80 90	This interval includes both 80 and 90.
$80 \leq a < 90$	 ●———○ 80 90	This interval includes 80 and excludes 90. Usually, an average in this interval means a grade of B.
$80 < a \leq 90$	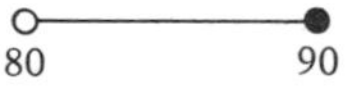 ○———● 80 90	This interval excludes 80 and includes 90.

Figure 1.21

Finally, consider Figure 1.22. In this case we wish to describe the collection of numbers that are either less than 80 or more than 90 and we write

$$a < 80 \quad \text{or} \quad a > 90$$

We do not write $80 > a > 90$. This statement means $a < 80$ and $a > 90$ simultaneously. No value for a satisfies both inequalities. Thus, it is important to distinguish between the mathematical meanings of "and" and "or." In an "and" statement both inequalities must be true simultaneously, whereas an "or" statement

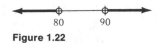

Figure 1.22

is true if at least one of the inequalities is true. Only an "and" statement can be written in compact form.

EXAMPLE 14: A company manufactures two products, A and B. Product A costs $20 per unit to produce and product B costs $60 per unit. The company has $90,000 to spend on the production. If x represents the number of units produced of product A and y represents the number of units produced of product B, write an inequality that expresses the restriction placed on the company because of available funds.

SOLUTION: x represents the number of units produced of product A and each unit costs $20 to produce. Thus, the company spends $20x$ dollars to produce x units of product A. Similarly, the company spends $60y$ dollars to produce y units of product B. Since the total cost cannot exceed $90,000, we have

$$20x + 60y \leq 90,000$$

Since the total cost cannot be negative, we also have

$$0 \leq 20x + 60y$$

Thus, the restriction placed on the company because of available funds is given by

$$0 \leq 20x + 60y \leq 90,000 \quad \text{with} \quad x \geq 0 \text{ and } y \geq 0$$

The question in Example 14 is designed to show the usefulness of inequalities in describing practical situations. Capital is only one of many restrictions that a company must consider because of limited resources. For example, machine time, man-hours of labor, and storage space are other considerations. Subject to these restrictions, the company must determine the number of units of each product that should be manufactured for maximum profit. The solution of this problem is found by techniques from a branch of applied mathematics called *linear programming*. Other problems that have been solved by linear programming include:

1. Determining the most economical mixture of ingredients that will result in a product with certain minimum requirements.
2. Determining the quickest and most economical route in distributing a product.
3. Determining the most efficient use of industrial machinery.
4. Determining the most efficient production schedule.

EXERCISES 1-8

In Exercises 1-20 use the proper symbol ($<$, $>$, $=$) between the pairs of numbers to indicate their correct order.

1. 6 _____ 9   2. -1 _____ -4
3. -7 _____ 1   4. 42.8 _____ -91

5. $|14|$ ___<___ $|-16|$

6. $|1.46|$ _____ $|-1.46|$

7. $|-16.2|$ ___<___ $|-18.1|$

8. $\frac{1}{2}$ _____ 0.5

9. -0.0001 ___<___ -0.00001

10. 0.0001 _____ 0.00001

11. 100% ___=___ 1

12. 5% _____ 0.5

13. $\frac{5}{8}$ ___<___ $\frac{2}{3}$

14. $\frac{3}{7}$ _____ $\frac{5}{9}$

15. 2 ___>___ $\sqrt{3}$

16. $\sqrt{77}$ _____ 8

17. $\frac{1}{3}$ _____ 0.33

18. $\frac{2}{3}$ _____ 0.67

19. π ___>___ 3.14

20. $\frac{1}{9}$ _____ $0.\overline{1}$

In Exercises 21–50 solve the inequality and graph each solution.

21. $x - 8 > 0$

22. $x + 5 < 0$

23. $4x > 8$

24. $-9x \le 81$

25. $4x \le 3x + 1$

26. $10x < 11x - 3$

27. $2x + 9 < 0$

28. $3 - 4x > 0$

29. $-4x - 5 \ge 6x + 5$

30. $4 - 6x \ge -7 - 9x$

31. $13 - 2x < 7 - 3x$

32. $16x - 7 \le 17x - 4$

33. $-2x + 5 > 5x - .7$

34. $1 - 4x \ge 2x + 3$

35. $7 - x > 5 - 2x$

36. $214 + 6x \le -4x + 6$

37. $2(x - 3) \ge 6(x + 1)$

38. $4(3 + x) > 8(x - 4)$

39. $\dfrac{x}{2} + 4 < 8 + x$

40. $24 - 13x \ge -12.5x + 26$

41. $x > x - 1$

42. $x < x - 1$

43. $3x < 4x$

44. $-x \ge 2x$

45. $2(x + 3) \ge 7 + 2x$

46. $7 - 4y \le 4(7 - y)$

47. $1 - 3y \le 1 - 4y$

48. $5x + 4 > x + 4$

49. $2(2 - x) + 1 > 5 - 2x$

50. $2(2 - x) + 1 \ge 5 - 2x$

In Exercises 51–62 answer true or false. If false, give an example for which the statement does not hold. (*Note:* True means that the statement holds for *all* values of x and y. False means there is at least one pair of values of x and y for which the statement is not true.)

51. If $x < 2$ and $y < 5$, then $x + y < 10$.

52. If $x < 1$ and $y < 2$, then $x < y$.

53. If $x < 0$ and $y < 0$, then $xy < 0$.

54. If $x < 0$ and $y < 0$, then $x + y < 0$.

55. If $x > 0$ and $y < 0$, then $x - y > 0$.

56. If $x < 2$ and $y < 3$, then $xy < 6$.

57. If $x > 2$ and $y > 3$, then $xy > 6$.

58. If $x < 0$ and $y > 1$, then $xy < -1$.

59. If $x < y$, then $x^2 < y^2$. true

60. If $x^2 > y^2$, then $x > y$.

61. If x and y are negative numbers and $x < y$, then $x^2 > y^2$.

62. If x and y are positive numbers and $x < y$, then $x^2 < y^2$.

In Exercises 63–66 graph each collection of numbers.

63. $x < 1$ or $x > 3$ **64.** $x > 1$ and $x < 3$

65. $1 \le x < 3$ **66.** $x < 0$ or $x \ge 2$

In Exercises 67–70 use inequalities to describe the collection of numbers illustrated. If possible, write the interval in compact form.

67.   2   4   **68.**   2   4

69.   2   4   **70.**   2   4

71. The most that you can deduct from your earnings for a retirement fund is 15 percent. If your salary is \$10,000, use inequalities to write an interval that contains the amount (a) that you might contribute to your retirement fund.

72. To the nearest inch, the height of a man is 69 in. Use inequalities to write an interval that contains the man's exact height. If we use a more accurate measuring device and record his height as 69.3 in., which interval now contains the man's exact height?

73. In Lobachevskian geometry the sum of the three angles in a triangle is always less than 180°. If the second angle is 20° more than the first and the third is twice the first, how many degrees might there be in the first angle?

74. A farmer has 50 acres on which to plant corn and potatoes. If x represents the number of acres on which he plants corn and y represents the number of acres on which he plants potatoes, find an inequality that expresses the restriction that is placed on him because of available land.

75. The same farmer as in Exercise 74 has \$10,000 in capital. It costs him \$150 per acre to plant and maintain corn and \$250 per acre to plant and maintain potatoes. If x represents the number of acres on which he plants corn and y represents the number of acres on which he plants potatoes, find an inequality that expresses the restriction that is placed on him because of capital.

76. The same farmer as in Exercise 74 has 1000 man-hours of labor available to him for the season. An acre of corn requires 24 hours of labor while an acre of potatoes requires 17 hours of labor. If x represents the number of acres on which he plants corn and y represents the number of acres on which he plants potatoes, find an inequality that expresses the restriction that is placed on him because of available manpower.

77. A man has \$2000 to invest in the stock market. Stock A sells for \$90 per share, stock B costs \$50 per share, and stock C costs \$300 per share. If x, y, and z represent the number of shares that he bought of each stock, respectively, find an inequality that expresses this relationship.

78. A TV manufacturer has available 10,000 man-hours of work per month. A console requires 70 man-hours of labor and a portable requires 50 man-hours of labor. If x represents the number of consoles produced and y represents the number of portables produced, find an inequality that expresses this relationship.

79. Government regulations limit the amount of a pollutant that a company can discharge into the water to A gal/day. If product 1 produces B gallons of pollutant per unit and product 2 produces C gallons of pollutant per unit, find an inequality that expresses the restriction placed on this company because of the government regulation.

80. Sad Sam has been informed by his doctor that he would be less sad if he obtained at least the minimum adult requirements of thiamine (1 mg) and of niacin (10 mg) per day. A trip to the supermarket reveals the following facts about his favorite cereals:

Cereal	Thiamine per ounce (mg)	Niacin per ounce (mg)
A	0.10	2.40
B	0.30	1.30

If x represents the number of ounces he eats of cereal A and y represents the number of ounces he eats of cereal B, find two inequalities that express what must be done if he is to satisfy his minimum daily requirement.

Sample Test: Chapter 1

(*Note:* At the end of each chapter there is a sample test. These questions will help you consolidate the concepts presented in the chapter and indicate which areas require further study. The test should be considered as only part of the review that is necessary for an exam.)

1. True or false (if false, give a specific counterexample)

   All real numbers are rational numbers.

2. Express the rational number $\frac{13}{11}$ as a repeating decimal.

3. Name the property of real numbers illustrated in the following statement:

   $(x + y) + z = z + (x + y)$

4. Evaluate: $1 - 3(2 - 4)$.

5. Evaluate the expression $-x^2 + 2xy^3$ if $x = -1$ and $y = 3$.

6. If $m = (y_2 - y_1)/(x_2 - x_1)$, evaluate m for $x_1 = 2$, $x_2 = -5$, $y_1 = -1$, and $y_2 = 4$.

7. Combine similar terms: $2(x - y) + (x + y) - (3 - y)$.

8. Solve: $3(x - 6) = -2(12 - 3x)$.

9. Solve: $1 - x = 1 + x$.

10. Find the value of t in the formula $a = p(1 + rt)$ if $a = 3000$, $p = 1500$, and $r = 0.05$.

11. The total cost (including tax) of a new car is \$5406. If the sales tax is 6 percent, how much did you pay in taxes?

12. You have 20 qt of a solution that is 10 percent antifreeze. How much pure antifreeze must be added to obtain a solution that is 30 percent antifreeze?

13. Solve the wave equation $n = v/L$ for the wavelength L.

14. Solve for y: $P = (x - y - z)/10$.

15. On a test you received a grade of 123 on a scale of 0–150. Convert your score to a scale of 0–100.

16. A map has a scale of 1.5 in. = 11 mi. To the nearest tenth of a mile, what distance is represented by 7 in. on the map?

17. Use the correct symbol ($<$, $>$, $=$) between the following pairs of numbers:

   a. -0.1 _____ -0.01 b. 2% _____ 0.2

18. Solve: $-x + 1 > 3$.

19. Solve: $1 - x \leq 1 + x$.

20. Use inequalities to describe the collection of numbers illustrated.

$$-x + 1 > 3$$
$$-x > 3 - 1$$
$$-x < 2$$

$$1 - x \leq 1 + x$$
$$-x - x \leq 1 - 1$$
$$-2x \leq 0$$

$$0 \leq x > 6$$

Chapter 2
Functions and graphs

2–1 DEFINING FUNCTIONS

One of the most important considerations in mathematics is determining the relationship between two variables. For example:

The postage required to mail a package is related to the weight of the package.
The bill from an electric company is related to the number of kilowatt-hours of electricity that are purchased.
The current in a circuit with a fixed voltage is related to the resistance in the circuit.
The demand for a product is related to the price charged for the product.
The perimeter of a square is related to the length of the side of the square.

In each of these examples we determine the relationship between the two variables by finding a rule that establishes a correspondence between values of each variable. For example, if we know the length of the side of a square, we can determine the perimeter by the formula $P = 4s$. In this case the rule is a formula or equation. Sometimes the rule is given in tabular form. For example, consider the formula table in Figure 2.1, which assigns to each final average (a) a final grade for the course.

Finally, the rule is sometimes given verbally and it is best to just make a list of the correspondences. For example, a telephone book is an enormous list of

$$\text{Final grade} = \begin{cases} A & \text{if} \quad 90 \le a \le 100 \\ B & \text{if} \quad 80 \le a < 90 \\ C & \text{if} \quad 70 \le a < 80 \\ D & \text{if} \quad 60 \le a < 70 \\ F & \text{if} \quad 0 \le a < 60 \end{cases}$$

Figure 2.1

correspondences which results from the rule "assign to each subscriber a telephone number."

In determining the relationship between two variables, the most useful rule is one that assigns to each value of the first variable *exactly one* value of the second variable. For example, the table in Figure 2.1 allows us to determine a final grade because the rule assigns to each final average exactly one final grade. We illustrate this concept in Figure 2.2, in which the assignments are represented as arrows from the points that represent final averages to the points that represent final grades.

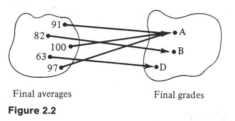

Final averages   Final grades

Figure 2.2

On the other hand, a rule that assigns to each final grade some final average is not as useful. We cannot determine an individual's final average if we know his final grade. For example, an A grade may correspond to any average between 90 and 100 (inclusive). This concept is illustrated in Figure 2.3.

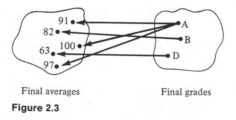

Final averages   Final grades

Figure 2.3

To summarize, we pick a particular value for a variable (which is then called. the *independent variable*) and some rule assigns one or more values for the other variable (which is called the *dependent variable*). We decided the rule is more useful if to each value of the independent variable there corresponds exactly one value for the dependent variable. This discussion leads us to the concept of a function.

DEFINITION OF A FUNCTION: A function consists of three features:
1. A collection of values that may be substituted for the independent variable, called the *domain* of the function.
2. A collection of assigned values for the dependent variable, called the *range* of the function.
3. A rule (usually represented by f) that assigns to each member of the domain *exactly one* member of the range.

It may help you to think of a function in terms of the computing machine in Figure 2.4. For convenience, we think of the "machine" f as transforming x values into y values. In goes an x value, out comes exactly one y value. With this in mind, you should have a clear image of the features of a function: (1) the domain (the input), (2) the range (the output), and (3) the rule f (the machine).

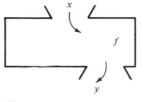

Figure 2.4

When working with functions it is common practice to refer to the function by stating only the rule, such as: "Consider the perimeter function $P = 4s$." The domain of the function is then assumed to be the collection of values for the independent variable that are interpretable in the problem. Thus, the domain of $P = 4s$ is the positive real numbers since the length of the side of a square must be positive. When working with functions defined by algebraic equations, such as $y = 2x^2 - 7$, the domain is the collection of all real numbers for which a real number exists in the range.

EXAMPLE 1: The formula $C = \pi d$ defines a function since to each value of the diameter (d) of a circle there corresponds exactly one circumference (C). The domain of the function is all the positive real numbers since the diameter of a circle must be positive.

EXAMPLE 2: The equation $y = 1/x$ defines a function since to each value of x there corresponds exactly one y value. The domain of the function is all the real numbers except zero. We exclude zero since when $x = 0$, $y = 1/0$, which is undefined.

EXAMPLE 3: The cost (c) for n radial tires if each tire costs \$50 is given by the formula $c = 50n$. This formula defines a function since to a given number of tires there corresponds exactly one cost. The domain of the function is the nonnegative integers since the number of tires purchased must be a whole number.

EXAMPLE 4: The deduction that motorists in New York may take on their federal income tax as a result of paying gasoline taxes is as follows for mileages (m) up to 4000 miles:

$$\text{Deduction} = \begin{cases} \$12 & \text{if} \quad 0 \le m < 3000 \\ \$19 & \text{if} \quad 3000 \le m < 3500 \\ \$22 & \text{if} \quad 3500 \le m < 4000 \end{cases}$$

This table defines a function since to each mileage there corresponds exactly one deduction. The domain is $0 \le m < 4000$. Remember this table is one rule that defines the relationship between mileage and tax deductions. It is not three separate functions. Notice also that we cannot determine an individual's mileage by knowing the deduction, so that if we reverse the assignments we do not have a function.

EXAMPLE 5: Consider the following rule:

$$y = \begin{cases} x + 1 & \text{if} \quad 0 \le x \le 2 \\ 3 & \text{if} \qquad x > 2 \end{cases}$$

This rule assigns to any value of x between 0 and 2 (inclusive) the number one greater than the x value. For example, when $x = 1$, $y = 2$, and when $x = \frac{1}{2}$, $y = \frac{3}{2}$. This rule also assigns the number 3 to each value of x greater than 2. The rule defines a function since to each value of x there corresponds exactly one y value. The domain of the function is symbolized by $x \ge 0$.

EXAMPLE 6: The total cost of producing a certain product consists of $350 per month in rent plus $2 per unit for material and labor. Find a rule that expresses the company's monthly total costs (c) as a function of the number of units (x) they produce.

SOLUTION: If each unit costs $2 to produce, the cost of producing x units is $2x$. To this number we must add the fixed cost of $350 that is paid in rent. Thus, the total cost may be found by the rule

$$C = 2x + 350$$

Since the company must produce a whole number of units, the domain of the function is the nonnegative integers.

EXAMPLE 7: Each month a salesman earns $500 plus 7 percent commission on sales above $2000. Find a rule that expresses the monthly earnings (e) of the salesman in terms of the amount (a) of merchandise he sells.

SOLUTION: If the salesman sells less than or equal to $2000 worth of merchandise for the month, he earns $500. Thus,

$$e = \$500 \quad \text{if} \quad \$0 \leq a \leq \$2000$$

If the salesman sells above $2000, he earns $500 plus 7 percent of the amount above $2000. Thus,

$$e = \$500 + 0.07(a - \$2000) \quad \text{if} \quad a > \$2000$$

The following rule may then be used to determine the monthly earnings of the salesman when we know the amount of merchandise that he sold:

$$e = \begin{cases} \$500 & \text{if} \quad \$0 \leq a \leq \$2000 \\ \$500 + 0.07(a - \$2000) & \text{if} \quad a > \$2000 \end{cases}$$

Domain of the function: $a \geq \$0$.

In mathematical notation we use *ordered pairs* to show the correspondences in a function. For example, consider the equation $y = 2x + 1$.

If x equals	then $y = 2x + 1$	thus the ordered pairs
2	$2(2) + 1 = 5$	$(2, 5)$
1	$2(1) + 1 = 3$	$(1, 3)$
0	$2(0) + 1 = 1$	$(0, 1)$
-1	$2(-1) + 1 = -1$	$(-1, -1)$
-2	$2(-2) + 1 = -3$	$(-2, -3)$

In the pairs that represent the correspondences in a function we list the values of the independent variable first and the values of the dependent variable second. Thus, the order of the numbers in the pair is significant. The pairing $(2, 5)$ indicates that when $x = 2$, $y = 5$; $(5, 2)$ means when $x = 5$, $y = 2$. In the equation $y = 2x + 1$, $(2, 5)$ is an ordered pair that makes the equation a true statement, so $(2, 5)$ is said to be a solution of the equation; $(5, 2)$ is not a solution of this equation.

EXAMPLE 8: Which of the following ordered pairs are solutions of the equation $y = 3x - 2$?
(a) $(-1, -5)$ 　　　　　　　　　　(b) $(-5, -1)$

SOLUTION:
(a) $(-1, -5)$ means that when $x = -1$, $y = -5$

$$y = 3x - 2$$
$$(-5) \stackrel{?}{=} 3(-1) - 2$$
$$-5 \stackrel{\vee}{=} -5$$

Thus, $(-1, -5)$ is a solution.

(b) $(-5, -1)$ means that when $x = -5$, $y = -1$

$$y = 3x - 2$$
$$(-1) \stackrel{?}{=} 3(-5) - 2$$
$$-1 \neq -17$$

Thus, $(-5, -1)$ is not a solution.

The representation of the correspondences in a function as ordered pairs gives us a different perspective of the function concept. In higher mathematics the following definition of a function is very popular.

> FUNCTION: A function is a collection of ordered pairs in which no two different ordered pairs have the same first component. The collection of all first components (of the ordered pairs) is called the *domain* of the function. The collection of all second components is called the *range* of the function.

EXAMPLE 9: Consider the ordered pairs (2, 1), (3, 1), and (4, −1). This collection is a function because the first component in the ordered pairs is always different. Domain: 2, 3, 4; range: 1, −1 (see Figure 2.5).

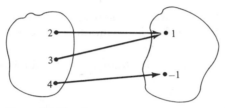

Figure 2.5 Function

EXAMPLE 10: Consider the ordered pairs (1, 2), (1, 3), and (−1, 4). This collection is not a function because the number 1 is the first component in two ordered pairs (see Figure 2.6).

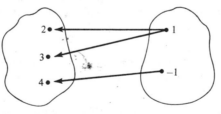

Figure 2.6 Not a Function

In this section we have defined a function in terms of (1) a rule and (2) ordered pairs. Since the function concept is so important, you should consider both definitions and satisfy yourself that these definitions are equivalent.

EXERCISES 2–1

In Exercises 1–10 find five ordered pairs that are solutions of each formula or equation.

1. $y = x$ **2.** $y = -x$

3. $y = 4 - x$ **4.** $y = 2x - 7$

5. $y = x^2 + 2x - 1$

6. $y = x^3 - 1$

7. $d = 60t$

8. $c = \pi d$

9. $C = \frac{5}{9}(F - 32)$

10. $F = \frac{9}{5}C + 32$

In Exercises 11–14 determine which of the ordered pairs are solutions of the equation.

Equation	Ordered pairs
11. $y = 3x + 1$	$(4, 1), (-2, -5), (1, 4), (-5, -2)$
12. $y = \|x\|$	$(-1, -1), (1, 1), (-1, 1), (1, -1)$
13. $y = x^2$	$(-1, -1), (1, 1), (-1, 1), (1, -1)$
14. $y = 1/x$	$(2, 0.5), (0.5, 2), (-1, -1), (0, 0)$

In Exercises 15–20 state which collections of ordered pairs are functions.

15. $(1, -3)(2, 0)(3, 1)$

16. $(-3, 1)(0, 2)(1, 3)$

17. $(2, -1)(3, 0)(4, -1)$

18. $(-1, 2)(0, 3)(-1, 4)$

19. $(2, 1)(2, 2)(2, 3)$

20. $(1, 2)(2, 2)(3, 2)$

In Exercises 21–30 find the domain of the function.

21. $C = 2\pi r$ (circumference of a circle)

22. $A = S^2$ (area of a square)

23. $y = x^2$

24. $y = 2x + 5$

25. $y = \dfrac{1}{x + 4}$

26. $y = \dfrac{2}{x(x - 1)}$

27. $(-2, 4)(-1, 1)(1, 1)(2, 4)$

28. $(0, 5)(1, 6)(2, 7)(3, 8)$

29. $y = \begin{cases} x & \text{if } 0 \le x \le 1 \\ -1 & \text{if } x > 1 \end{cases}$

30. $y = \begin{cases} x + 1 & \text{if } -2 \le x < 1 \\ x - 1 & \text{if } 1 \le x < 5 \end{cases}$

In Exercises 31–36 study each table and write a formula (rule) to define the relationship between the variables.

31.

If x is	1	2	3	4	5
y is	2	4	6	8	10

32.

If r is	15	20	25	30	35
d is	75	100	125	150	175

33.

If x is	6	7	8	9	10
y is	3	4	5	6	7

34.

If s is	12	17	22	27	32
z is	27	32	37	42	47

35.

If s is	1	2	3	4	5
a is	1	4	9	16	25

36.

If e is	1	2	3	4	5
V is	1	8	27	64	125

In Exercises 37–42 each collection of ordered pairs gives a few of the correspondences in some function. Write a formula to define the relationship. Use x for the independent variable and y for the dependent variable.

37. $(-1, -1)(0, 0)(1, 1)$

38. $(-1, 1)(0, 0)(1, 1)$

39. (1, 12)(2, 22)(3, 32)(4, 42)   **40.** (1, 7)(2, 13)(3, 19)(4, 25)

41. (6, 11)(7, 13)(8, 15)(9, 17)   **42.** (−2, 4)(0, 2)(2, 0)(4, −2)

In Exercises 43–62 find a formula or table that defines the functional relationship between the two variables; in each case indicate the domain of the function.

43. Express the area (A) of a square in terms of its side (s).

44. Express the area (A) of a rectangle of width 5 as a function of its length (L).

45. Express the length (L) of a rectangle of width 5 in terms of its area A.

46. Express the area (A) of a circle as a function of its radius (r).

47. Express the side (S) of a square in terms of its perimeter (P).

48. Express the area (A) of a square as a function of the perimeter (P) of the square.

49. Express the distance (d) that a car will travel in t hours if the car is always traveling at 40 mi/hour.

50. Express the cost (c) of y gal of gasoline if the gas cost 60 cents/gal.

51. Express the earnings (e) of an electrician in terms of the number (n) of hours worked if he makes $8 per hour.

52. Express the interest (i) earned on $1000 at 6 percent simple interest as a function of the time (t).

53. Express the earning (e) of a real estate agent who receives a 6 percent commission as a function of the sale price (p) of a house.

54. Express the monthly cost (c) for a checking account as a function of the number (n) of checks serviced that month if the bank charges 10 cents per check plus a 75 cents maintenance charge.

55. The total cost of producing a certain product consists of paying $400 per month rent plus $5 per unit for material. Express the company's monthly total costs (c) as a function of the number of units (x) they produce that month.

56. A company plans to purchase a machine to package its new product. If the company estimates that the machine will cost $5000 per year plus $1 to package each unit, express the yearly cost (c) of using the machine as a function of the number of units (x) packaged during the year.

57. A reservoir contains 10,000 gal of water. If water is being pumped from the reservoir at a rate of 50 gal/minute, write a formula expressing the amount (a) of water remaining in the reservoir as a function of the number (n) of minutes the water is being pumped.

58. Express the monthly cost of renting a computer as a function of the number (n) of hours the computer is used if the company charges $200 plus $100 for every hour the computer is used during the month.

59. Express the federal income tax (t) for a single person as a function of his taxable income (i) if the person's taxable income is between $2000 and $4000 inclusive and the tax rate is $310 plus 19 percent of the excess over $2000.

60. Express the monthly earnings (e) of a salesman in terms of the cash amount (a) of merchandise sold if the salesman earns $600 per month plus 8 percent commission on sales above $10,000.

61. Express the monthly cost (c) of an electric bill in terms of the number (n) of electrical units purchased (the unit of measure is the kilowatt-hour) if the person used no more than 48 units and the electric company has the following rate schedule:

	Amount	Charge
First	12 units or less	$1.75
Next	36 units at	3.82 cents/unit

62. A long strip of galvanized sheet metal 12 in. wide is to be formed into an open gutter by bending up the edges to form a gutter with rectangular cross section. Write the cross-sectional area of the gutter as a function of the depth (x).

2–2 VARIATION

In many scientific laws the functional relationship between variables is stated in the language of variation. The statement "y varies *directly* as x" means that there is some positive number k such that $y = kx$. The constant k is called the *variation constant*.

EXAMPLE 1: The perimeter of a square varies directly as the side. Algebraically, we write $P = ks$. In this case we know that $k = 4$.

EXAMPLE 2: The circumference of a circle varies directly as the diameter. Algebraically, we write $C = kd$. In this case we know that $k = \pi$.

EXAMPLE 3: The sales tax (T) on a purchase varies directly as the price (p) of the item. Algebraically, we write $T = kp$. The value of k depends on the sales tax rate in a given location.

EXAMPLE 4: Hooke's law states that the distance (d) a spring is stretched varies directly as the force (F) applied to the spring. Algebraically, we write $d = kF$. The value of k depends upon the spring in question (see Figure 2.7).

After expressing the functional relationship in the language of variation, we may compute the value of k if one correspondence between the variables is known. This value of k may then be used to find other correspondences.

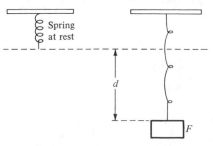

Figure 2.7

EXAMPLE 5: If y varies directly as x and $y = 21$ when $x = 3$, write y as a function of x and determine the value of y when $x = 10$.

SOLUTION: Since y varies directly as x, we have

$$y = kx$$

To find k, replace y by 21 and x by 3:

$$21 = k \cdot 3$$
$$7 = k$$

Thus, $y = 7x$. When $x = 10$, $y = 7(10) = 70$.

EXAMPLE 6: In a spring to which Hooke's law applies, a force of 20 lb stretches the spring 6 in. How far will the spring be stretched by a force of 13 lb?

SOLUTION: From Example 4 we have

$$d = kF$$

To find k, replace d by 6 and F by 20:

$$6 = k \cdot 20$$
$$0.3 = k$$

Thus, $d = 0.3F$. When $F = 13$, $d = 0.3(13) = 3.9$ in.   **Answer**

When the product of any correspondence between the variables is a constant, we say we have *inverse* variation. The statement "y varies inversely as x" means that there is some positive number k (variation constant) such that $xy = k$, or $y = k/x$. Note that with direct variation as x increases, y increases; while with inverse variation as x increases, y decreases.

EXAMPLE 7: If y varies inversely with x and $y = 27$ when $x = 2$, write y as a function of x and find y when $x = 9$.

SOLUTION: Since y varies inversely as x, we have

$$y = \frac{k}{x}$$

To find k, replace y by 27 and x by 2:

$$27 = \frac{k}{2}$$

$$54 = k$$

Thus, $y = 54/x$. When $x = 9$, $y = \frac{54}{9} = 6$.   **Answer**

EXAMPLE 8: The speed (S) of a gear varies inversely as the number (n) of teeth. Gear A, which has 12 teeth, makes 400 rpm. How many revolutions per minute are made by a gear with 32 teeth which is connected to gear A?

SOLUTION: Since S varies inversely with n, we have

$$S = \frac{k}{n}$$

To find k, replace S by 400 and n by 12:

$$400 = \frac{k}{12}$$

$$4800 = k$$

Thus, $S = 4800/n$. When $n = 32$, $S = 4800/32 = 150$ rpm.   **Answer**

We may extend the concept of variation to include direct and inverse variation of variables raised to specified powers and relationships that involve more than two variables.

EXAMPLE 9: If y varies directly as x^2 and inversely as z^3 and $y = 18$ when $x = 3$ and $z = 2$, determine the value of y when $x = 1$ and $z = 5$.

SOLUTION: Since y varies directly as x^2 and inversely as z^3, we have

$$y = \frac{kx^2}{z^3}$$

To find k, replace y by 18, x by 3, and z by 2:

$$18 = \frac{k(3)^2}{(2)^3}$$

$$18 = \frac{9k}{8}$$

$$16 = k$$

Thus, $y = 16x^2/z^3$. When $x = 1$ and $z = 5$,

$$y = \frac{16(1)^2}{(5)^3} = \frac{16}{125} \quad \textbf{Answer}$$

EXAMPLE 10: Newton's law of gravitation states that the gravitational attraction between two objects varies directly as the product of their masses and inversely as the square of the distance between their centers of mass. What will be the change in attraction between the two objects if both masses are doubled and the distance between their centers is cut in half?

SOLUTION: If F represents the gravitational attraction, m_1 and m_2 the masses of the objects, and d the distance between the centers of mass, then algebraically we write Newton's law as

$$F = \frac{km_1m_2}{d^2}$$

If the masses are doubled and the distance is cut in half, we have

$$F = \frac{k(2m_1)(2m_2)}{(\frac{1}{2}d)^2}$$

$$= \frac{4km_1m_2}{\frac{1}{4}d^2}$$

$$= \frac{16km_1m_2}{d^2}$$

Thus, the gravitational attraction becomes 16 times as great.   **Answer**

Finally, we point out that any variation problem may also be solved by using a proportion. For example, if y varies directly as x, we know that $y = kx$. Thus, $y/x = k$ and there is a constant ratio between any correspondence of x and y. This means that

$$\frac{y_1}{x_1} = \frac{y_2}{x_2} \quad \text{or} \quad \frac{y_1}{y_2} = \frac{x_1}{x_2}$$

The equation $y_1/y_2 = x_1/x_2$ is called a *proportion* (see Section 1-7) and may be used to solve the problem. For this reason the variation constant k is often called the constant of proportionality, and the expression "varies directly as" is often replaced by "is proportional to." However, the language of variation usually provides a more convenient and informative statement of a relationship.

EXERCISES 2–2

In Exercises 1–30 write an equation relating the variables and solve the problem.

1. If y varies directly as x and $y = 14$ when $x = 6$, write y as a function of x and determine the value of y when $x = 10$.

2. If y varies directly as x and $y = 2$ when $x = 5$, write y as a function of x and determine the value of y when $x = 11$.

3. If y varies directly as the square of x and $y = 3$ when $x = 2$, write y as a function of x and find y when $x = 4$.

4. If y varies directly as x^3 and $y = 5$ when $x = 3$, write y as a function of x and find y when $x = 2$.

5. If y varies inversely as x and $y = 9$ when $x = 8$, write y as a function of x and find y when $x = 24$.

6. If y varies inversely as x and $y = 3$ when $x = 7$, write y as a function of x and find y when $x = 21$.

7. If y varies inversely as x^3 and $y = 3$ when $x = 2$, write y as a function of x and find y when $x = 3$.

8. If y varies inversely as the square of x and $y = 2$ when $x = 4$, write y as a function of x and find y when $x = 8$.

9. If y varies directly as x and z and $y = 105$ when $x = 7$ and $z = 5$, find y when $x = 10$ and $z = 2$.

10. If y varies directly as x and inversely as z and $y = 10$ when $x = 4$ and $z = 3$, find y when $x = 7$ and $z = 15$.

11. If y varies directly as x and inversely as z^2 and $y = 36$ when $x = 4$ and $z = 7$, find y when $x = 9$ and $z = 9$.

12. If y varies inversely as x^2 and z^3 and $y = 0.5$ when $x = 3$ and $z = 2$, find y when $x = 2$ and $z = 3$.

13. In a spring to which Hooke's law applies (see Examples 4 and 6) a force of 15 lb stretches the spring 10 in. How far will the spring be stretched by a force of 6 lb?

14. The weight of an object on the moon varies directly with the weight of the object on earth. An object that weighs 114 lb on earth weighs 19 lb on the moon. How much will a man who weighs 174 lb on earth weigh on the moon?

15. The amount of garbage produced in a given location varies directly with the number of people living in the area. It is known that 25 tons of garbage is produced by 100 people in 1 year. If there are 8 million people in New York City, how much garbage is produced by New York City in 1 year?

16. Property tax varies directly as assessed valuation. The tax on property assessed at $12,000 is $400. What is the tax on property assessed at $40,000?

17. If the area of a rectangle remains constant, the length varies inversely as the width. The length of a rectangle is 9 in. when the width is 8 in. If the area of the rectangle remains constant, find the width when the length is 24 in.

18. The speed of a gear varies inversely as the number of teeth. Gear A with 48 teeth makes 40 rpm. How many revolutions per minute are made by a gear with 120 teeth that is connected to gear A?

19. The speed of a pulley varies inversely as the diameter of the pulley. The speed of pulley A, which has an 8-in. diameter is 450 rpm. What is the speed of a 6-in. diameter pulley connected to pulley A?

20. The time required to complete a certain job varies inversely as the number of machines that work on the job (assuming each machine does the same amount of work). It takes 5 machines 55 hours to complete an order. How long will it take 11 machines to complete the same job?

21. The volume of a sphere varies directly as the cube of its radius. The volume is 36π cubic units when the radius is 3 units. What is the volume when the radius is 5 units?

22. The distance an object falls due to gravity varies directly as the square of the time of fall. If an object falls 144 ft in 3 seconds, how far did it fall the first second?

23. The weight of an object varies inversely as the square of the distance from the object to the center of the earth. At sea level (4000 mi from the center of the earth) a man weighs 200 lb. Find his weight when he is 200 mi above the surface of the earth.

24. The intensity of light on a plane surface varies inversely as the square of the distance from the source of light. If we double the distance from the source to the plane, what happens to the intensity?

25. The exposure time for photographing an object varies inversely as the square of the lens diameter. What will happen to the exposure time if the lens diameter is cut in half?

26. The resistance of a wire to an electrical current varies directly as its length and inversely as the square of its diameter. If a wire 100 ft long with a diameter of 0.01 in. has a resistance of 10 ohms, what is the resistance of a wire of the same length and material but 0.03 in. in diameter?

27. The general gas law states that the pressure of an ideal gas varies directly as the absolute temperature and inversely as the volume. If $P = 4$ atm when $V = 10$ cm^3 and $T = 200°$ Kelvin, find P when $V = 30$ cm^3 and $T = 250°$ Kelvin.

28. The safe load of a beam (the amount it supports without breaking), which is supported at both ends, varies directly as the width and the square of the height and inversely as the distance between supports. If the width and height are doubled and the distance between supports remains the same, what is the effect on the safe load?

29. Coulomb's law states that the magnitude of the force that acts on two charges q_1 and q_2 varies directly as the product of the magnitude of q_1 and q_2 and inversely as the square of the distance between them. If the magnitude of q_1 is doubled, the magnitude of q_2 is tripled, and the distance between the charges is cut in half, what happens to the force?

30. Newton's law of gravitation states that the gravitational attraction between two objects varies directly as the product of their masses and inversely as the square of the distance between their centers of mass. What will be the change in attraction between the two objects if both masses are cut in half and the distance between their centers is doubled?

2–3 FUNCTIONAL NOTATION

A useful notation commonly used with functions allows us to represent more conveniently the value of the dependent variable for a particular value of the independent variable. In this notation we write an equation like $y = 2x$ as $f(x) = 2x$, where we replace the dependent variable y with the symbol $f(x)$. The symbol representing the independent variable appears in the parentheses. The term $f(x)$ is read "f at x" and means the value of the function (the y value) corresponding to some value of x. Similarly, $f(3)$ is read "f at 3" and means the value of the function when $x = 3$. To find $f(3)$ we substitute 3 for x in the equation $f(x) = 2x$.

$$f(3) = 2 \cdot 3 = 6 \quad \text{therefore} \quad f(3) = 6$$

EXAMPLE 1: If $P = f(S) = 4S$, find $f(1)$, $f(\tfrac{1}{4})$, and $f(8)$.

SOLUTION: $P_{\text{when } S=1} = f(1) = 4(1) = 4$

$P_{\text{when } S=\frac{1}{4}} = f(\tfrac{1}{4}) = 4(\tfrac{1}{4}) = 1$

$P_{\text{when } S=8} = f(8) = 4(8) = 32$

EXAMPLE 2: If $y = g(x) = (x + 2)/x$, find $g(1)$, $g(2)$, and $g(0)$.

SOLUTION: $y_{\text{when } x=1} = g(1) = \dfrac{1 + 2}{1} = \dfrac{3}{1} = 3$

$y_{\text{when } x=2} = g(2) = \dfrac{2 + 2}{2} = \dfrac{4}{2} = 2$

$y_{\text{when } x=0} = g(0) = \dfrac{0 + 2}{0} = \dfrac{2}{0}$ undefined

(Note that a function is not always labeled f. In this example the letter g was used to designate the function.)

EXAMPLE 3: If $f(x) = x - 1$ and $g(x) = x^2 + 1$, find $3f(-1) - 4g(2)$.

SOLUTION:

$f(x) = x - 1 \qquad\qquad g(x) = x^2 + 1$

$f(-1) = (-1) - 1 \qquad g(2) = (2)^2 + 1$

$f(-1) = -2 \qquad\qquad g(2) = 5$

Thus,

$3f(-1) - 4g(2) = 3(-2) - 4(5)$

$= -26$

EXAMPLE 4: If $f(x) = cx$, where c is a constant, show that $f(a + b) = f(a) + f(b)$.

SOLUTION:

$f(x) = cx$

$f(a + b) = c(a + b) = ca + cb$

$f(a) = ca$

$f(b) = cb$

Therefore, $f(a + b) = ca + cb = f(a) + f(b)$.

EXAMPLE 5: The state sales tax that single persons in Ohio can deduct from their income tax for incomes (x) of up to \$5000 is as follows; find $f(2000)$, $f(4010)$, $f(3500)$, and $f(4000)$:

$$\text{Deduction} = f(x) = \begin{cases} \$27 & \text{if} \quad \$0 \le x < \$3000 \\ \$36 & \text{if} \quad \$3000 \le x < \$4000 \\ \$44 & \text{if} \quad \$4000 \le x < \$5000 \end{cases}$$

SOLUTION:

$$\text{deduction}_{\text{when } x=\$2000} = f(2000) = \$27$$
$$\text{deduction}_{\text{when } x=\$4010} = f(4010) = \$44$$
$$\text{deduction}_{\text{when } x=\$3500} = f(3500) = \$36$$
$$\text{deduction}_{\text{when } x=\$4000} = f(4000) = \$44$$

When working with functions it is important to recognize that we frequently use the symbols f and x more out of custom than necessity, and that other symbols work just as well. The notations $f(x) = 2x$, $f(t) = 2t$, $g(y) = 2y$, and $h(z) = 2z$ all define exactly the same function if x, t, y, and z may be replaced by the same numbers.

EXERCISES 2–3

1. If $f(x) = x - 2$, find $f(0)$, $f(-1)$, and $f(4)$.
2. If $g(y) = y - 2$, find $g(0)$, $g(-1)$, and $g(4)$.
3. If $h(x) = -2x + 7$, find $h(-2)$, $h(1)$, and $h(5)$.
4. If $f(t) = t^2 + 1$, find $f(-1)$, $f(0)$, and $f(1)$.
5. If $g(x) = 2x^2 - x + 4$, find $g(3)$, $g(0)$, and $g(-1)$.
6. If $h(t) = -t^2$, find $h(5)$, $h(1)$, and $h(-5)$.
7. If $f(x) = (x + 1)/(x - 2)$, find $f(-3)$, $f(1)$, and $f(2)$.
8. If $h(y) = y/(y + 1)$, find $h(1)$, $h(0)$, and $h(-1)$.
9. If $f(x) = x - 1$, for what value of x will:
   $f(x) = 2$?   **b.** $f(x) = 0$?   **c.** $f(x) = -2$?
10. If $g(x) = 2x + 7$, for what value of x will:
    a. $g(x) = 5$?   **b.** $g(x) = 0$?   **c.** $g(x) = -3$?
11. If $f(x) = x + 1$ and $g(x) = x^2$, find:
    a. $f(1) + g(0)$ **b.** $f(-2) - g(-3)$ **c.** $4g(0) + 5f(1)$

    d. $2f(3) - 3g(2)$ **e.** $f(2) \cdot g(1)$ **f.** $\dfrac{f(1)}{g(2)}$

    g. $[f(1)]^2$ **h.** $g[f(1)]$ **i.** $f[g(1)]$ **j.** $f[g(x)]$
12. If $f(x) = x + 1$, find:
    a. $f(2)$ **b.** $f(3)$ **c.** $f(2 + 3)$
    d. Does $f(2 + 3) = f(2) + f(3)$?   **e.** $f(a)$
    f. $f(b)$ **g.** $f(a + b)$ **h.** Does $f(a + b) = f(a) + f(b)$?
13. If $f(x) = x + 1$, find:
    a. $f(2)$ **b.** $f(-2)$ **c.** Does $f(-2) = -f(2)$?
14. If $g(x) = x^2$, find:
    a. $g(2)$
    b. $g(-2)$
    c. Does $g(x) = g(-x)$ for all values of x?
15. Postage rates for first class mail were once as follows for weights (w) of up to 3 oz:

$$\text{Charge} = f(w) = \begin{cases} 10 \text{ cents} & \text{if } 0 \text{ oz} < w \leq 1 \text{ oz} \\ 20 \text{ cents} & \text{if } 1 \text{ oz} < w \leq 2 \text{ oz} \\ 30 \text{ cents} & \text{if } 2 \text{ oz} < w \leq 3 \text{ oz} \end{cases}$$

Find: **a.** $f(\frac{1}{2})$ **b.** $f(2.5)$ **c.** $f(2)$

16. The deduction that motorists in New York may take on their federal income tax for gasoline taxes is given in the following formula table for mileages (m) up to 4000 mi:

$$\text{Deduction} = f(m) = \begin{cases} \$11 & \text{if } 0 \leq m < 3000 \\ \$19 & \text{if } 3000 \leq m < 3500 \\ \$21 & \text{if } 3500 \leq m < 4000 \end{cases}$$

Find: **a.** $f(3275)$ **b.** $f(3500)$ **c.** $f(1212)$

17. If $f(x) = \begin{cases} 1 & \text{if } x \geq 0 \\ -1 & \text{if } x < 0, \text{ find:} \end{cases}$

a. $f(3)$ **b.** $f(0)$ **c.** $f(-3)$

18. If $f(x) = \begin{cases} -1 & \text{if } x > 0 \\ 0 & \text{if } x = 0 \\ 1 & \text{if } x < 0, \text{ find:} \end{cases}$

a. $f(1)$ **b.** $f(0)$ **c.** $f(-1)$ **d.** $f(0.0001)$

19. If $g(x) = \begin{cases} x & \text{if } x \geq 0 \\ -x & \text{if } x < 0, \text{ find:} \end{cases}$

a. $g(3)$ **b.** $g(0)$ **c.** $g(-3)$

20. If $h(x) = \begin{cases} x & \text{if } 0 \leq x \leq 1 \\ -1 & \text{if } x > 1, \text{ find:} \end{cases}$

a. $h(3)$ **b.** $h(1)$ **c.** $h(\frac{1}{2})$
d. $h(0)$ **e.** $h(-3)$

2–4 GRAPHS OF FUNCTIONS

One of the sources of information and insight about a relationship is a picture that describes the particular situation. Pictures or graphs are often used in business reports, laboratory reports, and newspapers to present data quickly and vividly. Similarly, it is useful to have a graph that describes the behavior of a particular function, for this picture helps us see the essential characteristics of the relationship.

We can pictorially represent a function by using the *Cartesian coordinate system*. This system was devised by the French mathematician and philosopher René Descartes and is formed from the intersection of two real number lines at right angles. The values for the independent variable (usually x) are represented on a horizontal number line and values for the dependent variable (usually y) on a vertical number line. These two lines are called *axes* and they intersect at their common zero point, which is called the *origin* (see Figure 2.8a).

This coordinate system divides the plane into four regions, called *quadrants*. The quadrant in which both x and y are positive is designated the first quadrant. The remaining quadrants are labeled in a counterclockwise direction. Figure 2.8b shows the name of the quadrants as well as the sign of x and y in that quadrant.

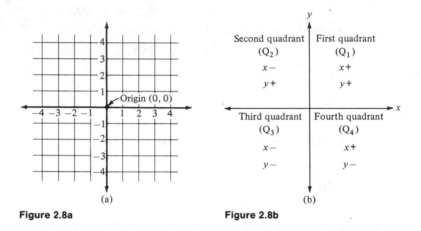

(a)

Figure 2.8a

Figure 2.8b

Any ordered pair can be represented as a point in this coordinate system. The first component indicates the distance of the point to the right or left of the vertical axis. The second component indicates the distance of the point above or below the horizontal axis. These components are called the *coordinates* of the point.

EXAMPLE 1: Represent the ordered pair (4, 2) as a point in the coordinate system.

SOLUTION: See Figure 2.9.

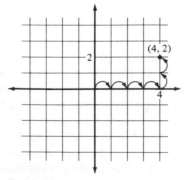

Figure 2.9

EXAMPLE 2: Represent the ordered pair $(2, 4)$ as a point in the coordinate system.

SOLUTION: See Figure 2.10. Notice that the location of the ordered pair $(2, 4)$ differs from $(4, 2)$.

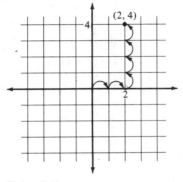

Figure 2.10

EXAMPLE 3: Represent the ordered pairs $(0, -3)$, $(-4, -2)$, $(\pi, -\frac{5}{2})$, and $(-\frac{4}{3}, \sqrt{2})$ as points in the coordinate system.

SOLUTION: See Figure 2.11.

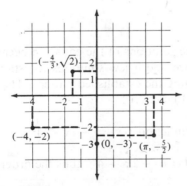

Figure 2.11

With this coordinate system we can represent any ordered pair of real numbers by a particular point in the system. This enables us to draw a geometric picture (graph) of a function. The *graph* of a function is the collection of all the points in the coordinate system that correspond to ordered pairs in the function.

EXAMPLE 4: Graph the function defined by the equation $y = 2x + 1$.

SOLUTION: The horizontal axis is used to represent the values of the independent variable and thus is called the *x axis*. The vertical axis will be named for the dependent variable and called the *y axis*.

If $x =$	then $y = 2x + 1$	thus the ordered-pair solutions
2	$2(2) + 1 = 5$	$(2, 5)$
1	$2(1) + 1 = 3$	$(1, 3)$
0	$2(0) + 1 = 1$	$(0, 1)$
-1	$2(-1) + 1 = -1$	$(-1, -1)$
-2	$2(-2) + 1 = -3$	$(-2, -3)$

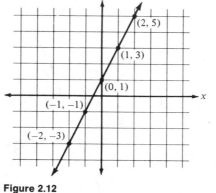

Figure 2.12

We cannot possibly list all the solutions of the equation because it is an infinite set of ordered pairs. Therefore, we find a few solutions to establish a trend and then complete the graph by following the established pattern. The graph of $y = 2x + 1$ is a straight line, as shown in Figure 2.12.

EXAMPLE 5: Graph the function $f(x) = -2x + 4$.

SOLUTION:

If $x =$	then $f(x) = -2x + 4$	thus the ordered-pair solutions
3	$-2(3) + 4 = -2$	$(3, -2)$
1	$-2(1) + 4 = 2$	$(1, 2)$
0	$-2(0) + 4 = 4$	$(0, 4)$
-1	$-2(-1) + 4 = 6$	$(-1, 6)$

The graph of $f(x) = -2x + 4$ is a straight line (Figure 2.13). In every case the graph of a function of the form $f(x) = mx + b$, where m and b are constants, is a straight line. We discuss these functions in detail in Section 8-2.

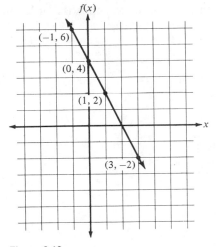

Figure 2.13

EXAMPLE 6: It costs a company $10 to manufacture one unit of a particular product. Therefore, the total cost (C) of manufacturing x units of the product is given by the formula $C = 10x$. Graph the function defined by this equation.

SOLUTION:

If $x =$	then $C = 10x$	thus the ordered-pair solutions
1	$10(1) = 10$	$(1, 10)$
3	$10(3) = 30$	$(3, 30)$
4	$10(4) = 40$	$(4, 40)$

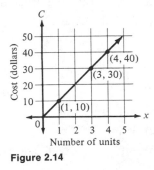

Figure 2.14

Note that it would not be practical to draw this graph with both variables plotted to the same scale. Thus, the two axes are scaled differently in this graph (Figure 2.14) as in many graphs that represent physical data. Technically, the graph should consist only of points above each of the positive integers since the company produces only a whole number of units. But in common practice we connect the points to obtain a better visual image of the relationship.

EXAMPLE 7: Graph the function $y = 4$ (or $y = 4 + 0x$).

SOLUTION:

If $x =$	then $y = 4 + 0 \cdot x$	thus the ordered-pair solutions
3	$4 + 0(3) = 4$	$(3, 4)$
1	$4 + 0(1) = 4$	$(1, 4)$
-2	$4 + 0(-2) = 4$	$(-2, 4)$
-3	$4 + 0(-3) = 4$	$(-3, 4)$

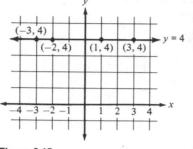

Figure 2.15

Note that no matter what number we substitute for x, the resulting value for y is 4 (see Figure 2.15). Since the value of the function does not change, $y = 4$ is called a *constant function*. A physical example of a constant function would be the acceleration of a falling body as a function of time ($a = 32$) since the acceleration is always 32 ft/second^2, regardless of the time.

EXAMPLE 8: Graph the function $y = x^2$.

SOLUTION:

If $x =$	then $y = x \cdot x = x^2$	thus the ordered-pair solutions
3	$y = 3(3) = 9$	$(3, 9)$
2	$y = 2(2) = 4$	$(2, 4)$
1	$y = 1(1) = 1$	$(1, 1)$
0	$y = 0(0) = 0$	$(0, 0)$
-1	$y = -1(-1) = 1$	$(-1, 1)$
-2	$y - -2(-2) = 4$	$(-2, 4)$
-3	$y = -3(-3) = 9$	$(-3, 9)$

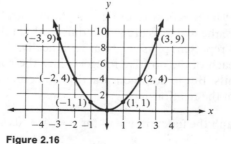

Figure 2.16

The cuplike curve in Figure 2.16 is called a *parabola*. In every case the graph of a function of the form $y = ax^2 + bx + c$, where a, b, and c are constants ($a \neq 0$), is a parabola. We discuss these functions in detail in Section 8-6.

EXAMPLE 9: Graph the function $f(x) = 4 - x^2$.

SOLUTION:

If $x =$	then $f(x) = 4 - x^2$	thus the ordered-pair solutions
3	$f(3) = 4 - (3)^2 = -5$	$(3, -5)$
2	$f(2) = 4 - (2)^2 = 0$	$(2, 0)$
1	$f(1) = 4 - (1)^2 = 3$	$(1, 3)$
0	$f(0) = 4 - (0)^2 = 4$	$(0, 4)$
−1	$f(-1) = 4 - (-1)^2 = 3$	$(-1, 3)$
−2	$f(-2) = 4 - (-2)^2 = 0$	$(-2, 0)$
−3	$f(-3) = 4 - (-3)^2 = -5$	$(-3, -5)$

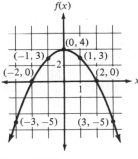

Figure 2.17

The graph of $f(x) = 4 - x^2$, shown in Figure 2.17, is a parabola.

EXAMPLE 10: The state sales tax that a single person in New York can deduct from his income tax for incomes (x) of up to $5000 is given in the formula table below. Graph this function.

SOLUTION:

$$\text{Deduction} = f(x) = \begin{cases} \$37 & \text{if} \quad 0 \le x < \$3000 \\ \$49 & \text{if} \quad \$3000 \le x < \$4000 \\ \$60 & \text{if} \quad \$4000 \le x < \$5000 \end{cases}$$

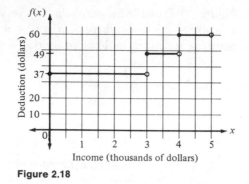

Figure 2.18

The deduction remains constant at $37 until the income reaches $3000. At an income of $3000 the deduction then jumps to $49 (see Figure 2.18). The fact that an income of exactly $3000 allows one a deduction of $49 and not $37 is indicated by filling in the circle above $3000 at a height of 49. The incomes between $3000 and $4000 all allow for the same deduction of $49 until the income of $4000 is reached. The graph then jumps to 60 and remains there until an income of $5000 is reached.

EXAMPLE 11: Graph the function defined as follows:

$$f(x) = \begin{cases} x & \text{if } x \geq 0 \\ -x & \text{if } x < 0 \end{cases}$$

SOLUTION:

If $x =$	then $f(x) = \begin{cases} x & \text{if } x \geq 0 \\ -x & \text{if } x < 0 \end{cases}$	thus the ordered-pair solutions
3	since $3 \geq 0, f(3) = 3$	(3, 3)
2	since $2 \geq 0, f(2) = 2$	(2, 2)
1	since $1 \geq 0, f(1) = 1$	(1, 1)
0	since $0 \geq 0, f(0) = 0$	(0, 0)
−1	since $-1 < 0, f(-1) = -(-1) = 1$	(−1, 1)
−2	since $-2 < 0, f(-2) = -(-2) = 2$	(−2, 2)
−3	since $-3 < 0, f(-3) = -(-3) = 3$	(−3, 3)

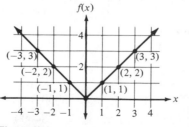

Figure 2.19

See Figure 2.19. (*Note:* This rule defines algebraically the absolute value function $f(x) = |x|$.)

In a function no two ordered pairs may have the same first component. Graphically, this means that a function cannot contain two points that lie on the same vertical line. This observation leads to the following simple test for determining if a graph represents a function.

VERTICAL LINE TEST: Imagine a vertical line sweeping across the graph. If the vertical line at any position intersects the graph in more than one point, the graph is not the graph of a function.

EXAMPLE 12: Use the vertical line test to determine which graphs in Figure 2.20 represent the graph of a function.

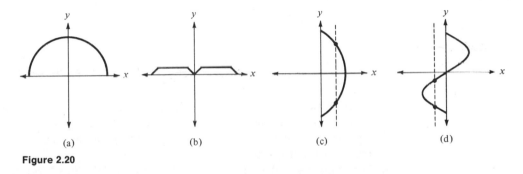

(a)   (b)   (c)   (d)

Figure 2.20

SOLUTION: By the vertical line test graphs (a) and (b) represent functions, whereas (c) and (d) do not.

In summary, coordinate (or analytical) geometry bridges the gap between algebra and geometry. The fundamental principle in this scheme is that every ordered pair which satisfies an equation corresponds to a point in its graph, and every point in the graph corresponds to an ordered pair that satisfies the equation. Thus, a picture may be drawn for each equation relating x and y, which helps us see the essential characteristics of the relationship. On the other hand, each graph can be represented as an equation, so that the powerful methods of algebra can be used to analyze these curves. This simple, yet brilliant, idea is one of the major foundations of mathematics.

EXERCISES 2–4

  1. Graph the following ordered pairs:

    a. $(3, 1)$ **b.** $(-3, 4)$

    c. $(0, 2)$ **d.** $(-3, 0)$

    e. $(-1, -2)$ **f.** $(2, -3)$

g. $(-2, -3)$ **h.** $(-1, 2)$
i. $(\sqrt{2}, 3)$ **j.** $(1, -\pi)$

2. Approximate (use integers) the ordered pairs corresponding to the points shown in the graph.

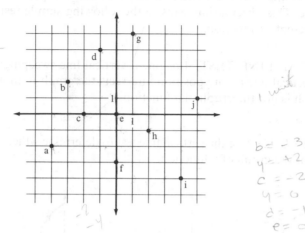

3. Three vertices of a square are $(3, 3)$, $(3, -1)$, and $(-1, 3)$. Find the ordered pair corresponding to the fourth vertex.

4. Three vertices of a rectangle are $(4, 2)$, $(-5, 2)$, and $(-5, -4)$. Find the ordered pair corresponding to the fourth vertex.

5. The points $(0, 0)$, $(-2, -4)$, and $(1, 2)$ lie on a straight line. Find the coordinates of three other points on this line.

6. The points $(-3, 8)$, $(0, 2)$, and $(1, 0)$ lie on a straight line. Find the coordinates of three other points on this line.

7. If we designate the horizontal axis as the x axis and the vertical axis as the y axis:

    a. What is the x value of all the points on the y axis?

    b. What is the y value of all the points on the x axis?

8. Graph the points $A(-5, 2)$ and $B(1, 2)$ and draw a straight line through them. Use the following information to plot the points C, D, and E that lie on this line, and find their coordinates.

    a. Point C is halfway between A and B.

    b. Point B is halfway between A and D.

    c. The distance from point E to point A is twice the distance from point E to point B.

In Exercises 9–29 graph the function defined by the equation.

9. $y = x$ **10.** $y = -3x$

11. $y = x + 3$ **12.** $f(x) = x - 1$

13. $f(x) = 2x - 4$ **14.** $y = 5x - 5$

15. $y = -x + 2$ **16.** $f(x) = -2x + 3$

17. $f(x) = 1$ **18.** $y = -3$

19. $y = x^2 + 1$ **20.** $f(x) = x^2 - 4$

21. $f(x) = 1 - x^2$ **22.** $y = x^2 - 2x + 1$
23. $f(x) = x^3$ **24.** $y = x^3 - 2$
25. $y = |x - 1|$ **26.** $y = |x + 3|$
27. $f(x) = -|x|$ **28.** $y = 1/x$
29. $y = 1/(x - 2)$

30. A projectile fired vertically up from the ground with an initial velocity of 128 ft/second will hit the ground 8 seconds later and the speed of the projectile in terms of the elapsed time t equals $|128 - 32t|$. Graph the function: $s = |128 - 32t|$.

31. The state sales tax that a single person in Ohio can deduct from his income tax for incomes (x) of up to $5000 is as follows:

$$\text{Deduction} = \begin{cases} \$27 & \text{if } \ 0 \le x < \$3000 \\ \$36 & \text{if } \ \$3000 \le x < \$4000 \\ \$44 & \text{if } \ \$4000 \le x < \$5000 \end{cases}$$

Graph this function.

32. The postage rates for first class mail were once given by the following formula table for weights (w) of up to 3 oz:

$$\text{Postage charge} = \begin{cases} 10 \text{ cents} & \text{if } \ 0 < w \le 1 \text{ oz} \\ 20 \text{ cents} & \text{if } \ 1 \text{ oz} < w \le 2 \text{ oz} \\ 30 \text{ cents} & \text{if } \ 2 \text{ oz} < w \le 3 \text{ oz} \end{cases}$$

Graph this function.

33. The charge for the first 3 minutes of a Monday station-to-station call from New York to Los Angeles is dependent upon the time of day that the call is made. The following formula table indicates the charge in terms of military time (that is, 1700 hours is equivalent to 5 P.M.):

$$\text{Charge} = \begin{cases} \$0.85 & \text{if } \ 0 \text{ hours} \le t < 0800 \text{ hours} \\ \$1.45 & \text{if } \ 0800 \text{ hours} \le t \le 1700 \text{ hours} \\ \$0.85 & \text{if } \ 1700 \text{ hours} < t \le 2400 \text{ hours} \end{cases}$$

Graph this function.

34. The union dues of a faculty member in terms of his salary (s) are given in the following formula table:

$$\text{Union dues} = \begin{cases} \$50 & \text{if } \ 0 \le s \le \$2{,}500 \\ \$70 & \text{if } \ \$2{,}500 < s \le \$6{,}000 \\ \$95 & \text{if } \ \$6{,}000 < s \le \$8{,}500 \\ \$110 & \text{if } \ \$8{,}500 < s \le \$15{,}000 \\ \$125 & \text{if } \ \$15{,}000 < s \end{cases}$$

Graph this function.

35. Because of the energy crisis the local power company advertises to "save a watt." To this end they calculate the efficiency of air conditioners by finding the ratio between the cooling ability of the machine (Btu) and the amount of watts of electricity that it requires (that is, efficiency = Btu/watt). Their rating in terms of the efficiency (e) of the air conditioner is as follows:

$$\text{Rating} = \begin{cases} 0 \text{ (flunk)} & \text{if } 0 \le e < 6 \\ 1 \text{ (pass)} & \text{if } 6 \le e < 8 \\ 2 \text{ (good)} & \text{if } 8 \le e < 10 \\ 3 \text{ (very good)} & \text{if } 10 \le e \end{cases}$$

Graph this function.

36. Graph the function defined as follows:

$$f(x) = \begin{cases} 1 & \text{if } x > 2 \\ 0 & \text{if } x \le 2 \end{cases}$$

37. Graph the function defined as follows:

$$g(x) = \begin{cases} 1 & \text{if } x \ge 0 \\ -1 & \text{if } x < 0 \end{cases}$$

38. Graph the function defined as follows:

$$h(x) = \begin{cases} x & \text{if } x \ge 0 \\ 1 & \text{if } x < 0 \end{cases}$$

39. Graph the function defined as follows:

$$f(x) = \begin{cases} 1 & \text{if } x < 0 \\ 0 & \text{if } x = 0 \\ -1 & \text{if } x > 0 \end{cases}$$

40. Graph the function defined as follows:

$$y = \begin{cases} -x & \text{if } x > 0 \\ 2 & \text{if } x = 0 \\ x + 1 & \text{if } x < 0 \end{cases}$$

41. The formula for the side (S) of a square in terms of the perimeter (p) of the square is $S = p/4$. Graph the function defined by this equation.

42. If a car is traveling at a rate of 50 mi/hour, the distance (d) traveled by the car is given by the formula $d = 50t$, where t represents the number of hours traveled. Graph the function defined by this equation.

43. A formula for the amount of work (w) done in lifting a 10-lb object x ft in the air would be $w = 10x$. Graph the function defined by this equation.

44. A formula for the cost (C) of n radial tires if each tire costs $50 would be $C = 50n$. Graph the function defined by this equation for up to six tires.

45. Which of the graphs shown represent the graph of a function?

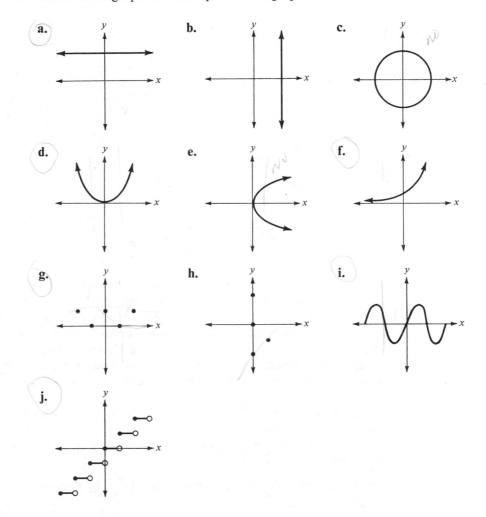

46. You may have noticed in graphs where the two axes were scaled differently that your graph looked different from another's graph. That is, although both graphs were correct, the behavior exhibited in the graphs seemed to be somewhat different. This is common when two people use different scales and since graphing is such a popular method of presenting information, it is important that you be aware of the dramatic effect that scaling can have on the shape of a curve. For example, consider the graph shown at the left, which illustrates the rate at which President Nixon was withdrawing troops from Vietnam during 1969. Notice that the behavior of the curve is horizontal in nature and leaves the impression that our forces were leaving at a very slow rate. If you worked in the Administration, this graph is totally inappropriate, so you present a second graph, as shown on the right.

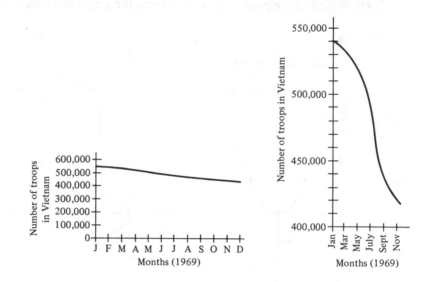

Now that's better! The behavior of this curve is almost vertical in nature and leaves the impression that the planes cannot take off fast enough to take home all the men waiting at the airport. The point is that although both graphs are correct, they leave distinctly different impressions because they were not scaled in the same manner. Thus, when you are trying to obtain information from a graph, do not be overwhelmed by the strong and immediate visual impression given by the curve; rather, examine the scales closely and form your judgment from the information presented.

a. How does this discussion relate to the topic of graphing algebraic functions? Illustrate your answer by using different scales to sketch two graphs of the function $y = x^2$. Consider both graphs and determine the essential characteristics of the function.

In b–e consider the graphs shown and explain why they might produce a deceptive impression.

b.

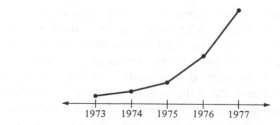

Land values have increased dramatically over the last 5 years!

c.

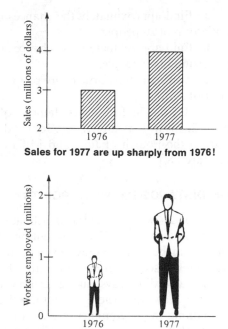

Sales for 1977 are up sharply from 1976!

d.

Over the last year the number of workers employed has risen sharply in our state!

e.

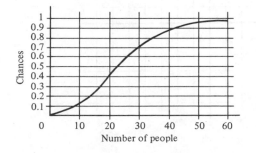

If the President's welfare reform program is passed, the equivalent of all of the people in the shaded states will be on welfare!

47. The graph shown illustrates the surprisingly high chances of two or more persons in a group sharing a birthday. For example, among 30 people the chances are about 7 out of 10 (expressed on the graph as the decimal 0.7) that at least two of them will have the same birthday.

a. Find approximately the chances of two or more persons sharing a birthday in a group of 20 people.

b. Find approximately the chances of two or more persons sharing a birthday in a group of 60 people.

c. Approximately how many people must be in the group for the chances to be 1 out of 2?

[*Note:* Of the 34 Presidents, James Polk (Nov. 2, 1795) and Warren G. Harding (Nov. 2, 1865) share a birthday. John Adams, James Monroe, and Thomas Jefferson died on the Fourth of July, and Millard Fillmore and William Howard Taft died on March 8.]

2–5 DISTANCE BETWEEN POINTS

In Section 2-4 we saw that the coordinate system enables us to bridge the gap between algebra and geometry. By representing the ordered pairs that satisfy some equation as points in the coordinate system, we generate a geometric picture. This interpretation can then be used to better understand the behavior of various functions.

The coordinate system has another important use in the development of trigonometry. However, before we can use the coordinate system as a frame of reference for defining the trigonometric functions, we must find a method to determine the distance between any two points in the system.

It is easiest to calculate the distance between two points on a horizontal or vertical line. If the points lie on the same horizontal line, we find the distance between them by taking the absolute value of the difference between their x coordinates.

EXAMPLE 1: Find the distance between $(-4, 3)$ and $(7, 3)$.

SOLUTION: See Figure 2.21.

$$d = |x_2 - x_1| = |7 - (-4)| = |11| = 11$$

If two points lie on the same vertical line, we calculate the distance between them by taking the absolute value of the difference between their y coordinates.

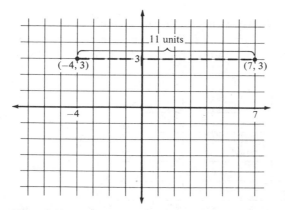

Figure 2.21

EXAMPLE 2: Find the distance between $(1, 1)$ and $(1, -3)$.

SOLUTION: See Figure 2.22.

$$d = |y_2 - y_1| = |-3 - 1| = |-4| = 4$$

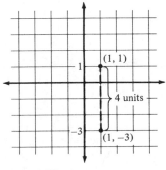

Figure 2.22

If two points do not lie on the same horizontal or vertical line, we can find the distance between them by drawing a horizontal line through one point and a vertical line through the other, as illustrated in Figure 2.23. The x coordinate at the point where the two lines intersect will be x_2. The y coordinate will be y_1. The distance between (x_1, y_1) and (x_2, y_1) will be $|x_2 - x_1|$ because they lie on the same horizontal line. The distance between (x_2, y_2) and (x_2, y_1) will be $|y_2 - y_1|$ because they lie on the same vertical line. The vertical and horizontal lines meet at a right angle $(90°)$ and thus the resulting triangle is a right triangle.

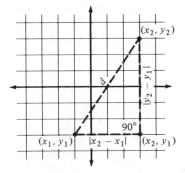

Figure 2.23

The distance d is the length of the hypotenuse in the right triangle, and by the Pythagorean theorem we get

$$d^2 = |x_2 - x_1|^2 + |y_2 - y_1|^2$$
$$d = \sqrt{(x_2 - x_1)^2 + (y_2 - y_1)^2}$$

Note that we no longer need the absolute value symbols since the square of any real number is never negative.

EXAMPLE 3: Find the distance between (2, 1) and (6, 4).

SOLUTION:

$$d = \sqrt{(x_2 - x_1)^2 + (y_2 - y_1)^2}$$
$$= \sqrt{(6 - 2)^2 + (4 - 1)^2}$$
$$= \sqrt{4^2 + 3^2}$$
$$= \sqrt{16 + 9}$$
$$= \sqrt{25}$$
$$= 5 \text{ (see Figure 2.24)}$$

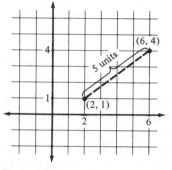

Figure 2.24

EXAMPLE 4: Find the distance between (−3, −2) and (−1, 4).

SOLUTION:

$$d = \sqrt{(x_2 - x_1)^2 + (y_2 - y_1)^2}$$
$$= \sqrt{[-1 - (-3)]^2 + [4 - (-2)]^2}$$
$$= \sqrt{(2)^2 + (6)^2}$$
$$= \sqrt{4 + 36}$$
$$= \sqrt{40} \text{ (see Figure 2.25)}$$

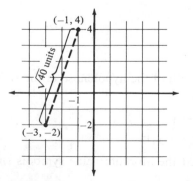

Figure 2.25

EXAMPLE 5: Find the distance between $(0, 0)$ and $(-6, 5)$.

SOLUTION:

$$d = \sqrt{(x_2 - x_1)^2 + (y_2 - y_1)^2}$$
$$= \sqrt{(-6 - 0)^2 + (5 - 0)^2}$$
$$= \sqrt{(-6)^2 + (5)^2}$$
$$= \sqrt{36 + 25}$$
$$= \sqrt{61} \text{ (see Figure 2.26)}$$

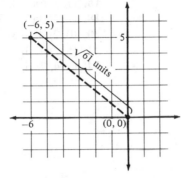

Figure 2.26

EXAMPLE 6: Find the distance between $(-4, 3)$ and $(7, 3)$.

SOLUTION:

$$d = \sqrt{(x_2 - x_1)^2 + (y_2 - y_1)^2}$$
$$= \sqrt{[7 - (-4)]^2 + (3 - 3)^2}$$
$$= \sqrt{11^2 + 0^2}$$
$$= \sqrt{121}$$
$$= 11 \text{ (see Figure 2.27)}$$

Note that the distance formula also works if the two points are on the same horizontal or vertical line.

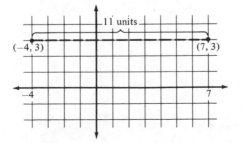

Figure 2.27

EXERCISES 2–5

Represent each ordered pair as points in the coordinate system and find the distance between them.

 1. $(4, 1)(0, 1)$

 2. $(-3, -2)(2, -2)$

 3. $(-1, 3)(-1, 1)$

 4. $(2, -4)(2, -1)$

 5. $(0, 0)(4, 0)$

 6. $(0, -3)(0, 0)$

 7. $(0, 0)(4, 3)$

 8. $(-3, -4)(0, 0)$

 9. $(-1, 1)(0, 0)$

10. $(0, 0)(2, -3)$

11. $(-12, 5)(0, 0)$

12. $(0, 0)(-4, 1)$

13. $(1, 2)(-2, 6)$

14. $(-2, 5)(4, -3)$

15. $(6, -2)(-6, 3)$

16. $(1, -5)(-4, 7)$

17. $(-4, -3)(4, 3)$

18. $(-3, 0)(1, -3)$

19. $(4, -1)(-2, -3)$

20. $(0, 2)(4, -3)$

21. $(-1, -1)(-2, -2)$

22. $(5, 2)(1, 3)$

23. $(4, 0)(-3, -5)$

24. $(-4, 7)(6, 3)$

25. $(3, -1)(-1, 3)$

Sample Test: Chapter 2

1. Fill in the missing component in each of the following ordered pairs so that they are solutions of the equation $y = -2x + 4$:

   (0,)   (, -1)   (-1,)   (, 0)   (3,)

2. Give three correspondences in the function $f(x) = 2$.

3. Is the collection of ordered pairs $(-1, 3)(0, 3)(1, 3)$ a function?

4. What is the domain of the function $y = 1/[x(x + 1)]$?

5. Consider the following table, which gives a few of the correspondences in some function:

If x is	1	2	3	4	5
y is	2	5	10	17	26

   $y = x \cdot 5$

   Write a formula to define this relationship.

6. Write a formula expressing the circumference of a circle as a function of its radius.

7. Write a formula that expresses the area of a square as a function of the diagonal. (*Hint:* Use the drawing and the Pythagorean theorem).

8. Is the graph shown the graph of a function?

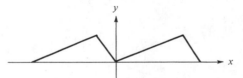

9. If $f(x) = x - 1/x$, find $f(0)$.

10. If $f(x) = x - 1$ and $g(x) = 2x$, find $3f(3) - 4g(-2)$.

11. If $g(x) = 2x + 1$, for what value of x does $g(x) = 0$?

In questions 12–14 consider the function

$$f(x) = \begin{cases} -x & \text{if } 0 \le x \le 3 \\ 3 & \text{if } x > 3 \end{cases}$$

12. Find $f(3)$.

13. Graph the function.

14. What is the domain of the function?

15. Three vertices of a rectangle are $(-4, -3)$, $(-4, 2)$, and $(3, 2)$. Determine the ordered pair corresponding to the fourth vertex.

16. Graph the function: $y = x^2 + x$.

17. If y varies directly as x and $y = 5$ when $x = 12$, write y as a function of x and determine the value of y when $x = 20$.

18. The speed of a pulley varies inversely as the diameter of the pulley. The speed of pulley A, which has a 7-in. diameter, is 320 rpm. What is the speed of a 5-in.-diameter pulley connected to pulley A?

19. Consider the drawing, which illustrates that the intensity of light on a surface is inversely proportional to the square of the distance from the source of light. You are reading a book 6 ft from an electric light. If you move 2 ft closer, how many times greater is the intensity of light on the book?

20. Find the distance between $(-2, 1)$ and $(3, -1)$.

Chapter 3
The trigonometric functions

3–1 DEFINING THE TRIGONOMETRIC FUNCTIONS

Historically, trigonometry involved the study of triangles for the purpose of measuring angles and distances in surveying and astronomy. Although this application of trigonometry is still important, a more general approach emphasizing the repetitive characteristics of the trigonometric functions is more useful today. The rhythmic motion of the heart, the sound waves produced by a musical instrument, and alternating electrical current are but a few of the physical phenomena that occur in cycles and can be studied through the aid of the trigonometric functions. Thus, we will study the more general approach to trigonometry and later we will show that the study of triangles can be considered as a special case.

A *ray* is a half-line that begins at a point and extends indefinitely in some direction. Two rays that share a common end point (or vertex) form an angle. If we designate one ray as the *initial ray* and the other ray as the *terminal ray* (see Figure 3.1), the measure of the angle is the amount of rotation needed to make the initial ray coincide with the terminal ray. Notice that there are many rotations that will make the rays coincide since there is no limitation on the number of revolutions made by the initial ray. In fact, it is useful to allow the initial ray to rotate through many revolutions, since the rotating initial ray will demonstrate a cyclic behavior that can serve as a model to simulate physical phenomena that occur in cycles. Also, the initial ray can rotate in two possible directions. To show the direction of the rotation

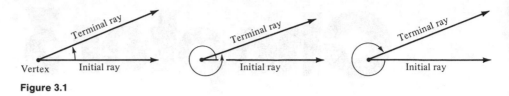

Figure 3.1

we define the measure of an angle to be positive if the rotation is counterclockwise, negative if the rotation is clockwise (see Figure 3.2).

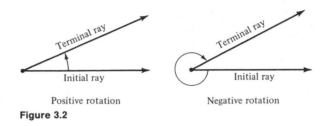

Positive rotation   Negative rotation

Figure 3.2

The frame of reference that we use to study the trigonometric functions is the coordinate system. First we put an angle θ (theta) in standard position (see Figure 3.3) by:

1. Place the vertex of the angle at the origin (0, 0).
2. Placing the initial ray of the angle along the positive x axis.

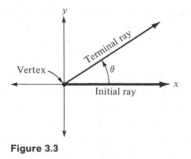

Figure 3.3

When we have the angle in standard position we can define the trigonometric functions of an angle by considering any point on the terminal ray of θ [except (0, 0)]. Three numbers can be associated with the location of this point (see Figure 3.4):

1. The x coordinate of the point.
2. The y coordinate of the point.
3. The distance r between the point and the origin.

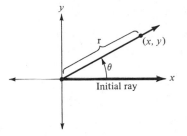

Figure 3.4

Since r represents the distance from the origin $(0, 0)$ to the point (x, y), we can find the relationship between x, y, and r by using the distance formula:

$$r = \sqrt{(x - 0)^2 + (y - 0)^2}$$
$$r = \sqrt{x^2 + y^2} \quad \text{or} \quad r^2 = x^2 + y^2$$

If we consider the number of ratios that can be obtained from the three variables x, y, and r, we find that there are six. It is these six ratios that define the six *trigonometric functions*.

DEFINITION OF THE TRIGONOMETRIC FUNCTIONS: If θ is an angle in standard position and if (x, y) is any point on the terminal ray of θ [except $(0, 0)$], then:

Name of function	*Abbreviation*	*Ratio*	
sine of angle θ	$\sin \theta$	$= \dfrac{y}{r}$	reciprocal functions
cosecant of angle θ	$\csc \theta$	$= \dfrac{r}{y} \, (y \neq 0)$	
cosine of angle θ	$\cos \theta$	$= \dfrac{x}{r}$	reciprocal functions
secant of angle θ	$\sec \theta$	$= \dfrac{r}{x} \, (x \neq 0)$	
tangent of angle θ	$\tan \theta$	$= \dfrac{y}{x} \, (x \neq 0)$	reciprocal functions
cotangent of angle θ	$\cot \theta$	$= \dfrac{x}{y} \, (y \neq 0)$	

EXAMPLE 1: Find the values of the six trigonometric functions of an angle θ if $(-6, 8)$ is a point on the terminal ray of θ (see Figure 3.5).

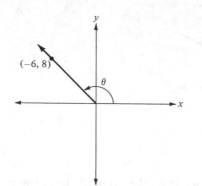

Figure 3.5

SOLUTION:

$$x = -6 \qquad y = 8$$
$$r = \sqrt{x^2 + y^2}$$
$$= \sqrt{(-6)^2 + (8)^2}$$
$$= \sqrt{36 + 64} = \sqrt{100} = 10$$

$\sin \theta = \dfrac{y}{r} = \dfrac{8}{10} = \dfrac{4}{5}$ ←reciprocals→ $\csc \theta = \dfrac{r}{y} = \dfrac{10}{8} = \dfrac{5}{4}$

$\cos \theta = \dfrac{x}{r} = \dfrac{-6}{10} = \dfrac{-3}{5}$ ←reciprocals→ $\sec \theta = \dfrac{r}{x} = \dfrac{10}{-6} = \dfrac{5}{-3}$

$\tan \theta = \dfrac{y}{x} = \dfrac{8}{-6} = \dfrac{4}{-3}$ ←reciprocals→ $\cot \theta = \dfrac{x}{y} = \dfrac{-6}{8} = \dfrac{-3}{4}$

Note that if we say $\sin \theta = \frac{4}{5}$, we do not necessarily mean that $y = 4$ and $r = 5$. In this problem $y = 8$ and $r = 10$, but the ratio reduced to $\frac{4}{5}$.

EXAMPLE 2: Find the value of the six trigonometric functions of an angle θ if $(2, -5)$ is a point on the terminal ray of θ (see Figure 3.6).

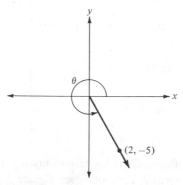

Figure 3.6

SOLUTION: $x = 2 \qquad y = -5$

$$r = \sqrt{x^2 + y^2}$$
$$= \sqrt{(2)^2 + (-5)^2}$$
$$= \sqrt{4 + 25} = \sqrt{29} \qquad (Note: \sqrt{4 + 25} \neq \sqrt{4} + \sqrt{25})$$

$$\sin \theta = \frac{y}{r} = \frac{-5}{\sqrt{29}} \leftarrow \text{reciprocals} \rightarrow \csc \theta = \frac{r}{y} = \frac{\sqrt{29}}{-5}$$

$$\cos \theta = \frac{x}{r} = \frac{2}{\sqrt{29}} \leftarrow \text{reciprocals} \rightarrow \sec \theta = \frac{r}{x} = \frac{\sqrt{29}}{2}$$

$$\tan \theta = \frac{y}{x} = \frac{-5}{2} \leftarrow \text{reciprocals} \rightarrow \cot \theta = \frac{x}{y} = \frac{2}{-5}$$

To demonstrate why these relationships define functions, let us consider the sine function. The independent variable in this function is the amount of rotation made by the initial ray. The ratio y/r, corresponding to *any* point on the terminal ray of θ, is the dependent variable. The rule in the sine function assigns to each angle measure exactly one real number. To show that this relationship is a function we must demonstrate that the ratios y/r, corresponding to a given rotation, are equal.

Consider Figure 3.7, in which we pick any two points on the terminal ray of an angle θ, say (x_1, y_1) and (x_2, y_2). Observe that we can form the two triangles in Figure 3.8 that contain the same angles. Two triangles with the same angles are called *similar triangles* and one of the properties of similar triangles is that corresponding sides are proportional. Therefore, $y_1/r_1 = y_2/r_2$ and the ratios y/r will always be equal for all the points on the terminal ray of θ. This method could also be used to show that the remaining five trigonometric relations define functions.

The coordinate system divides the plane into four quadrants. To determine the sign of the ratios for the trigonometric functions of an angle whose terminal ray is in a particular quadrant, we examine the sign of x and y in that quadrant. Since r

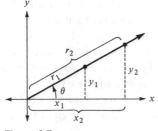

Figure 3.7

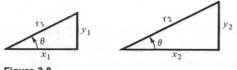

Figure 3.8

is a distance represented by some positive number, r does not affect the sign of a ratio. Consider an angle whose terminal ray is in the second quadrant, where x is negative and y is positive.

$$\sin \theta = \frac{y}{r} = \frac{+}{+} = + \ \leftarrow\text{reciprocals}\rightarrow \ \csc \theta = \frac{r}{y} = \frac{+}{+} = +$$

$$\cos \theta = \frac{x}{r} = \frac{-}{+} = - \ \leftarrow\text{reciprocals}\rightarrow \ \sec \theta = \frac{r}{x} = \frac{+}{-} = -$$

$$\tan \theta = \frac{y}{x} = \frac{+}{-} = - \ \leftarrow\text{reciprocals}\rightarrow \ \cot \theta = \frac{x}{y} = \frac{-}{+} = -$$

Therefore, the sin θ, and its reciprocal csc θ, will be positive if the terminal ray of θ is in the second quadrant. The values of the remaining four functions are negative.

If we repeat this procedure for all four quadrants, we obtain the results summarized in Figure 3.9. We shall use this important chart frequently. Notice that in each

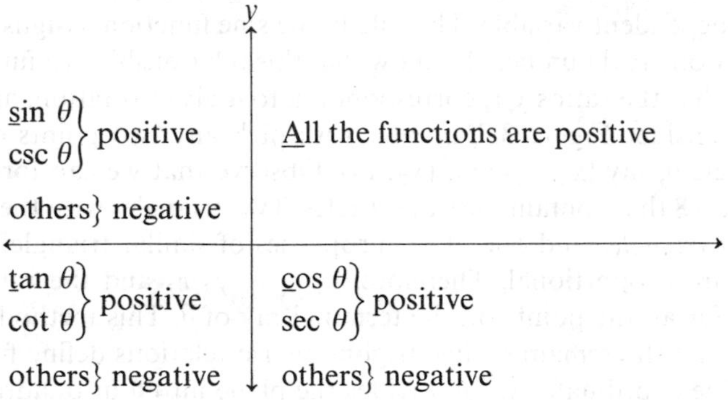

Figure 3.9

quadrant, except for the first quadrant, the value of only two functions is positive; and in each case the two functions are reciprocal functions. Thus, if we can remember in which quadrant sin θ, tan θ, and cos θ are positive, we can generate the chart. A sentence that is useful in remembering this chart is: "All students take chemistry." Notice that the underlined first letter in each of the words corresponds to the underlined first letters in the chart. Finally, the chart does not mention what will happen if the terminal ray of θ coincides with one of the axes. This is a special case where θ is not contained in any quadrant and we will deal with it later.

To see how this chart can be useful, consider each of the following examples.

EXAMPLE 3: In which quadrant is the terminal ray of θ if sin θ is positive and cos θ is negative?

SOLUTION: sin θ is positive in quadrant one (Q_1) or quadrant two (Q_2); cos θ is negative in quadrant two (Q_2) or quadrant three (Q_3). Only quadrant two (Q_2) satisfies both conditions. Therefore, the terminal ray of θ is in quadrant two (Q_2).

EXAMPLE 4: In which quadrant is the terminal ray of θ if $\csc \theta < 0$ and $\cot \theta > 0$?

SOLUTION: $\csc \theta < 0$ means that $\csc \theta$ is negative and θ is in Q_3 or Q_4; $\cot \theta > 0$ means that $\cot \theta$ is positive and θ is in Q_1 or Q_3. Only Q_3 satisfies both conditions. Therefore, the terminal ray of θ is in quadrant three (Q_3).

EXAMPLE 5: Find the values of the remaining trigonometric functions if $\sin \theta = \frac{4}{5}$ and $\cos \theta$ is negative.

SOLUTION: First, determine the quadrant that contains the terminal ray of θ. $\sin \theta = \frac{4}{5}$, which is positive. Therefore, θ is in Q_1 or Q_2; $\cos \theta$ is negative in Q_2 or Q_3. Only quadrant two (Q_2) satisfies both conditions. Therefore, the terminal ray of θ is in Q_2, where x is negative and y is positive.

Second, determine appropriate values for x, y, and r.

$$\sin \theta = \frac{y}{r} = \frac{4}{5}$$

Since y is positive in Q_2, let $y = 4$ and $r = 5$; then find x.

$$r^2 = x^2 + y^2$$
$$(5)^2 = x^2 + (4)^2$$
$$25 = x^2 + 16$$
$$9 = x^2$$
$$x = 3 \quad \text{or} \quad -3$$

Since x is negative in Q_2, let $x = -3$.

Third, find the values of the remaining trigonometric ratios. If $x = -3$, $y = 4$, and $r = 5$, then

$$\sin \theta = \frac{y}{r} = \frac{4}{5} \quad \leftarrow\text{reciprocals}\rightarrow \quad \csc \theta = \frac{r}{y} = \frac{5}{4}$$

$$\cos \theta = \frac{x}{r} = \frac{-3}{5} \quad \leftarrow\text{reciprocals}\rightarrow \quad \sec \theta = \frac{r}{x} = \frac{5}{-3}$$

$$\tan \theta = \frac{y}{x} = \frac{4}{-3} \quad \leftarrow\text{reciprocals}\rightarrow \quad \cot \theta = \frac{x}{y} = \frac{-3}{4}$$

EXAMPLE 6: Find the value of the remaining trigonometric functions if $\tan \theta = \frac{2}{3}$ and $\sec \theta < 0$.

SOLUTION: First, determine the quadrant that contains the terminal ray of θ. $\tan \theta = \frac{2}{3}$, which is positive. Therefore, θ is in Q_1 or Q_3; $\sec \theta < 0$ means that $\sec \theta$ is negative and θ is in Q_2 or Q_3. Only Q_3 satisfies both conditions. Therefore, the terminal ray of θ is in Q_3, where x is negative and y is negative.

Second, determine appropriate values for x, y, and r.

$$\tan \theta = \frac{y}{x} = \frac{2}{3}$$

Since both x and y are negative in Q_3, let $x = -3$ and $y = -2$; then find r.

$$r = \sqrt{x^2 + y^2}$$
$$= \sqrt{(-3)^2 + (-2)^2}$$
$$= \sqrt{9 + 4}$$
$$= \sqrt{13}$$

Third, calculate the values of the different trigonometric functions. If $x = -3$, $y = -2$, and $r = \sqrt{13}$, then

$$\sin \theta = \frac{y}{r} = \frac{-2}{\sqrt{13}} \leftarrow \text{reciprocals} \rightarrow \csc \theta = \frac{r}{y} = \frac{\sqrt{13}}{-2}$$

$$\cos \theta = \frac{x}{r} = \frac{-3}{\sqrt{13}} \leftarrow \text{reciprocals} \rightarrow \sec \theta = \frac{r}{x} = \frac{\sqrt{13}}{-3}$$

$$\tan \theta = \frac{y}{x} = \frac{-2}{-3} = \frac{2}{3} \leftarrow \text{reciprocals} \rightarrow \cot \theta = \frac{x}{y} = \frac{-3}{-2} = \frac{3}{2}$$

Note once again that the value of a trigonometric function is a ratio. $\tan \theta = \frac{2}{3}$ did not mean that $y = 2$ and $x = 3$.

EXERCISES 3–1

In Exercises 1–10 find the values of the six trigonometric functions of the angle θ, which is in standard position if:
1. $(3, -4)$ is on the terminal ray of θ.
2. $(-1, -1)$ is on the terminal ray of θ.
3. $(-12, 5)$ is on the terminal ray of θ.
4. $(2, 4)$ is on the terminal ray of θ.
5. $(9, -12)$ is on the terminal ray of θ.
6. $(-7, -13)$ is on the terminal ray of θ.
7. The terminal ray of θ lies in Q_1 on the line $y = x$.
8. The terminal ray of θ lies in Q_3 on the line $y = x$.
9. The terminal ray of θ lies on the line $y = -2x$ and θ is in Q_2.
10. The terminal ray of θ lies on the line $y = -2x$ and θ is in Q_4.

In Exercises 11–16 determine in which quadrant the terminal ray of θ lies.
11. $\sin \theta$ is negative, $\cos \theta$ is positive.
12. $\cot \theta$ is positive, $\csc \theta$ is positive.
13. $\tan \theta$ is positive, $\sec \theta$ is negative.

14. $\sec \theta > 0$, $\csc \theta < 0$

15. $\tan \theta < 0$, $\sin \theta > 0$

16. $\cos \theta < 0$, $\cot \theta < 0$

In Exercises 17–25 find the values of the remaining trigonometric functions of θ.

17. $\sin \theta = -\frac{3}{5}$, terminal ray of θ is in Q_3.

18. $\sec \theta = \frac{13}{12}$, terminal ray of θ is in Q_4.

19. $\cos \theta = \frac{1}{2}$, $\tan \theta$ is negative.

20. $\tan \theta = \frac{2}{3}$, $\sin \theta$ is negative.

21. $\csc \theta = -\frac{4}{3}$, $\cot \theta$ is positive.

22. $\cot \theta = 1$, $\csc \theta < 0$

23. $\sin \theta = \frac{1}{3}$, $\sec \theta > 0$

24. $\csc \theta = 2$, $\cos \theta < 0$

25. $\sec \theta = -3$, $\sin \theta > 0$

Use the reciprocal relations among the functions to answer Exercises 26–40. (*Note:* Reciprocal ratios hold true only for the same angle.)

26. If $\sin \theta = \frac{1}{2}$, find $\csc \theta$.

27. If $\cos \theta = \frac{3}{4}$, find $\sec \theta$.

28. If $\tan \theta = 1$, find $\cot \theta$.

29. If $\sec \theta = 1.4$, find $\cos \theta$.

30. If $\csc \theta = 1.7$, find $\sin \theta$.

31. If $\cot \theta = 0$, find $\tan \theta$.

32. What is the value of $\sin \theta \cdot \csc \theta$ for all values of θ for which both functions are defined?

33. What is the value of $3 \cos \theta \cdot \sec \theta$ for all values of θ for which both functions are defined?

34. What is the value of $\tan \theta \cdot \cot \theta$ for all values of θ for which both functions are defined?

35. Express $\sin \theta$ in terms of $\csc \theta$.

36. Express $\cos \theta$ in terms of $\sec \theta$.

37. Express $\tan \theta$ in terms of $\cot \theta$.

38. Express $\sec \theta$ in terms of $\cos \theta$.

39. Express $\cot \theta$ in terms of $\tan \theta$.

40. Express $\csc \theta$ in terms of $\sin \theta$.

3–2 TRIGONOMETRIC FUNCTIONS OF ACUTE ANGLES

In Section 3-1 we were able to find the values of the trigonometric functions from the coordinates of a point on the terminal ray of the angle (θ). We use another way of finding these values if we know the measure of angle θ. We defined the measure of an angle to be the amount of rotation needed to make the initial ray coincide with the terminal ray. A common unit used to describe the measure of an angle is a degree, written $1°$.

DEGREE: We define one degree (1°) to be $\frac{1}{360}$ of a complete counterclockwise rotation. Equivalently, this means that there are 360 degrees in a complete counterclockwise rotation.

Since precise measurements are often needed, the degree is subdivided into 60 smaller units called *minutes*. [That is, there are 60 minutes (written 60′) in 1°.] Similarly, a minute is subdivided into 60 seconds. [That is, there are 60 seconds (written 60″) in 1′.] The measure of an angle to the nearest 10 minutes will generally be sufficient for our purposes.

In the beginning we will work with acute angles. These are angles between 0° and 90° and their terminal rays lie in the first quadrant. To find the values of the trigonometric functions for acute angles, we consult Table 1 in the Appendix. These values are computed through the aid of calculus and it should be understood that these values are only *approximations* that are rounded off to four significant digits. Angles from 0° to 45° are listed in the column on the left side of the page, and the measures of these angles increase as we read down the column. Angles from 45° to 90° are listed in the column on the right side of the page and the angle measures increase as we read up the column. To use the table to find the approximate value of a particular trigonometric function for an acute angle, do the following:

1. Locate the row containing the angle measure.
2. Read at the intersection of this row and the column corresponding to the particular trigonometric function. The function names are taken from the top row if the angle is between 0° and 45°. Function names in the bottom row apply to angles between 45° and 90°.

EXAMPLE 1: Use Table 1 to approximate sin 22° and cos 68°

SOLUTION: Consider the excerpt from Table 1 in Figure 3.10.

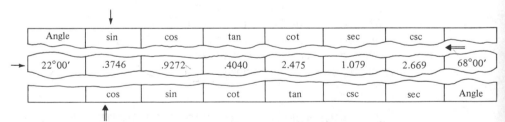

Angle	sin	cos	tan	cot	sec	csc	
22°00′	.3746	.9272	.4040	2.475	1.079	2.669	68°00′
	cos	sin	cot	tan	csc	sec	Angle

Figure 3.10

Thus,

$$\sin 22° \approx 0.3746$$
$$\cos 68° \approx 0.3746$$

(*Note:* The symbol $\approx$ means approximately.)

EXAMPLE 2: Use Table 1 to approximate cot 4°, cos 17°10′, tan 72°50′, sec 89°50′, and csc 0°.

SOLUTION: From Table 1, we read the following values:

$$cot\ 4° \approx 14.30$$
$$cos\ 17°10′ \approx 0.9555$$
$$tan\ 72°50′ \approx 3.237$$
$$sec\ 89°50′ \approx 343.8$$
$$csc\ 0°\ \text{is undefined}$$

If we wish to find the value of a trigonometric function for angles that are given more accurately than to the nearest 10′, we must approximate by a method called *interpolation*. This method assumes that we will commit only a small error if we consider the function to be changing uniformly during any 10′ interval. For example, to approximate sin 34°35′, find

$$sin\ 34°30′ \approx 0.5664$$

and find

$$sin\ 34°40′ \approx 0.5688$$

Then since the angle 34°35′ is halfway between 34°30′ and 34°40′, we assume that the sin 34°35′ is halfway between the sin 34°30′ and the sin 34°40′. Halfway between 0.5664 and 0.5688 is 0.5676. Therefore, sin 34°35′ ≈ 0.5676. This method is generally summarized in the following chart:

$$
10′\left\{5′\left\{\begin{array}{ll} \theta & \sin\theta \\ 34°30′ & 0.5664 \\ 34°35′ & n \approx 0.56\_\_ \\ 34°40′ & 0.5688 \end{array}\right\}x\right\}\ 5688 - 5664 = 24
$$

Note: Although 0.5688 − 0.5664 actually equals 0.0024, it is convenient to forget about the decimal for now and treat the numbers as integers. We know the value of the function begins 0.56, so we can put the decimal in the appropriate spot at the end of the problem.

The chart leads to the proportion

$$\frac{5}{10} = \frac{x}{24}$$

Solving the proportion for x, we get

$$24 \cdot \tfrac{1}{2} = x$$
$$12 = x$$

Add the value for x to 5664:

$$5664 + 12 = 5676$$

Replace the decimal in the appropriate spot:

$$n \approx 0.5676$$

Therefore, $\sin 34°35' \approx 0.5676$.

EXAMPLE 3: Approximate $\cos 67°42'$.

SOLUTION: Find in Table 1 the values of the cosine function for the angles closest to $67°42'$, then proceed as in the last example.

$$
\begin{array}{ccc}
\theta & \cos\theta & \\
10'\left\{2'\left\{\begin{matrix}67°40'\\67°42'\\67°50'\end{matrix}\right.\right. & \left.\begin{matrix}0.3800\\n \approx 0.37\_\_\\0.3773\end{matrix}\right\}x & \left.\vphantom{\begin{matrix}0.3800\\0.37\\0.3773\end{matrix}}\right\}3773 - 3800 = -27
\end{array}
$$

Note: As the angle increased from $67°40'$ to $67°50'$, the value of the cosine function decreased from 0.3800 to 0.3773. This means that we must use a negative number to express the difference between 0.3773 and 0.3800 (that is, $3773 - 3800 = -27$). A negative number will be needed whenever we interpolate with the *co*sine, *co*tangent, or *co*secant function.

This chart leads to the proportion

$$\frac{2}{10} = \frac{x}{-27}$$

Solving the proportion for x, we get

$$(-27) \cdot \frac{2}{10} = x$$

$$\frac{-54}{10} = x$$

$$-5.4 = x$$

Round off the value for x to the nearest integer and add the result to 3800:

$$3800 + (-5) = 3795$$

Replace the decimal in the appropriate place:

$$n \approx 0.3795$$

Therefore, $\cos 67°42' \approx 0.3795$.

EXAMPLE 4: Approximate the value of $\sec 88°47'$.

SOLUTION:

$$10' \left\{ 7' \left\{ \begin{array}{ll} 88°40' \\ 88°47' \\ 88°50' \end{array} \right. \begin{array}{l} \quad 42.98 \\ n \approx 4\_\_\_\_ \\ \quad 49.11 \end{array} \left. \begin{array}{l} \end{array} \right\} x \right\} 4911 - 4298 = 613$$

(θ and sec θ column headers above)

The chart leads to the proportion

$$\frac{7}{10} = \frac{x}{613}$$

Solving the proportion for x, we get

$$(613) \cdot \frac{7}{10} = x$$

$$\frac{4291}{10} = x$$

$$429.1 = x$$

Round off the value for x to the nearest integer and add the result to 4298:

$$4298 + 429 = 4727$$

Replace the decimal in the appropriate place:

$$n \approx 47.27$$

Therefore, sec $88°47' \approx 47.27$.

A similar procedure can be used if we wish to approximate the measure of an acute angle to the nearest minute when the value of one of the trigonometric functions is known.

EXAMPLE 5: Approximate the acute angle θ if tan $\theta \approx 0.4357$.

SOLUTION: Read down the column labeled "tan" in Table 1 until you find the closest values to 0.4357; then use the angles that are in the same row with these values to set up the chart.

$$35 \left\{ 9 \left\{ \begin{array}{ll} 0.4348 \\ 0.4357 \\ 0.4383 \end{array} \right. \begin{array}{l} 23°30' \\ 23°3\_' \\ 23°40' \end{array} \left. \begin{array}{l} \end{array} \right\} x \right\} 23°40' - 23°30' = 10'$$

(tan θ and θ column headers above)

The chart leads to the proportion

$$\frac{9}{35} = \frac{x}{10}$$

Solving the proportion, we get

$$(10) \cdot \frac{9}{35} = x$$

$$\frac{90}{35} = x$$

$$2.6 = x$$

Round off the value for x to the nearest integer and add the result to $23°30'$:

$$23°30' + 3' = 23°33'$$

Therefore, $\theta \approx 23°33'$.

EXAMPLE 6: Approximate the acute angle θ if $\csc \theta \approx 19.75$.

SOLUTION: Read down the column labeled "csc" in Table 1 until you find the closest values to 19.75; then use the angles that are in the same row with these values to set up the chart:

$$112 \left\{ 64 \left\{ \begin{array}{ll} \csc \theta & \theta \\ 19.11 & 3°00' \\ 19.75 & 2°5\_' \\ 20.23 & 2°50' \end{array} \right\} x \right. \right\} 2°50' - 3°00' = -10'$$

Note: $3°00'$ is equivalent to $2°60'$.

The chart leads to the proportion

$$\frac{64}{112} = \frac{x}{-10}$$

Solving the proportion for x, we get

$$(-10) \cdot \frac{64}{112} = x$$

$$\frac{-640}{112} = x$$

$$-5.7 = x$$

Round off the value for x to the nearest integer and add the result to $3°00'$ (or $2°60'$):

$$3°00' + (-6') = 2°54'$$

Therefore, $\theta \approx 2°54'$.

EXERCISES 3–2

In Exercises 1–13 approximate each using Table 1; interpolate in Exercises 14–25.

1. sin $42°50'$ **2.** cos $10°20'$ **3.** tan $79°30'$

4. cot $16°40'$ **5.** sec $65°10'$ **6.** csc $80°$

7. sin $30°$ **8.** cot $0°$ **9.** tan $0°$

10. tan $89°50'$ **11.** tan $45°$ **12.** cos $90°$

13. sec $60°$ **14.** sin $7°14'$ **15.** cos $81°58'$

16. cot $44°22'$ **17.** tan $12°05'$ **18.** csc $89°45'$

19. sec $1°17'$ **20.** tan $87°52'$ **21.** cos $0°45'$

22. sin $52°21'$ **23.** sec $75°47'$ **24.** cot $2°07'$

25. cos $19°19'$

In Exercises 26–38 approximate the measure of angle θ from Table 1; interpolate in Exercises 39–50.

26. sin $\theta \approx 0.7071$ **27.** cos $\theta \approx 0.8660$ **28.** tan $\theta \approx 0.7907$

29. cot $\theta \approx 2.699$ **30.** sec $\theta \approx 1.781$ **31.** csc $\theta \approx 49.11$

32. sin $\theta \approx 0.1765$ **33.** sec $\theta \approx 6.392$ **34.** tan $\theta \approx 0.5581$

35. cos $\theta \approx 1.963$ **36.** sec $\theta \approx 0.6128$ **37.** csc $\theta \approx 0.4157$

38. sin $\theta \approx 2.671$ **39.** sin $\theta \approx 0.5551$ **40.** cos $\theta \approx 0.9409$

41. cot $\theta \approx 0.7651$ **42.** tan $\theta \approx 0.0402$ **43.** csc $\theta \approx 16.00$

44. sec $\theta \approx 1.549$ **45.** sin $\theta \approx 0.9973$ **46.** cot $\theta \approx 0.0711$

47. tan $\theta \approx 0.4000$ **48.** cos $\theta \approx 0.9513$ **49.** csc $\theta \approx 1.403$

50. sin $\theta \approx 0.7244$

51. a. Approximate sin $45°$ by use of Table 1.

  b. Approximate sin $45°$ by interpolation; use sin $0°$ and sin $90°$.

  c. Explain the difference in your results from parts a and b.

52. Explain why the values of sin θ range between 0 and 1 when θ is an acute angle.

53. Explain why the values of csc θ are always greater than or equal to 1 when θ is an acute angle.

54. Complete the following table by determining on the basis of the appropriate ratio among x, y, and r if the function is increasing or decreasing in the given interval. Use Table 1 to verify your answer for $0° < \theta < 90°$.

	sin	csc	cos	sec	tan	cot
$0° < \theta < 90°$						
$90° < \theta < 180°$						
$180° < \theta < 270°$						
$270° < \theta < 360°$						

55. As was mentioned, the tabular values of the trigonometric functions are *approximations* that are computed by the aid of calculus. A simpler method to obtain a rough estimate of their values is by construction. Start by using a compass to draw a circle with a radius of 20 spaces on a piece of graph paper. (*Note:* Do not be concerned if the circle goes slightly off the paper.) Label the graph so that a distance of two spaces corresponds to one-tenth of a unit, and the radius of 20 spaces corresponds to one unit (that is, $r = 1$). Now use a protractor to find the point on the circle that corresponds to the following rotations, and on the basis of your estimate for their x or y components, approximate to two significant digits the value of the given trigonometric functions.

    a. sin 30°, sin 150°, sin 210°, sin 330°
    b. cos 30°, cos 150°, cos 210°, cos 330°
    c. cos 45°, cos 135°, cos 225°, cos 315°
    d. sin 45°, sin 135°, sin 225°, sin 315°
    e. sin 62°, sin 118°, sin 242°, sin 298°
    f. cos 62°, cos 118°, cos 242°, cos 298°
    g. sec 15°, sec 165°, sec 195°, sec 345°
    h. csc 15°, csc 165°, csc 195°, csc 345°
    i. tan 78°, tan 102°, tan 258°, tan 282°
    j. cot 78°, cot 102°, cot 258°, cot 282°

56. The formula for the horizontal distance traveled by a projectile is

$$d = V^2 \frac{\sin A \sin B}{16}$$

where V is the initial velocity, A the angle of elevation, and $B = 90° - A$. In the 16-lb shot-putting event an athlete releases the ball at an angle of elevation of 42° with an initial velocity of 47 ft/second. Determine the distance of his throw. (*Note:* The maximum distance is attained when the release angle is 45°.)

57. In Exercise 31 in Section 1-7 we considered the law of reflection for the important phenomenon of light. The second major principle in the theory of this subject is the law of refraction. Refraction is the bending of light as it passes from one medium to another. For example, consider this figure, which shows a ray of light bending toward the perpendicular as it passes from air to water. The bending is caused by the change in speed of the light ray as it slows down in the water, which is the denser medium. The theory of refraction is extremely important; cameras, eyeglasses, and other optical instruments are designed by analyzing the bending of light rays by a lens. Historically, the contribution to science of telescopes and microscopes cannot be overemphasized. The mathematical relation between the angle of incidence (i) and the angle of refraction (r) is a trigonometric ratio discovered by Rene Descartes. The law is

$$\frac{\sin i}{\sin r} = \frac{v_i}{v_r}$$

where v_i/v_r is the ratio between the velocities of light in the two mediums.

    a. As light passes from air to water v_i/v_r is about $\frac{4}{3}$. Find the angle of refraction if $i = 28°30'$.

b. When the sun is near the horizon, the angle of incidence approaches 90° and the angle at which the sun's rays penetrate the water approaches a limiting value. Determine this value of r. It is interesting to note that this restriction on the angle of refraction causes an optical illusion for a diver under water, as shown in the figure. As he looks up, the world above the surface appears to be in the shape of a cone. This distorted perspective is called the fish-eye view of the world.

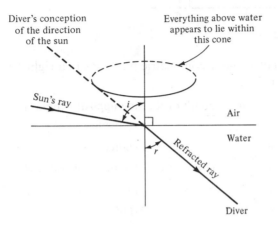

3–3 RIGHT TRIANGLES

A right triangle is a triangle that contains a 90° angle. Generally the angles of the triangle are labeled A, B, and C, with C designating the right angle. The side opposite angle A is labeled a; the side opposite angle B is labeled b; and the side opposite angle C (the 90° angle) is called the *hypotenuse* of the triangle and is labeled c (see Figure 3.11). To *solve* a right triangle means to find the measures of the two acute angles (A and B) and the lengths of the three sides (a, b, and c) of the triangle. To accomplish this, at least two of the five values above must be known, one of which must be a side.

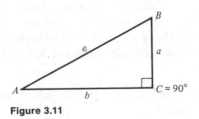

Figure 3.11

The trigonometric functions can be used to solve a right triangle if we place the triangle so that the vertex of angle A is at the origin $(0, 0)$ and side b lies along the positive x axis, as shown in Figure 3.12. We can now consider the vertex of angle B to be a point on the terminal ray of angle A. The x coordinate of this point is b; the y coordinate of this point is a; and the distance r of this point from the origin

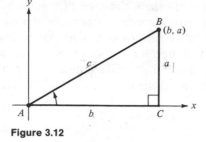

Figure 3.12

is c. Consequently, the trigonometric functions in a right triangle are defined in the following manner:

$$\sin A = \frac{y}{r} = \frac{a}{c} = \frac{\text{length of the side \textbf{opposite} angle } A}{\textbf{hypotenuse}}$$

$$\csc A = \frac{r}{y} = \frac{c}{a} = \frac{\textbf{hypotenuse}}{\text{length of the side \textbf{opposite} angle } A}$$

$$\cos A = \frac{x}{r} = \frac{b}{c} = \frac{\text{length of the side \textbf{adjacent} to angle } A}{\textbf{hypotenuse}}$$

$$\sec A = \frac{r}{x} = \frac{c}{b} = \frac{\textbf{hypotenuse}}{\text{length of the side \textbf{adjacent} to angle } A}$$

$$\tan A = \frac{y}{x} = \frac{a}{b} = \frac{\text{length of the side \textbf{opposite} angle } A}{\text{length of the side \textbf{adjacent} to angle } A}$$

$$\cot A = \frac{x}{y} = \frac{b}{a} = \frac{\text{length of the side \textbf{adjacent} to angle } A}{\text{length of the side \textbf{opposite} angle } A}$$

Thus, right triangle trigonometry is a special case of our more general definition of the trigonometric functions for it arises when we are dealing with an acute angle whose terminal ray lies in the first quadrant. Once we have used the general definition of the trigonometric functions to obtain the ratios in terms of the sides of the triangle, we no longer have to consider the triangle with reference to the coordinate system. Rather, we use the ratios stated above to "solve" the triangle. Before attempting these problems it is suggested that you consider Sections A-1 and A-2 in the Appendix concerning significant digits.

EXAMPLE 1: Solve the right triangle ABC in which angle $A = 30°$ and the hypotenuse $c = 1\overline{0}0$ ft (Figure 3.13).

SOLUTION: First, we can find angle B since the sum of angle A and angle B is $90°$. (*Note:* Two acute angles whose sum is $90°$ are called *complementary.*)

$$A + B = 90°$$
$$30° + B = 90°$$
$$B = 60°$$

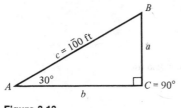

Figure 3.13

Second, we can find side a by using the sine function.

$$\sin A = \frac{a}{c}$$

$$\sin 30° = \frac{a}{100}$$

$$100(\sin 30°) = a$$

$$100(0.5000) = a$$

$$5\overline{0} \text{ ft} = a \quad \text{(two significant digits)}$$

Third, we can find side b by using the cosine function.

$$\cos A = \frac{b}{c}$$

$$\cos 30° = \frac{b}{100}$$

$$100(\cos 30°) = b$$

$$100(0.8660) = b$$

$$87 \text{ ft} = b \quad \text{(two significant digits)}$$

To summarize, we found that $B = 60°$, $a = 5\overline{0}$ ft, and $b = 87$ ft.

In applications of trigonometry, our values are only as accurate as the devices we use to measure the data. However, although all our answers are approximations, the symbol for equality (=) is generally used instead of the more precise symbol for approximation (≈). Computed results should not be used to determine other parts of a triangle since the given data produce more accurate answers. The results are usually rounded off as follows:

Accuracy of Sides	*Accuracy of Angle*
Two significant digits	Nearest degree
Three significant digits	Nearest 10 minutes
Four significant digits	Nearest minute

EXAMPLE 2: Solve the triangle ABC in which angle $B = 47°10'$ and side $a = 45.6$ ft (Figure 3.14).

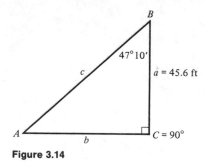

Figure 3.14

SOLUTION : First, we find angle A.

$$A + B = 90°$$
$$A + 47°10' = 90°$$
$$A = 42°50'$$

Second, we can find the hypotenuse c by using the secant function.

$$\sec B = \frac{c}{a}$$

$$\sec 47°10' = \frac{c}{45.6}$$

$$45.6(\sec 47°10') = c$$
$$45.6(1.471) = c$$
$$67.1 \text{ ft} = c \quad \text{(three significant digits)}$$

Third, we can find side b by using the tangent function.

$$\tan B = \frac{b}{a}$$

$$\tan 47°10' = \frac{b}{45.6}$$

$$45.6(\tan 47°10') = b$$
$$45.6(1.079) = b$$
$$49.2 \text{ ft} = b \quad \text{(three significant digits)}$$

To summarize, we determined that $A = 42°50'$, $c = 67.1$ ft, and $b = 49.2$ ft. Note that we could have picked different trigonometric functions to compute sides b and c. For example, to compute the hypotenuse c we could have used $\cos B = a/c$. The reason we avoided this is that it leads to a difficult division,

$$c = \frac{45.6}{\cos 47°10'} = \frac{45.6}{0.6799}$$

To avoid this type of division always pick a function that has the side you wish to find in the numerator of its ratio.

EXAMPLE 3: Solve the triangle ABC in which side $a = 11$ ft and side $b = 5.0$ ft (Figure 3.15).

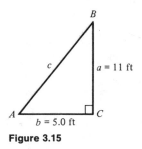

Figure 3.15

SOLUTION: First, we can find the hypotenuse c by the Pythagorean theorem.

$$c^2 = a^2 + b^2$$
$$= (11)^2 + (5.0)^2$$
$$= 121 + 25$$
$$= 146$$
$$c = \sqrt{146}$$
$$= 12 \text{ ft} \quad \text{(two significant digits)}$$

Second, we can find angle A by using the tangent function.

$$\tan A = \frac{a}{b}$$

$$= \frac{11}{5} = 2.2$$

$$A = 66° \quad \text{(from Table 1 to the nearest degree)}$$

Third, we can find angle B since angle A and angle B are complementary.

$$A + B = 90°$$
$$66° + B = 90°$$
$$B = 24°$$

To summarize, we found that $c = 12$ ft, $A = 66°$, and $B = 24°$.

In many practical applications of right triangles, an angle is measured with respect to a horizontal line. This measurement is accomplished by use of a transit.

(By centering a bubble of air in a water chamber, the table of this instrument may be horizontally leveled.) The sighting tube of the transit is then tilted upward or downward until the desired object is sighted. This measuring technique will result in an angle that is described as either an angle of elevation or an angle of depression (see Figure 3.16). The measurement results in an angle of elevation if the object being measured is above the observer; and the measurement results in an angle of depression if the object being measured is below the observer.

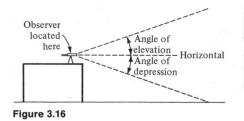

Figure 3.16

EXAMPLE 4: It is necessary to determine the height of a smokestack in order to estimate the cost of painting. At a point $2\overline{0}0$ ft from the base of the stack, the angle of elevation is 30°. How high is the smokestack? (See Figure 3.17.)

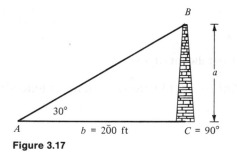

Figure 3.17

SOLUTION: In right triangle ABC we can find the length of side a by using the tangent function.

$$\tan A = \frac{a}{b}$$

$$\tan 30° = \frac{a}{200}$$

$$200(\tan 30°) = a$$

$$200(0.5774) = a$$

$$120 \text{ ft} = a \quad \text{(two significant digits)}$$

EXAMPLE 5: The measure of the angle of depression of a buoy from the platform of a radar tower that is 85 ft above the ocean is 15°. Find the distance of the buoy from the base of the tower. (See Figure 3.18.)

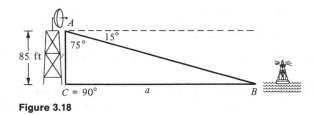

Figure 3.18

SOLUTION: In right triangle ABC we can find angle A since angle A and the angle of depression are complementary.

$$A + 15° = 90°$$
$$A = 75°$$

We can find the length of side a by using the tangent function.

$$\tan A = \frac{a}{b}$$

$$\tan 75° = \frac{a}{85}$$

$$85(\tan 75°) = a$$
$$85(3.732) = a$$
$$320 \text{ ft} = a \quad \text{(two significant digits)}$$

EXERCISES 3-3

In Exercises 1–19 solve each right triangle ABC ($C = 90°$) for the given data.

1. $A = 30°$, $a = 5\overline{0}$ ft
2. $B = 45°$, $a = 85$ ft
3. $A = 60°$, $c = 15$ ft
4. $A = 22°$, $b = 62$ ft
5. $B = 71°$, $c = 25$ ft
6. $A = 19°$, $a = 17$ ft
7. $A = 55°$, $c = 25$ ft
8. $B = 10°20'$, $a = 24.5$ ft
9. $A = 45°30'$, $a = 86.6$ ft
10. $A = 84°50'$, $c = 12.4$ ft
11. $B = 52°40'$, $c = 625$ ft
12. $A = 31°30'$, $b = 29.7$ ft
13. $B = 88°10'$, $a = 31.2$ ft
14. $a = 5\overline{0}$ ft, $b = 120$ ft
15. $a = 6.0$ ft, $c = 15$ ft
16. $b = 1.0$ ft, $c = 2.0$ ft
17. $a = 1.0$ ft, $b = 1.0$ ft
18. $a = 7.00$ ft, $c = 11.0$ ft
19. $b = 12.0$ ft, $c = 26.0$ ft

20. The three sides of a right triangle are 5, 12, and 13. Find the trigonometric functions of the smaller acute angle.

21. In a right triangle the hypotenuse is 17 and the shorter side is 8. Find the trigonometric functions of the larger acute angle.

22. In right triangle ABC, $\cos B = \frac{4}{5}$ and $a = 16$. Find c.

23. In right triangle ABC, $\sin A = \frac{12}{13}$ and $c = 39$. Find a.

24. In right triangle ABC, $\tan A = \frac{7}{12}$ and $a = 21$. Find b.

25. In right triangle ABC, $\csc B = 2$ and $c = 8$. Find b.

In Exercises 26–57 solve each problem by making a careful diagram and using right triangles.

26. A ladder leans against the side of a building and makes an angle of 60° with the ground. If the foot of the ladder is $1\overline{0}$ ft from the building, find the height the ladder reaches on the building.

27. An escalator from the first floor to the second floor of a building is $5\overline{0}$ ft long and makes an angle of 30° with the floor. Find the vertical distance between the floors.

28. If the angle of elevation of the sun at a certain time is 40°, find the height of a tree that casts a shadow of 45 ft.

29. A road has a uniform elevation of 6°. Find the increase in elevation in driving $5\overline{0}0$ yd along the road.

30. The distance from ground level to the underside of a cloud is called the "ceiling." At an airport, a ceiling light projector throws a spotlight vertically on the underside of a cloud. At a distance of $6\overline{0}0$ ft from the projector, the angle of elevation of the spot of light on the cloud is 58°. What is the ceiling?

31. To find the width of a river, a surveyor sets up his transit at C and sights across the river to point B (both B and C are at the water's edge). He then measures off $2\overline{0}0$ ft from C to A such that C is a right angle. If he determines that angle A is 24°, how wide is the river?

32. A pilot in an airplane at an altitude of $4\overline{0}00$ ft observes the angle of depression of an airport to be 12°. How far is the airport from a point on the ground directly below the plane?

33. A lighthouse built at sea level is 180 ft high. From its top, the angle of depression of a buoy is 24°. Find the distance from the buoy to the foot of the lighthouse.

34. A surveyor stands on a cliff $5\overline{0}$ ft above the water of the river below. If the angle of depression to the water's edge on the opposite bank is 10°, how wide is the river at this point?

35. At an airport, cars drive down a ramp 85 ft long to reach the lower baggage claim area 15 ft below the main level. What angle does the ramp make with the ground at the lower level?

36. A road rises 25 ft in a horizontal distance of $4\overline{0}0$ ft. Find the angle that the road makes with the horizontal.

37. From the top of a building $3\overline{0}$ ft tall, the angle of depression to the foot of a building across the street is 60° and the angle of elevation to the top of the same building is 70°. How tall is the building?

38. A carpenter has to build a stairway. The total rise is 8 ft 6 in. and the angle of rise is 30°, as shown. What is the shortest piece of 2- by 12-in. stock that can be used to make the stringer?

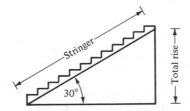

39. In building a stairway a carpenter determines that the total rise must be 7 ft 6 in. Because of a support column near the base of the stairs, the total run cannot exceed 6 ft 3 in. Determine the angle of rise. (*Note:* To make climbing stairs as easy as possible, the angle of rise should be from 25° to 35°.)

40. For maximum safety the distance between the base of a ladder and a building should be one-fourth of the length of the ladder. Find the angle that the ladder makes with the ground when it is set up in the safest position.

41. In building a warehouse, a carpenter checks the drawings and finds the roof span to be $4\overline{0}$ ft, as shown in the sketch. If the slope of the roof is 17°, what length of 2- by 6-in. stock will he need to make rafters if a 12-in. overhang is desired?

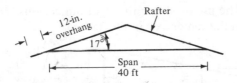

42. A welder is to weld supports for a 25-ft conveyor so that it will operate at a 14° angle. What is the length of the supports?

43. A carpenter is to build a concrete ramp with an 18° slope leading up to a platform. If the platform is 38 in. above the ground, how far from the base of the platform should he start to lay the concrete forms?

44. A tinsmith forms an angle of 120° for a $1\overline{0}$-in. flashing, as shown. Give the height of the bend in inches.

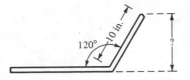

45. In putting in a pipeline, a steamfitter has to run a 45° offset. The perpendicular distance between the two pipelines is 3 ft 6 in. (see the figure). What length of pipe will he need for the connection between the two lines?

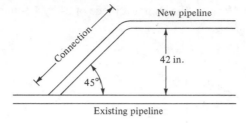

46. A steamfitter is to connect a pipeline to a tank. The run (forward distance) from the pipeline to the tank is 36 in., and the set (side distance) is 27 in., as shown. Will the steamfitter be able to use standard connections of $22\frac{1}{2}°$, $30°$, $45°$, $60°$, or $90°$?

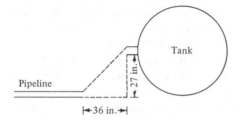

47. A machinist is given a 5.00 in. diameter steel rod with instructions to make a tapered pin 12.0 in. long. The pin must have diameters of 4.00 in. and 2.00 in. What angle of taper should he use to obtain the right dimensions? (*Hint:* make use of the dashed line parallel to the center axis in the figure.)

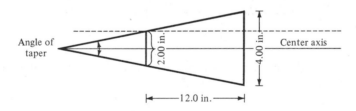

48. A circular disc 24.0 in. in diameter is to have five equally spaced holes as shown. Determine the correct setting (x) for the dividers to space these holes.

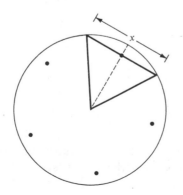

49. In an isosceles triangle, each of the equal sides is $7\overline{0}$ in. and the base is $8\overline{0}$ in. Find the angles of the triangle.

50. The sides of a rectangle are 18 and 31 ft. Find the angle that the diagonal makes with the shorter side.

51. An observer on the third floor of a building determines that the angle of depression of the foot of a building across the street is 28° and that the angle of elevation of the top of the same building is 51°. If the distance between the two buildings is $5\overline{0}$ ft, find the height of the observed building.

52. An observer on the top of a hill 350 ft above the level of a road spots two cars directly in line. Find the distance between the cars if the angles of depression noted by the observer were 16° and 27°.

53. A right triangle, called an *impedance triangle*, is used to analyze alternating current (a.c.) circuits. Consider the first figure, which shows a resistor and an inductor in series. As current flows through these components it encounters some resistance, and the total effective resistance is called the *impedance*. We determine the impedance by making the resistances of the two circuit components the legs of a right triangle. As shown in the second figure, the hypotenuse of the triangle is then the impedance. The degree to which the voltage and current are in phase is given by angle θ, which is called the *phase angle*. For the data in this figure determine the impedance and the phase angle.

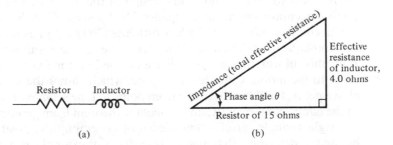

(a)   (b)

54. An important principle in the mathematical analysis of light is the law of reflection. This law states that a ray of light which strikes a reflecting surface is reflected so that the angle of incidence (i) equals the angle of reflection (r). In the figure a photographer positions his flash at A. For a better lighting effect he aims this flash at position P on a reflecting surface to take a picture of a subject at B. If $AC = 4.0$ ft, $CD = 12$ ft, and $i = 41°$, find the length of BD.

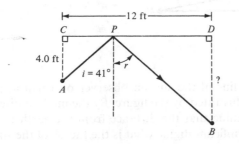

55. Consider the figure shown, which illustrates that twilight lasts until the sun is 18° below the horizon. From this, estimate the height (*h*) of the atmosphere.

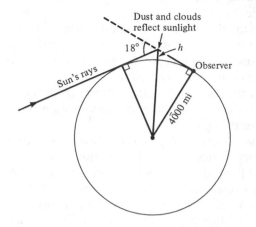

56. Historically, trigonometry was developed to analyze observations in astronomy. In early times this science was the primary concern of learned men who sought aesthetically to understand God's design of the universe, and practically, to obtain a more accurate system of navigation. In Exercise 32 in Section 1-7 we discussed the ingenious method used by Eratosthenes (200 B.C.) to estimate the circumference, and consequently the radius, of the earth. The correct result for the radius is 4000 mi. With this information we now discuss a method for finding the distance from the earth to the moon. Consider this figure, which shows the moon (*M*) to be directly above point *A* and barely visible from point *B*. Line segment *BM* is said to be tangent to the circle; consequently, an elementary theorem from geometry tells us that *OBM* is a right triangle. From a knowledge of geography the positions of *A* and *B* can be used to determine that angle 0 is 89°04′. For example, if *A* and *B* were points on the equator, then angle 0 is the difference in their longitudes. Complete the line of reasoning and estimate the distance from the earth to the moon (segment *AM*) to two significant digits.

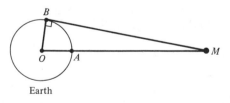

57. To determine the radius of the sun an observer on earth at point 0 measures angle *AOC* to be 16′, as illustrated by the figure. By the method discussed in Exercise 56 it is possible to determine that the distance from the earth to the sun is about 93,000,000 mi. To two significant digits, what is the radius of the sun?

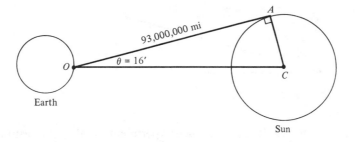

3–4 VECTORS

Another application of right triangles occurs with the study of physical quantities that act in a definite direction. For example, when the weatherman describes wind he mentions both the speed of the wind and the direction from which the wind is blowing. Similarly, quantities such as forces, weights, and velocities must be described in such a way that both the strength (magnitude) and the direction of the quantity can be determined. Mathematically, we represent such a quantity by a line segment with an arrowhead at one end, and this directed line segment is called a *vector*. The direction in which the arrowhead is pointing represents the direction in which the quantity is acting, while the length of the line segment is proportional to the magnitude of the quantity.

EXAMPLE 1: Use a vector to represent graphically a wind that is blowing due north at a speed of 40 mi/hour.

SOLUTION: If 1 unit represents 10 mi/hour, the directed line segment **OA**, which is 4 units long and pointing due north, is the appropriate vector (Figure 3.19). (*Note:* A vector that starts at O and ends at A is labeled **OA**; a vector that starts at A and ends at O is a different vector and is labeled **AO**.)

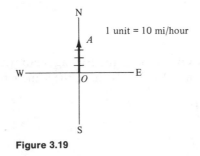

Figure 3.19

Frequently, there are two (or more) forces acting on a body from different directions, and their net effect is a third force with a new direction. This new force is called the *resultant*, or vector sum, of the given forces. For example, in Figure 3.20 two forces of 20 lb and 30 lb are both acting on a body with an angle of 30° between

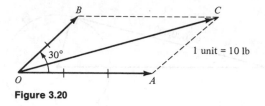

Figure 3.20

the two forces. It can be shown experimentally that the resultant of vectors **OA** and **OB** is vector **OC**, which is the diagonal of a parallelogram formed from the given vectors. The magnitude of the resultant **OC** can be determined by the length of the line segment from O to C. The direction of the resultant is the same as the direction of the arrowhead on vector **OC** and is described in terms of the original vectors by finding either angle AOC or angle BOC. In this example we can only approximate the magnitude and direction of vector **OC** from the diagram since it is necessary to solve a triangle that is not a right triangle to find their values. We will study oblique triangles in Sections 3-8 and 3-9. Here we will confine ourselves to problems that rely on right triangles for their solution.

EXAMPLE 2: A force of 3.0 lb and a force of 4.0 lb are acting on a body with an angle of 90° between the two forces. Find (a) the magnitude of the resultant and (b) the angle between the resultant and the larger force.

SOLUTION: Let **OA** represent the 4.0-lb force and **OB** the 3.0-lb force. Then vector **OC** represents the resultant (see Figure 3.21).

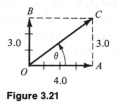

Figure 3.21

a. We can find the length of **OC** by using the Pythagorean theorem in right triangle OAC. Notice that line segments OB and AC have the same length.

$$(OA)^2 + (AC)^2 = (OC)^2$$
$$(4)^2 + (3)^2 = (OC)^2$$
$$16 + 9 = (OC)^2$$
$$25 = (OC)^2$$
$$5.0 = OC$$

Thus, the magnitude of the resultant is 5.0 lb.

b. We can find the angle between the resultant and the 4.0 lb force by using the tangent function in right triangle OAC.

$$\tan \theta = \frac{AC}{OA}$$

$$= \frac{3}{4} = 0.7500$$

$$\theta = 37° \quad \text{(to the nearest degree)}$$

Thus, the angle between the resultant and the larger force is $37°$.

EXAMPLE 3: A ship is headed due east at $2\bar{0}$ knots (nautical miles per hour) while the current carries the ship due south at 5.0 knots. Find (a) the speed of the ship and (b) the direction (course) of the ship.

SOLUTION: Let vector **OA** represent the velocity of the ship in the easterly direction. Let vector **OB** represent the velocity of the ship in the southerly direction due to the current. Let vector **OC** represent the actual velocity of the ship. (See Figure 3.22.)

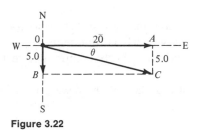

Figure 3.22

a. We can find the length of **OC** by using the Pythagorean theorem in right triangle OAC.

$$(OA)^2 + (AC)^2 = (OC)^2$$
$$(20)^2 + (5)^2 = (OC)^2$$
$$400 + 25 = (OC)^2$$
$$425 = (OC)^2$$
$$\sqrt{425} = OC$$
$$21 = OC \quad \text{(two significant digits)}$$

Thus, the speed of the ship is 21 knots.

b. We can find the angle between the resultant and vector **OA**, which is due east by using the tangent function in right triangle OAC.

$$\tan \theta = \frac{AC}{OA}$$

$$= \frac{5}{20} = 0.2500$$

$$\theta = 14° \quad \text{(to the nearest degree)}$$

Thus, the ship is heading in a direction that is 14° south of east.

In the previous examples we considered how the net effect of having two forces acting on a body produced a third force, the resultant. However, the reverse situation often arises where we are given a single force that we think of as the resultant, and we need to calculate two forces, which are called *components*, that produce the resultant. This process of expressing a single force in terms of two components, which are usually at right angles to each other, is called *resolving a vector*. The following examples illustrate the usefulness of resolving a vector into components.

EXAMPLE 4: An airplane, pointed due west, is traveling 10° north of west at a rate of $4\overline{0}0$ mi/hour. This resultant course is due to a wind blowing north. Find (a) the velocity of the plane if there were no wind (that is, the vector pointing due west) and (b) the velocity of the wind (that is, the vector pointing due north).

SOLUTION: Let vector **OC** represent the resultant velocity of the plane. The resolution of vector **OC** results in a westerly component vector **OA** (the velocity of the plane if there were no wind) and a northerly component vector **OB** (the velocity of the wind) (see Figure 3.23).

a. We can find the velocity of the plane if there were no wind by using the cosine function in right triangle OAC.

$$\cos 10° = \frac{x}{400}$$

$$400(\cos 10°) = x$$

$$400(0.9848) = x$$

$$390 = x \quad \text{(two significant digits)}$$

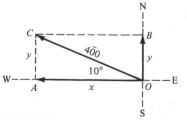

Figure 3.23

Thus, the velocity of the plane if there were no wind would be 390 mi/hour due west.
b. We can find the velocity of the wind by using the sine function in right triangle OAC.

$$\sin 10° = \frac{y}{400}$$

$$400(\sin 10°) = y$$

$$400(0.1736) = y$$

$$69 = y \quad \text{(two significant digits)}$$

Thus, the velocity of the wind would be 69 mi/hour due north.

EXAMPLE 5: Find the horizontal and vertical components of a vector that has a magnitude of 25 lb and makes an angle of 40° with the positive x axis.

SOLUTION: Let vector **OA** be the given vector. The resolution of vector **OA** results in the horizontal component, vector **OB**, and the vertical component, vector **OC** (see Figure 3.24).

We can find the magnitude of the horizontal component, vector **OB**, by using the cosine function in right triangle OBA.

$$\cos 40° = \frac{x}{25}$$

$$25(\cos 40°) = x$$

$$25(0.7660) = x$$

$$19 = x \quad \text{(two significant digits)}$$

We can find the magnitude of the vertical component, vector **OC**, by using the sine function in right triangle OBA.

$$\sin 40° = \frac{y}{25}$$

$$25(\sin 40°) = y$$

$$25(0.6428) = y$$

$$16 = y \quad \text{(two significant digits)}$$

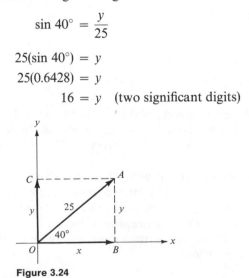

Figure 3.24

Thus, the magnitude of the horizontal component is 19 lb, while the magnitude of the vertical component is 16 lb.

EXAMPLE 6: A man pulls with a force of $4\overline{0}$ lb on a window pole in an effort to lower a window. What part of the man's force lowers the window if the pole makes an angle of 20° with the window?

SOLUTION: Let vector **OA** represent the pull on the window pole of $4\overline{0}$ lb. The resolution of vector **OA** results in the vertical component vector **OC** (the force lowering the window) and the horizontal component vector **OB** (wasted force) (see Figure 3.25).

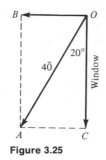

Figure 3.25

We can find the magnitude of vector **OC** by using the cosine function in right triangle ACO

$$\cos 20° = \frac{OC}{40}$$

$$40(\cos 20°) = OC$$
$$40(0.9397) = OC$$
$$38 = OC \quad \text{(two significant digits)}$$

Thus, the force lowering the window is 38 lb.

EXAMPLE 7: A car weighing $3\overline{0}00$ lb is parked in a driveway that makes an angle of 10° with the horizontal. Find the minimum brake force that is needed to keep the car from rolling down the driveway. (Assume no friction.)

SOLUTION: Let vector **OA** drawn vertically down represent the weight of the car, $3\overline{0}00$ lb. The resolution of vector **OA** into components that are parallel and perpendicular to the inclined plane results in vector **OB** (the force tending to pull the object down the plane) and vector **OC** (the pressure of the car on the driveway). The vector that we seek is vector **OD**, which has the same magnitude, but the opposite direction, of vector **OB** (see Figure 3.26).

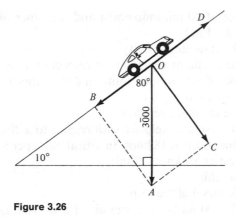

Figure 3.26

We can find the length of **OB** by using the cosine function in right triangle *OBA*. Notice that vector **OA** makes a right angle with the horizontal. Thus, if the plane is inclined 10°, angle *BOA* is 80°.

$$\cos 80° = \frac{OB}{3000}$$

$$3000(\cos 80°) = OB$$

$$3000(0.1736) = OB$$

$$520 = OB \quad \text{(two significant digits)}$$

Thus, the minimum brake force required is 520 lb.

EXERCISES 3–4

In Exercises 1–20 find the magnitude of all forces to two significant digits and all angles to the nearest degree.

1. A force of 5.0 lb and a force of 12 lb are acting on a body with an angle of 90° between the two forces. Find:
    a. The magnitude of the resultant.
    b. The angle between the resultant and the larger force.

2. A force of 12 lb and a force of 16 lb are acting on a body with an angle of 90° between the two forces. Find:
    a. The magnitude of the resultant.
    b. The angle between the resultant and the smaller force.

3. Two forces, one of $1\overline{0}$ lb and the other of 15 lb, act on the same object at right angles to each other. Find:
    a. The magnitude of the resultant.
    b. The angle between the resultant and the larger force.

4. Two velocities, one of $2\overline{0}$ mi/hour north and the other of $3\overline{0}$ mi/hour east are acting on the same body. Find:

a. The speed of the resultant velocity.

b. The angle of the resultant velocity with respect to a direction of due east.

5. Two velocities, one of 5.0 mi/hour south and the other of 15 mi/hour west are acting on the same body. Find:

a. The speed of the resultant velocity.

b. The angle of the resultant velocity with respect to a direction of due west.

6. A ship is headed due south at 18 knots (nautical miles per hour) while the current carries the ship due west at 5.0 knots. Find:

a. The speed of the ship.

b. The direction (course) of the ship.

7. An airplane can fly $5\overline{0}0$ mi/hour in still air. If it is heading due north in a wind that is blowing due east at a rate of $8\overline{0}$ mi/hour, find:

a. The distance the plane can fly in 1 hour.

b. The angle that the resultant velocity will make with respect to due east.

8. An object is dropped from a plane that is moving horizontally at a speed of $3\overline{0}0$ ft/second. If the vertical velocity of the object in terms of time is given by the formula $v = 32t$, 5 seconds later:

a. What is the speed of the object?

b. What angle does the direction of the object make with the horizontal?

9. A pilot wishes to fly due east at $3\overline{0}0$ mi/hour when a $7\overline{0}$ mi/hour south wind is blowing.

a. How many degrees south of east should the pilot point the plane to attain the desired course?

b. What airspeed should the pilot maintain?

10. A ship wishes to travel due south at 25 knots in an easterly current of 7.0 knots.

a. How many degrees west of south should the navigator direct the ship?

b. What speed must the ship maintain?

In Exercises 11–14 find the horizontal and vertical component of each vector.

11. A magnitude of $1\overline{0}0$ lb and makes an angle of 30° with the positive x axis.

12. A magnitude of 75 lb and makes an angle of 45° with the positive x axis.

13. A magnitude of 18 lb and makes an angle of 27° with the positive x axis.

14. A magnitude of 125 lb and makes an angle of 72° with the positive x axis.

15. An airplane pointed due east is traveling 15° north of east at a rate of 350 mi/hour. The resultant course is due to a wind blowing north. Find:

a. The velocity of the plane if there were no wind (that is, the vector pointing due east).

b. The velocity of the wind (that is, the vector pointing due north).

16. A man pulls with a force of 25 lb on a window pole in an effort to lower a window. How much force is wasted if the pole makes an angle of 15° with the window?

17. A man pushes with a force of $4\overline{0}$ lb on the handle of a lawn mower that makes an angle of 33° with the ground. How much force pushes the lawn mower forward?

18. A car weighing 3500 lb is parked in a driveway that makes an angle of 5° with the horizontal. Find the minimum brake force that is needed to keep the car from rolling down the driveway (assume no friction).

19. Find the force needed to keep a barrel weighing $1\overline{0}0$ lb from rolling down a ramp that makes an angle of 15° with the horizontal (assume no friction).

20. A box is resting on a ramp that makes an angle of 18° with the horizontal. What is the force of friction between the box and the ramp if the box weighs $8\overline{0}$ lb?

3–5 SPECIAL ANGLES AND COFUNCTIONS

The values of the trigonometric functions that appear in Table 1 are computed through the aid of calculus. Although these values can be determined to any desired accuracy, they will always be *approximations*. However, for certain angles we can determine the *exact* values of the trigonometric functions. These values were determined geometrically long before calculus was developed and they played a central role in the development of trigonometry.

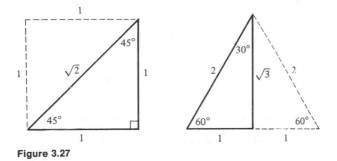

Figure 3.27

The first angles whose trigonometric values are exactly known are the acute angles 30°, 45°, and 60°. Geometrically, these values may be found by considering the right triangles in Figure 3.27. The diagonal in a square with a side length of 1 unit forms the right triangle with a 45° angle; while the altitude in an equilateral triangle with a side length of 2 units forms the right triangle with angles of 30° and 60°. Using these two triangles we may determine the value of any of their trigonometric functions. The results are tabulated in the following table:

θ	$\sin \theta$	$\csc \theta$	$\cos \theta$	$\sec \theta$	$\tan \theta$	$\cot \theta$
30°	$\dfrac{1}{2}$	2	$\dfrac{\sqrt{3}}{2}$	$\dfrac{2}{\sqrt{3}}$	$\dfrac{1}{\sqrt{3}}$	$\sqrt{3}$
60°	$\dfrac{\sqrt{3}}{2}$	$\dfrac{2}{\sqrt{3}}$	$\dfrac{1}{2}$	2	$\sqrt{3}$	$\dfrac{1}{\sqrt{3}}$
45°	$\dfrac{1}{\sqrt{2}}$	$\sqrt{2}$	$\dfrac{1}{\sqrt{2}}$	$\sqrt{2}$	1	1

Note the similarity between the values of the trigonometric functions for 30° and 60°. For example:

$$\sin 30° = \frac{1}{2} \quad \text{and} \quad \cos 60° = \frac{1}{2}$$

$$\tan 30° = \frac{1}{\sqrt{3}} \quad \text{and} \quad \cot 60° = \frac{1}{\sqrt{3}}$$

$$\sec 30° = \frac{2}{\sqrt{3}} \quad \text{and} \quad \csc 60° = \frac{2}{\sqrt{3}}$$

This similarity results from the fact that in the right triangle the side opposite the 30° angle is adjacent to the 60° angle. Thus,

$$\sin 30° = \frac{\text{length of side opposite 30° angle}}{\text{hypotenuse}} = \frac{1}{2}$$

$$= \frac{\text{length of side adjacent to 60° angle}}{\text{hypotenuse}} = \cos 60°$$

$$\tan 30° = \frac{\text{length of side opposite 30° angle}}{\text{length of side adjacent to 30° angle}} = \frac{1}{\sqrt{3}}$$

$$= \frac{\text{length of side adjacent to 60° angle}}{\text{length of side opposite 60° angle}} = \cot 60°$$

$$\sec 30° = \frac{\text{hypotenuse}}{\text{length of side adjacent to 30° angle}} = \frac{2}{\sqrt{3}}$$

$$= \frac{\text{hypotenuse}}{\text{length of side opposite 60° angle}} = \csc 60°$$

Observe that in each case a trigonometric function of 30° is equal to the corresponding cofunction of 60°. The corresponding cofunction is easy to remember since the *co*function of the sine is the *co*sine, the cofunction of the tangent is the *co*tangent, and the cofunction of the secant is the *co*secant. We can generalize from these examples concerning 30° and 60° to any two angles A and B that are complementary (that is, $A + B = 90°$), since in any right triangle the side opposite angle A is adjacent to angle B. Thus, *a trigonometric function of any acute angle is equal to the corresponding cofunction of the complementary angle.*

EXAMPLE 1: sin 23° = cos 67° because 23° and 67° are complementary angles (that is, 23° + 67° = 90°) and the cofunction of the sine is the cosine.

EXAMPLE 2: cot 15°20′ = tan 74°40′ since 15°20′ and 74°40′ are complementary angles (that is, 15°20′ + 74°40′ = 90°) and the cofunction of the cotangent is the tangent.

EXAMPLE 3: sec 64°07′ = csc 25°53′ since 64°07′ and 25°53′ are complementary angles (that is, 64°07′ + 25°53′ = 90°) and the cofunction of the secant is the cosecant.

EXAMPLE 4: Find an acute angle θ for which sin θ = cos(θ + 40°).

SOLUTION: Since the sine and the cosine are cofunctions, we know that the angle θ and the angle (θ + 40°) are complementary. Thus,

$$\theta + (\theta + 40°) = 90°$$
$$2\theta + 40° = 90°$$
$$2\theta = 50°$$
$$\theta = 25°$$

Check: Substitute 25° for θ in the original equation.

$$\sin 25° \overset{?}{=} \cos(25° + 40°)$$
$$\sin 25° \overset{\checkmark}{=} \cos 65°$$

EXAMPLE 5: Find an acute angle θ for which sec θ = csc 5θ.

SOLUTION: Since the secant and the cosecant are cofunctions, we know that the angle θ and the angle 5θ are complementary. Thus,

$$\theta + 5\theta = 90°$$
$$6\theta = 90°$$
$$\theta = 15°$$

Check: Substitute 15° for θ in the original equation.

$$\sec 15° \overset{?}{=} \csc 5(15°)$$
$$\sec 15° \overset{\checkmark}{=} \csc 75°$$

A second collection of angles whose trigonometric values are exactly known are angles whose terminal ray coincides with one of the axes. These angles are called *quadrantal angles,* and any quadrantal angle can always be expressed as the product of 90° and some integer. For example, 270° = 90° · 3 and −180° = 90° · (−2). To find the values of the trigonometric functions for quadrantal angles, we merely pick any point on the terminal ray of the angle (θ) and calculate the various ratios. Since we have shown earlier that it does not matter which point we pick on the terminal ray of θ, it is convenient to let $r = 1$ and pick the point that is 1 unit from the origin (0, 0). For example, the terminal ray of a 90° angle is the positive y axis. We can calculate the trigonometric ratios by picking any point on the positive y axis, but it is easier to let $r = 1$ and pick the point (0, 1), as shown in Figure 3.28. Then since $x = 0$, $y = 1$, and $r = 1$, we can calculate

$$\sin 90° = \frac{y}{r} = \frac{1}{1} = 1 \leftarrow\text{reciprocals}\rightarrow \csc 90° = \frac{r}{y} = \frac{1}{1} = 1$$

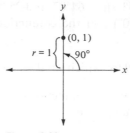

Figure 3.28

$$\cos 90° = \frac{x}{r} = \frac{0}{1} = 0 \;\leftarrow\text{reciprocals}\rightarrow\; \sec 90° = \frac{r}{x} = \frac{1}{0} \quad \text{undefined}$$

$$\tan 90° = \frac{y}{x} = \frac{1}{0} \quad \text{undefined} \;\leftarrow\text{reciprocals}\rightarrow\; \cot 90° = \frac{x}{y} = \frac{0}{1} = 0$$

We can repeat this procedure for any angle whose terminal ray coincides with one of the axes. Figure 3.29 shows the four possible positions for the terminal ray of a quadrantal angle.

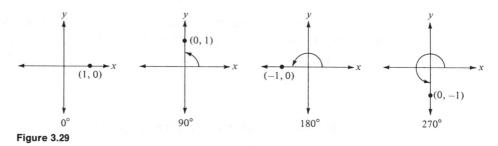

Figure 3.29

There are other quadrantal angles besides 0°, 90°, 180°, and 270°, but their trigonometric values must be the same as one of the four listed. For example, the trigonometric values for 360° will be the same as the trigonometric values for 0°, since the terminal ray for both angles is the same (positive x axis). In general, *if two angles have the same terminal ray, they are called coterminal and the trigonometric functions of coterminal angles are equal.* The following table summarizes the values of the trigonometric functions for quadrantal angles:

θ	$\sin \theta$	$\csc \theta$	$\cos \theta$	$\sec \theta$	$\tan \theta$	$\cot \theta$
0°	0	undefined	1	1	0	undefined
90°	1	1	0	undefined	undefined	0
180°	0	undefined	−1	−1	0	undefined
270°	−1	−1	0	undefined	undefined	0

EXAMPLE 6: Find the exact value of sin 90° + cos 180°.

SOLUTION:

$$\sin 90° = 1$$
$$\cos 180° = -1$$

Therefore, $\sin 90° + \cos 180° = (1) + (-1) = 0.$

EXAMPLE 7: Find the exact value of $3 \cos 0° - \csc 270°.$

SOLUTION:

$$\cos 0° = 1$$
$$\csc 270° = -1$$

Therefore, $3 \cos 0° - \csc 270° = 3(1) - (-1) = 3 + 1 = 4.$

EXAMPLE 8: Find the exact value of $\sin 540°.$

SOLUTION: 540° is a quadrantal angle that is coterminal with 180° (see Figure 3.30). Therefore, $\sin 540° = \sin 180° = 0.$

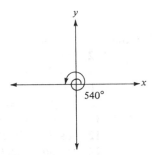

Figure 3.30

EXAMPLE 9: Find the exact value of $\cos(-270°).$

SOLUTION: $-270°$ is a quadrantal angle that is coterminal with 90° (see Figure 3.31). Therefore, $\cos(-270°) = \cos 90° = 0.$

(*Note:* Remember that a negative angle means a rotation in a clockwise direction.)

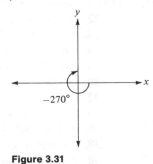

Figure 3.31

EXAMPLE 10: Find the exact value of $3 \tan 0° - 2 \sin 30°$.

SOLUTION:

$\tan 0° = 0$

$\sin 30° = \frac{1}{2}$

Therefore, $3 \tan 0° - 2 \sin 30° = 3(0) - 2(\frac{1}{2}) = 0 - 1 = -1$.

EXAMPLE 11: Find the exact value of $(\sec \theta)^2 - (\tan \theta)^2$ if $\theta = 180°$.

SOLUTION:

$\sec 180° = -1$

$\tan 180° = 0$

Therefore, $(\sec 180°)^2 - (\tan 180°)^2 = (-1)^2 - (0)^2 = 1$.

EXERCISES 3–5

In Exercises 1–14 find an acute angle θ that satisfies the equation.

1. $\sin 20° = \cos \theta$
2. $\cot 54° = \tan \theta$
3. $\csc \theta = \sec 5°$
4. $\cos \theta = \sin 71°$
5. $\tan 33°10' = \cot \theta$
6. $\sec \theta = \csc 64°40'$
7. $\sin \theta = \cos 40°30'$
8. $\tan 51°17' = \cot \theta$
9. $\sin \theta = \cos(\theta + 20°)$
10. $\cot \theta = \tan(\theta - 15°)$
11. $\sec \theta = \csc 2\theta$
12. $\cos \theta = \sin \theta$
13. $\cot 2\theta = \tan 2\theta$
14. $\csc 3\theta = \sec(2\theta - 40°)$

In Exercises 15–20 express each term as a function of the angle complementary to the given angle.

15. $\sin 17°$
16. $\cos 64°$
17. $\tan 1°$
18. $\sec 33°10'$
19. $\cot 78°09'$
20. $\csc 13°58'$

In Exercises 21–45 find the exact value of the expression.

21. $\sin 30° + \tan 45°$
22. $\cos 60° + \csc 30°$
23. $\cot 45° + \sec 60°$
24. $3 \tan 45° - 2 \cos 60°$
25. $4 \csc 30° - \sin 30°$
26. $\sin 0° + \cos 90°$
27. $\tan 180° - \sec 0°$
28. $\csc 270° + \cot 90°$
29. $2 \sec 180° - 3 \sin 270°$
30. $\cos 0° + 4 \tan 0°$
31. $\tan(-180°) + \sin(-270°)$
32. $\cos(-90°) - \sin(-360°)$
33. $\csc 450° + \tan 540°$
34. $5 \cot 630° - 3 \sin 720°$
35. $\cot(-450°) - 2 \cos(-630°)$
36. $3 \sin(-540°) + \tan(-720°)$
37. $\sin 45° + \tan 180°$
38. $\csc 60° + \cot 90°$
39. $\tan 30° + \sec 90°$
40. $\cos 30° - \csc 180°$

41. $2 \sin \theta$ if $\theta = 30°$

42. $\sin 2\theta$ if $\theta = 30°$

43. $4 \tan \theta \cdot \cot \theta$ if $\theta = 45°$

44. $(\sin \theta)^2 + (\cos \theta)^2$ if $\theta = 90°$

45. $(\csc \theta)^2 - (\cot \theta)^2$ if $\theta = 270°$

3–6 REDUCING FUNCTIONS OF ANGLES IN ANY QUADRANT

At present we have developed methods that enable us to calculate approximate trigonometric values for acute angles (Table 1) and exact trigonometric values for special angles (30°, 45°, 60°, and quadrantal angles). We now need to develop a method for finding the trigonometric values of an angle whose terminal ray lies in quadrant 2, quadrant 3, or quadrant 4, for then we will be able to compute the values of the trigonometric functions for *any* angle.

The procedure for finding these values is to relate any nonacute angle to an angle that appears in Table 1 through what is called a *reference angle*. The reference angle for an angle θ is defined to be the *positive acute angle formed by the terminal ray of θ and the horizontal axis*. For example, the reference angle for 150° is 30° since the closest segment of the horizontal axis (negative x axis) may correspond to a rotation of 180° and $|180° - 150°| = 30°$, as shown in Figure 3.32.

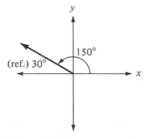

Figure 3.32

EXAMPLE 1: The reference angle for 225° is 45° since the closest segment of the horizontal axis (negative x axis) may correspond to a rotation of 180° and $|225° - 180°| = 45°$ (see Figure 3.33).

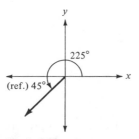

Figure 3.33

EXAMPLE 2: The reference angle for 333°17′ is 26°43′ since the closest segment of the horizontal axis (positive x axis) may correspond to a rotation of 360° and $|360° - 333°17′| = 26°43′$ (see Figure 3.34).

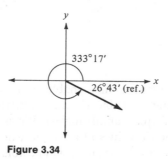

Figure 3.34

EXAMPLE 3: The reference angle for 600° is 60° since the closest segment of the horizontal axis (negative x axis) may correspond to a rotation of 360° + 180° or 540° and $|600° - 540°| = 60°$ (see Figure 3.35).

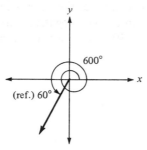

Figure 3.35

EXAMPLE 4: The reference angle for $-230°$ is 50° since the closest segment of the horizontal axis (negative x axis) may correspond to $-180°$ and $|-180° - (-230°)| = 50°$ (see Figure 3.36).

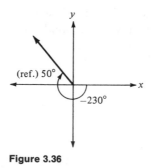

Figure 3.36

Now consider the angles 30°, 150°, 210°, and 330°, where the reference angle for each of these angles is 30°. Note that the points on the terminal ray for each of these

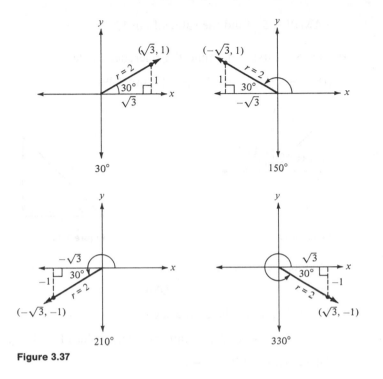

Figure 3.37

angles differ only in the sign of the x coordinate or the y coordinate. Therefore, the values of the trigonometric functions (which are ratios among x, y, and r) of these angles can differ only in their sign. For instance, by considering the points in Figure 3.37, we can determine that $\sin 30° = 1/2$, $\sin 150° = 1/2$, $\sin 210° = -1/2$, and $\sin 330° = -1/2$. Thus, the reference angle provides us with a method for relating an angle in any quadrant to some acute angle whose trigonometric values can be found. Although the trigonometric values of an angle and its reference angle may differ in sign, we can determine the correct sign according to the quadrant containing the terminal ray of the angle. In Section 3-1 we developed the chart in Figure 3.38, which summarizes the sign of the different trigonometric functions (<u>A</u>ll <u>s</u>tudents <u>t</u>ake <u>c</u>hemistry) in the various quadrants.

$\left.\begin{array}{l}\underline{\text{s}}\text{in } \theta \\ \text{csc } \theta\end{array}\right\}\ +$	<u>A</u>ll the functions are positive
others$\}\ -$	
$\left.\begin{array}{l}\underline{\text{t}}\text{an } \theta \\ \text{cot } \theta\end{array}\right\}\ +$	$\left.\begin{array}{l}\underline{\text{c}}\text{os } \theta \\ \text{sec } \theta\end{array}\right\}\ +$
others$\}\ -$	others$\}\ -$

Figure 3.38

EXAMPLE 5: Find the value of tan 135°.

SOLUTION: First, determine the reference angle.

$$|180° - 135°| = 45° \qquad \text{(see Figure 3.39)}$$

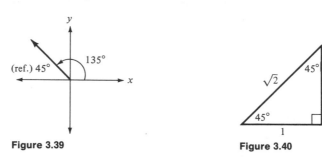

Figure 3.39   **Figure 3.40**

Second, determine tan 45°.

tan 45° = 1   (see Figure 3.40)

Third, determine the correct sign.

135° is in the second quadrant where the value of the tangent function is negative

Therefore, tan 135° = −1.

In general, we can find the trigonometric value of any angle by doing the following:

1. Find the reference angle for the given angle.
2. Find the trigonometric value of the reference angle using the appropriate function. Since reference angles are always positive acute angles, this value can always be determined from Table 1. (*Note:* If the reference angle is 30°, 45°, or 60°, the exact answer is preferable.)
3. Determine the correct sign according to the terminal ray of the angle.

EXAMPLE 6: Approximate cos 227°20′.

SOLUTION: First, determine the reference angle.

$$|227°20′ - 180°| = 47°20′ \qquad \text{(see Figure 3.41)}$$

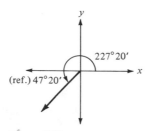

Figure 3.41

Second, determine cos 47°20′.

cos 47°20′ ≈ 0.6777   (Table 1)

Third, determine the correct sign.

227°20′ is in Q_3 where the cosine function is negative

Therefore, cos 227°20′ = −cos 47°20′ ≈ −0.6777.

EXAMPLE 7: Find the exact value of cos 300°.

SOLUTION: First, determine the reference angle.

$|360° − 300°| = 60°$ (see Figure 3.42)

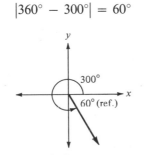

Figure 3.42

Figure 3.43

Second, determine cos 60°.

cos 60° = ½   (see Figure 3.43)

Third, determine the correct sign.

300° is in Q_4 where the cosine function is positive

Therefore, cos 300° = cos 60° = ½.

EXAMPLE 8: Evaluate sec 510°.

SOLUTION: First, determine the reference angle.

$|540° − 510°| = 30°$ (see Figure 3.44)

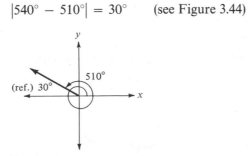

Figure 3.44

Second, determine sec 30°.

$$\sec 30° = \frac{2}{\sqrt{3}} \qquad \text{(see Figure 3.45)}$$

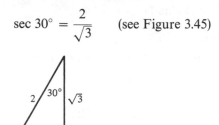

Figure 3.45

Third, determine the correct sign.

510° is in Q_2 where the secant function is negative

Therefore, $\sec 510° = -\sec 30° = -2/\sqrt{3}$.

EXAMPLE 9: Approximate $\cot(-110°)$.

SOLUTION: First, determine the reference angle.

$$\left|-180° - (-110°)\right| = 70° \qquad \text{(see Figure 3.46)}$$

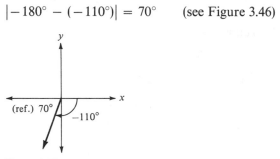

Figure 3.46

Second, determine $\cot 70°$.

$\cot 70° \approx 0.3640$ (Table 1)

Third, determine the correct sign.

$-110°$ is in Q_3 where the cotangent function is positive

Therefore, $\cot(-110°) = \cot 70° \approx 0.3640$.

EXAMPLE 10: Approximate $\csc(-412°50')$.

SOLUTION: First, determine the reference angle.

$$\left|-360 - (-412°50')\right| = 52°50' \qquad \text{(see Figure 3.47)}$$

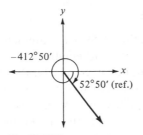

Figure 3.47

Second, determine csc 52° 50′.

csc 52° 50′ ≈ 1.255   (Table 1)

Third, determine the correct sign.

−412° 50′ is in Q₄ where the cosecant function is negative

Therefore, csc (−412° 50′) = −csc 52° 50′ ≈ −1.255.

Note from the last two examples that negative angles (clockwise rotations) have trigonometric values that may be either positive or negative.

EXERCISES 3–6

In Exercises 1–30 find the exact value of each expression.

1. sin 210°	**2.** tan 225°	**3.** sec 330°
4. cos 135°	**5.** cot 315°	**6.** csc 120°
7. sin 150°	**8.** tan 300°	**9.** cos 225°
10. cot 240°	**11.** cos 330°	**12.** sin 315°
13. sec 390°	**14.** tan 420°	**15.** cot 690°
16. cos 840°	**17.** sin 1035°	**18.** csc 675°
19. cos 495°	**20.** sin 570°	**21.** sin(−60°)
22. cos(−45°)	**23.** tan(−120°)	**24.** sec(−225°)
25. cot(−315°) ·	**26.** csc(−210°)	**27.** sec(−330°)
28. cos(−480°)	**29.** cot(−495°)	**30.** sin(−1050°)

In Exercises 31–60 find the approximate value of each expression.

31. sin 212°	**32.** cos 307°	**33.** tan 254°
34. cot 115°	**35.** sec 301° 20′	**36.** csc 163° 40′
37. cos 148° 50′	**38.** sin 354° 30′	**39.** cot 298° 10′
40. tan 190° 40′	**41.** sin 177° 10′	**42.** cos 252° 20′
43. csc 672° 40′	**44.** cot 392° 10′	**45.** sin 626° 40′
46. sin 531° 30′	**47.** cos 952° 20′	**48.** sec 452° 20′
49. tan 738° 30′	**50.** sin 521° 50′	**51.** cos(−81°)
52. sin(−25°)	**53.** tan(−131°)	**54.** csc(−322° 40′)
55. cos(−61°)	**56.** sin(−251° 30′)	**57.** tan(−214° 10′)
58. sec(−400°)	**59.** cot(−512°)	**60.** csc(−938° 20′)

3–7 TRIGONOMETRIC EQUATIONS

An equation (as we discussed in Section 1-5) is a statement that two expressions have the same value, and the truth of this mathematical assertion depends upon the number that is substituted for the unknown. Similarly, the validity of an equation that involves the trigonometric functions depends upon the number of degrees that is substituted for the unknown angle. Thus, in solving a trigonometric equation, we are looking for the collection of *all* the angles that make the given equation valid. However, before we attempt to find a procedure for writing all the solutions to a particular equation, let us first establish a method for finding solutions that are between $0°$ and $360°$.

EXAMPLE 1: Find the exact values of θ ($0° \leq \theta < 360°$) for which the equation $\sin \theta = \frac{1}{2}$ is a true statement.

SOLUTION: First, determine the quadrant that contains the terminal ray of θ.

$\sin \theta = \frac{1}{2}$ which is a positive number

The terminal ray of θ could be in either quadrant 1 or 2 since the sine function is positive in both quadrants.

Second, determine the reference angle.

$\sin 30° = \frac{1}{2}$

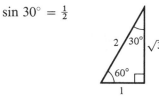

Figure 3.48

Therefore, the reference angle is $30°$ (see Figure 3.48).

Third, determine the appropriate values of θ (see Figure 3.49).

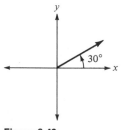 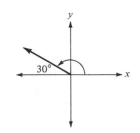

Figure 3.49

$30°$ is the angle in Q_1 with a reference angle of $30°$

$150°$ is the angle in Q_2 with a reference angle of $30°$

Therefore, $30°$ and $150°$ make the equation a true statement.

EXAMPLE 2: Approximate the values of θ ($0° \leq \theta < 360°$) for which the equation $\cos \theta = -0.7969$ is a true statement.

SOLUTION : First, determine the quadrant that contains the terminal ray of θ.

$\cos \theta = -0.7969$ which is a negative number

The terminal ray of θ could be in either quadrant 2 or 3 since the cosine function is negative in both quadrants.

Second, determine the reference angle.

$\cos 37° 10' \approx 0.7969$ (from Table 1)

Therefore, the reference angle is $37° 10'$.

Third, determine the appropriate values of θ (see Figure 3.50).

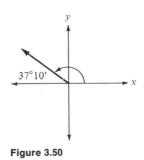
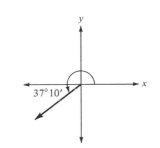

Figure 3.50

$142° 50'$ is the angle in Q_2 with a reference angle of $37° 10'$

$217° 10'$ is the angle in Q_3 with a reference angle of $37° 10'$

Therefore, $142° 50'$ and $217° 10'$ make the equation a true statement.

EXAMPLE 3: Approximate the values of θ ($0° \leq \theta < 360°$) that make the equation $4 \tan \theta - 1 = 0$ a true statement.

SOLUTION : First, solve the equation for $\tan \theta$.

$4 \tan \theta - 1 = 0$
$4 \tan \theta = 1$
$\tan \theta = \frac{1}{4} = 0.2500$

Second, determine the quadrant that contains the terminal ray of θ.

$\tan \theta = 0.2500$ which is a positive number

The terminal ray of θ could be in either Q_1 or Q_3 since the tangent function is positive in both quadrants.

Third, determine the reference angle.

$$\tan 14°00' \approx 0.2500 \qquad \text{(from Table 1)}$$

Therefore, $14°00'$ is the reference angle.

Fourth, determine the appropriate values of θ (see Figure 3.51).

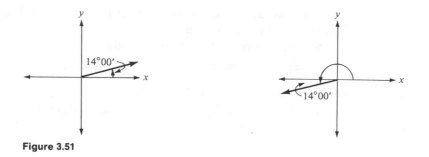

Figure 3.51

$14°00'$ is the angle in Q_1 with a reference angle of $14°00'$

$194°00'$ is the angle in Q_3 with a reference angle of $14°00'$

Therefore, $14°00'$ and $194°00'$ make the equation a true statement.

EXAMPLE 4: Find the exact values of θ ($0° \le \theta < 360°$) for which the equation $2 \sin \theta + \sqrt{3} = 0$ is a true statement.

SOLUTION: First, solve the equation for $\sin \theta$.

$$2 \sin \theta + \sqrt{3} = 0$$
$$2 \sin \theta = -\sqrt{3}$$
$$\sin \theta = \frac{-\sqrt{3}}{2}$$

Second, determine the quadrant that contains the terminal ray of θ.

$$\sin \theta = \frac{-\sqrt{3}}{2} \qquad \text{which is a negative number}$$

The terminal ray of θ could be in either Q_3 or Q_4 since the sine function is negative in both quadrants.

Third, determine the reference angle.

$$\sin 60° = \frac{\sqrt{3}}{2}$$

Therefore, $60°$ is the reference angle.

Fourth, determine the appropriate values of θ (see Figure 3.52).

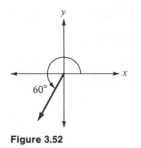

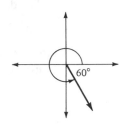

Figure 3.52

240° is the angle in Q_3 with a reference angle of 60°

300° is the angle in Q_4 with a reference angle of 60°

Therefore, 240° and 300° make the equation a true statement.

Once we are able to find the solutions of a trigonometric equation that are between 0° and 360°, we can determine *all* the solutions by finding the angles that are coterminal with our results. That is, we wish to find all the angles that have the same terminal ray with the solutions that are between 0° and 360°. The problem arises that we cannot possibly list all the angles that are coterminal with a given angle. Therefore, we indicate some rule by which coterminal angles can be found. We can generate all the angles coterminal to a given angle of say 30° by adding to 30° the multiples of 360°. Thus,

$$30° + (0)360° = 30°$$
$$30° + (1)360° = 390°$$
$$30° + (2)360° = 750°$$
$$30° + (-1)360° = -330°$$
$$30° + (-2)360° = -690° \quad \text{etc.}$$

In general, $30° + k360°$, where k is an integer, will generate all the angles that have the same terminal ray as 30°.

EXAMPLE 5: Approximate all of the solutions to the equation

$$10 \cot \theta = 3$$

SOLUTION: First, solve the equation for $\cot \theta$.

$$10 \cot \theta = 3$$
$$\cot \theta = \tfrac{3}{10} = 0.3000$$

Second, determine which quadrant contains the terminal ray of θ.

$\cot \theta$ is a positive number

The terminal ray of θ could be in either Q_1 or Q_3 since the cotangent function is positive in both quadrants.

Third, determine the reference angle.

cot 73°20′ ≈ 0.3000   (from Table 1)

Therefore, 73°20′ is the reference angle.

Fourth, determine the solutions between 0° and 360° (see Figure 3.53).

Figure 3.53

73°20′ is the angle in Q_1 with a refer-   253°20′ is the angle in Q_3 with a reference angle of 73°20′   ence angle of 73°20′

Therefore, 73°20′ and 253°20′ make the equation a true statement.

Fifth, indicate how angles coterminal to the above angles may be generated.

$$\left.\begin{array}{l} 73°20′ + k360° \\ 253°20′ + k360° \end{array}\right\} \quad \text{or equivalently} \quad 73°20′ + k180°$$

where k is an integer, generates all the solutions to the equation.

EXAMPLE 6: Find the exact values of θ for which the equation $2 \sec \theta + 5 = 1$ is a true statement.

SOLUTION: First, solve the equation for $\sec \theta$.

$$2 \sec \theta + 5 = 1$$
$$2 \sec \theta = -4$$
$$\sec \theta = -2$$

Second, determine the quadrant that contains the terminal ray of θ.

$\sec \theta$ is a negative number

The terminal ray of θ could be in either Q_2 or Q_3 since the secant function is negative in both quadrants.

Third, determine the reference angle.

$$\sec 60° = 2$$

Therefore, 60° is the reference angle.

Fourth, determine the solutions between 0° and 360° (see Figure 3.54).

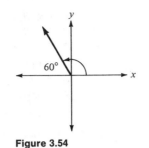
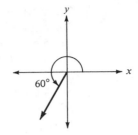

Figure 3.54

120° is the angle in Q_2 with a reference angle of 60°

240° is the angle in Q_3 with a reference angle of 60°

Therefore, 120° and 240° make the equation a true statement.

Fifth, indicate how angles coterminal to the above angles may be generated.

$$120° + k360°$$
$$240° + k360°$$

where k is an integer, generates all the solutions to the equation.

EXERCISES 3–7

In Exercises 1–20 find the exact values of θ between 0° and 360° that make the equation a true statement.

1. $\cos \theta = \dfrac{1}{2}$

2. $\cos \theta = \dfrac{-1}{2}$

3. $\tan \theta = -1$

4. $\tan \theta = 1$

5. $\sin \theta = \dfrac{1}{\sqrt{2}}$

6. $\sin \theta = \dfrac{-1}{\sqrt{2}}$

7. $\sec \theta = \dfrac{-2}{\sqrt{3}}$

8. $\sec \theta = \dfrac{2}{\sqrt{3}}$

9. $\cot \theta = \sqrt{3}$

10. $\cot \theta = -\sqrt{3}$

11. $\csc \theta = -\sqrt{2}$

12. $\csc \theta = \sqrt{2}$

13. $2 \tan \theta = 2\sqrt{3}$

14. $-2 \cos \theta = \sqrt{3}$

15. $2 \sin \theta + \sqrt{3} = 0$

16. $\csc \theta - 2 = 0$

17. $4 \sin \theta + 3 = 1$

18. $3 \sec \theta - 7 = -1$

19. $2 \tan \theta + 5 = 7$

20. $3 \cot \theta + 4 = 1$

In Exercises 21–40 find the approximate values of θ between 0° and 360° that make the equation a true statement.

21. $\sin \theta = 0.1219$

22. $\sin \theta = -0.1219$

23. $\cos \theta = -0.5125$

24. $\cos \theta = 0.5125$

25. $\cot \theta = 2.457$

26. $\cot \theta = -2.457$

27. $\sec \theta = -1.058$

28. $\sec \theta = 1.058$

29. $\tan \theta = 3.145$

30. $\tan \theta = -3.145$

31. $5 \cot \theta = -1$

32. $7 \sin \theta = 2$

33. $3 \cos \theta + 2 = 0$

34. $3 \tan \theta - 1 = 0$

35. $2 \tan \theta - 7 = 0$

36. $5 \sin \theta + 2 = 0$

37. $4 \csc \theta + 9 = 0$

38. $\cos \theta - 3 = 0$

39. $\sin \theta + 2 = 0$

40. $3 \sec \theta - 1 = 1$

In Exercises 41–50 find five angles that are coterminal with the given angle.

41. $45°$

42. $90°$

43. $22°10'$

44. $84°40'$

45. $120°$

46. $215°$

47. $-30°$

48. $-60°$

49. $-100°50'$

50. $-312°20'$

In Exercises 51–60 find all the values of θ for which the given trigonometric equation is a true statement. Where possible, find exact solutions.

51. $\sqrt{2} \sin \theta = 1$

52. $\sqrt{3} \csc \theta = 2$

53. $3 \sin \theta + 2 = 1$

54. $5 \tan \theta + 2 = -3$

55. $10 \cos \theta + 7 = 3$

56. $2 \cot \theta - 3 = 0$

57. $5 \sec \theta + 1 = 3$

58. $\sin \theta = 1$

59. $\cos \theta = -1$

60. $\cos \theta = 2$

3–8 LAW OF SINES

In Section 3-3 we learned to solve right triangles. We now wish to extend our ability to solve triangles by considering the solution of general triangles, which may or may not be right triangles. Remember that we "solve" a triangle by finding the values of the three angles and three sides of the triangle. In order to accomplish this, at least three of these six values must be known, one of which must be a side.

The first technique that we will use to solve general triangles is called the *law of sines*. We can illustrate the origin of this method in the case where the three angles of the triangle are acute angles (that is, less than 90°) by considering triangle ABC in Figure 3.55. A different approach is used to prove this law for oblique triangles. The altitude of the triangle BD, of length h, divides the triangle into two right triangles, ADB and CDB. In right triangle ADB,

$$\sin A = \frac{h}{c}$$

Thus,

$$h = c \sin A$$

In right triangle CDB,

$$\sin C = \frac{h}{a}$$

Thus,

$$h = a \sin C$$

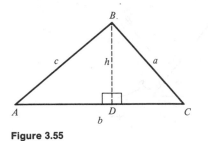

Figure 3.55

Setting these two expressions for h equal to each other, we have

$$c \sin A = a \sin C$$

or

$$\frac{\sin A}{a} = \frac{\sin C}{c}$$

Similarly, by drawing the altitude from vertex A we can show that

$$\frac{\sin B}{b} = \frac{\sin C}{c}$$

Combining these results, we have

$$\frac{\sin A}{a} = \frac{\sin B}{b} = \frac{\sin C}{c}$$

This relationship, the *law of sines*, states the following:

> **LAW OF SINES**   The sines of the angles in a triangle are proportional to the lengths of the opposite sides.

Note that if C is a right angle, $\sin C = \sin 90° = 1$ and the law of sines yields the right-triangle relationships

$$\sin A = \frac{a}{c} \quad \text{and} \quad \sin B = \frac{b}{c}$$

The law of sines can be used to solve any triangle in the following two cases:

1. If we know two angles and one side of the triangle.
2. If we know two sides of the triangle and the angle opposite one of them.

The following example illustrates how the law of sines can be used to solve a triangle in the first case in which two angles and one side of the triangle are known. Note that in computing results, the symbol for equality ($=$) is generally used even though the symbol for approximation ($\approx$) may be more appropriate.

EXAMPLE 1: Approximate the missing parts of triangle ABC, shown in Figure 3.56, in which $A = 35°$, $B = 50°$, and $a = 12$ ft.

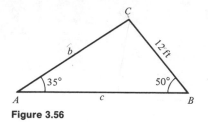

Figure 3.56

SOLUTION: First, we find angle C since the sum of the angles in a triangle is 180°.

$$A + B + C = 180°$$
$$35° + 50° + C = 180°$$
$$C = 95°$$

Second, we find side b by applying the law of sines.

$$\frac{\sin A}{a} = \frac{\sin B}{b}$$

$$\frac{\sin 35°}{12} = \frac{\sin 50°}{b}$$

$$b = \frac{12 \sin 50°}{\sin 35°}$$

$$= \frac{12(0.7660)}{0.5736}$$

$$= 16 \text{ ft}$$

Third, we find side c by applying the law of sines.

$$\frac{\sin A}{a} = \frac{\sin C}{c}$$

$$\frac{\sin 35°}{12} = \frac{\sin 95°}{c}$$

$$c = \frac{12 \sin 95°}{\sin 35°}$$

$$= \frac{12(0.9962)}{0.5736} \quad (Note: \sin 95° = \sin 85° = 0.9962)$$

$$= 21 \text{ ft}$$

Thus, the solution to the triangle is shown in Figure 3.57.

(*Note:* Remember that our computed results cannot be more accurate than the data that are given. Guidelines for the desired accuracy in a solution can be found in Section 3-3. The arithmetic involved in these calculations is quite tedious, so you are encouraged to use a calculator.)

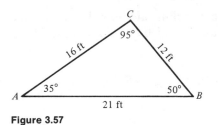

Figure 3.57

EXAMPLE 2: Two observation towers, A and B, are located $1\overline{0}$ mi apart. A fire is sighted at point C and the observer in tower A measures angle CAB to be $80°$. At the same time the observer in tower B measures angle CBA to be $40°$. How far is the fire from tower A?

SOLUTION: First, draw a diagram picturing the data (Figure 3.58).

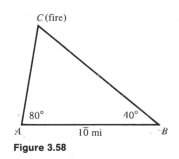

Figure 3.58

Second, find angle C.

$$A + B + C = 180°$$
$$80° + 40° + C = 180°$$
$$C = 60°$$

Third, we find side b by applying the law of sines.

$$\frac{\sin B}{b} = \frac{\sin C}{c}$$

$$\frac{\sin 40°}{b} = \frac{\sin 60°}{10}$$

$$b = \frac{10 \sin 40°}{\sin 60°}$$

$$= \frac{10(0.6428)}{0.8660}$$

$$= 7.4$$

Thus, the fire is about 7.4 mi from station A.

The following examples illustrate how the law of sines can be used to solve a triangle in the second case, in which two sides of the triangle and the angle opposite one of them is known.

EXAMPLE 3: Solve the triangle ABC, in which $B = 60°, b = 5\bar{0}$ ft, and $c = 3\bar{0}$ ft (see Figure 3.59).

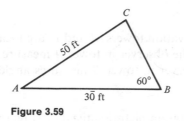

Figure 3.59

SOLUTION: First, we find angle C by applying the law of sines.

$$\frac{\sin B}{b} = \frac{\sin C}{c}$$

$$\frac{\sin 60°}{50} = \frac{\sin C}{30}$$

$$\frac{30(\sin 60°)}{50} = \sin C$$

$$\frac{30(0.8660)}{50} = \sin C$$

$$0.5196 = \sin C$$

We now have two possibilities for angle C since the sine of both first and second quadrant angles is positive.

Case 1 (acute angle in Q_1):	*Case 2* (Obtuse angle in Q_2):
$\sin C = 0.5196$	$\sin C = 0.5196$

reference angle $= 31°20'$ (from Table 1)

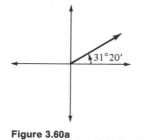

Figure 3.60a

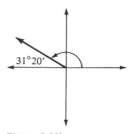

Figure 3.60b

$31°20'$ is the angle in Q_1 with a reference angle of $31°20'$ (Figure 3.60a)

Therefore, $C = 31°20'$ or

$148°40'$ is the angle in Q_2 with a reference angle of $31°20'$ (Figure 3.60b)

$C = 148°40'$.

Second, we find angle A.

$$A + B + C = 180°$$
$$A + 60° + 31°20' = 180°$$
$$A = 88°40'$$

$$A + B + C = 180°$$
$$B = 60° \qquad C = 148°40'$$

Here we find that $B + C = 209°40'$, so regardless of the value of A, the sum of the angles of the triangle exceeds $180°$. Therefore, we reject $C = 148°40'$ as a solution.

Third, we find side a by applying the law of sines.

$$\frac{\sin A}{a} = \frac{\sin B}{b}$$

$$\frac{\sin 88°40'}{a} = \frac{\sin 60°}{50}$$

$$\frac{50(\sin 88°40')}{\sin 60°} = a$$

$$\frac{50(0.9997)}{0.8660} = a$$

$$58 \text{ ft} = a$$

When we round off the angles to the nearest degree, the solution to the triangle is as shown in Figure 3.61.

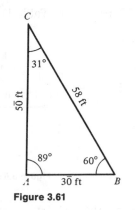

Figure 3.61

EXAMPLE 4: Approximate the missing parts of triangle ABC in which $A = 37°20'$, $a = 12.5$ ft, and $c = 20.1$ ft.

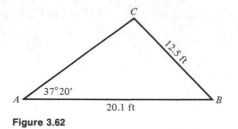

Figure 3.62

SOLUTION: First, sketch Figure 3.62. We find angle C by applying the law of sines.

$$\frac{\sin A}{a} = \frac{\sin C}{c}$$

$$\frac{\sin 37°20'}{12.5} = \frac{\sin C}{20.1}$$

$$\frac{20.1(\sin 37°20')}{12.5} = \sin C$$

$$\frac{20.1(0.6065)}{12.5} = \sin C$$

$$0.9753 = \sin C$$

We now have two possibilities:

Case 1 (acute angle in Q_1):   *Case 2* (obtuse angle in Q_2):

   $\sin C = 0.9753$ $\sin C = 0.9753$

reference angle $= 77°10'$ (from Table 1)

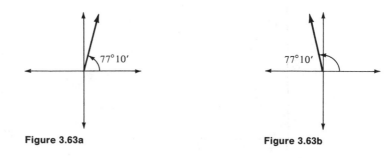

Figure 3.63a   **Figure 3.63b**

$77°10'$ is the angle in Q_1 with a refer- $102°50'$ is the angle in Q_2 with a reference angle of $77°10'$ ence angle of $77°10'$
                Therefore, $C = 77°10'$ or $C = 102°50'$.

Second, we find angle *B*.

$$A + B + C = 180°$$
$$37°20' + B + 77°10' = 180°$$
$$B + 114°30' = 180°$$
$$B = 65°30'$$

$$A + B + C = 180°$$
$$37°20' + B + 102°50' = 180°$$
$$B + 140°10' = 180°$$
$$B = 39°50'$$

Third, we find side *b* by applying the law of sines.

$$\frac{\sin A}{a} = \frac{\sin B}{b}$$

$$\frac{\sin 37°20'}{12.5} = \frac{\sin 65°30'}{b}$$

$$b = \frac{12.5(\sin 65°30')}{\sin 37°20'}$$

$$= \frac{12.5(0.9100)}{0.6065}$$

$$= 18.8 \text{ ft}$$

$$\frac{\sin A}{a} = \frac{\sin B}{b}$$

$$\frac{\sin 37°20'}{12.5} = \frac{\sin 39°50'}{b}$$

$$b = \frac{12.5(\sin 39°50')}{\sin 37°20'}$$

$$= \frac{12.5(0.6406)}{0.6065}$$

$$= 13.2 \text{ ft}$$

The two possible solutions from the given data are shown in Figure 3.64.

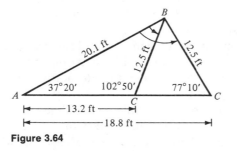

Figure 3.64

Note that when we attempt to solve a triangle in which two sides of the triangle and the angle opposite one of them are given, there may be one triangle that fits the data (as in Example 3) or there may be two triangles that fit the data (as in Example 4). If there is no triangle that can be constructed from the data, this implies that the measurements were not taken accurately.

EXERCISES 3–8

In Exercises 1–20 approximate the remaining parts of the triangles for the data given.
  1. $A = 30°$, $a = 25$ ft, and $B = 45°$.
  2. $C = 60°$, $c = 4\overline{0}$ ft, and $A = 80°$.

3. $A = 15°$, $C = 87°$, and $b = 42$ ft.
4. $B = 68°$, $C = 72°$, and $a = 18$ ft.
5. $B = 120°$, $C = 40°$, and $a = 55$ ft.
6. $C = 135°$, $c = 98$ ft, and $B = 15°$.
7. $A = 62°10'$, $a = 31.5$ ft, and $B = 76°30'$.
8. $C = 44°50'$, $B = 86°20'$, and $a = 62.7$ ft.
9. $B = 111°20'$, $C = 35°40'$, and $a = 142$ ft.
10. $A = 98°30'$, $B = 6°10'$, and $a = 415$ ft.
11. $A = 45°$, $a = 8\overline{0}$ ft, and $b = 5\overline{0}$ ft.
12. $C = 60°$, $c = 75$ ft, and $a = 45$ ft.
13. $B = 30°$, $b = 3\overline{0}$ ft, and $a = 4\overline{0}$ ft.
14. $B = 22°$, $b = 78$ ft, and $a = 86$ ft.
15. $C = 150°$, $c = 92$ ft, and $b = 69$ ft.
16. $A = 95°$, $a = 54$ ft, and $c = 38$ ft.
17. $B = 7°10'$, $b = 74.8$ ft, and $c = 92.4$ ft.
18. $B = 152°50'$, $b = 13\overline{0}$ ft, and $c = 45.0$ ft.
19. $A = 81°$, $a = 11$ ft, and $c = 35$ ft.
20. $C = 105°30'$, $c = 46.1$ ft, and $b = 75.2$ ft.

21. Two surveyors establish a baseline AB on a level field. The surveyor at point A is $20\overline{0}$ ft from the surveyor at point B. Each one sights a stake at point C. The surveyor at A measures angle CAB to be $72°30'$, while the surveyor at B measures angle CBA to be $81°20'$. Find the distance from B to C.

22. Engineers wish to build a bridge across a river to join point A on one side to either point B or point C on the other side. This distance from B to C is $40\overline{0}$ ft, angle ABC is $67°20'$, and angle ACB is $84°30'$. By how many feet does the distance from A to B exceed the distance from A to C?

23. Airport A is $30\overline{0}$ mi due north of airport B. Their radio stations receive a distress signal from a ship located at point C. It is determined that point C is located $54°$ south of east with respect to airport A and $76°$ north of east from airport B. How far is the ship from airport A?

24. One gun is located at point A, while a second gun at point B is located 5.0 mi directly east of A. From point A the direction to the target is $27°$ north of east. From point B the direction to the target is $72°$ north of east. For what firing range should the guns be set?

25. Two points, A and B are $10\overline{0}$ yd apart. A point C across a canyon is located so that angle CAB is $70°$ and CBA is $80°$. Compute the distance BC across the canyon.

26. Two engineers are located on points A and B on the opposite sides of a hill. They are both able to see a stake at point C which is at a distance of $80\overline{0}$ ft from A and $70\overline{0}$ ft from B. If angle ABC is $25°$, find the distance AB through the hill.

27. A and B are two points located on opposite edges of a swamp. A third point C is located so that AC is 421 ft and BC is 376 ft. Angle ABC is measured to be $65°30'$. Compute the distance AB across the swamp.

28. A force $\mathbf{A}$ of $40\overline{0}$ lb and a force $\mathbf{B}$ of $60\overline{0}$ lb act at a point. Their resultant, $\mathbf{R}$, makes an angle of $42°$ with $\mathbf{A}$. Find the magnitude of $\mathbf{R}$.

29. Two forces act on a body to produce a resultant of 75 lb. If the angle between the two forces is 56° and one of the forces is $6\overline{0}$ lb, find the magnitude of the other force.

30. City B is located 50° north of east of city A. There is a $3\overline{0}$-mi/hour wind from the west and the pilot wishes to maintain an airspeed of 450 mi/hour. How many degrees north of east should the pilot head the plane in order to arrive directly at city B from city A?

3–9 LAW OF COSINES

In Section 3-8 we found that the law of sines can be used to solve any triangle in the following two cases:

1. If we know two angles and one side of the triangle.
2. If we know two sides of the triangle and the angle opposite one of them.

However, there exists two other cases for which the law of sines cannot be applied. They are:

3. If we know two sides of the triangle and the angle between these two sides.
4. If we know three sides of the triangle.

In these cases we use the *law of cosines*, which states the following:

> **LAW OF COSINES**   In any triangle, the square of any side equals the sum of the squares of the other two sides, minus twice the product of these other two sides and the cosine of their included angle.

Thus, for triangle ABC in Figure 3.65 we have

$$a^2 = b^2 + c^2 - 2bc \cos A$$
$$b^2 = a^2 + c^2 - 2ac \cos B$$
$$c^2 = a^2 + b^2 - 2ab \cos C$$

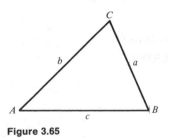

Figure 3.65

If we know two sides of the triangle and the included angle we find the third side by substituting in one of these formulas. After finding this part, we complete the solution by use of the law of sines. To obtain accuracy in the angles the computed side should be carried in the calculations to at least one more significant digit then stated in the solution.

If the three forms of the law of cosines are solved for the cosine of the angle, we have

$$\cos A = \frac{b^2 + c^2 - a^2}{2bc}$$

$$\cos B = \frac{a^2 + c^2 - b^2}{2ac}$$

$$\cos C = \frac{a^2 + b^2 - c^2}{2ab}$$

These formulas are used to find the angles in a triangle when we know the three sides. In this case we do not use the law of sines to complete the solution because results are more accurate when they are computed from the data given.

EXAMPLE 1: Approximate the missing parts of triangle ABC in which $A = 60°$, $b = 25$ ft, and $c = 42$ ft (Figure 3.66).

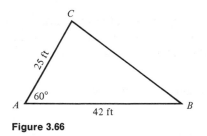

Figure 3.66

SOLUTION: First, we find side a by applying the law of cosines.

$$\begin{aligned}
a^2 &= b^2 + c^2 - 2bc \cos A \\
&= (25)^2 + (42)^2 - 2(25)(42)\cos 60° \\
&= 625 + 1764 - 2100(0.5000) \\
&= 2389 - 1050 \\
&= 1339 \\
a &= \sqrt{1339} \approx 36.6 \\
a &= 37 \text{ ft}
\end{aligned}$$

Second, we find the *smaller* of the remaining angles, angle B, by applying the law of sines. This angle must be acute. (Why?)

$$\frac{\sin A}{a} = \frac{\sin B}{b}$$

$$\frac{\sin 60°}{36.6} = \frac{\sin B}{25}$$

$$\frac{25 \sin 60°}{36.6} = \sin B$$

$$\frac{25(0.8660)}{36.6} = \sin B$$

$$0.5915 = \sin B$$

$$36° = B$$

(*Note:* $\sin B = 0.5915$ is true if $B = 36°$ or if $B = 144°$. We eliminate $144°$ as a possible solution since we know that angle B must be acute.)

Third, we find angle C.

$$A + B + C = 180°$$

$$60° + 36° + C = 180°$$

$$C = 84°$$

Thus, the solution to the triangle is as shown in Figure 3.67.

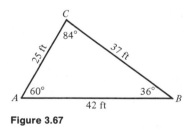

Figure 3.67

EXAMPLE 2: Approximate the missing parts of triangle ABC in which $a = 23.5$ ft, $b = 44.2$ ft, and $c = 30.1$ ft (Figure 3.68).

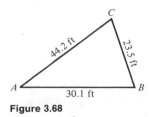

Figure 3.68

SOLUTION: First, we find angle A by applying the law of cosines.

$$\cos A = \frac{b^2 + c^2 - a^2}{2bc}$$

$$= \frac{(44.2)^2 + (30.1)^2 - (23.5)^2}{2(44.2)(30.1)}$$

$$= 0.8672$$

$$A = 29°50'$$

(*Note:* Remember that if the cosine of the angle is positive, the angle is acute; if the cosine of the angle is negative, the angle is in Q_2 and is obtuse.)

Second, we find the smaller of the remaining angles, acute angle C, by applying the law of cosines.

$$\cos C = \frac{a^2 + b^2 - c^2}{2ab}$$

$$= \frac{(44.2)^2 + (23.5)^2 - (30.1)^2}{2(44.2)(23.5)}$$

$$= 0.7701$$

$$C = 39°40'$$

Third, we find angle C.

$$A + B + C = 180°$$

$$29°50' + B + 39°40' = 180°$$

$$B = 110°30'$$

Thus, the solution to the triangle is as shown in Figure 3.69.

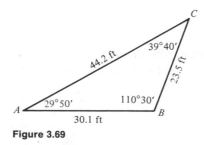

Figure 3.69

EXAMPLE 3: A body is acted upon by two forces with magnitudes of 78 lb and 42 lb which act at an angle of 25° with each other, as shown in Figure 3.70. Find:
a. The magnitude of the resultant.
b. The angle between the resultant and the larger force.

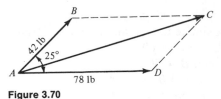

Figure 3.70

SOLUTION:

a. The resultant of the two given forces is the diagonal of the parallelogram $ABCD$ formed from the given vectors. (*Note:* Refer to Section 3-4.) In the parallelogram $AB = DC = 42$ and the sum of angle A and angle D equals $180°$. Thus,

$$A + D = 180°$$
$$25° + D = 180°$$
$$D = 155°$$

We find the magnitude of the resultant **AC** by applying the law of cosines:

$$(AC)^2 = (AD)^2 + (DC)^2 - 2(AD)(DC)\cos D$$
$$(AC)^2 = (78)^2 + (42)^2 - (2(78)(42)\cos 155°$$

Since $\cos 155° = -\cos 25° = -0.9063$, we have

$$(AC)^2 = (78)^2 + (42)^2 - 2(78)(42)(-0.9063)$$
$$AC \approx 117$$
$$AC = 120$$

Thus, the magnitude of the resultant is 120 lb.

b. We find the angle between the resultant and the larger force, angle CAD, by applying the law of sines. Note that since angle D is greater than $90°$, angle CAD must be acute.

$$\frac{\sin CAD}{DC} = \frac{\sin D}{AC}$$

$$\frac{\sin CAD}{42} = \frac{\sin 155°}{117}$$

$$\sin CAD = \frac{42 \sin 155°}{117}$$

Since $\sin 155° = \sin 25° = 0.4226$, we have

$$\sin CAD = \frac{42(0.4226)}{117}$$

$$= 0.1517$$
$$CAD = 9°$$

Thus, the angle between the resultant and the larger force is $9°$

EXERCISES 3–9

In Exercises 1–16 approximate the remaining parts of triangle ABC.

1. $a = 12$ ft, $b = 15$ ft, and $C = 60°$.

2. $a = 2\overline{0}$ ft, $c = 3\overline{0}$ ft, and $B = 30°$.

3. $c = 19.2$ ft, $a = 46.1$ ft, and $B = 10°20'$

4. $b = 126$ ft, $c = 92.1$ ft, and $A = 72°50'$.

5. $a = 4\overline{0}$ ft, $b = 5\overline{0}$ ft, and $C = 120°$.

6. $b = 36$ ft, $c = 75$ ft, and $A = 98°$.

7. $b = 11.1$ ft, $a = 19.2$ ft, and $C = 95°40'$

8. $c = 127$ ft, $b = 315$ ft, and $A = 162°30'$.

9. $a = 4.0$ ft, $b = 2.0$ ft, and $c = 3.0$ ft.

10. $a = 11$ ft, $b = 15$ ft, and $c = 19$ ft.

11. $a = 12$ ft, $b = 5.2$ ft, and $c = 8.1$ ft.

12. $a = 4.9$ ft, $b = 5.3$ ft, and $c = 2.6$ ft.

13. $a = 15\overline{0}$ ft, $b = 175$ ft, and $c = 20\overline{0}$ ft.

14. $a = 84.8$ ft, $b = 36.8$ ft, and $c = 76.5$ ft.

15. $a = 34.4$ ft, $b = 56.1$ ft, and $c = 42.3$ ft.

16. $a = 45.0$ ft, $b = 108$ ft, and $c = 117$ ft.

17. A body is acted upon by two forces with magnitudes of 16 lb and 25 lb which act at an angle of 30° with each other. Find:

    a. The magnitude of the resultant.

    b. The angle between the resultant and the larger force.

18. Two forces with magnitudes of 45 lb and $9\overline{0}$ lb are applied to the same point. If the angle between them measures 72°, find:

    a. The magnitude of the resultant.

    b. The angle between the resultant and the smaller force.

19. Forces with magnitudes of 126 lb and 198 lb act simultaneously upon a body in such a way that the angle between the forces is 14°50'. Find:

    a. The magnitude of the resultant.

    b. The angle between the resultant and the larger force.

20. Two forces with magnitudes of 15 lb and $2\overline{0}$ lb act on a body in such a way that the magnitude of the resultant is 28 lb. Find the angles that the three forces make with each other.

21. Two forces with magnitudes of 42 lb and 71 lb are applied to the same object. If the magnitude of the resultant is 85 lb, find the angles that the three forces make with each other.

22. A surveyor at point C sights two points A and B on opposite sides of a lake. If C is 760 ft from A and 920 ft from B and angle ACB is 96°, how wide is the lake?

23. A baseball diamond is a square that is $9\overline{0}$ ft long on each side. If the pitcher's mound is $6\overline{0}$ ft from home plate on the diagonal from home to second base, how far is the pitcher's mound from first base?

24. A draftsman drew to scale (1 in. = $5\overline{0}$ yd) a map of a development that includes a triangular recreation area with sides of lengths 75 yd, 125 yd, and 150 yd. On the map, what are the angles of the triangle representing the recreation area?

25. A ship sails due east for $4\overline{0}$ mi and then changes direction and sails 20° north of east for $6\overline{0}$ mi. How far is the ship from its starting point?

26. A plane's airspeed is 350 mi/hour on a heading that is 30° north of east. If a wind is blowing from the east with a speed of $4\overline{0}$ mi/hour, what is the ground speed of the plane?

27. If ABC is a right triangle ($C = 90°$), show that the law of cosines simplifies to the relationship $c^2 = a^2 + b^2$.

1. If $\tan \theta = \frac{3}{4}$ and $\cos \theta < 0$, find $\sin \theta$.

2. If the point $(-1, 1)$ is on the terminal ray of θ, give one possible measure of angle θ.

3. Interpolate and find $\cos 32°47'$.

4. What is the exact value of $3 \cos 720° + \tan(-540°)$?

5. Determine the exact value of $\sec 135°$.

6. Determine the exact value of $\sin(-210°)$.

7. What is the approximate value of $\cos 626°40'$?

8. Solve the equation $3 \csc \theta - 7 = -13$ for $0° \le \theta < 360°$.

9. Determine all the solutions to the equation $3 \cot \theta - 1 = 0$.

10. Give five angles that are coterminal with $100°$. Include at least two negative angles.

11. Determine the acute angle θ for which $\sin 2\theta = \cos(\theta + 60°)$.

12. Multiple choice: As angle θ increases from $180°$ to $360°$, the sine of that angle:
    a. Increases.
    b. Decreases.
    c. Increases, then decreases.
    d. Decreases, then increases.

13. In a right triangle the hypotenuse is 17 and the longer side is 15. Determine the sine of the smaller acute angle.

14. In right triangle ABC with $C = 90°$ if $b = 5.00$ and $c = 8.00$, determine angle B.

15. What is the vertical component of a vector with a magnitude of $5\overline{0}$ lb which makes an angle of $25°$ with the positive x axis?

16. In triangle ABC if $c = 6.0$, $A = 40°$, and $B = 60°$, find b.

17. True or false: If $B = 30°$, $a = 39$ ft, and $b = 57$ ft, there are two triangles that satisfy the data.

18. In triangle ABC if $a = 5.0$, $b = 9.0$, and $c = 6.0$, determine angle B.

19. In triangle ABC if $a = 3.0$, $b = 4.0$, and $C = 120°$, find c.

20. The first major attempt to measure the relative distances of the sun and the moon from the earth was made in about 280 B.C. by the Alexandrian astronomer Aristarchus. Basically he reasoned that the moon has no light of its own since we do not always see a "full moon." Therefore, the moon, like the earth, must receive its light from the sun and then reflect this light toward earth. As illustrated

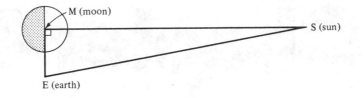

in the figure, he also correctly reasoned that at a quarter phase of the moon, when it is half light and half dark, angle *EMS* is a right angle.

a. Using primitive instruments, Aristarchus measured angle *SEM* to be 29/30 of a right angle (or 87°). About how many times more distant is the sun than the moon in Aristarchus's estimate?

b. Your answer in part a, which is Aristarchus's estimate, is not close to the true ratio. The error is caused by his measurement for angle *SEM*, which should be about 89° 50′. The absence of a refined instrument that could provide such a precise measurement was a problem that plagued the science of astronomy until modern times. On the basis of the correct measurement for angle *SEM*, about how many times farther away is the sun than the moon?

(*Note:* Actually Aristarchus computed his estimate by using Euclidean geometry. Trigonometry was developed later to solve precisely this type of problem. Aristarchus is often referred to as the Copernicus of antiquity because he was the first to propose the heliocentric hypothesis that the earth and other planets revolve around a fixed sun.)

Chapter 4

Products and factoring

4–1 LAWS OF EXPONENTS

In Chapter 1 we discussed some essential topics in algebra that have served our purposes to this point. However, in order to continue our study of trigonometric and other functions, we need more algebraic techniques. We will start by reconsidering the meaning of an exponent that is a positive integer.

5^3 is shorthand for $5 \cdot 5 \cdot 5$

$(-4)^2$ is shorthand for $(-4) \cdot (-4)$

x^4 is shorthand for $x \cdot x \cdot x \cdot x$

x^n, where n is a positive integer, is shorthand for $\underbrace{x \cdot x \cdot x \cdots x}_{n \text{ factors}}$

Thus, by x^n, where n is a positive integer, we mean to use x as a factor n times.

The first rule of exponents concerns the product of two expressions that contain exponents. For instance, what is the product of 5^2 and 5^4?

5^2 means $5 \cdot 5$

5^4 means $5 \cdot 5 \cdot 5 \cdot 5$

Therefore,

$$5^2 \cdot 5^4 \text{ means } \underbrace{(5 \cdot 5)}_{2 \text{ factors}} \underbrace{(5 \cdot 5 \cdot 5 \cdot 5)}_{4 \text{ factors}} = \underbrace{5 \cdot 5 \cdot 5 \cdot 5 \cdot 5 \cdot 5}_{6 \text{ factors}} = 5^6$$

Note that the exponent of 6 in the resulting product is the sum of the original exponents (2 and 4). Thus, in general, if m and n denote positive integers,

$$x^m \cdot x^n = x^{m+n}$$

EXAMPLE 1: Simplify the product $2^2 \cdot 2^3$ and check your result.

$$2^2 \cdot 2^3 = 2^{2+3} = 2^5$$
$$\downarrow \quad \downarrow \qquad\qquad \downarrow$$
$$4 \cdot 8 = \qquad\qquad 32$$

EXAMPLE 2: Simplify the product $3^3 \cdot 3$ and check your result.

$$3^3 \cdot 3 = 3^{3+1} = 3^4$$
$$\downarrow \quad \downarrow \qquad\qquad \downarrow$$
$$27 \cdot 3 = \qquad\qquad 81$$

(*Note:* If no exponent is indicated, the exponent is 1. Thus, 3 is equivalent to 3^1.)

EXAMPLE 3: Simplify the following products:

a. $c^5 \cdot c^2$

$\qquad = c^{5+2} = c^7$

b. $(\cos \theta)^5 (\cos \theta)^2$

$\qquad = (\cos \theta)^{5+2} = (\cos \theta)^7$

EXAMPLE 4: Simplify the following products:

a. $(2s^3)(7s^2)$

$\qquad = 2 \cdot 7 s^{3+2}$

$\qquad = 14s^5$

b. $2(\sin \theta)^3 \cdot 7(\sin \theta)^2$

$\qquad = 2 \cdot 7(\sin \theta)^{3+2}$

$\qquad = 14(\sin \theta)^5$

EXAMPLE 5: Simplify the following products:

a. $(6t^5 c^3)(-2tc^7)$

$\qquad = 6(-2)t^{5+1}c^{3+7}$

$\qquad = -12t^6 c^{10}$

b. $6(\tan \theta)^5(\cos \theta)^3 \cdot (-2 \tan \theta)(\cos \theta)^7$

$\qquad = 6(-2)(\tan \theta)^{5+1}(\cos \theta)^{3+7}$

$\qquad = -12(\tan \theta)^6(\cos \theta)^{10}$

Note in the above examples how trigonometric expressions are handled in the same manner as algebraic expressions. Throughout the chapter, we will present problems in two parallel forms: the algebraic form on the left side of the page, and a similar trigonometric expression on the right side of the page. The techniques for both problems are the same since a symbol in algebra, such as s, represents a real number; while an expression in trigonometry, such as $\sin \theta$, represents a ratio that is also a real number. Thus, the rules that govern both algebraic and trigonometric expressions are the rules that govern the real numbers.

When using exponents with functions, it is common notation to avoid parentheses and to put the exponent between the function and the symbol for the independent variable. For example, $(\sin \theta)^3$, which means $(\sin \theta)(\sin \theta)(\sin \theta)$, is written as $\sin^3 \theta$. We will use both notations interchangeably and, for the sake of comparison, the following example solves the same problem using both notations.

EXAMPLE 6: Simplify the following product:

$$[2(\sin \theta)^2(\cos \theta)^3][-5 \sin \theta(\cos \theta)^5] \quad \text{or} \quad [2 \sin^2 \theta \cos^3 \theta][-5 \sin \theta \cos^5 \theta]$$
$$= 2(-5)(\sin \theta)^{2+1}(\cos \theta)^{3+5} \quad \text{or} \quad 2(-5)\sin^{2+1} \theta \cos^{3+5} \theta$$
$$= -10(\sin \theta)^3(\cos \theta)^8 \quad \text{or} \quad -10 \sin^3 \theta \cos^8 \theta$$

The second rule of exponents concerns the quotient of two expressions that contain exponents. For instance, what is the result when 5^6 is divided by 5^2?

$$\frac{5^6}{5^2} \text{ means } \frac{\overset{1}{\cancel{5}} \cdot \overset{1}{\cancel{5}} \cdot 5 \cdot 5 \cdot 5 \cdot 5}{\underset{1}{\cancel{5}} \cdot \underset{1}{\cancel{5}}} = 5^4$$

Note that the exponent of 4 in the result is the difference of the original exponents $(6 - 2)$. Similarly, the result when 5^2 is divided by 5^6 is $1/5^4$.

$$\frac{5^2}{5^6} = \frac{\overset{1}{\cancel{5}} \cdot \overset{1}{\cancel{5}}}{\underset{1}{\cancel{5}} \cdot \underset{1}{\cancel{5}} \cdot 5 \cdot 5 \cdot 5 \cdot 5} = \frac{1}{5^4}$$

Thus, in general, if m and n are positive integers,

$$\frac{x^m}{x^n} = x^{m-n} \quad \text{if} \quad m > n$$

$$\frac{x^m}{x^n} = \frac{1}{x^{n-m}} \quad \text{if} \quad n > m$$

EXAMPLE 7: Simplify the quotient $2^5/2^2$ and check your result.

$$\frac{2^5}{2^2} = 2^{5-2} = 2^3$$
$$\downarrow \qquad\qquad \downarrow$$
$$\frac{32}{4} = \qquad\quad 8$$

EXAMPLE 8: Simplify the following quotients:

a. $\dfrac{-4s^3}{32s^{12}} = \dfrac{-4}{32s^{12-3}}$

b. $\dfrac{-4(\sin \theta)^3}{32(\sin \theta)^{12}} = \dfrac{-4}{32(\sin \theta)^{12-3}}$

$$= \frac{-1}{8s^9} \qquad\qquad\qquad\qquad = \frac{-1}{8(\sin \theta)^9}$$

EXAMPLE 9: Simplify the following quotients:

a. $\dfrac{18s^3tc^2}{12st^4c}$

$= \dfrac{18s^{3-1}c^{2-1}}{12t^{4-1}}$

$= \dfrac{3s^2c}{2t^3}$

b. $\dfrac{18 \sin^3 \theta \tan \theta \cos^2 \theta}{12 \sin \theta \tan^4 \theta \cos \theta}$

$= \dfrac{18 \sin^{3-1} \theta \cos^{2-1} \theta}{12 \tan^{4-1} \theta}$

$= \dfrac{3 \sin^2 \theta \cos \theta}{2 \tan^3 \theta}$

The third rule of exponents concerns raising to a power an expression that contains exponents. For example, what is the result when 5^2 is raised to the fourth power?

$(5^2)^4$ means $5^2 \cdot 5^2 \cdot 5^2 \cdot 5^2 = 5^{2+2+2+2} = 5^8$

Note that the exponent 8 in the result is the product of the original exponents (2 and 4). Thus, in general, if m and n denote positive integers,

$(x^m)^n = x^{mn}$

EXAMPLE 10: Simplify $(2^3)^2$ and check your result.

$(2^3)^2 = 2^{3 \cdot 2} = 2^6$
$\downarrow$
$(8)^2 = 64 = 2 \cdot 2 \cdot 2 \cdot 2 \cdot 2 \cdot 2$

EXAMPLE 11: Simplify:
a. $(t^3)^5 = t^{3 \cdot 5} = t^{15}$

b. $(\tan^3 \theta)^5 = \tan^{3 \cdot 5} \theta = \tan^{15} \theta$

This rule of multiplying exponents may be generalized to include raising to a power expressions that contain more than one factor. For example, what is the result when $2x^2y^5$ is raised to the third power?

$(2x^2y^5)^3$ means $(2x^2y^5)(2x^2y^5)(2x^2y^5) = 2^{1+1+1}x^{2+2+2}y^{5+5+5} = 2^3x^6y^{15}$

Note that the exponents in the result can be found by multiplying each of the original exponents by 3. Thus, in general, the exponents in the result are found by multiplying each of the original exponents by the desired power.

EXAMPLE 12: Simplify $(3 \cdot 4)^2$ and check your result.

$(3 \cdot 4)^2 = 3^{1 \cdot 2} \cdot 4^{1 \cdot 2} = 3^2 \cdot 4^2$
$\downarrow \qquad\qquad\qquad \downarrow$
$(12)^2 = 144 = 9 \cdot 16$

EXAMPLE 13: Simplify:
a. $(-2s^5)^2$

$\quad = (-2)^{1\cdot2}s^{5\cdot2}$

$\quad = (-2)^2 s^{10}$ or $4s^{10}$

b. $(-2\sin^5\theta)^2$

$\quad = (-2)^{1\cdot2}\sin^{5\cdot2}\theta$

$\quad = (-2)^2\sin^{10}\theta$ or $4\sin^{10}\theta$

EXAMPLE 14: Simplify:
a. $(-t^4c^2)^3$

$\quad = (-1)^3 t^{4\cdot3}c^{2\cdot3}$

$\quad = -1t^{12}c^6$ or $-t^{12}c^6$

b. $[-(\tan\theta)^4(\cos\theta)^2]^3$

$\quad = (-1)^3(\tan\theta)^{4\cdot3}(\cos\theta)^{2\cdot3}$

$\quad = -1(\tan\theta)^{12}(\cos\theta)^6$

or $\quad -(\tan\theta)^{12}(\cos\theta)^6$

EXAMPLE 15: Simplify:
a. $(3s^5t)^2(-2s^3t^4)^3$

$\quad = 3^{1\cdot2}s^{5\cdot2}t^{1\cdot2}(-2)^{1\cdot3}s^{3\cdot3}t^{4\cdot3}$

$\quad = 3^2 s^{10}t^2(-2)^3 s^9 t^{12}$

$\quad = 3^2(-2)^3 s^{10+9}t^{2+12}$

$\quad = 3^2(-2)^3 s^{19}t^{14}$

or $\quad -72s^{19}t^{14}$

b. $(3\sin^5\theta\tan\theta)^2(-2\sin^3\theta\tan^4\theta)^3$

$\quad = 3^{1\cdot2}\sin^{5\cdot2}\theta\tan^{1\cdot2}\theta(-2)^{1\cdot3}\sin^{3\cdot3}\theta\tan^{4\cdot3}\theta$

$\quad = 3^2\sin^{10}\theta\tan^2\theta(-2)^3\sin^9\theta\tan^{12}\theta$

$\quad = 3^2(-2)^3\sin^{10+9}\theta\tan^{2+12}\theta$

$\quad = 3^2(-2)^3\sin^{19}\theta\tan^{14}\theta$

or $\quad -72\sin^{19}\theta\tan^{14}\theta$

EXERCISES 4–1

In Exercises 1–10 simplify each expression and check your result.

1. $2^4 \cdot 2$

2. $3^2 \cdot 3^3$

3. $\dfrac{3^5}{3^2}$

4. $\dfrac{2}{2^4}$

5. $(3^2)^3$

6. $(2^3)^2$

7. $(2\cdot5)^2$

8. $(6\cdot2)^2$

9. $(2^3\cdot3)^2$

10. $(3^2\cdot4)^3$

In Exercises 11–20 simplify each expression. Variables as exponents denote positive integers.

11. $2^3 \cdot 2^7 \cdot 2$

12. $(-3)^2(-3)^5(-3)^3$

13. $2^x \cdot 2^5 \cdot 2^y$

14. $5^a \cdot 5$

15. $(4^a)^b$

16. $(5^x)^{2x}$

17. $\dfrac{2}{2^n}$

18. $\dfrac{5^{2x}}{5^x}$

19. $\dfrac{3^5 \cdot 5^3 \cdot 7^5}{3^2 \cdot 5^4 \cdot 7^7}$

20. $\dfrac{2^4 \cdot 5 \cdot 3^2}{2 \cdot 5^3 \cdot 3}$

In Exercises 21–55 simplify each expression. Variables as exponents denote positive integers; variables in the denominator do not denote zero.

21. a. $c^5 \cdot c^7$ **b.** $(\cos\theta)^5(\cos\theta)^7$

22. a. $s^3 \cdot s$ **b.** $(\sin\theta)^3(\sin\theta)$

23. a. $t \cdot t^4$ **b.** $\tan\theta \cdot \tan^4\theta$

24. a. $s^9 \cdot s^4$ **b.** $\sin^9\theta \cdot \sin^4\theta$

25. a. $c^a \cdot c^b$ **b.** $\cos^a\theta \cdot \cos^b\theta$

26. a. $t^{2x} \cdot t^5$ **b.** $\tan^{2x}\theta \cdot \tan^5\theta$

27. a. $s^{2a} \cdot s^{5a}$ **b.** $(\sin\theta)^{2a}(\sin\theta)^{5a}$

28. a. $s^{a+2}s^{2a}$ **b.** $(\sin\theta)^{a+2}(\sin\theta)^{2a}$

29. a. $c^8 \div c^3$ **b.** $(\cos\theta)^8 \div (\cos\theta)^3$

30. a. $c^2 \div c^7$ **b.** $(\cos\theta)^2 \div (\cos\theta)^7$

31. a. $t \div t^4$ **b.** $(\tan\theta) \div (\tan\theta)^4$

32. a. $t^9 \div t$ **b.** $(\tan\theta)^9 \div (\tan\theta)$

33. a. $\dfrac{s^a}{s}$ **b.** $\dfrac{\sin^a\theta}{\sin\theta}$

34. a. $\dfrac{s^x}{s^{4x}}$ **b.** $\dfrac{\sin^x\theta}{\sin^{4x}\theta}$

35. a. $\dfrac{c^x}{c^{x+1}}$ **b.** $\dfrac{\cos^x\theta}{\cos^{x+1}\theta}$

36. a. $\dfrac{c^{2n+2}}{c^2}$ **b.** $\dfrac{\cos^{2n+2}\theta}{\cos^2\theta}$

37. a. $(t^2)^4$ **b.** $(\tan^2\theta)^4$

38. a. $(t^a)^2$ **b.** $(\tan^a\theta)^2$

39. a. $(3s)^2$ **b.** $(3\sin\theta)^2$

40. a. $(-2s)^2$ **b.** $(-2\sin\theta)^2$

41. a. $(-t)^4$ **b.** $(-\tan\theta)^4$

42. a. $(sc)^2$ **b.** $(\sin\theta\cos\theta)^2$

43. a. $(3t^3s^2)^4$ **b.** $(3\tan^3\theta\sin^2\theta)^4$

44. a. $(-5c^5t)^3$ **b.** $(-5\cos^5\theta\tan\theta)^3$

45. a. $(2t^3)(-5t)^2$ **b.** $[2(\tan\theta)^3](-5\tan\theta)^2$

46. a. $(-4c)^3(5c^3)$ **b.** $(-4\cos\theta)^3(5\cos^3\theta)$

47. a. $(2st)^2(3tc)^3$ **b.** $(2\sin\theta\tan\theta)^2(3\tan\theta\cos\theta)^3$

48. a. $(-4s^2c)^3(3sc^2)^5$ **b.** $(-4\sin^2\theta\cos\theta)^3(3\sin\theta\cos^2\theta)^5$

49. a. $(6t^2c^4)^3(-tc^3)^2$ **b.** $(6\tan^2\theta\cos^4\theta)^3(-\tan\theta\cos^3\theta)^2$

50. a. $(-c^2)(-c)^2$ **b.** $(-\cos^2\theta)(-\cos\theta)^2$

51. a. $(-3t^2)(-3t)^2$

b. $(-3 \tan^2 \theta)(-3 \tan \theta)^2$

52. a. $\dfrac{(4s^2ct^3)^4}{(-20s^6ct^2)^2}$

b. $\dfrac{(4 \sin^2 \theta \cos \theta \tan^3 \theta)^4}{(-20 \sin^6 \theta \cos \theta \tan^2 \theta)^2}$

53. a. $\dfrac{(3s^2t)^2}{(-9st)^3}$

b. $\dfrac{(3 \sin^2 \theta \tan \theta)^2}{(-9 \sin \theta \tan \theta)^3}$

54. a. $\dfrac{(2t)^2(4s^3)}{(5t^3)(-6s)^4}$

b. $\dfrac{(2 \tan \theta)^2(4 \sin^3 \theta)}{(5 \tan^3 \theta)(-6 \sin \theta)^4}$

55. a. $\dfrac{(-c^3)^2(4st)}{(-2c^2)(2s^2t^3)^3}$

b. $\dfrac{(-\cos^3 \theta)^2(4 \sin \theta \tan \theta)}{(-2 \cos^2 \theta)(2 \sin^2 \theta \tan^3 \theta)^3}$

56. The formula for the surface area of a sphere is $A = \pi d^2$. Determine a formula for the area in terms of the radius r.

57. The formula for the volume of a sphere is $V = \frac{4}{3}\pi r^3$. Determine a formula for the volume in terms of the diameter d.

4–2 PRODUCTS OF ALGEBRAIC EXPRESSIONS

In Chapter 1 we often made use of the distributive property of real numbers, which provides an alternative method for multiplying a number by an expression that contains more than one term. For example:

Method 1   or   **Method 2 (Distributive Property)**

$2(3 + 4)$ $2(3 + 4) = 2(3) + 2(4)$

$= 2(7)$ $= 6 + 8$

$= 14$ $= 14$

Note that method 1 can be used only if we are able to combine the terms that are in the parentheses. In most algebraic cases, it is not possible to combine these terms into a single term and thus method 2 must be employed. The following examples illustrate the use of the distributive property in multiplying a number by an expression that contains more than one term.

EXAMPLE 1: Multiply:
a. $5(s + 2)$

$= 5 \cdot s + 5 \cdot 2$

$= 5s + 10$

b. $5(\sin \theta + 2)$

$= 5 \sin \theta + 5 \cdot 2$

$= 5 \sin \theta + 10$

EXAMPLE 2: Multiply:
a. $3t(t^2 - 2t + 1)$

$= (3t)(t^2) + (3t)(-2t) + (3t)(1)$

$= 3t^3 - 6t^2 + 3t$

b. $3 \tan \theta(\tan^2 \theta - 2 \tan \theta + 1)$

$= (3 \tan \theta)(\tan^2 \theta)$

$\quad + (3 \tan \theta)(-2 \tan \theta) + (3 \tan \theta)(1)$

$= 3 \tan^3 \theta - 6 \tan^2 \theta + 3 \tan \theta$

EXAMPLE 3: Multiply:

a. $-cs(4cs^2 - 5c^2s)$

$= (-cs)(4cs^2) + (-cs)(-5c^2s)$

$= -4c^2s^3 + 5c^3s^2$

b. $-\cos\theta\sin\theta(4\cos\theta\sin^2\theta - 5\cos^2\theta\sin\theta)$

$= (-\cos\theta\sin\theta)(4\cos\theta\sin^2\theta)$

$\quad + (-\cos\theta\sin\theta)(-5\cos^2\theta\sin\theta)$

$= -4\cos^2\theta\sin^3\theta + 5\cos^3\theta\sin^2\theta$

EXAMPLE 4: Multiply:

a. $(2c^2 + t)ct$

$= (2c^2)(ct) + (t)(ct)$

$= 2c^3t + ct^2$

b. $(2\cos^2\theta + \tan\theta)(\cos\theta\tan\theta)$

$= (2\cos^2\theta)(\cos\theta\tan\theta)$

$\quad + (\tan\theta)(\cos\theta\tan\theta)$

$= 2\cos^3\theta\tan\theta + \cos\theta\tan^2\theta$

Since division is defined in terms of multiplication, we can use the distributive property to simplify similar problems involving division.

EXAMPLE 5: Simplify:

a. $(12c^5 + 6c^2) \div 3c$

$= (12c^5 + 6c^2) \cdot \dfrac{1}{3c}$

$= (12c^5)\dfrac{1}{3c} + (6c^2)\dfrac{1}{3c}$

$= 4c^4 + 2c$

b. $(12\cos^5\theta + 6\cos^2\theta) \div 3\cos\theta$

$= (12\cos^5\theta + 6\cos^2\theta) \cdot \dfrac{1}{3\cos\theta}$

$= (12\cos^5\theta)\dfrac{1}{3\cos\theta} + (6\cos^2\theta)\dfrac{1}{3\cos\theta}$

$= 4\cos^4\theta + 2\cos\theta$

If both factors in the multiplication contain more than one term, the distributive property must be used more than once. For example, no matter what expression is inside the parentheses

$(\quad)(x + 2)$ means $(\quad)x + (\quad)2$

Thus,

$(x + 3)(x + 2)$ means $(x + 3)x + (x + 3)2$

Using the distributive property the second time, we get

$(x + 3)x + (x + 3)2 = x^2 + 3x + 2x + 6 = x^2 + 5x + 6$

Therefore,

$(x + 3)(x + 2) = x^2 + 5x + 6$

EXAMPLE 6: Multiply:

a. $(s - 7)(s + 4)$

$= (s - 7)(s) + (s - 7)(4)$

$= s^2 - 7s + 4s - 28$

$= s^2 - 3s - 28$

b. $(\sin\theta - 7)(\sin\theta + 4)$

$= (\sin\theta - 7)(\sin\theta) + (\sin\theta - 7)(4)$

$= \sin^2\theta - 7\sin\theta + 4\sin\theta - 28$

$= \sin^2\theta - 3\sin\theta - 28$

EXAMPLE 7: Multiply:

a. $(2t + 3)(5t - 1)$

$\qquad = (2t + 3)(5t) + (2t + 3)(-1)$

$\qquad = 10t^2 + 15t - 2t - 3$

$\qquad = 10t^2 + 13t - 3$

b. $(2 \tan \theta + 3)(5 \tan \theta - 1)$

$\qquad = (2 \tan \theta + 3)(5 \tan \theta)$

$\qquad\quad + (2 \tan \theta + 3)(-1)$

$\qquad = 10 \tan^2 \theta + 15 \tan \theta - 2 \tan \theta - 3$

$\qquad = 10 \tan^2 \theta + 13 \tan \theta - 3$

Notice that this method of multiplication is equivalent to multiplying each term of the first factor by each term of the second factor and then combining similar terms. This observation leads to the following arrangement (of Example 7), which is good for multiplying longer expressions since similar terms are placed under each other:

$$
\begin{array}{l}
2t + 3 \\
5t - 1 \\
\hline
10t^2 + 15t \\
\;- 2t - 3 \\
\hline
10t^2 + 13t - 3
\end{array}
$$

add

this line equals $5t(2t + 3)$
this line equals $-1(2t + 3)$

EXAMPLE 8: Multiply:

a. $(c^2 - 5c + 25)(c + 3)$

SOLUTION:

$$
\begin{array}{l}
c^2 - 5c + 25 \\
c + 3 \\
\hline
c^3 - 5c^2 + 25c \\
\quad\;\; 3c^2 - 15c + 75 \\
\hline
c^3 - 2c^2 + 10c + 75
\end{array}
$$

add

this line equals $c(c^2 - 5c + 25)$
this line equals $3(c^2 - 5c + 25)$

b. $(\cos^2 \theta - 5 \cos \theta + 25)(\cos \theta + 3)$

SOLUTION:

$$
\begin{array}{l}
\cos^2 \theta - 5 \cos \theta + 25 \\
\cos \theta + 3 \\
\hline
\cos^3 \theta - 5 \cos^2 \theta + 25 \cos \theta \\
\quad\;\; 3 \cos^2 \theta - 15 \cos \theta + 75 \\
\hline
\cos^3 \theta - 2 \cos^2 \theta + 10 \cos \theta + 75
\end{array}
$$

add

this line equals $\cos \theta(\cos^2 \theta - 5 \cos \theta + 25)$
this line equals $3(\cos^2 \theta - 5 \cos \theta + 25)$

EXAMPLE 9: Multiply:

a. $(3s - 2t)^2$ or $(3s - 2t)(3s - 2t)$

SOLUTION:

$$3s - 2t$$
$$3s - 2t$$

add $\dfrac{\begin{aligned}9s^2 - 6st \\ - 6st + 4t^2\end{aligned}}{9s^2 - 12st + 4t^2}$ $\quad$ this line equals $3s(3s - 2t)$
this line equals $-2t(3s - 2t)$

[*Note*: $(3s - 2t)^2 \neq (3s)^2 + (-2t)^2$, for this result leaves out the middle term, $-12st$.]

b. $(3 \sin \theta - 2 \tan \theta)^2$ or $(3 \sin \theta - 2 \tan \theta)(3 \sin \theta - 2 \tan \theta)$

SOLUTION:

$$3 \sin \theta - 2 \tan \theta$$
$$3 \sin \theta - 2 \tan \theta$$

add $\dfrac{\begin{aligned}9 \sin^2 \theta - 6 \sin \theta \tan \theta \\ -6 \sin \theta \tan \theta + 4 \tan^2 \theta\end{aligned}}{9 \sin^2 \theta - 12 \sin \theta \tan \theta + 4 \tan^2 \theta}$ $\quad$ this line equals $3 \sin \theta (3 \sin \theta - 2 \tan \theta)$
this line equals $-2 \tan \theta (3 \sin \theta - 2 \tan \theta)$

Finally, when multiplying expressions that contain two terms, a mental shortcut is often used. To illustrate this shortcut, let's first multiply $(x + 5)(x + 2)$ the long way:

$$
\begin{aligned}
x &+ 5 \\
x &+ 2 \\
\hline
x^2 &+ 5x \\
&+ 2x + 10 \\
\hline
x^2 &+ 7x + 10
\end{aligned}
$$

Now let's consider how we arrived at each term of the result $x^2 + 7x + 10$.

$$
\begin{aligned}
x &+ 5 \\
x &+ 2 \quad \text{First} \\
\hline
x^2 &+ 5x \\
&2x + 10 \\
\hline
x^2 &+ 7x + 10
\end{aligned}
$$
$(x + 5)(x + 2)$
First terms

$$
\begin{aligned}
x &+ 5 \\
x &+ 2 \quad \text{Outer} \\
\hline
x^2 &+ 5x \\
&2x + 10 \\
\hline
x^2 &+ 7x + 10
\end{aligned}
$$
$(x + 5)(x + 2)$
Outer terms

$$
\begin{aligned}
x &+ 5 \\
x &+ 2 \quad \text{Inner} \\
\hline
x^2 &+ 5x \\
&2x + 10 \\
\hline
x^2 &+ 7x + 10
\end{aligned}
$$
$(x + 5)(x + 2)$
Inner terms

$$
\begin{aligned}
x &+ 5 \\
x &+ 2 \quad \text{Last} \\
\hline
x^2 &+ 5x \\
&2x + 10 \\
\hline
x^2 &+ 7x + 10
\end{aligned}
$$
$(x + 5)(x + 2)$
Last terms

First + Outer + Inner + Last

$$x^2 + 2x + 5x + 10$$
$$= x^2 + 7x + 10$$

This method is easily remembered because the combination of the first letters in the above words spells FOIL. Note that it is usually possible to combine the outer term and the inner term to form the middle term in the product.

EXAMPLE 10: Multiply using the FOIL method:
a. $(c - 7)(c + 4)$ **b.** $(\cos \theta - 7)(\cos \theta + 4)$

SOLUTION:

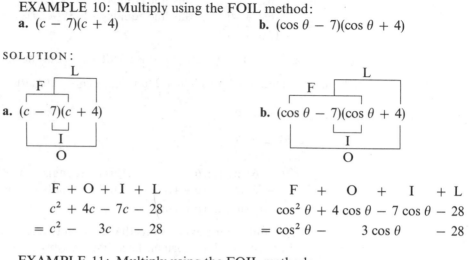

$$F + O + I + L$$
$$c^2 + 4c - 7c - 28$$
$$= c^2 - \quad 3c \quad - 28$$

$$F + O + I + L$$
$$\cos^2 \theta + 4 \cos \theta - 7 \cos \theta - 28$$
$$= \cos^2 \theta - \quad 3 \cos \theta \quad - 28$$

EXAMPLE 11: Multiply using the FOIL method:
a. $(2t + 5)(3t - 4)$.   **b.** $(2 \tan \theta + 5)(3 \tan \theta - 4)$.

SOLUTION:

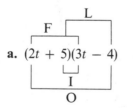

$$F + O + I + L$$
$$6t^2 - 8t + 15t - 20$$
$$= 6t^2 + \quad 7t \quad - 20$$

$$F + O + I + L$$
$$6 \tan^2 \theta - 8 \tan \theta + 15 \tan \theta - 20$$
$$= 6 \tan^2 \theta + \quad 7 \tan \theta \quad - 20$$

EXERCISES 4–2

In Exercises 1–44 perform the multiplication or division and combine similar terms.

1. a. $5(s + t)$ **b.** $5(\sin \theta + \tan \theta)$
2. a. $4(t - c)$ **b.** $4(\tan \theta - \cos \theta)$
3. a. $-2(c - s)$ **b.** $-2(\cos \theta - \sin \theta)$
4. a. $-6(t + s)$ **b.** $-6(\tan \theta + \sin \theta)$
5. a. $2s(s^3 - 7s + 1)$ **b.** $2 \sin \theta(\sin^3 \theta - 7 \sin \theta + 1)$
6. a. $3c(4c^2 + 5c - 2)$ **b.** $3 \cos \theta(4 \cos^2 \theta + 5 \cos \theta - 2)$

7. a. $-5t(t^3 - t^2 - t)$ **b.** $-5\tan\theta(\tan^3\theta - \tan^2\theta - \tan\theta)$

8. a. $-s(3s^2 + 5s - 6)$ **b.** $-\sin\theta(3\sin^2\theta + 5\sin\theta - 6)$

9. a. $c^2t(2c - 3t + 4)$ **b.** $\cos^2\theta\tan\theta(2\cos\theta - 3\tan\theta + 4)$

10. a. $5st^2(s^2 + 2st + t^2)$ **b.** $5\sin\theta\tan^2\theta(\sin^2\theta + 2\sin\theta\tan\theta + \tan^2\theta)$

11. a. $-2stc(4s - t + 7c)$ **b.** $-2\sin\theta\tan\theta\cos\theta(4\sin\theta - \tan\theta + 7\cos\theta)$

12. a. $-4s^2tc^3(3s^3 - 2tc + 5sc^4)$

     b. $-4\sin^2\theta\tan\theta\cos^3\theta(3\sin^3\theta - 2\tan\theta\cos\theta + 5\sin\theta\cos^4\theta)$

13. a. $(4s - 5t)st$ **b.** $(4\sin\theta - 5\tan\theta)\sin\theta\tan\theta$

14. a. $(c - 4s)c^3s^2$ **b.** $(\cos\theta - 4\sin\theta)\cos^3\theta\sin^2\theta$

15. a. $(t^2 + s^2)t^2s^2$ **b.** $(\tan^2\theta + \sin^2\theta)\tan^2\theta\sin^2\theta$

16. a. $(t^2 - c^2)tc$ **b.** $(\tan^2\theta - \cos^2\theta)\tan\theta\cos\theta$

17. a. $(2c^2 - 7c) \div c$ **b.** $(2\cos^2\theta - 7\cos\theta) \div \cos\theta$

18. a. $(8s^2 - 16s) \div -4s$ **b.** $(8\sin^2\theta - 16\sin\theta) \div -4\sin\theta$

19. a. $(24s^2t + 18st^2) \div 6st$

     b. $(24\sin^2\theta\tan\theta + 18\sin\theta\tan^2\theta) \div 6\sin\theta\tan\theta$

20. a. $(6tc^2 - 12t^2c) \div 2tc$

     b. $(6\tan\theta\cos^2\theta - 12\tan^2\theta\cos\theta) \div 2\tan\theta\cos\theta$

21. a. $\dfrac{4s^3 - 6s^2 + 8s}{2s}$ **b.** $\dfrac{4\sin^3\theta - 6\sin^2\theta + 8\sin\theta}{2\sin\theta}$

22. a. $\dfrac{cs^2 - 2c^3s}{cs}$ **b.** $\dfrac{\cos\theta\sin^2\theta - 2\cos^3\theta\sin\theta}{\cos\theta\sin\theta}$

23. a. $(t + 3)(t + 4)$ **b.** $(\tan\theta + 3)(\tan\theta + 4)$

24. a. $(s - 2)(s - 6)$ **b.** $(\sin\theta - 2)(\sin\theta - 6)$

25. a. $(c + 4)(c - 4)$ **b.** $(\cos\theta + 4)(\cos\theta - 4)$

26. a. $(2s + 5)(s - 3)$ **b.** $(2\sin\theta + 5)(\sin\theta - 3)$

27. a. $(3t - 4)(2t - 1)$ **b.** $(3\tan\theta - 4)(2\tan\theta - 1)$

28. a. $(5c - 4s)(3c + s)$ **b.** $(5\cos\theta - 4\sin\theta)(3\cos\theta + \sin\theta)$

29. a. $(2t + 5c)(6t - c)$ **b.** $(2\tan\theta + 5\cos\theta)(6\tan\theta - \cos\theta)$

30. a. $(s + 5)^2$ **b.** $(\sin\theta + 5)^2$

31. a. $(c - 2)^2$ **b.** $(\cos\theta - 2)^2$

32. a. $(2s + 3t)^2$ **b.** $(2\sin\theta + 3\tan\theta)^2$

33. a. $(4c - s)^2$ **b.** $(4\cos\theta - \sin\theta)^2$

34. a. $(t + 3)(t^2 + 2t + 3)$ **b.** $(\tan\theta + 3)(\tan^2\theta + 2\tan\theta + 3)$

35. a. $(s - 4)(s^2 + 5s - 1)$ **b.** $(\sin\theta - 4)(\sin^2\theta + 5\sin\theta - 1)$

36. a. $(2c - 1)(c^3 + c + 1)$ **b.** $(2\cos\theta - 1)(\cos^3\theta + \cos\theta + 1)$

37. a. $(3c + 2)(c^3 - c - 1)$ **b.** $(3\cos\theta + 2)(\cos^3\theta - \cos\theta - 1)$

38. a. $(s + t)(s^2 - st + t^2)$ **b.** $(\sin\theta + \tan\theta)(\sin^2\theta - \sin\theta\tan\theta + \tan^2\theta)$

39. a. $(t - c)(t^2 + tc + c^2)$ **b.** $(\tan\theta - \cos\theta)(\tan^2\theta + \tan\theta\cos\theta + \cos^2\theta)$

40. a. $(2s + 3c)(s^2 - 2sc + c^2)$

     b. $(2\sin\theta + 3\cos\theta)(\sin^2\theta - 2\sin\theta\cos\theta + \cos^2\theta)$

41. a. $(4t - c)(2t^2 + tc - 4c^2)$

     b. $(4\tan\theta - \cos\theta)(2\tan^2\theta + \tan\theta\cos\theta - 4\cos^2\theta)$

42. a. $(c - 3)(c + 2)(c - 4)$ **b.** $(\cos\theta - 3)(\cos\theta + 2)(\cos\theta - 4)$

43. a. $(2t - 1)(3t + 2)(1 - t)$ **b.** $(2 \tan \theta - 1)(3 \tan \theta + 2)(1 - \tan \theta)$
44. a. $(s + 1)^3$ **b.** $(\sin \theta + 1)^3$

In Exercises 45–61, multiply using the FOIL method.

45. a. $(c + 5)(c + 2)$ **b.** $(\cos \theta + 5)(\cos \theta + 2)$
46. a. $(c - 3)(c + 4)$ **b.** $(\cos \theta - 3)(\cos \theta + 4)$
47. a. $(s + 7)(s - 6)$ **b.** $(\sin \theta + 7)(\sin \theta - 6)$
48. a. $(s - 4)(s - 11)$ **b.** $(\sin \theta - 4)(\sin \theta - 11)$
49. a. $(t - 6)(t + 6)$ **b.** $(\tan \theta - 6)(\tan \theta + 6)$
50. a. $(t - 1)(t + 1)$ **b.** $(\tan \theta - 1)(\tan \theta + 1)$
51. a. $(2s + 5)(2s - 5)$ **b.** $(2 \sin \theta + 5)(2 \sin \theta - 5)$
52. a. $(5c - 3)(5c + 3)$ **b.** $(5 \cos \theta - 3)(5 \cos \theta + 3)$
53. a. $(3t + 7)(t - 4)$ **b.** $(3 \tan \theta + 7)(\tan \theta - 4)$
54. a. $(6s - 1)(5s - 2)$ **b.** $(6 \sin \theta - 1)(5 \sin \theta - 2)$
55. a. $(2 - t)(6 - 5t)$ **b.** $(2 - \tan \theta)(6 - 5 \tan \theta)$
56. a. $(1 - 2s)(3 - 5s)$ **b.** $(1 - 2 \sin \theta)(3 - 5 \sin \theta)$
57. a. $(4 + 3s)(7 - s)$ **b.** $(4 + 3 \sin \theta)(7 - \sin \theta)$
58. a. $(c + 4)^2$ **b.** $(\cos \theta + 4)^2$
59. a. $(s - 3)^2$ **b.** $(\sin \theta - 3)^2$
60. a. $(2t - 1)^2$ **b.** $(2 \tan \theta - 1)^2$
61. a. $(3 - 4t)^2$ **b.** $(3 - 4 \tan \theta)^2$

62. Consider the square with side length $a + b$ in the figure.

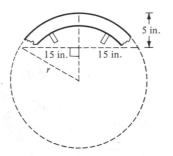

a. Algebraically, calculate the area: $(a + b)^2$.
b. Find the area by adding the areas of regions 1, 2, 3, and 4.
c. Does $(a + b)^2 = a^2 + b^2$?

63. Consider this figure, which shows a piece of a broken chariot wheel found by a group of archaeologists. What is the radius of the wheel?

64. Consider the square with side length $a + b$, shown here. Calculate the area of the square in two ways and establish the Pythagorean relation: $c^2 = a^2 + b^2$. (*Note:* It must be shown that $\theta = 90°$.)

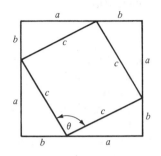

4–3 BINOMIAL THEOREM

A leading candidate for the most common mistake in algebra appears when students are asked to find the product or expansion of $(x + y)^2$. Too frequently they write the expansion as $x^2 + y^2$, which leaves out the middle term $2xy$. The problem becomes worse when expanding other expressions of the form $(x + y)^n$, where n is a positive integer, since there are usually many more terms than just x^n and y^n. We now discuss a method for finding these middle terms which is much easier than repeated multiplication.

We start by trying to find some patterns in the following expansions of the powers of $a + b$. You can verify each expansion by direct multiplication.

$$(a + b)^1 = a + b \qquad \text{(2 terms)}$$
$$(a + b)^2 = a^2 + 2ab + b^2 \qquad \text{(3 terms)}$$
$$(a + b)^3 = a^3 + 3a^2b + 3ab^2 + b^3 \qquad \text{(4 terms)}$$
$$(a + b)^4 = a^4 + 4a^3b + 6a^2b^2 + 4ab^3 + b^4 \qquad \text{(5 terms)}$$
$$(a + b)^5 = a^5 + 5a^4b + 10a^3b^2 + 10a^2b^3 + 5ab^4 + b^5 \qquad \text{(6 terms)}$$

Observe that in all the above cases the expansion of $(a + b)^n$ behaved as follows:

1. The number of terms in the expansion is $n + 1$. For example, in the expression $(a + b)^5$ we have $n = 5$ and there are six terms in the expansion.
2. The first term is a^n and the last term is b^n.
3. The second term is $na^{n-1}b$ and the nth term is nab^{n-1}.
4. The exponent of a decreases by 1 in each successive term while the exponent of b increases by 1.
5. The sum of the exponents of a and b in any term is n.

Thus, if we assume that this pattern continues when n is a positive integer greater than 5 (and it can be proved that it does), we need only find a method for determining the constant coefficients if we wish to expand such expressions. Our observations are generally summarized in the following formula, which is called the *binomial theorem*. The word "binomial" is used because we are expanding the power of an expression with two terms.

> **BINOMIAL THEOREM**: For any positive integer n,
>
> $$(a + b)^n = a^n + na^{n-1}b + (\text{constant})a^{n-2}b^2$$
> $$+ (\text{constant})a^{n-3}b^3 + \cdots + nab^{n-1} + b^n$$

A more complete statement of the binomial theorem also gives a formula to obtain the constant coefficients. Although this formula is very efficient (particularly for large values of n), it is too complicated for our present purpose. Therefore, we have summarized the coefficients in the expansion of $(a + b)^n$ for $n \le 20$ in Table 7 in the Appendix. This table is known as *Pascal's triangle*. Use of the table is simple. For example, the constant coefficients in the expansion of $(a + b)^7$ are found by reading across the row marked $n = 7$. The coefficients for the eight terms are 1, 7, 21, 35, 21, 7, and 1, respectively. To determine the coefficient of the eleventh term in the expansion of $(a + b)^{14}$, merely read 1001, which is the number at the intersection of row 14 with column 11. We are now ready to try some problems.

EXAMPLE 1: Expand $(x + y)^8$ by the binomial theorem.

SOLUTION: In our statement of the binomial theorem we substitute x for a and y for b. From Table 7 we determine the coefficients in the nine terms of the expansion as 1, 8, 28, 56, 70, 56, 28, 8, and 1, respectively. Thus, we have

$$(x + y)^8 = x^8 + 8x^7y + 28x^6y^2 + 56x^5y^3$$
$$+ 70x^4y^4 + 56x^3y^5 + 28x^2y^6 + 8xy^7 + y^8$$

EXAMPLE 2: Expand $(x + 2y)^4$ by the binomial theorem.

SOLUTION: Substitute x for a and $2y$ for b in our statement of the binomial theorem. From Table 7 we determine the coefficients of the five terms as 1, 4, 6, 4, and 1, respectively. Thus, we have

$$(x + 2y)^4 = x^4 + 4x^3(2y) + 6x^2(2y)^2 + 4x(2y)^3 + (2y)^4$$
$$= x^4 + 8x^3y + 24x^2y^2 + 32xy^3 + 16y^4$$

EXAMPLE 3: Expand $(2x - y)^3$ by the binomial theorem.

SOLUTION: First rewrite $(2x - y)^3$ as $[2x + (-y)]^3$. In our statement of the binomial theorem we now substitute $2x$ for a and $-y$ for b. From Table 7 we determine the coefficients of the four terms as 1, 3, 3, and 1, respectively. Thus, we have

$$(2x - y)^3 = (2x)^3 + 3(2x)^2(-y) + 3(2x)(-y)^2 + (-y)^3$$
$$= 8x^3 - 12x^2y + 6xy^2 - y^3$$

Note that when the binomial is the difference of two terms, the terms in the expansion alternate in sign.

Finally, we can write any single term of the binomial expansion without producing the rest of the terms. Note in the binomial theorem that the exponent of b is 1 in the second term, 2 in the third term, and in general $r - 1$ in the rth term.

Since the sum of the exponents of a and b is always n, the exponent above a in the rth term must be $n - (r - 1)$. Thus, the formula for the rth term of $(a + b)^n$ is

(number in nth row and rth column in Table 7)$a^{n-(r-1)}b^{r-1}$

EXAMPLE 4: Write the fourteenth term in the expansion of $(3x - y)^{15}$.

SOLUTION: In this case $n = 15$, $r = 14$, $a = 3x$, and $b = -y$. The number at the intersection of row 15 and column 14 in Table 7 is 105. Thus, we have

14th term $= 105(3x)^2(-y)^{13} = -945x^2y^{13}$

EXERCISES 4–3

In Exercises 1–12 expand each expression by the binomial theorem.

1. $(x + y)^6$ 2. $(x + y)^7$
3. $(x - y)^5$ 4. $(x - y)^4$
5. $(x + h)^4$ 6. $(x + h)^3$
7. $(x - 1)^7$ 8. $(y + 1)^5$
9. $(2x + y)^3$ 10. $(x - 2y)^6$
11. $(3c - 4d)^4$ 12. $(4c + 3d)^3$

In Exercises 13–16 write the first four terms in the expansion of the given expression.

13. $(x + y)^{15}$ 14. $(x - y)^{10}$
15. $(x - 3y)^{12}$ 16. $(x + 3)^{17}$

In Exercises 17–20 write the indicated term of the expansion.

17. Second term of $(x + y)^{18}$ 18. Fifth term of $(x - y)^7$
19. Twelfth term of $(3x - y)^{13}$ 20. Sixth term of $(x + 3y)^{10}$
21. Look for the pattern in Pascal's triangle (Table 7) and extend it by writing the first five entries for $n = 21$.

4–4 FACTORING

In Sections 4-2 and 4-3 we considered how to multiply expressions and change a product into a sum. For instance,

$$\underbrace{5(t + c)}_{\text{product}} = \underbrace{5t + 5c}_{\text{sum}}$$

Each of the expressions that is to be multiplied in the product is called a *factor*; each of the expressions that is to be added in the sum is called a *term*. Thus, in the above example, 5 and $t + c$ are factors, and $5t$ and $5c$ are terms. It is important to be able to transform products into sums and sums into products. In different situations one form is more useful than the other.

If we change a product into a sum, we are multiplying:
$5(t + c) \xrightarrow{\text{multiplying}} 5t + 5c$
If we change a sum into a product, we are factoring:
$5(t + c) \xleftarrow{\text{factoring}} 5t + 5c$

Thus, in factoring an expression, we will try to reverse the process of multiplication that we learned in Section 4-2. In general, the procedures for factoring are not as straightforward as those for multiplication and often a great deal of trial and error is necessary.

The first method that should always be employed in factoring a sum is to attempt to find a factor that is common to each of the terms. For example:

Sum	Common factor
$7x + 7y$	7
$9s + 6t$	3
$15x^2 + (-5x)$ (or $15x^2 - 5x$)	$5x$
$2a^2b + 8ab^2$	$2ab$
$12c^2 + (-20ct)$ (or $12c^2 - 20ct$)	$4c$

Note that in each case we attempt to pick the greatest common factor that will divide into each term of the sum. Therefore, although 5 is a common factor of $15x^2 - 5x$, a preferable common factor is $5x$ since $5x$ is the largest factor that divides both terms. After we determine the greatest common factor, if we divide this factor into the original sum, we obtain a second factor of our original expression. The product of these two factors represents our expression in factored form.

EXAMPLE 1: Factor completely:
(*Note:* To factor an expression completely means that none of the factors can be factored again. Unless otherwise specified, we are interested only in factors having integer coefficients.)
a. $9s + 6t$ **b.** $9 \sin \theta + 6 \tan \theta$

SOLUTION:

a. greatest common factor: 3

$$\frac{9s + 6t}{3} = 3s + 2t$$

Therefore, $9s + 6t = 3(3s + 2t)$.

b. greatest common factor: 3

$$\frac{9 \sin \theta + 6 \tan \theta}{3} = 3 \sin \theta + 2 \tan \theta$$

Therefore, $9 \sin \theta + 6 \tan \theta$
$$= 3(3 \sin \theta + 2 \tan \theta).$$

EXAMPLE 2: Factor completely:
a. $21c^2 - 7c$ **b.** $21 \cos^2 \theta - 7 \cos \theta$

SOLUTION:

a. greatest common factor: $7c$

$$\frac{21c^2 - 7c}{7c} = 3c - 1$$

Therefore, $21c^2 - 7c = 7c(3c - 1)$.

b. greatest common factor: $7 \cos \theta$

$$\frac{21 \cos^2 \theta - 7 \cos \theta}{7 \cos \theta} = 3 \cos \theta - 1$$

Therefore, $21 \cos^2 \theta - 7 \cos \theta$
$$= 7 \cos \theta(3 \cos \theta - 1).$$

EXAMPLE 3: Factor completely:

a. $12ct^2 + 9c^2t$

b. $12 \cos \theta \tan^2 \theta + 9 \cos^2 \theta \tan \theta$

SOLUTION:

a. greatest common factor: $3ct$

$$\frac{12ct^2 + 9c^2t}{3ct} = 4t + 3c$$

b. greatest common factor: $3 \cos \theta \tan \theta$

$$\frac{12 \cos \theta \tan^2 \theta + 9 \cos^2 \theta \tan \theta}{3 \cos \theta \tan \theta}$$

$$= 4 \tan \theta + 3 \cos \theta$$

Therefore,

$12ct^2 + 9c^2t = 3ct(4t + 3c).$

Therefore,

$12 \cos \theta \tan^2 \theta + 9 \cos^2 \theta \tan \theta$

$= 3 \cos \theta \tan \theta(4 \tan \theta + 3 \cos \theta).$

EXAMPLE 4: Factor completely:

a. $3s^3 + 12s^2 - 9s$

b. $3 \sin^3 \theta + 12 \sin^2 \theta - 9 \sin \theta$

SOLUTION:

a. greatest common factor: $3s$

$$\frac{3s^3 + 12s^2 - 9s}{3s} = s^2 + 4s - 3$$

b. greatest common factor: $3 \sin \theta$

$$\frac{3 \sin^3 \theta + 12 \sin^2 \theta - 9 \sin \theta}{3 \sin \theta}$$

$$= \sin^2 \theta + 4 \sin \theta - 3$$

Therefore,

$3s^3 + 12s^2 - 9s = 3s(s^2 + 4s - 3).$

Therefore,

$3 \sin^3 \theta + 12 \sin^2 \theta - 9 \sin \theta$

$= 3 \sin \theta(\sin^2 \theta + 4 \sin \theta - 3).$

EXAMPLE 5: Factor completely:

a. $s(c + 2) - t(c + 2)$

b. $\sin \theta(\cos \theta + 2) - \tan \theta(\cos \theta + 2)$

SOLUTION:

a. greatest common factor: $c + 2$

$$\frac{s(c + 2) - t(c + 2)}{c + 2} = s - t$$

b. greatest common factor: $\cos \theta + 2$

$$\frac{\sin \theta(\cos \theta + 2) - \tan \theta(\cos \theta + 2)}{\cos \theta + 2}$$

$$= \sin \theta - \tan \theta$$

Therefore,

$s(c + 2) - t(c + 2) = (c + 2)(s - t).$

Therefore,

$\sin \theta(\cos \theta + 2) - \tan \theta(\cos \theta + 2)$

$= (\cos \theta + 2)(\sin \theta - \tan \theta).$

A second method of factoring comes from reversing the following product:

$$(x + y)(x - y) \;\overset{\text{multiplying}}{\underset{\text{factoring}}{\longrightarrow}}\; x^2 - y^2$$

Note that the factors on the left differ only in the operation between x and y, while the result of the multiplication is two perfect squares with a minus sign between them. (Remember that x^2, x^4, x^6, x^{even} and 1, 4, 9, 16, 25, etc., are perfect squares.) Thus, in factoring two perfect squares with a minus sign between them, we obtain two factors that consist of the sum and the difference of the square roots of each of the squared terms.

EXAMPLE 6: Factor completely:

a. $c^2 - 81$ **b.** $\cos^2 \theta - 81$

SOLUTION:

a. $c^2 = c \cdot c$

$81 = 9 \cdot 9$

Therefore, $(c^2 - 81) = (c + 9)(c - 9)$.

b. $\cos^2 \theta = \cos \theta \cdot \cos \theta$

$81 = 9 \cdot 9$

Therefore, $(\cos^2 \theta - 81)$
$$= (\cos \theta + 9)(\cos \theta - 9).$$

EXAMPLE 7: Factor completely:

a. $16 - 9t^2$ **b.** $16 - 9 \tan^2 \theta$

SOLUTION:

a. $16 = 4 \cdot 4$

$9t^2 = (3t)(3t)$

Therefore, $(16 - 9t)^2 = (4 + 3t)(4 - 3t)$.

b. $16 = 4 \cdot 4$

$9 \tan^2 \theta = (3 \tan \theta)(3 \tan \theta)$

Therefore, $(16 - 9 \tan^2 \theta)$
$$= (4 + 3 \tan \theta)(4 - 3 \tan \theta).$$

EXAMPLE 8: Factor completely:

a. $s^4 - c^4$ **b.** $\sin^4 \theta - \cos^4 \theta$

SOLUTION:

a. $s^4 = s^2 \cdot s^2$

$c^4 = c^2 \cdot c^2$

Therefore,

$(s^4 - c^4) = (s^2 + c^2)(s^2 - c^2)$

But

$(s^2 - c^2) = (s + c)(s - c)$

Therefore,

$(s^4 - c^4) = (s^2 + c^2)(s + c)(s - c)$

b. $\sin^4 \theta = \sin^2 \theta \cdot \sin^2 \theta$

$\cos^4 \theta = \cos^2 \theta \cdot \cos^2 \theta$

Therefore,

$(\sin^4 \theta - \cos^4 \theta)$
$$= (\sin^2 \theta + \cos^2 \theta)(\sin^2 \theta - \cos^2 \theta)$$

But

$(\sin^2 \theta - \cos^2 \theta)$
$$= (\sin \theta + \cos \theta)(\sin \theta - \cos \theta)$$

Therefore,

$(\sin^4 \theta - \cos^4 \theta)$
$$= (\sin^2 \theta + \cos^2 \theta)$$
$$\times (\sin \theta + \cos \theta)(\sin \theta - \cos \theta)$$

The final method of factoring that we will discuss attempts to reverse FOIL multiplication that was illustrated in Section 4-2. For example,

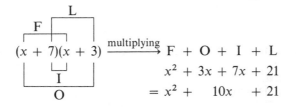

$$
\begin{array}{l}
(x + 7)(x + 3) \xrightarrow{\text{multiplying}} F + O + I + L \\
\qquad\qquad\qquad\quad x^2 + 3x + 7x + 21 \\
\qquad\qquad\qquad = x^2 + \quad 10x \quad + 21
\end{array}
$$

To reverse this process and factor $x^2 + 10x + 21$, we first note that the factors on the left consist of two terms and are in the form

$$(? + ?)(? + ?)$$

The first term, x^2, is the result of multiplying the first terms in the FOIL method. Thus,

$$x^2 + 10x + 21 = (x + ?)(x + ?)$$

The last term, 21, is the result of multiplying the last terms in the FOIL method. Since 21 equals $(21)(1)$, $(-21)(-1)$, $(7)(3)$, and $(-7)(-3)$, we have four possibilities:

$$
x^2 + 10x + 21 \overset{?}{=}
\begin{array}{c}
(x + 21)(x + 1) \\
\text{or} \\
(x - 21)(x - 1) \\
\text{or} \\
(x + 7)(x + 3) \\
\text{or} \\
(x - 7)(x - 3)
\end{array}
$$

The middle term, $10x$, is the sum of the inner and the outer terms:

$(x + 21)(x + 1)$ $\qquad$ $(x - 21)(x - 1)$

$x + 21x = 22x$ $\qquad$ $-x - 21x = -22x$

$(x + 7)(x + 3)$ $\qquad$ $(x - 7)(x - 3)$

$3x + 7x = 10x$ $\qquad$ $-3x - 7x = -10x$

Thus, $x^2 + 10x + 21 = (x + 7)(x + 3)$ is the correct choice. Note that since the middle term, $10x$, is positive, we could have immediately eliminated $(-21)(-1)$ and $(-7)(-3)$, for these choices result in a negative middle term. Also, once we find the correct combination, $(x + 7)(x + 3)$, we can stop because no other combinations will work. Since there are often many combinations of factors for the first and last terms, much trial and error can be involved in these problems. The experience that you will gain from practice often helps to eliminate some of the possibilities.

EXAMPLE 9: Factor completely:
a. $s^2 - 9s + 18$
b. $\sin^2 \theta - 9 \sin \theta + 18$

SOLUTION:
a. The first term, s^2, is the result of multiplying the first terms in the FOIL method. Thus,

$$s^2 - 9s + 18 = (s + ?)(s + ?)$$

The last term, 18, is the result of multiplying the last terms in the FOIL method. Since 18 equals $(-18)(-1)$, $(-9)(-2)$, and $(-6)(-3)$, we have three possibilities. We eliminate $(18)(1)$, $(9)(2)$, and $(6)(3)$ since the middle term is negative.

$$(s - 18)(s - 1)$$
$$\text{or}$$
$$s^2 - 9s + 18 \overset{?}{=} (s - 9)(s - 2)$$
$$\text{or}$$
$$(s - 6)(s - 3)$$

The middle term, $-9s$, is the sum of the inner and the outer terms:

$$(s - 18)(s - 1) \qquad (s - 9)(s - 2) \qquad (s - 6)(s - 3)$$
$$-s - 18s = -19s \qquad -2s - 9s = -11s \qquad -3s - 6s = \underline{-9s}$$

Thus, $s^2 - 9s + 18 = (s - 6)(s - 3)$ is the correct choice.
b. In a similar way, $\sin^2 \theta - 9 \sin \theta + 18 = (\sin \theta - 6)(\sin \theta - 3)$.

EXAMPLE 10: Factor completely:
a. $\tan^2 \theta - 5 \tan \theta - 6$
b. $t^2 - 5t - 6$

SOLUTION:
a. The first term, $\tan^2 \theta$, is the result of multiplying the first terms in the FOIL method. Thus,

$$(\tan^2 \theta - 5 \tan \theta - 6) = (\tan \theta + ?)(\tan \theta + ?)$$

The last term, -6, is the result of multiplying the last terms in the FOIL method. Since -6 equals $(-6)(1)$, $(6)(-1)$, $(-3)(2)$, and $(3)(-2)$, we have four possibilities.

$$\tan^2 \theta - 5 \tan \theta - 6 \overset{?}{=} \begin{array}{l} (\tan \theta - 6)(\tan \theta + 1) \\ (\tan \theta + 6)(\tan \theta - 1) \\ (\tan \theta - 3)(\tan \theta + 2) \\ (\tan \theta + 3)(\tan \theta - 2) \end{array}$$

The middle term, $-5 \tan \theta$, is the sum of the inner and the outer terms:

$$(\tan \theta - 6)(\tan \theta + 1)$$

$$\tan \theta + (-6 \tan \theta) = -5 \tan \theta \qquad \text{stop here}$$

Thus, $\tan^2 \theta - 5 \tan \theta - 6 = (\tan \theta - 6)(\tan \theta + 1)$ is the correct choice.

b. In a similar way, $t^2 - 5t - 6 = (t - 6)(t + 1)$.

Note from the last three examples that if the coefficient of the squared term is 1, the correct factors of the last term when added together result in the coefficient of the middle term.

$$x^2 + 10x + 21 = (x + 7)(x + 3)$$

$$s^2 - 9s + 18 = (s - 6)(s - 3)$$

$$\tan^2 \theta - 5 \tan \theta - 6 = (\tan \theta - 6)(\tan \theta + 1)$$

EXAMPLE 11: Factor completely:
a. $4c^2 - 12c + 5$
b. $4 \cos^2 \theta - 12 \cos \theta + 5$

SOLUTION:

a. The first term, $4c^2$, is the result of multiplying the first terms in the FOIL method. Thus,

$$4c^2 - 12c + 5 \overset{?}{=} \begin{array}{c} (4c + ?)(c + ?) \\ \text{or} \\ (2c + ?)(2c + ?) \end{array}$$

The last term, 5, is the result of multiplying the last terms in the FOIL method. There is only one possibility, $(-5)(-1)$, for 5 since we can eliminate $(5)(1)$ because the middle term is negative:

$$4c^2 - 12c + 5 \overset{?}{=} \begin{array}{c} (4c - 5)(c - 1) \\ \text{or} \\ (2c - 5)(2c - 1) \end{array}$$

The middle term, $-12c$, is the sum of the inner and the outer terms:

$$(4c - 5)(c - 1) \qquad\qquad (2c - 5)(2c - 1)$$

$$-4c - 5c = -9c \qquad\qquad -2c - 10c = -12c$$

Thus, $4c^2 - 12c + 5 = (2c - 5)(2c - 1)$ is the correct choice.

b. In a similar way, $4 \cos^2 \theta - 12 \cos \theta + 5 = (2 \cos \theta - 5)(2 \cos \theta - 1)$.

Remember that, although there are often many combinations to consider in factoring these expressions, experience and practice will help you to eliminate *mentally* many of the possibilities.

EXAMPLE 12: Factor completely:
a. $9 \sin^2 \theta + 12 \sin \theta + 4$
b. $9s^2 + 12s + 4$

SOLUTION:
a. The first term, $9 \sin^2 \theta$ is the result of multiplying the first terms in the FOIL method. Thus,

$$9 \sin^2 \theta + 12 \sin \theta + 4 \overset{?}{=} \begin{matrix} (9 \sin \theta + ?)(\sin \theta + ?) \\ \text{or} \\ (3 \sin \theta + ?)(3 \sin \theta + ?) \end{matrix}$$

The last term, 4, is the result of multiplying the last terms in the FOIL method. Since 4 equals (4)(1) and (2)(2), we have two possibilities. We eliminate negative factors of 4 since the middle term is positive.

$$9 \sin^2 \theta + 12 \sin \theta + 4 \overset{?}{=} \begin{matrix} (9 \sin \theta + 4)(\sin \theta + 1) \\ (9 \sin \theta + 2)(\sin \theta + 2) \\ (3 \sin \theta + 4)(3 \sin \theta + 1) \\ (3 \sin \theta + 2)(3 \sin \theta + 2) \end{matrix}$$

The middle term, $12 \sin \theta$, is the sum of the inner and the outer terms:

$(9 \sin \theta + 4)(\sin \theta + 1)$

$9 \sin \theta + 4 \sin \theta = 13 \sin \theta$

$(9 \sin \theta + 2)(\sin \theta + 2)$

$18 \sin \theta + 2 \sin \theta = 20 \sin \theta$

$(3 \sin \theta + 4)(3 \sin \theta + 1)$

$3 \sin \theta + 12 \sin \theta = 15 \sin \theta$

$(3 \sin \theta + 2)(3 \sin \theta + 2)$

$6 \sin \theta + 6 \sin \theta = 12 \sin \theta$

Thus, $9 \sin^2 \theta + 12 \sin \theta + 4 = (3 \sin \theta + 2)(3 \sin \theta + 2)$ or $(3 \sin \theta + 2)^2$.
b. In a similar way, $9s^2 + 12s + 4 = (3s + 2)(3s + 2)$ or $(3s + 2)^2$. (*Note:* Very often when the first term and the last term of the expression to be factored are perfect squares, the factored form is also a perfect square. Try this possibility first.)

Finally, it should be noted that there are many expressions that cannot be factored. For example,

$x^2 + 4$
$x^2 - 7$
$x^2 - x + 4$
$3x^2 + 4x - 2$

cannot be factored with integer coefficients. A quick test of whether an expression of the form $ax^2 + bx + c$ can be factored is to determine the value of $b^2 - 4ac$.

If this value is a perfect square, the expression can be factored. For instance, in the expression $3x^2 + 4x - 2$, we have $a = 3$, $b = 4$, and $c = -2$. The value of $b^2 - 4ac$ is $(4)^2 - 4(3)(-2) = 16 + 24 = 40$. Since 40 is not a perfect square, the expression cannot be factored.

EXERCISES 4–4

In Exercise 1–70 factor each expression completely. First factor out any common factors.

1. **a.** $7s + 7c$ **b.** $7 \sin \theta + 7 \cos \theta$
2. **a.** $3t - 3s$ **b.** $3 \tan \theta - 3 \sin \theta$
3. **a.** $sc + st$ **b.** $\sin \theta \cos \theta + \sin \theta \tan \theta$
4. **a.** $ct - cs$ **b.** $\cos \theta \tan \theta - \cos \theta \sin \theta$
5. **a.** $21s^2 - 14s$ **b.** $21 \sin^2 \theta - 14 \sin \theta$
6. **a.** $7t - 35t^3$ **b.** $7 \tan \theta - 35 \tan^3 \theta$
7. **a.** $9s^2t^2 + 3st$ **b.** $9 \sin^2 \theta \tan^2 \theta + 3 \sin \theta \tan \theta$
8. **a.** $22s^3c^3 - 2s^2c^2$ **b.** $22 \sin^3 \theta \cos^3 \theta - 2 \sin^2 \theta \cos^2 \theta$
9. **a.** $15ct^2 - 21c^2t$ **b.** $15 \cos \theta \tan^2 \theta - 21 \cos^2 \theta \tan \theta$
10. **a.** $20t^2 - 15ts$ **b.** $20 \tan^2 \theta - 15 \tan \theta \sin \theta$
11. **a.** $2s^2t^2 + 4sc - 5s^2c^2$ **b.** $2 \sin^2 \theta \tan^2 \theta + 4 \sin \theta \cos \theta - 5 \sin^2 \theta \cos^2 \theta$
12. **a.** $3s^{10}t^6 - 9s^7t^7 + 6s^9t^4$ **b.** $3 \sin^{10} \theta \tan^6 \theta - 9 \sin^7 \theta \tan^7 \theta + 6 \sin^9 \theta \tan^4 \theta$
13. **a.** $c(s - 5) + t(s - 5)$ **b.** $\cos \theta (\sin \theta - 5) + \tan \theta (\sin \theta - 5)$
14. **a.** $s(2t + 1) - c(2t + 1)$ **b.** $\sin \theta (2 \tan \theta + 1) - \cos \theta (2 \tan \theta + 1)$
15. **a.** $t(c - s) + s(c - s)$ **b.** $\tan \theta (\cos \theta - \sin \theta) + \sin \theta (\cos \theta - \sin \theta)$
16. **a.** $2s(t + c) - 5c(t + c)$ **b.** $2 \sin \theta (\tan \theta + \cos \theta) - 5 \cos \theta (\tan \theta + \cos \theta)$
17. **a.** $s^2 - 64$ **b.** $\sin^2 \theta - 64$
18. **a.** $9 - t^2$ **b.** $9 - \tan^2 \theta$
19. **a.** $36c^2 - 1$ **b.** $36 \cos^2 \theta - 1$
20. **a.** $100 - 81s^2$ **b.** $100 - 81 \sin^2 \theta$
21. **a.** $4t^2 - 25s^2$ **b.** $4 \tan^2 \theta - 25 \sin^2 \theta$
22. **a.** $49s^2 - 16c^2$ **b.** $49 \sin^2 \theta - 16 \cos^2 \theta$
23. **a.** $36s^6 - t^4$ **b.** $36 \sin^6 \theta - \tan^4 \theta$
24. **a.** $25t^{12} - 36c^{10}$ **b.** $25 \tan^{12} \theta - 36 \cos^{10} \theta$
25. **a.** $9s^2 - t^2c^2$ **b.** $9 \sin^2 \theta - \tan^2 \theta \cos^2 \theta$
26. **a.** $25s^2c^2 - 16t^4$ **b.** $25 \sin^2 \theta \cos^2 \theta - 16 \tan^4 \theta$
27. **a.** $s^2 + 5s + 4$ **b.** $\sin^2 \theta + 5 \sin \theta + 4$
28. **a.** $c^2 + 7c + 10$ **b.** $\cos^2 \theta + 7 \cos \theta + 10$
29. **a.** $t^2 - 3t + 2$ **b.** $\tan^2 \theta - 3 \tan \theta + 2$
30. **a.** $t^2 - 8t + 12$ **b.** $\tan^2 \theta - 8 \tan \theta + 12$
31. **a.** $s^2 + s - 12$ **b.** $\sin^2 \theta + \sin \theta - 12$
32. **a.** $c^2 - 3c - 18$ **b.** $\cos^2 \theta - 3 \cos \theta - 18$
33. **a.** $t^2 - 9t + 20$ **b.** $\tan^2 \theta - 9 \tan \theta + 20$
34. **a.** $s^2 - 8s - 20$ **b.** $\sin^2 \theta - 8 \sin \theta - 20$
35. **a.** $6 + s - s^2$ **b.** $6 + \sin \theta - \sin^2 \theta$
36. **a.** $9 - 8c - c^2$ **b.** $9 - 8 \cos \theta - \cos^2 \theta$
37. **a.** $3s^2 - 8s - 3$ **b.** $3 \sin^2 \theta - 8 \sin \theta - 3$

38. a. $2t^2 - 7t + 6$ **b.** $2 \tan^2 \theta - 7 \tan \theta + 6$

39. a. $7c^2 - 13c - 2$ **b.** $7 \cos^2 \theta - 13 \cos \theta - 2$

40. a. $5c^2 - 4c - 12$ **b.** $5 \cos^2 \theta - 4 \cos \theta - 12$

41. a. $6s^2 + 7s + 2$ **b.** $6 \sin^2 \theta + 7 \sin \theta + 2$

42. a. $4c^2 - 13c + 3$ **b.** $4 \cos^2 \theta - 13 \cos \theta + 3$

43. a. $9t^2 - 25t - 6$ **b.** $9 \tan^2 \theta - 25 \tan \theta - 6$

44. a. $12t^2 + 19t - 18$ **b.** $12 \tan^2 \theta + 19 \tan \theta - 18$

45. a. $20s^2 - 43s - 12$ **b.** $20 \sin^2 \theta - 43 \sin \theta - 12$

46. a. $18c^2 - 57c + 24$ **b.** $18 \cos^2 \theta - 57 \cos \theta + 24$

47. a. $s^2 + 5sc + 4c^2$ **b.** $\sin^2 \theta + 5 \sin \theta \cos \theta + 4 \cos^2 \theta$

48. a. $2t^2 - 7tc + 6c^2$ **b.** $2 \tan^2 \theta - 7 \tan \theta \cos \theta + 6 \cos^2 \theta$

49. a. $4c^2 - 9ct + 2t^2$ **b.** $4 \cos^2 \theta - 9 \cos \theta \tan \theta + 2 \tan^2 \theta$

50. a. $2s^2 + 5sc - 7c^2$ **b.** $2 \sin^2 \theta + 5 \sin \theta \cos \theta - 7 \cos^2 \theta$

51. a. $16t^2 - 24t + 9$ **b.** $16 \tan^2 \theta - 24 \tan \theta + 9$

52. a. $9c^2 + 24c + 16$ **b.** $9 \cos^2 \theta + 24 \cos \theta + 16$

53. a. $4s^2 - 4s + 1$ **b.** $4 \sin^2 \theta - 4 \sin \theta + 1$

54. a. $4t^2 + 20t + 25$ **b.** $4 \tan^2 \theta + 20 \tan \theta + 25$

55. a. $9s^2 - 12sc + 4c^2$ **b.** $9 \sin^2 \theta - 12 \sin \theta \cos \theta + 4 \cos^2 \theta$

56. a. $16s^2t^2 + 8st + 1$ **b.** $16 \sin^2 \theta \tan^2 \theta + 8 \sin \theta \tan \theta + 1$

57. a. $7s^2 - 63$ **b.** $7 \sin^2 \theta - 63$

58. a. $3t^2 - 75s^2$ **b.** $3 \tan^2 \theta - 75 \sin^2 \theta$

59. a. $s - st^2$ **b.** $\sin \theta - \sin \theta \tan^2 \theta$

60. a. $c^3 - c$ **b.** $\cos^3 \theta - \cos \theta$

61. a. $t^4 - 1$ **b.** $\tan^4 \theta - 1$

62. a. $c^4 - t^4$ **b.** $\cos^4 \theta - \tan^4 \theta$

63. a. $3s^2 - 6s - 24$ **b.** $3 \sin^2 \theta - 6 \sin \theta - 24$

64. a. $5t^2 - 15t - 50$ **b.** $5 \tan^2 \theta - 15 \tan \theta - 50$

65. a. $s^3 - 6s^2 + 8s$ **b.** $\sin^3 \theta - 6 \sin^2 \theta + 8 \sin \theta$

66. a. $t^4 - 5t^3 + 6t^2$ **b.** $\tan^4 \theta - 5 \tan^3 \theta + 6 \tan^2 \theta$

67. a. $3c^3 - 12c^2 + 9c$ **b.** $3 \cos^3 \theta - 12 \cos^2 \theta + 9 \cos \theta$

68. a. $2s^3c - 18s^2c + 40sc$ **b.** $2 \sin^3 \theta \cos \theta - 18 \sin^2 \theta \cos \theta + 40 \sin \theta \cos \theta$

69. a. $t^4 + 4t^2 - 5$ **b.** $\tan^4 \theta + 4 \tan^2 \theta - 5$

70. a. $4s^4 - 5s^2 + 1$ **b.** $4 \sin^4 \theta - 5 \sin^2 \theta + 1$

In Exercises 71–80 determine whether each expression can be factored (with integer coefficients) by calculating $b^2 - 4ac$.

71. $6s^2 + 7s + 2$ **72.** $3 \sin^2 \theta + \sin \theta - 1$

73. $t^2 - 5t + 3$ **74.** $2 \tan^2 \theta - \tan \theta - 3$

75. $2 \cos^2 \theta - 5 \cos \theta - 7$ **76.** $c^2 - 3c - 4$

77. $4 \tan^2 \theta - 2 \tan \theta + 3$ **78.** $12s^2 - s + 1$

79. $\cos^2 \theta + 1$ **80.** $2 - \sin^2 \theta$

4–5 SOLVING QUADRATIC EQUATIONS BY FACTORING

An equation that can be put in the form $ax^2 + bx + c = 0$, where a, b, and c represent constants with $a \neq 0$, is called a *second-degree* or *quadratic equation*. For example,

$$2x^2 - 7x + 6 = 0$$
$$4t^2 = 9t - 2$$
$$\sin^2 \theta - 1 = 0$$

are quadratic equations

Remember that an equation is a statement that two expressions have the same value, and the truth of this mathematical assertion depends upon the number that is substituted for the variable. In solving an equation, we are attempting to find the numbers that make the equation true.

A technique that can often be used to solve a second-degree equation relies on factoring. For example, let us try to find the numbers that make the equation $x^2 + 5x + 4 = 0$ a true statement.

First, factor the expression on the left-hand side of the equation:

$$(x + 4)(x + 1) = 0$$

We now have the situation where the product of two factors is zero. In multiplication, zero is a very special number, for the product of two numbers can be zero only if at least one of the factors is zero. That is, if $a \cdot b = 0$, either $a = 0$ or $b = 0$. Zero is the only number that has this property. Thus, if either factor of the left-hand side of our equation is zero, the product will be zero.

$$(x + 4)(x + 1) = 0$$

$$x + 4 = 0 \qquad x + 1 = 0$$
$$\text{when } x = -4 \qquad \text{when } x = -1$$

The solutions of $x^2 + 5x + 4 = 0$ are -4 and -1. To check, substitute our solutions -4 and -1 in the original equation:

$$x^2 + 5x + 4 = 0 \qquad\qquad x^2 + 5x + 4 = 0$$
$$(-4)^2 + 5(-4) + 4 \overset{?}{=} 0 \qquad (-1)^2 + 5(-1) + 4 \overset{?}{=} 0$$
$$16 - 20 + 4 \overset{?}{=} 0 \qquad\qquad 1 - 5 + 4 \overset{?}{=} 0$$
$$0 \overset{\checkmark}{=} 0 \qquad\qquad\qquad 0 \overset{\checkmark}{=} 0$$

EXAMPLE 1: For what values of t and $\tan \theta$ will the following equations be true?

a. $t^2 - 3t = 18$ \qquad\qquad **b.** $\tan^2 \theta - 3 \tan \theta = 18$

SOLUTION:

a. \qquad $t^2 - 3t = 18$ \qquad\qquad **b.** \qquad $\tan^2 \theta - 3 \tan \theta = 18$

$$t^2 - 3t - 18 = 0 \qquad\qquad \tan^2 \theta - 3 \tan \theta - 18 = 0$$
$$(t - 6)(t + 3) = 0 \qquad\qquad (\tan \theta - 6)(\tan \theta + 3) = 0$$
$$t - 6 = 0 \qquad\qquad\qquad \tan \theta - 6 = 0$$
$$t = 6 \qquad\qquad\qquad\qquad \tan \theta = 6$$
$$t + 3 = 0 \qquad\qquad\qquad \tan \theta + 3 = 0$$
$$t = -3 \qquad\qquad\qquad\qquad \tan \theta = -3$$

Check: Substitute the solutions in the original equations:

$$t^2 - 3t = 18 \qquad\qquad \tan^2 \theta - 3 \tan \theta = 18$$
$$(6)^2 - 3(6) \stackrel{?}{=} 18 \qquad\qquad (6)^2 - 3(6) \stackrel{?}{=} 18$$
$$36 - 18 \stackrel{?}{=} 18 \qquad\qquad 36 - 18 \stackrel{?}{=} 18$$
$$18 \stackrel{\checkmark}{=} 18 \qquad\qquad 18 \stackrel{\checkmark}{=} 18$$

$$t^2 - 3t = 18 \qquad\qquad \tan^2 \theta - 3 \tan \theta = 18$$
$$(-3)^2 - 3(-3) \stackrel{?}{=} 18 \qquad\qquad (-3)^2 - 3(-3) \stackrel{?}{=} 18$$
$$9 + 9 \stackrel{?}{=} 18 \qquad\qquad 9 + 9 \stackrel{?}{=} 18$$
$$18 \stackrel{\checkmark}{=} 18 \qquad\qquad 18 \stackrel{\checkmark}{=} 18$$

Note that, in general, the following procedure is used:

1. If necessary, change the form of the equation so that one side is 0.
2. Factor the nonzero side of the equation.
3. Set each factor equal to zero and obtain two solutions by solving the resulting equations.
4. Check each solution by substituting it in the original equation.

EXAMPLE 2: For what values of s and $\sin \theta$ will the following equations be true?
a. $4s^2 = s$ **b.** $4 \sin^2 \theta = \sin \theta$

SOLUTION:

a.
$$4s^2 = s$$
$$4s^2 - s = 0$$
$$s(4s - 1) = 0$$
$$s = 0$$
$$4s - 1 = 0$$
$$s = \tfrac{1}{4}$$

b.
$$4 \sin^2 \theta = \sin \theta$$
$$4 \sin^2 \theta - \sin \theta = 0$$
$$\sin \theta(4 \sin \theta - 1) = 0$$
$$\sin \theta = 0$$
$$4 \sin \theta - 1 = 0$$
$$\sin \theta = \tfrac{1}{4}$$

Check:

$$4s^2 = s \qquad\qquad 4 \sin^2 \theta = \sin \theta$$
$$4(0)^2 \stackrel{?}{=} 0 \qquad\qquad 4(0)^2 \stackrel{?}{=} 0$$
$$4(0) \stackrel{?}{=} 0 \qquad\qquad 4(0) \stackrel{?}{=} 0$$
$$0 \stackrel{\checkmark}{=} 0 \qquad\qquad 0 \stackrel{\checkmark}{=} 0$$

$$4s^2 = s \qquad\qquad 4 \sin^2 \theta = \sin \theta$$
$$4(\tfrac{1}{4})^2 \stackrel{?}{=} \tfrac{1}{4} \qquad\qquad 4(\tfrac{1}{4})^2 \stackrel{?}{=} \tfrac{1}{4}$$
$$4(\tfrac{1}{16}) \stackrel{?}{=} \tfrac{1}{4} \qquad\qquad 4(\tfrac{1}{16}) \stackrel{?}{=} \tfrac{1}{4}$$
$$\tfrac{1}{4} \stackrel{\checkmark}{=} \tfrac{1}{4} \qquad\qquad \tfrac{1}{4} \stackrel{\checkmark}{=} \tfrac{1}{4}$$

EXAMPLE 3: For what values of c and $\cos \theta$ will the following equations be true?
a. $6c^2 = 5c - 1$ **b.** $6 \cos^2 \theta = 5 \cos \theta - 1$

SOLUTION:

a.
$$6c^2 = 5c - 1$$
$$6c^2 - 5c + 1 = 0$$
$$(3c - 1)(2c - 1) = 0$$
$$3c - 1 = 0$$
$$c = \tfrac{1}{3}$$
$$2c - 1 = 0$$
$$c = \tfrac{1}{2}$$

b.
$$6 \cos^2 \theta = 5 \cos \theta - 1$$
$$6 \cos^2 \theta - 5 \cos \theta + 1 = 0$$
$$(3 \cos \theta - 1)(2 \cos \theta - 1) = 0$$
$$3 \cos \theta - 1 = 0$$
$$\cos \theta = \tfrac{1}{3}$$
$$2 \cos \theta - 1 = 0$$
$$\cos \theta = \tfrac{1}{2}$$

Check:

$$6c^2 = 5c - 1$$
$$6(\tfrac{1}{3})^2 \overset{?}{=} 5(\tfrac{1}{3}) - 1$$
$$6(\tfrac{1}{9}) \overset{?}{=} \tfrac{5}{3} - \tfrac{3}{3}$$
$$\tfrac{2}{3} \overset{\checkmark}{=} \tfrac{2}{3}$$

$$6 \cos^2 \theta = 5 \cos \theta - 1$$
$$6(\tfrac{1}{3})^2 \overset{?}{=} 5(\tfrac{1}{3}) - 1$$
$$6(\tfrac{1}{9}) \overset{?}{=} \tfrac{5}{3} - \tfrac{3}{3}$$
$$\tfrac{2}{3} \overset{\checkmark}{=} \tfrac{2}{3}$$

$$6c^2 = 5c - 1$$
$$6(\tfrac{1}{2})^2 \overset{?}{=} 5(\tfrac{1}{2}) - 1$$
$$6(\tfrac{1}{4}) \overset{?}{=} \tfrac{5}{2} - \tfrac{2}{2}$$
$$\tfrac{3}{2} \overset{\checkmark}{=} \tfrac{3}{2}$$

$$6 \cos^2 \theta = 5 \cos \theta - 1$$
$$6(\tfrac{1}{2})^2 \overset{?}{=} 5(\tfrac{1}{2}) - 1$$
$$6(\tfrac{1}{4}) \overset{?}{=} \tfrac{5}{2} - \tfrac{2}{2}$$
$$\tfrac{3}{2} \overset{\checkmark}{=} \tfrac{3}{2}$$

EXAMPLE 4: For what values of t and $\tan \theta$ will the following equations be true?
a. $t^2 - 6t + 9 = 0$ **b.** $\tan^2 \theta - 6 \tan \theta + 9 = 0$

SOLUTION:

a.
$$t^2 - 6t + 9 = 0$$
$$(t - 3)(t - 3) = 0$$
$$t - 3 = 0$$
$$t = 3$$
$$t - 3 = 0$$
$$t = 3$$

b.
$$\tan^2 \theta - 6 \tan \theta + 9 = 0$$
$$(\tan \theta - 3)(\tan \theta - 3) = 0$$
$$\tan \theta - 3 = 0$$
$$\tan \theta = 3$$
$$\tan \theta - 3 = 0$$
$$\tan \theta = 3$$

(*Note:* When the same number makes both factors zero, we think of the equation as having two solutions that are equal.)

Check:

$$t^2 - 6t + 9 = 0$$
$$(3)^2 - 6(3) + 9 \overset{?}{=} 0$$
$$9 - 18 + 9 \overset{?}{=} 0$$
$$0 \overset{\checkmark}{=} 0$$

$$\tan^2 \theta - 6 \tan \theta + 9 = 0$$
$$(3)^2 - 6(3) + 9 \overset{?}{=} 0$$
$$9 - 18 + 9 \overset{?}{=} 0$$
$$0 \overset{\checkmark}{=} 0$$

EXAMPLE 5: For what values of θ ($0° \leq \theta < 360°$) will the equation $2 \sin^2 \theta - 5 \sin \theta = 3$ be a true statement?

SOLUTION:

$$2 \sin^2 \theta - 5 \sin \theta = 3$$
$$2 \sin^2 \theta - 5 \sin \theta - 3 = 0$$
$$(2 \sin \theta + 1)(\sin \theta - 3) = 0$$

$2 \sin \theta + 1 = 0$	$\sin \theta - 3 = 0$
$\sin \theta = -\frac{1}{2}$	$\sin \theta = 3$

1. Since $\sin \theta$ is a negative number, θ could be in Q_3 or Q_4.
2. Since $\sin 30° = \frac{1}{2}$, the reference angle is 30° (Figure 4.1a).
3. 210° is the angle in Q_3 with a reference angle of 30° (Figure 4.1a). 330° is the angle in Q_4 with a reference angle of 30° (Figure 4.1b).

Since the value of $\sin \theta$ never exceeds 1, there is no solution to this equation.

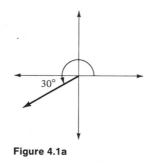

Figure 4.1a

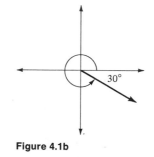

Figure 4.1b

Thus, 210° and 330° are the solutions of the equation ($0° \leq \theta < 360°$).

EXAMPLE 6: The height (y) of a ball thrown vertically up from a roof 64 ft high with an initial velocity of 48 ft/second is given by the formula $y = 64 + 48t - 16t^2$. When will the ball strike the ground?

SOLUTION: When the ball strikes the ground, the height (y) of the ball is 0. Therefore, we want to find the values of t for which $64 + 48t - 16t^2$ equals 0.

$$0 = 64 + 48t - 16t^2$$
$$= 16(4 + 3t - t^2)$$
$$= 16(4 - t)(1 + t)$$

Since the first factor 16 cannot be 0, there are two ways that $0 = 16(4 - t)(1 + t)$ can be true:

$4 - t = 0$		$1 + t = 0$
$4 = t$	or	$t = -1$

The ball hits the ground 4 seconds later. We reject -1 as a solution since for this problem a negative time has no physical significance.

Note that this method cannot be used to solve all second-degree equations since, in many cases, an expression of the form $ax^2 + bx + c$ cannot be factored. We shall deal with this problem in Chapter 8.

Finally, this factoring technique can be used to solve higher degree equations if these equations can be factored.

EXAMPLE 7: For what values of s and $\sin \theta$ will the equations $s^3 = s$ and $\sin^3 \theta = \sin \theta$ be a true statement?

a. $s^3 = s$ \qquad\qquad\qquad\qquad **b.** $\sin^3 \theta = \sin \theta$

SOLUTION:

a.
$$s^3 = s$$
$$s^3 - s = 0$$
$$s(s^2 - 1) = 0$$
$$s(s + 1)(s - 1) = 0$$
$$s = 0$$
$$s + 1 = 0$$
$$s = -1$$
$$s - 1 = 0$$
$$s = 1$$

b.
$$\sin^3 \theta = \sin \theta$$
$$\sin^3 \theta - \sin \theta = 0$$
$$\sin \theta(\sin^2 \theta - 1) = 0$$
$$\sin \theta(\sin \theta + 1)(\sin \theta - 1) = 0$$
$$\sin \theta = 0$$
$$\sin \theta + 1 = 0$$
$$\sin \theta = -1$$
$$\sin \theta - 1 = 0$$
$$\sin \theta = 1$$

Check:

$$s^3 = s$$
$$0^3 \overset{?}{=} 0$$
$$0 \overset{\checkmark}{=} 0$$
$$s^3 = s$$
$$(-1)^3 \overset{?}{=} -1$$
$$-1 \overset{\checkmark}{=} -1$$
$$s^3 = s$$
$$(1)^3 \overset{?}{=} 1$$
$$1 \overset{\checkmark}{=} 1$$

$$\sin^3 \theta = \sin \theta$$
$$0^3 \overset{?}{=} 0$$
$$0 \overset{\checkmark}{=} 0$$
$$\sin^3 \theta = \sin \theta$$
$$(-1)^3 \overset{?}{=} -1$$
$$-1 \overset{\checkmark}{=} -1$$
$$\sin^3 \theta = \sin \theta$$
$$(1)^3 \overset{?}{=} 1$$
$$1 \overset{\checkmark}{=} 1$$

EXERCISES 4-5

In Exercises 1–38 solve each equation. For the trigonometric equations, find the values of either $\sin \theta$, $\cos \theta$, or $\tan \theta$ that make the equation a true statement. Eliminate any values that do not lead to solutions for θ.

1. a. $(s - 4)(s + 1) = 0$ \qquad\qquad **b.** $(\sin \theta - 4)(\sin \theta + 1) = 0$

2. a. $(t + 2)(t - 3) = 0$ \qquad\qquad **b.** $(\tan \theta + 2)(\tan \theta - 3) = 0$

3. **a.** $t(t + 5) = 0$ **b.** $\tan \theta(\tan \theta + 5) = 0$

4. **a.** $s(s - 1) = 0$ **b.** $\sin \theta(\sin \theta - 1) = 0$

5. **a.** $(2c - 5)(5c + 3) = 0$ **b.** $(2 \cos \theta - 5)(5 \cos \theta + 3) = 0$

6. **a.** $(4t + 1)(3t - 5) = 0$ **b.** $(4 \tan \theta + 1)(3 \tan \theta - 5) = 0$

7. **a.** $c^2 - 5c = 0$ **b.** $\cos^2 \theta - 5 \cos \theta = 0$

8. **a.** $s^2 = 2s$ **b.** $\sin^2 \theta = 2 \sin \theta$

9. **a.** $4s^2 = 5s$ $s(4s - 5)$ **b.** $4 \sin^2 \theta = 5 \sin \theta$

10. **a.** $14t^2 + 7t = 0$ **b.** $14 \tan^2 \theta + 7 \tan \theta = 0$

11. **a.** $t^2 - 4 = 0$ **b.** $\tan^2 \theta - 4 = 0$

12. **a.** $5s^2 - 5 = 0$ **b.** $5 \sin^2 \theta - 5 = 0$

13. **a.** $c^2 - 3c + 2 = 0$ **b.** $\cos^2 \theta - 3 \cos \theta + 2 = 0$

14. **a.** $s^2 + 13s + 12 = 0$ **b.** $\sin^2 \theta + 13 \sin \theta + 12 = 0$

15. **a.** $t^2 - 2t - 8 = 0$ **b.** $\tan^2 \theta - 2 \tan \theta - 8 = 0$

16. **a.** $t^2 - 3t - 18 = 0$ **b.** $\tan^2 \theta - 3 \tan \theta - 18 = 0$

17. **a.** $s^2 = 8s + 20$ **b.** $\sin^2 \theta = 8 \sin \theta + 20$

18. **a.** $t^2 + t = 12$ **b.** $\tan^2 \theta + \tan \theta = 12$

19. **a.** $t^2 - 3t = 4$ **b.** $\tan^2 \theta - 3 \tan \theta = 4$

20. **a.** $t^2 + 3t = 4$ **b.** $\tan^2 \theta + 3 \tan \theta = 4$

21. **a.** $c^2 = 2c - 1$ **b.** $\cos^2 \theta = 2 \cos \theta - 1$

22. **a.** $t^2 + 6t = -9$ **b.** $\tan^2 \theta + 6 \tan \theta = -9$

23. **a.** $3s^2 - 16s + 5 = 0$ **b.** $3 \sin^2 \theta - 16 \sin \theta + 5 = 0$

24. **a.** $5c^2 - 11c + 2 = 0$ **b.** $5 \cos^2 \theta - 11 \cos \theta + 2 = 0$

25. **a.** $5t^2 - 16t + 3 = 0$ **b.** $5 \tan^2 \theta - 16 \tan \theta + 3 = 0$

26. **a.** $3c^2 + 10c - 8 = 0$ **b.** $3 \cos^2 \theta + 10 \cos \theta - 8 = 0$

27. **a.** $3s^2 = 5s - 2$ **b.** $3 \sin^2 \theta = 5 \sin \theta - 2$

28. **a.** $2c^2 = 7c + 4$ **b.** $2 \cos^2 \theta = 7 \cos \theta + 4$

29. **a.** $4t^2 = -12t - 9$ **b.** $4 \tan^2 \theta = -12 \tan \theta - 9$

30. **a.** $9s^2 = 6s - 1$ **b.** $9 \sin^2 \theta = 6 \sin \theta - 1$

31. **a.** $2(s^2 - 1) = 3s$ **b.** $2(\sin^2 \theta - 1) = 3 \sin \theta$

32. **a.** $3(c^2 + 1) = 10c$ **b.** $3(\cos^2 \theta + 1) = 10 \cos \theta$

33. **a.** $c^3 - 4c = 0$ **b.** $\cos^3 \theta - 4 \cos \theta = 0$

34. **a.** $9t^3 = 25t$ **b.** $9 \tan^3 \theta = 25 \tan \theta$

35. **a.** $t^3 + 9t^2 + 14t = 0$ **b.** $\tan^3 \theta + 9 \tan^2 \theta + 14 \tan \theta = 0$

36. **a.** $4c^3 - 13c^2 = -3c$ **b.** $4 \cos^3 \theta - 13 \cos^2 \theta = -3 \cos \theta$

37. **a.** $s^4 = s^2$ **b.** $\sin^4 \theta = \sin^2 \theta$

38. **a.** $t^4 - 5t^2 + 4 = 0$ **b.** $\tan^4 \theta - 5 \tan^2 \theta + 4 = 0$

In Exercises 39–44 find the values of θ ($0° \le \theta < 360°$), to the nearest 10 minutes, which make the equation a true statement.

39. $\tan^2 \theta + 4 \tan \theta - 21 = 0$ 40. $\sin^2 \theta = 4 \sin \theta + 5$

41. $6 \sin^2 \theta + \sin \theta - 2 = 0$ 42. $2 \cos^2 \theta = 3 \cos \theta - 1$

43. $\cos^3 \theta = \cos \theta$ 44. $\tan^3 \theta = \tan \theta$

In Exercises 45–50 find to the nearest 10 minutes all the values of θ that make the equation a true statement.

45. $\tan^2 \theta + 3 \tan \theta - 10 = 0$ 46. $\cos^2 \theta = -4 \cos \theta - 3$

47. $\cos^2 \theta = \cos \theta$
48. $\sin^2 \theta = -\sin \theta$
49. $12 \tan^2 \theta - 16 \tan \theta - 3 = 0$
50. $12 \sin^2 \theta = 5 \sin \theta + 2$

51. The length of a rectangle is 2 ft more than the width, and the area is 24 ft². Find the dimensions of the rectangle.

52. The area of a rectangular plate is to be 28 in.², and the sides of the plate are to differ in length by 3 in. Find the dimensions of the plate.

53. When a uniform border is added to a rug that is 6 ft by 4 ft, the area increases by 11 ft². Find the width of the border.

54. One side of a right triangle is 3 ft longer than the other, and the hypotenuse is 15 ft. What is the length of the shortest side of the triangle?

55. The height (y) after t seconds of a diver who jumps from a cliff that is 64 ft above water is given by the formula $y = 64 - 16t^2$. When will the diver hit the water?

56. The height (y) of a ball that is thrown down from a roof that is 160 ft high with an initial velocity of 48 ft/second is given by the formula $y = 160 - 48t - 16t^2$. When will the ball hit the ground?

57. The height (y) of a projectile that is shot directly up from the ground with an initial velocity of 96 ft/second is given by the formula $y = 96t - 16t^2$.

  a. When will the projectile strike the ground?
  b. At what time after firing is the projectile 80 ft off the ground?
  c. Physically, why are there two solutions to part b?

58. The height (y) of a projectile that is shot directly up from a height of 144 ft with an initial velocity of 96 ft/second is given by the formula $y = 144 + 96t - 16t^2$. How long after firing is the projectile 288 ft off the ground?

59. The total sales revenue for a company is estimated by the formula $S = 8x - x^2$, where x is the unit selling price (in dollars) and S is the total sales revenue (1 unit equals $10,000). What should be the unit selling price if the company wishes the total revenue from sales to be $150,000 (that is, $S = 15$)? Why are there two solutions to the problem?

60. The total cost (in dollars) to a company of producing x units of a product in 1 month is estimated by the formula $C = 10x^2 + 150x + 200$.

  a. What are the fixed costs of the company (that is, what costs does the company experience even if they do not produce any units)?

  b. If the total cost during a month is $2700, how many units did the company produce?

Sample Test: Chapter 4

In questions 1–9 perform the indicated operations and express your results in simplest form.

1. $3 \cdot 3^3$

2. $\dfrac{2^{3x}}{2^x}$

3. $(2 \sin^3 \theta)^2$

4. $\dfrac{(3x)^2}{y} \left(\dfrac{2zy}{x^2}\right)^3$

5. $ct^2(3c - 2t + 4)$

6. $\dfrac{\sin \theta \cos^3 \theta - 4 \sin^2 \theta \cos \theta}{\sin \theta \cos \theta}$

7. $(t - 3)^2$

8. $(2 \sin \theta + 1)(3 \sin \theta - 2)$

9. $(2x + 4)(4 + x - x^2)$

10. Expand $(x + h)^6$ by the binomial theorem.

11. Write the tenth term in the expansion of $(2x - y)^{11}$.

In questions 12–15 factor each of the expressions completely.

12. $x^2 - 8x + 12$

13. $\cos^2 \theta + 8 \cos \theta + 16$

14. $4c^2 - 2ct - 12t^2$

15. $t^3 - t^3x^2$

16. Calculate $b^2 - 4ac$ for the expression $2c^2 - 3c + 7$. Can the expression be factored?

17. Solve for s: $s(s + 2) = 0$.

18. Solve for x: $x^2 + 3x = 4$.

19. Solve for $\tan \theta$: $3 \tan^3 \theta + 10 \tan^2 \theta - 8 \tan \theta = 0$.

20. Solve to the nearest degree for $0° \leq \theta \leq 180°$: $\tan^2 \theta - 3 \tan \theta - 4 = 0$.

Chapter 5
Fractions

5-1 ALGEBRAIC FRACTIONS

In arithmetic, a fraction is used to indicate the quotient or ratio of two numbers. Since different ratios often have the same value, there are many fractions that are equivalent. For example, $\frac{1}{2}$, $\frac{2}{4}$, and $\frac{4}{8}$ are equivalent fractions since, although they appear in different forms, they have the same value. In working with fractions, it is essential that you understand the following principle, which shows how a fraction can change its form without changing its value: *The value of a fraction does not change when both the numerator and denominator of the fraction are multiplied by the same number (except zero).* For example, since

$$\frac{2}{4} = \frac{2(2)}{4(2)} = \frac{4}{8}$$

the fractions $\frac{2}{4}$ and $\frac{4}{8}$ are equivalent. Similarly, since

$$\frac{2}{4} = \frac{2(\frac{1}{2})}{4(\frac{1}{2})} = \frac{1}{2}$$

the fractions $\frac{2}{4}$ and $\frac{1}{2}$ are equivalent.

A fraction is said to be in simplest form when its numerator and denominator have no common factor other than 1. For example, to simplify the fraction $\frac{6}{9}$ we note that 3 is a factor of 6 and 3 is a factor of 9. That is,

$$\frac{6}{9} = \frac{3 \cdot 2}{3 \cdot 3}$$

We can now cancel the common factor 3 by multiplying the top and bottom of the fraction by the reciprocal of 3, which is $\frac{1}{3}$:

$$\frac{6}{9} = \frac{(\frac{1}{3}) \cdot 3 \cdot 2}{(\frac{1}{3}) \cdot 3 \cdot 3} = \frac{1 \cdot 2}{1 \cdot 3} = \frac{2}{3}$$

Thus, by *cancellation* we mean that if the same *factor* appears in the numerator and denominator, we can change this factor into a factor of 1 by multiplying the top and bottom of the fraction by the reciprocal of the factor. In common practice, when the same factor appears in the numerator and denominator, one merely crosses out the factor and puts a 1 above it, as follows:

$$\frac{\overset{1}{\cancel{3}} \cdot 2}{\underset{1}{\cancel{3}} \cdot 3} = \frac{2}{3}$$

EXAMPLE 1: In the expression $(3 + 2)/3$ may we cancel the 3's to obtain

$$\frac{\overset{1}{\cancel{3}} + 2}{\underset{1}{\cancel{3}}}?$$

SOLUTION:

$$\frac{3 + 2}{3} = \frac{\frac{1}{3}(3 + 2)}{\frac{1}{3} \cdot 3} \qquad \text{multiply the numerator and denominator by } \tfrac{1}{3}$$

$$= \frac{\frac{1}{3} \cdot 3 + \frac{1}{3} \cdot 2}{\frac{1}{3} \cdot 3} \qquad \text{distributive property}$$

$$= \frac{1 + \frac{2}{3}}{1}$$

We cannot cancel since

$$\frac{1 + \frac{2}{3}}{1} \quad \text{does not equal} \quad \frac{\overset{1}{\cancel{3}} + 2}{\underset{1}{\cancel{3}}}$$

Notice that we can cancel the 3's in the expression $(3 \cdot 2)/3$ since 3 is a *factor* in the numerator. We *cannot* cancel the 3's in the expression $(3 + 2)/3$ since 3 is a *term* in the numerator.

The same principles that govern a fraction in arithmetic will also apply if the numerator and denominator contain either algebraic or trigonometric expressions. For example, let us try to express in simplest form the fraction

$$\frac{3x + 9}{5x + 15}$$

First, we factor completely the numerator and denominator in an attempt to find a factor that is common to both.

$$\frac{3x + 9}{5x + 15} = \frac{3(x + 3)}{5(x + 3)}.$$

Since the numerator and denominator both contain the *factor* $x + 3$, we can cancel this factor.

$$\frac{3\cancel{(x + 3)}}{5\cancel{(x + 3)}} = \frac{3}{5}$$

Thus, the simplest form of the fraction

$$\frac{3x + 9}{5x + 15} \text{ is } \frac{3}{5}$$

EXAMPLE 2: Express in simplest form:

a. $\dfrac{5s + 10}{s^2 - 4}$

b. $\dfrac{5 \sin \theta + 10}{\sin^2 \theta - 4}$

SOLUTION:

a. $\dfrac{5s + 10}{s^2 - 4} = \dfrac{5\cancel{(s + 2)}}{\cancel{(s + 2)}(s - 2)}$

$= \dfrac{5}{s - 2}$

b. $\dfrac{5 \sin \theta + 10}{\sin^2 \theta - 4} = \dfrac{5\cancel{(\sin \theta + 2)}}{\cancel{(\sin \theta + 2)}(\sin \theta - 2)}$

$= \dfrac{5}{\sin \theta - 2}$

Note that the general procedure for expressing a fraction in simplest form is to:

1. Factor completely the numerator and denominator of the fraction.
2. Cancel *factors* that are common to the numerator and denominator.

EXAMPLE 3: Express in simplest form:

a. $\dfrac{c^2 - 3c - 4}{c^2 - 2c - 8}$

b. $\dfrac{\cos^2 \theta - 3 \cos \theta - 4}{\cos^2 \theta - 2 \cos \theta - 8}$

SOLUTION:

a. $\dfrac{c^2 - 3c - 4}{c^2 - 2c - 8} = \dfrac{\cancel{(c - 4)}(c + 1)}{(c + 2)\cancel{(c - 4)}}$

$= \dfrac{c + 1}{c + 2}$

b. $\dfrac{\cos^2 \theta - 3 \cos \theta - 4}{\cos^2 \theta - 2 \cos \theta - 8}$

$= \dfrac{\cancel{(\cos \theta - 4)}(\cos \theta + 1)}{(\cos \theta + 2)\cancel{(\cos \theta - 4)}}$

$= \dfrac{\cos \theta + 1}{\cos \theta + 2}$

EXAMPLE 4: Express in simplest form:

a. $\dfrac{s^2 - c^2}{c - s}$

b. $\dfrac{\sin^2 \theta - \cos^2 \theta}{\cos \theta - \sin \theta}$

SOLUTION:

a. $\dfrac{s^2 - c^2}{c - s} = \dfrac{(s + c)(s - c)}{c - s}$ [Note: $(-1)(s - c) = c - s$]

$= \dfrac{(s + c)(s - c)}{(-1)(s - c)}$

$= \dfrac{(s + c)\overset{1}{\cancel{(s - c)}}}{(-1)\underset{1}{\cancel{(s - c)}}}$

$= \dfrac{s + c}{-1}$ or $-(s + c)$

b. $\dfrac{\sin^2 \theta - \cos^2 \theta}{\cos \theta - \sin \theta} = \dfrac{(\sin \theta + \cos \theta)(\sin \theta - \cos \theta)}{\cos \theta - \sin \theta}$ [Note: $(-1)(\sin \theta - \cos \theta)$ $= (\cos \theta - \sin \theta)]$

$= \dfrac{(\sin \theta + \cos \theta)(\sin \theta - \cos \theta)}{(-1)(\sin \theta - \cos \theta)}$

$= \dfrac{(\sin \theta + \cos \theta)\overset{1}{\cancel{(\sin \theta - \cos \theta)}}}{(-1)\underset{1}{\cancel{(\sin \theta - \cos \theta)}}}$

$= \dfrac{\sin \theta + \cos \theta}{-1}$ or $-(\sin \theta + \cos \theta)$

EXAMPLE 5: Express in simplest form:

a. $\dfrac{t^3 - 5t^2 + 6t}{3t^3 - 12t}$

b. $\dfrac{\tan^3 \theta - 5 \tan^2 \theta + 6 \tan \theta}{3 \tan^3 \theta - 12 \tan \theta}$

SOLUTION:

a. $\dfrac{t^3 - 5t^2 + 6t}{3t^3 - 12t} = \dfrac{t(t^2 - 5t + 6)}{3t(t^2 - 4)}$

$= \dfrac{\overset{1}{\cancel{t}}(t - 3)\overset{1}{\cancel{(t - 2)}}}{3\underset{1}{\cancel{t}}(t + 2)\underset{1}{\cancel{(t - 2)}}}$

$= \dfrac{t - 3}{3(t + 2)}$

b. $\dfrac{\tan^3 \theta - 5 \tan^2 \theta + 6 \tan \theta}{3 \tan^3 \theta - 12 \tan \theta}$

$= \dfrac{\tan \theta(\tan^2 \theta - 5 \tan \theta + 6)}{3 \tan \theta(\tan^2 \theta - 4)}$

$= \dfrac{\overset{1}{\cancel{\tan \theta}}(\tan \theta - 3)\overset{1}{\cancel{(\tan \theta - 2)}}}{3 \underset{1}{\cancel{\tan \theta}}(\tan \theta + 2)\underset{1}{\cancel{(\tan \theta - 2)}}}$

$= \dfrac{\tan \theta - 3}{3(\tan \theta + 2)}$

$(t^2 - 3t)(t - 2)$

t^3

EXAMPLE 6: *The following are examples of fractions that cannot be simplified.* In each case a *mistake* that students commonly make is illustrated. Note that the reason behind each mistake is that the students did not understand that only *factors* that are common to the numerator and denominator may be canceled.

a. $\dfrac{\overset{1}{\cancel{4}} + s}{\underset{1}{\cancel{4}c}} = \dfrac{1+s}{c}$
   b. $\dfrac{\overset{\sin\theta}{\cancel{\sin^2\theta}} + \cos^2\theta}{\underset{1}{\cancel{\sin\theta}}} = \sin\theta + \cos^2\theta$

c. $\dfrac{\overset{1}{\cancel{\sin\theta}}}{3\,\underset{1}{\cancel{\sin\theta}} + \cos\theta} = \dfrac{1}{3+\cos\theta}$
   d. $\dfrac{3\overset{1}{\cancel{s}} + \cancel{c}}{\underset{1}{\cancel{s}} + \cancel{c}} = 3$

e. $\dfrac{\overset{1}{\cancel{\sin\theta}} + \overset{1}{\cancel{\cos\theta}} + \tan\theta}{\underset{1}{\cancel{\sin\theta}} + \underset{1}{\cancel{\cos\theta}}} = 1 + \tan\theta$
   f. $\dfrac{\overset{s}{\cancel{s^2}} + \overset{c}{\cancel{c^2}}}{\underset{1}{\cancel{s}} + \underset{1}{\cancel{c}}} = \dfrac{s+c}{2}$

Finally, it is important to remember that the denominator of a fraction cannot be zero since division by zero is undefined. Therefore in working with fractions, you should always be aware of values for the variable that will make the denominator zero.

EXAMPLE 7: For what values of s is $(5s + 10) \div (s^2 - 4)$ undefined (Example 2)?

SOLUTION:

$$\frac{5s + 10}{s^2 - 4} = \frac{5s + 10}{(s + 2)(s - 2)}$$

The denominator is zero when either factor in the denominator is zero. Therefore, $(5s + 10) \div (s^2 - 4)$ is undefined when $s = -2$ and when $s = 2$.

EXAMPLE 8: For what values of x does $(x^2 - 4) \div (x - 2) = x + 2$?

SOLUTION:

$$\frac{x^2 - 4}{x - 2} = \frac{(x + 2)\overset{1}{\cancel{(x - 2)}}}{\underset{1}{\cancel{(x - 2)}}} = x + 2$$

The two fractions are equivalent for all values of x except $x = 2$. When $x = 2$, we have

$$\frac{(2)^2 - 4}{(2) - 2} \overset{?}{=} (2) + 2$$

$$\frac{0}{0} \overset{?}{=} 4$$

undefined $\neq 4$

EXERCISES 5–1

In Exercises 1–30 express each fraction in simplest form.

1. a. $\dfrac{3s + 3c}{4s + 4c}$ $= \dfrac{3(s+c)}{4(s+c)}$

b. $\dfrac{3 \sin \theta + 3 \cos \theta}{4 \sin \theta + 4 \cos \theta}$

2. a. $\dfrac{5t - 5s}{2t - 2s}$ $=$

b. $\dfrac{5 \tan \theta - 5 \sin \theta}{2 \tan \theta - 2 \sin \theta}$

3. a. $\dfrac{4s - 8}{11s - 22}$ $= \dfrac{4(s-2)}{11(s-2)}$

b. $\dfrac{4 \sin \theta - 8}{11 \sin \theta - 22}$

4. a. $\dfrac{3t + 9}{15 + 5t}$ $= \dfrac{3(t+3)}{5(3+t)}$

b. $\dfrac{3 \tan \theta + 9}{15 + 5 \tan \theta}$

5. a. $\dfrac{3s - 5c}{6s^2 - 10sc}$ $=$

b. $\dfrac{3 \sin \theta - 5 \cos \theta}{6 \sin^2 \theta - 10 \sin \theta \cos \theta}$

6. a. $\dfrac{t^2 + tc}{t + c}$

b. $\dfrac{\tan^2 \theta + \tan \theta \cos \theta}{\tan \theta + \cos \theta}$

7. a. $\dfrac{s^2 + 2s}{s^2 + 3s}$

b. $\dfrac{\sin^2 \theta + 2 \sin \theta}{\sin^2 \theta + 3 \sin \theta}$

8. a. $\dfrac{2c + 4s}{2c - 4s}$

b. $\dfrac{2 \cos \theta + 4 \sin \theta}{2 \cos \theta - 4 \sin \theta}$

9. a. $\dfrac{s - 1}{s^2 - 1}$

b. $\dfrac{\sin \theta - 1}{\sin^2 \theta - 1}$

10. a. $\dfrac{c^2 - 4}{c - 2}$

b. $\dfrac{\cos^2 \theta - 4}{\cos \theta - 2}$

11. a. $\dfrac{4t}{12t^2 + 4t}$ $= \dfrac{4t}{4t(3t+1)} = 1$

b. $\dfrac{4 \tan \theta}{12 \tan^2 \theta + 4 \tan \theta}$

12. a. $\dfrac{20s^2 - 5s}{10s}$

b. $\dfrac{20 \sin^2 \theta - 5 \sin \theta}{10 \sin \theta}$

13. a. $\dfrac{t - 2}{2 - t}$

b. $\dfrac{\tan \theta - 2}{2 - \tan \theta}$

14. a. $\dfrac{s - t}{t - s}$

b. $\dfrac{\sin \theta - \tan \theta}{\tan \theta - \sin \theta}$

15. a. $\dfrac{1 - s}{s^2 - 1}$

b. $\dfrac{1 - \sin \theta}{\sin^2 \theta - 1}$

16. a. $\dfrac{(t - 4)^2}{4 - t}$

b. $\dfrac{(\tan \theta - 4)^2}{4 - \tan \theta}$

17. a. $\dfrac{64 - c^2}{2c + 16}$ **b.** $\dfrac{64 - \cos^2 \theta}{2 \cos \theta + 16}$

18. a. $\dfrac{6(s - c)^2}{9c - 9s}$ **b.** $\dfrac{6(\sin \theta - \cos \theta)^2}{9 \cos \theta - 9 \sin \theta}$

19. a. $\dfrac{s^2 - 3s - 4}{s^2 - 8s + 16}$ **b.** $\dfrac{\sin^2 \theta - 3 \sin \theta - 4}{\sin^2 \theta - 8 \sin \theta + 16}$

20. a. $\dfrac{t^2 - 2t - 3}{t^2 + 3t + 2}$ **b.** $\dfrac{\tan^2 \theta - 2 \tan \theta - 3}{\tan^2 \theta + 3 \tan \theta + 2}$

21. a. $\dfrac{c^2 - 8c + 16}{c^2 - 16}$ **b.** $\dfrac{\cos^2 \theta - 8 \cos \theta + 16}{\cos^2 \theta - 16}$

22. a. $\dfrac{9 - c^2}{6 - c - c^2}$ **b.** $\dfrac{9 - \cos^2 \theta}{6 - \cos \theta - \cos^2 \theta}$

23. a. $\dfrac{49 - 14t + t^2}{49 - t^2}$ **b.** $\dfrac{49 - 14 \tan \theta + \tan^2 \theta}{49 - \tan^2 \theta}$

24. a. $\dfrac{2s^2 - 5s + 2}{4s - 2}$ **b.** $\dfrac{2 \sin^2 \theta - 5 \sin \theta + 2}{4 \sin \theta - 2}$

25. a. $\dfrac{5c^2 - 5}{2c^2 - 14c + 12}$ **b.** $\dfrac{5 \cos^2 \theta - 5}{2 \cos^2 \theta - 14 \cos \theta + 12}$

26. a. $\dfrac{3s^2 - 10s + 3}{s^2 + s - 12}$ **b.** $\dfrac{3 \sin^2 \theta - 10 \sin \theta + 3}{\sin^2 \theta + \sin \theta - 12}$

27. a. $\dfrac{6t^3 - 3t^2 - 30t}{2t^2 - 4t - 16}$ **b.** $\dfrac{6 \tan^3 \theta - 3 \tan^2 \theta - 30 \tan \theta}{2 \tan^2 \theta - 4 \tan \theta - 16}$

28. a. $\dfrac{s^3 - s^2 - 6s}{s^3 + s^2 - 12s}$ **b.** $\dfrac{\sin^3 \theta - \sin^2 \theta - 6 \sin \theta}{\sin^3 \theta + \sin^2 \theta - 12 \sin \theta}$

29. a. $\dfrac{t^2 + 4t - 5}{5 - 4t - t^2}$ **b.** $\dfrac{\tan^2 \theta + 4 \tan \theta - 5}{5 - 4 \tan \theta - \tan^2 \theta}$

30. a. $\dfrac{(t - 1)(s - 2)(c - 3)}{(1 - t)(2 - s)(3 - c)}$ **b.** $\dfrac{(\tan \theta - 1)(\sin \theta - 2)(\cos \theta - 3)}{(1 - \tan \theta)(2 - \sin \theta)(3 - \cos \theta)}$

In Exercises 31–36 for what value(s) of s or $\tan \theta$ is the fraction undefined?

31. $\dfrac{4s - 8}{11s - 22}$ **34.** $\dfrac{1 - s}{s^2 - 1}$

32. $\dfrac{\tan \theta - 2}{2 - \tan \theta}$ **35.** $\dfrac{s^2 - 3s - 4}{s^2 - 8s + 16}$

33. $\dfrac{4 \tan \theta}{12 \tan^2 \theta + 4 \tan \theta}$

36. $\dfrac{\tan^2 \theta - 9}{6 - \tan \theta - \tan^2 \theta}$

37. For what values of x does $\dfrac{3x + 9}{5x + 15} = \dfrac{3}{5}$?

38. For what values of x does $\dfrac{3x - 6}{4x - 8} = \dfrac{3}{4}$?

39. For what values of x does $\dfrac{x^2 - 1}{x - 1} = x + 1$?

40. For what values of x does $\dfrac{12x^2 - 4x}{2x} = 6x - 2$?

41. Explain the following paradox:

$$a = b$$
$$a^2 = b^2 \qquad \text{square both sides of the equation}$$
$$a^2 - b^2 = 0 \qquad \text{subtract } b^2 \text{ from both sides of the equation}$$
$$(a + b)(a - b) = 0 \qquad \text{factor}$$
$$\dfrac{(a + b)\overset{1}{\cancel{(a - b)}}}{\underset{1}{\cancel{a - b}}} = \dfrac{0}{a - b} \qquad \text{divide both sides of the equation by the same number}$$
$$a + b = 0$$
$$a = -b \qquad \text{subtract } b \text{ from both sides of the equation}$$

Thus, a equals both b and the negative of b!

42. Explain the following paradox, which is obtained in solving the equation $x - 2 = 1$ in the following roundabout way:

$$x - 2 = 1$$
$$x^2 - 8x + 12 = x - 6 \qquad \text{multiply both sides of the equation by } x - 6$$
$$x^2 - 8x + 15 = x - 3 \qquad \text{add 3 to both sides of the equation}$$
$$\dfrac{(x - 5)\overset{1}{\cancel{(x - 3)}}}{\underset{1}{\cancel{x - 3}}} = \dfrac{\overset{1}{\cancel{x - 3}}}{\underset{1}{\cancel{x - 3}}} \qquad \text{divide both sides of the equation by } x - 3$$
$$x - 5 = 1$$
$$x = 6 \qquad \text{add 5 to both sides of the equation}$$

But 6 is not the correct solution since it does not satisfy the original equation $x - 2 = 1$ (that is, $6 - 2 \neq 1$).

5–2 MULTIPLICATION AND DIVISION

In arithmetic we know that the product of two or more fractions is the product of their numerators divided by the product of their denominators. For example,

$$\frac{2}{3} \cdot \frac{4}{5} = \frac{2 \cdot 4}{3 \cdot 5} = \frac{8}{15}$$

$$4 \cdot \frac{2}{7} \cdot \frac{3}{11} = \frac{4 \cdot 2 \cdot 3}{1 \cdot 7 \cdot 11} = \frac{24}{77}$$

$$\frac{2}{3} \cdot \frac{3}{4} \cdot \frac{4}{5} = \frac{2 \cdot \overset{1}{\cancel{3}} \cdot \overset{1}{\cancel{4}}}{\cancel{3} \cdot \cancel{4} \cdot 5} = \frac{2}{5}$$

Note from the last example that if the same factor appears in both the numerator and denominator, it is usually easier to cancel before you multiply.

Similarly, we will use the above procedure for multiplying algebraic and trigonometric fractions.

EXAMPLE 1: Express the following products in simplest form:

a. $\dfrac{4s}{5c} \cdot \dfrac{15c^3}{16s^2}$

b. $\dfrac{4 \sin \theta}{5 \cos \theta} \cdot \dfrac{15 \cos^3 \theta}{16 \sin^2 \theta}$

SOLUTION:

a. $\dfrac{4s}{5c} \cdot \dfrac{15c^3}{16s^2} = \dfrac{\overset{1}{\cancel{4}} \cdot \overset{3}{\cancel{15}} \overset{1}{\cancel{s}} \overset{c^2}{\cancel{c^3}}}{\underset{1}{\cancel{5}} \cdot \underset{4}{\cancel{16}} \underset{s}{\cancel{s^2}} \underset{1}{\cancel{c}}}$

$= \dfrac{3c^2}{4s}$

b. $\dfrac{4 \sin \theta}{5 \cos \theta} \cdot \dfrac{15 \cos^3 \theta}{16 \sin^2 \theta}$

$= \dfrac{\overset{1}{\cancel{4}} \cdot \overset{3}{\cancel{15}} \overset{1}{\cancel{\sin \theta}} \overset{\cos^2 \theta}{\cancel{\cos^3 \theta}}}{\underset{1}{\cancel{5}} \cdot \underset{4}{\cancel{16}} \underset{\sin \theta}{\cancel{\sin^2 \theta}} \underset{1}{\cancel{\cos \theta}}}$

$= \dfrac{3 \cos^2 \theta}{4 \sin \theta}$

EXAMPLE 2: Express the following products in simplest form:

a. $\dfrac{4c - 16}{9c} \cdot \dfrac{12c}{c^2 - 3c - 4}$

b. $\dfrac{4 \cos \theta - 16}{9 \cos \theta} \cdot \dfrac{12 \cos \theta}{\cos^2 \theta - 3 \cos \theta - 4}$

SOLUTION:

a. $\dfrac{4c - 16}{9c} \cdot \dfrac{12c}{c^2 - 3c - 4}$

$= \dfrac{4(c - 4)}{9c} \cdot \dfrac{12c}{(c - 4)(c + 1)}$

$= \dfrac{4 \cdot \overset{4}{\cancel{12}} \overset{1}{\cancel{c}} \cancel{(c - 4)}}{\underset{3}{\cancel{9}} \cdot \underset{1}{\cancel{c}} (c + 1) \underset{1}{\cancel{(c - 4)}}}$

$= \dfrac{16}{3(c + 1)}$

b. $\dfrac{4 \cos \theta - 16}{9 \cos \theta} \cdot \dfrac{12 \cos \theta}{\cos^2 \theta - 3 \cos \theta - 4}$

$= \dfrac{4(\cos \theta - 4)}{9 \cos \theta} \cdot \dfrac{12 \cos \theta}{(\cos \theta - 4)(\cos \theta + 1)}$

$= \dfrac{4 \cdot \overset{4}{\cancel{12}} \overset{1}{\cancel{\cos \theta}} \cancel{(\cos \theta - 4)}}{\underset{3}{\cancel{9}} \cdot \underset{1}{\cancel{\cos \theta}} (\cos \theta + 1) \underset{1}{\cancel{(\cos \theta - 4)}}}$

$= \dfrac{16}{3(\cos \theta + 1)}$

Note that the general procedure for multiplying two or more fractions is to:

1. Completely factor the numerator and denominator.
2. Cancel common factors.
3. Multiply the remaining factors in the numerator to obtain the numerator of the product.
4. Multiply the remaining factors in the denominator to obtain the denominator of the product.

EXAMPLE 3: Express the following products in simplest form:

a. $\dfrac{5t^2 + 15t}{4 - t^2} \cdot \dfrac{12 - 6t}{(3 + t)^2}$

b. $\dfrac{5 \tan^2 \theta + 15 \tan \theta}{4 - \tan^2 \theta} \cdot \dfrac{12 - 6 \tan \theta}{(3 + \tan \theta)^2}$

SOLUTION:

a. $\dfrac{5t^2 + 15t}{4 - t^2} \cdot \dfrac{12 - 6t}{(3 + t)^2} = \dfrac{5t(t + 3)}{(2 - t)(2 + t)} \cdot \dfrac{6(2 - t)}{(3 + t)^2}$

$$= \frac{5 \cdot 6t(t + 3)(2 - t)}{(2 + t)(3 + t)^2(2 - t)}$$

$$= \frac{30t}{(2 + t)(3 + t)}$$

b. $\dfrac{5 \tan^2 \theta + 15 \tan \theta}{4 - \tan^2 \theta} \cdot \dfrac{12 - 6 \tan \theta}{(3 + \tan \theta)^2} = \dfrac{5 \tan \theta(\tan \theta + 3)}{(2 - \tan \theta)(2 + \tan \theta)} \cdot \dfrac{6(2 - \tan \theta)}{(3 + \tan \theta)^2}$

$$= \frac{5 \cdot 6 \tan \theta(\tan \theta + 3)(2 - \tan \theta)}{(2 + \tan \theta)(3 + \tan \theta)^2(2 - \tan \theta)}$$

$$= \frac{30 \tan \theta}{(2 + \tan \theta)(3 + \tan \theta)}$$

In order to divide two fractions, we must recall that division is defined in terms of multiplication. That is, to divide a by b we multiply a by the reciprocal of b.

$$a \div b = a \cdot \frac{1}{b} \quad \text{for} \quad b \neq 0$$

The reciprocal of a fraction can be found by inverting the fraction. For example, the reciprocal of $\frac{2}{7}$ is $\frac{7}{2}$. Thus, to divide two fractions, we invert the fraction by which we are dividing to find its reciprocal, and then multiply.

EXAMPLE 4: Express the following quotients in simplest form:

a. $\dfrac{s^2 - 4}{3s} \div \dfrac{6s - 12}{5s^2}$

b. $\dfrac{\sin^2 \theta - 4}{3 \sin \theta} \div \dfrac{6 \sin \theta - 12}{5 \sin^2 \theta}$

SOLUTION:

a. $\dfrac{s^2 - 4}{3s} \div \dfrac{6s - 12}{5s^2} = \dfrac{s^2 - 4}{3s} \cdot \dfrac{5s^2}{6s - 12}$

$$= \dfrac{(s + 2)(s - 2)}{3s} \cdot \dfrac{5s^2}{6(s - 2)}$$

$$= \dfrac{5s^{\overset{s}{2}}(s + 2)(s - 2)}{3 \cdot 6s(s - 2)}$$

$$= \dfrac{5s(s + 2)}{18}$$

b. $\dfrac{\sin^2 \theta - 4}{3 \sin \theta} \div \dfrac{6 \sin \theta - 12}{5 \sin^2 \theta} = \dfrac{\sin^2 \theta - 4}{3 \sin \theta} \cdot \dfrac{5 \sin^2 \theta}{6 \sin \theta - 12}$

$$= \dfrac{(\sin \theta + 2)(\sin \theta - 2)}{3 \sin \theta} \cdot \dfrac{5 \sin^2 \theta}{6(\sin \theta - 2)}$$

$$= \dfrac{5 \sin^2 \theta(\sin \theta + 2)(\sin \theta - 2)}{3 \cdot 6 \sin \theta(\sin \theta - 2)}$$

$$= \dfrac{5 \sin \theta(\sin \theta + 2)}{18}$$

EXAMPLE 5: Express the following quotients in simplest form:

a. $\dfrac{t - 4}{t + 4} \div \dfrac{4 - t}{4 + t}$ **b.** $\dfrac{\tan \theta - 4}{\tan \theta + 4} \div \dfrac{4 - \tan \theta}{4 + \tan \theta}$

SOLUTION:

a. $\dfrac{t - 4}{t + 4} \div \dfrac{4 - t}{4 + t} = \dfrac{t - 4}{t + 4} \cdot \dfrac{4 + t}{4 - t}$

$$= \dfrac{(4 + t)(t - 4)}{(t + 4)(4 - t)}$$

$$= -1$$

b. $\dfrac{\tan \theta - 4}{\tan \theta + 4} \div \dfrac{4 - \tan \theta}{4 + \tan \theta}$

$$= \dfrac{\tan \theta - 4}{\tan \theta + 4} \cdot \dfrac{4 + \tan \theta}{4 - \tan \theta}$$

$$= \dfrac{(4 + \tan \theta)(\tan \theta - 4)}{(\tan \theta + 4)(4 - \tan \theta)}$$

$$= -1$$

EXERCISES 5-2

Express each product or quotient in simplest form.

1. $\frac{3}{7} \cdot \frac{5}{8}$

2. $\frac{4}{5} \cdot \frac{15}{4}$

3. a. $\dfrac{t}{15} \cdot 10$ **b.** $\dfrac{\tan \theta}{15} \cdot 10$

4. a. $3s^2 \cdot \dfrac{1}{6s^3}$

b. $3\sin^2\theta \cdot \dfrac{1}{6\sin^3\theta}$

5. a. $\dfrac{2s}{5c} \cdot \dfrac{5c^2}{8s^3}$

b. $\dfrac{2\sin\theta}{5\cos\theta} \cdot \dfrac{5\cos^2\theta}{8\sin^3\theta}$

6. a. $\dfrac{14t}{10c} \cdot \dfrac{15c}{35t}$

b. $\dfrac{14\tan\theta}{10\cos\theta} \cdot \dfrac{15\cos\theta}{35\tan\theta}$

7. a. $\dfrac{30s^2c^2}{7t} \cdot \dfrac{21t^3}{5sc}$

b. $\dfrac{30\sin^2\theta\cos^2\theta}{7\tan\theta} \cdot \dfrac{21\tan^3\theta}{5\sin\theta\cos\theta}$

8. a. $\dfrac{24tc}{5s} \cdot \dfrac{10s^2}{c}$

b. $\dfrac{24\tan\theta\cos\theta}{5\sin\theta} \cdot \dfrac{10\sin^2\theta}{\cos\theta}$

9. a. $\dfrac{5c}{c^2-9} \cdot \dfrac{6c+18}{15c^2}$

b. $\dfrac{5\cos\theta}{\cos^2\theta-9} \cdot \dfrac{6\cos\theta+18}{15\cos^2\theta}$

10. a. $\dfrac{t^2-t-12}{t^2} \cdot \dfrac{t}{t-4}$

b. $\dfrac{\tan^2\theta-\tan\theta-12}{\tan^2\theta} \cdot \dfrac{\tan\theta}{\tan\theta-4}$

11. a. $\dfrac{s^2+2s}{6t} \cdot \dfrac{t^2}{s^2-4}$

b. $\dfrac{\sin^2\theta+2\sin\theta}{6\tan\theta} \cdot \dfrac{\tan^2\theta}{\sin^2\theta-4}$

12. a. $\dfrac{s^2-st}{t} \cdot \dfrac{t^2}{3s-3t}$

b. $\dfrac{\sin\theta-\sin\theta\tan\theta}{\tan\theta} \cdot \dfrac{\tan^2\theta}{3\sin\theta-3\tan\theta}$

13. a. $\dfrac{(c-1)^2}{(c+2)^2} \cdot \dfrac{c^2-4}{c^2-1}$

b. $\dfrac{(\cos\theta-1)^2}{(\cos\theta+2)^2} \cdot \dfrac{\cos^2\theta-4}{\cos^2\theta-1}$

14. a. $(t+4)^2 \cdot \dfrac{t}{16-t^2}$

b. $(\tan\theta+4)^2 \cdot \dfrac{\tan\theta}{16-\tan^2\theta}$

15. a. $\dfrac{s^2-3s+2}{(s-2)^2} \cdot \dfrac{s^2-5s+4}{(s-1)^2}$

b. $\dfrac{\sin^2\theta-3\sin\theta+2}{(\sin\theta-2)^2} \cdot \dfrac{\sin^2\theta-5\sin\theta+4}{(\sin\theta-1)^2}$

16. a. $\dfrac{3c+6}{c^2+4c} \cdot \dfrac{c^2+2c}{3c^2-12}$

b. $\dfrac{3\cos\theta+6}{\cos^2\theta+4\cos\theta} \cdot \dfrac{\cos^2\theta+2\cos\theta}{3\cos^2\theta-12}$

17. a. $\dfrac{t^2+3t-4}{t^2-3t-4} \cdot \dfrac{t^2-5t+4}{t^2+5t+4}$

b. $\dfrac{\tan^2\theta+3\tan\theta-4}{\tan^2\theta-3\tan\theta-4} \cdot \dfrac{\tan^2\theta-5\tan\theta+4}{\tan^2\theta+5\tan\theta+4}$

18. a. $\dfrac{2t^2-7t+6}{t^2-t-6} \cdot \dfrac{5t-15}{t^2-4}$

b. $\dfrac{2\tan^2\theta-7\tan\theta+6}{\tan^2\theta-\tan\theta-6} \cdot \dfrac{5\tan\theta-15}{\tan^2\theta-4}$

19. a. $\dfrac{4}{7-s} \cdot (s^2-5s-14)$

b. $\dfrac{4}{7-\sin\theta} \cdot (\sin^2\theta-5\sin\theta-14)$

20. a. $\dfrac{4-s^2}{s^2-9} \cdot \dfrac{9+3s}{8s-16}$

b. $\dfrac{4-\sin^2\theta}{\sin^2\theta-9} \cdot \dfrac{9+3\sin\theta}{8\sin\theta-16}$

21. $\frac{2}{5} \div \frac{4}{9}$

22. $\frac{7}{4} \div \frac{21}{16}$

23. a. $\dfrac{5}{s^2} \div \dfrac{5}{s}$ **b.** $\dfrac{5}{\sin^2 \theta} \div \dfrac{5}{\sin \theta}$

24. a. $\dfrac{21s}{20c} \div \dfrac{3s}{5c}$ **b.** $\dfrac{21 \sin \theta}{20 \cos \theta} \div \dfrac{3 \sin \theta}{5 \cos \theta}$

25. a. $\dfrac{sc}{t} \div \dfrac{s^2c}{t}$ **b.** $\dfrac{\sin \theta \cos \theta}{\tan \theta} \div \dfrac{\sin^2 \theta \cos \theta}{\tan \theta}$

26. a. $\dfrac{st^2}{2c} \div \dfrac{s^2t}{10}$ **b.** $\dfrac{\sin \theta \tan^2 \theta}{2 \cos \theta} \div \dfrac{\sin^2 \theta \tan \theta}{10}$

27. a. $\dfrac{4s - 6}{5} \div \dfrac{6s - 9}{25}$ **b.** $\dfrac{4 \sin \theta - 6}{5} \div \dfrac{6 \sin \theta - 9}{25}$

28. a. $\dfrac{t - 4}{t} \div \dfrac{t^2 - 16}{t^2}$ **b.** $\dfrac{\tan \theta - 4}{\tan \theta} \div \dfrac{\tan^2 \theta - 16}{\tan^2 \theta}$

29. a. $\dfrac{c^2 + c}{c - 3} \div (c + 2)$ **b.** $\dfrac{\cos^2 \theta + \cos \theta - 2}{\cos \theta - 3} \div (\cos \theta + 2)$

30. a. $\dfrac{4t^2 - 9}{4t - 9} \div (2t - 3)$ **b.** $\dfrac{4 \tan^2 \theta - 9}{4 \tan \theta - 9} \div (2 \tan \theta - 3)$

31. a. $\dfrac{(s - 1)^2}{3s^2} \div \dfrac{s^2 - 1}{4s}$ **b.** $\dfrac{(\sin \theta - 1)^2}{3 \sin^2 \theta} \div \dfrac{\sin^2 \theta - 1}{4 \sin \theta}$

32. a. $\dfrac{4s + 18}{14s + 35} \div \dfrac{4s^2 - 25}{6 + 3s}$ **b.** $\dfrac{4 \sin \theta + 18}{14 \sin \theta + 35} \div \dfrac{4 \sin^2 \theta - 25}{6 + 3 \sin \theta}$

33. a. $\dfrac{t^4 - 1}{7t + 7} \div \dfrac{t^3 + t}{t^2 - 1}$ **b.** $\dfrac{\tan^4 \theta - 1}{7 \tan \theta + 7} \div \dfrac{\tan^3 \theta + \tan \theta}{\tan^2 \theta - 1}$

34. a. $\dfrac{3s^2 + 6s - 24}{s^2 - 7s + 10} \div \dfrac{3s^2 + 4s}{s^3 - 5s^2}$ **b.** $\dfrac{3 \sin^2 \theta + 6 \sin \theta - 24}{\sin^2 \theta - 7 \sin \theta + 10} \div \dfrac{3 \sin^2 \theta + 4 \sin \theta}{\sin^3 \theta - 5 \sin^2 \theta}$

35. a. $\dfrac{-4}{3 - s} \div \dfrac{12}{s - 3}$ **b.** $\dfrac{-4}{3 - \sin \theta} \div \dfrac{12}{\sin \theta - 3}$

36. a. $\dfrac{s - t}{c} \div \dfrac{t - s}{c}$ **b.** $\dfrac{\sin \theta - \tan \theta}{\cos \theta} \div \dfrac{\tan \theta - \sin \theta}{\cos \theta}$

37. a. $\dfrac{1 - s}{c^2 + c} \div \dfrac{s^2 - 1}{c^2 - 1}$ **b.** $\dfrac{1 - \sin \theta}{\cos^2 \theta + \cos \theta} \div \dfrac{\sin^2 \theta - 1}{\cos^2 \theta - 1}$

38. a. $\dfrac{t^2 - 7t + 10}{5t - 25} \div \dfrac{4 - 2t}{25 - t^2}$ **b.** $\dfrac{\tan^2 \theta - 7 \tan \theta + 10}{5 \tan \theta - 25} \div \dfrac{4 - 2 \tan \theta}{25 - \tan^2 \theta}$

39. a. $\dfrac{t-1}{2-2t} \div \left(\dfrac{t-1}{3-3t} \div \dfrac{t-1}{4-4t} \right)$ **b.** $\dfrac{\tan\theta - 1}{2 - 2\tan\theta} \div \left(\dfrac{\tan\theta - 1}{3 - 3\tan\theta} \div \dfrac{\tan\theta - 1}{4 - 4\tan\theta} \right)$

40. a. $\dfrac{s-t}{4c} \div \left(\dfrac{s-t}{2c} \cdot \dfrac{s-t}{c} \right)$

b. $\dfrac{\sin\theta - \tan\theta}{4\cos\theta} \div \left(\dfrac{\sin\theta - \tan\theta}{2\cos\theta} \cdot \dfrac{\sin\theta - \tan\theta}{\cos\theta} \right)$

5–3 LEAST COMMON MULTIPLE

For the remainder of our work with fractions, it is important to determine a number that is exactly divisible by the denominators in a given problem. For convenience, we generally try to find the smallest such number and it is called the *least common multiple*, or LCM.

> LEAST COMMON MULTIPLE: The least common multiple of two or more integers is the smallest positive integer that is exactly divisible by each of the given integers.

For example:

1. The LCM of 2 and 3 is 6 since 6 is the smallest positive integer that is exactly divisible by 2 and 3.
2. The LCM of 6 and 12 is 12 since 12 is the smallest positive integer that is exactly divisible by 6 and 12.
3. The LCM of 12 and 15 is 60 since 60 is the smallest positive integer that is exactly divisible by 12 and 15.

To calculate the least common multiple, we must express each of the given integers as the product of factors that are prime numbers. A prime number is a positive integer that is exactly divisible only by itself and 1; the first 10 prime numbers are 2, 3, 5, 7, 11, 13, 17, 19, 23, and 29. In most cases, these are the only prime numbers we will need. The following examples indicate how to express an integer as the product of its prime factors.

EXAMPLE 1: Express 36 as the product of its prime factors.

SOLUTION: Divide the smallest prime, 2, into 36.

$$\begin{array}{r} 18 \\ 2\overline{)\,36} \end{array}$$

Divide the result (18) by 2 again.

$$\begin{array}{r} 9 \\ 2\overline{)\,18} \end{array}$$

Since the result (9) is not exactly divisible by 2, try the next highest prime, 3.

$$\begin{array}{r} 3 \\ 3\overline{)\,9} \end{array}$$

Stop when the result of the division is a prime number, for we can now express the original number in terms of its prime factors.

$$36 = 2 \cdot 18$$

$$= 2 \cdot \overbrace{2 \cdot 9}$$

$$= 2 \cdot 2 \cdot \overbrace{3 \cdot 3}$$

Thus,

$$36 = 2 \cdot 2 \cdot 3 \cdot 3 \quad \text{or} \quad 2^2 \cdot 3^2$$

EXAMPLE 2: Express 350 as the product of its prime factors.

SOLUTION: Perform successive divisions until the result of the division is a prime number.

$$\begin{array}{r} 2\overline{)\,350} \\ 5\overline{)\,175} \\ 5\overline{)\,\;35} \\ 7 \end{array}$$ divide 350 by 2
   divide the result (175) by 5
   divide the result (35) by 5

Thus,

$$350 = 2 \cdot 5 \cdot 5 \cdot 7 \quad \text{or} \quad 2 \cdot 5^2 \cdot 7$$

EXAMPLE 3: Express 114 as the product of its prime factors.

SOLUTION: Perform successive divisions until the result of the division is a prime number.

$$\begin{array}{r} 2\overline{)\,114} \\ 3\overline{)\,\;57} \\ 19 \end{array}$$

Thus,

$$114 = 2 \cdot 3 \cdot 19$$

We can now determine the least common multiple of two or more integers since the LCM is the product of all the different prime factors with each prime raised to the highest power to which it appears in any one of the products. The LCM is used in addition and subtraction of fractions.

EXAMPLE 4: Find the LCM of 10, 12, and 18.

SOLUTION: Express each number as the product of its prime factors.

$$10 = 2 \cdot 5$$
$$12 = 2^2 \cdot 3$$
$$18 = 2 \cdot 3^2$$

The LCM will contain the factors 2, 3 and 5; the highest exponent of 2 is 2, the highest exponent of 3 is 2, and the highest exponent of 5 is 1. Thus,

$$\text{LCM} = 2^2 \cdot 3^2 \cdot 5 \quad \text{or} \quad 180$$

Note that 180 is the smallest number that is exactly divisible by all three numbers, 10, 12, and 18.

EXAMPLE 5: Find the LCM of 21, 27, and 49.

SOLUTION: Express each number as the product of its prime factors.

$$21 = 3 \cdot 7$$
$$27 = 3^3$$
$$49 = 7^2$$

The LCM will contain the factors 3 and 7; the highest exponent of 3 is 3, and the highest exponent of 7 is 2. Thus,

$$\text{LCM} = 3^3 \cdot 7^2 \quad \text{or} \quad 1323$$

Note from this example that although the LCM can often be found through trial and error, in some cases that method would be inappropriate.

The least common multiple of algebraic and trigonometric expressions is handled in a similar way.

EXAMPLE 6: Find the least common multiple of $6s^2t$, $9st$, and $18st^3$.

SOLUTION: Write each expression as the product of its prime factors.

$$6s^2t = 2 \cdot 3 \cdot s^2 \cdot t$$
$$9st = 3^2 \cdot s \cdot t$$
$$18st^3 = 2 \cdot 3^2 \cdot s \cdot t^3$$

The LCM will contain the factors 2, 3, s, and t; the highest exponent of 2 is 1, the highest exponent of 3 is 2, the highest exponent of s is 2, and the highest exponent of t is 3. Thus,

$$\text{LCM} = 2 \cdot 3^2 \cdot s^2 \cdot t^3 \quad \text{or} \quad 18s^2t^3$$

EXAMPLE 7: Find the least common multiple of $10 \sin^2 \theta \tan^3 \theta$, $15 \sin^3 \theta \tan^5 \theta$, and $20 \sin \theta \tan^4 \theta$.

SOLUTION : Write the expression as the product of its prime factors.

$$10 \sin^2 \theta \tan^3 \theta = 2 \cdot 5 \sin^2 \theta \tan^3 \theta$$
$$15 \sin^3 \theta \tan^5 \theta = 3 \cdot 5 \sin^3 \theta \tan^5 \theta$$
$$20 \sin \theta \tan^4 \theta = 2^2 \cdot 5 \sin \theta \tan^4 \theta$$

The LCM will contain the factors 2, 3, 5, $\sin \theta$, and $\tan \theta$; the highest exponent of 2 is 2, the highest exponent of 3 is 1, the highest exponent of 5 is 1, the highest exponent of $\sin \theta$ is 3, and the highest exponent of $\tan \theta$ is 5. Thus,

$$\text{LCM} = 2^2 \cdot 3 \cdot 5 \cdot \sin^3 \theta \cdot \tan^5 \theta \quad \text{or} \quad 60 \sin^3 \theta \tan^5 \theta$$

EXAMPLE 8: Find the least common multiple of $2c + 4$ and $c^2 - 4$.

SOLUTION : Factor each expression completely.

$$2c + 4 = 2(c + 2)$$
$$c^2 - 4 = (c + 2)(c - 2)$$

The LCM will contain the factors 2, $c + 2$, and $c - 2$; the highest exponent of 2 is 1, the highest exponent of $c + 2$ is 1, and the highest exponent of $c - 2$ is 1. Thus,

$$\text{LCM} = 2(c + 2)(c - 2)$$

EXAMPLE 9: Find the least common multiple of $\tan^2 \theta - 1$ and $\tan^2 \theta + 2 \tan \theta + 1$.

SOLUTION : Factor each expression completely.

$$\tan^2 \theta - 1 = (\tan \theta + 1)(\tan \theta - 1)$$
$$\tan^2 \theta + 2 \tan \theta + 1 = (\tan \theta + 1)^2$$

The LCM will contain the factors $\tan \theta + 1$ and $\tan \theta - 1$; the highest exponent of $\tan \theta + 1$ is 2, and the highest exponent of $\tan \theta - 1$ is 1. Thus,

$$\text{LCM} = (\tan \theta + 1)^2(\tan \theta - 1)$$

EXAMPLE 10: Find the least common multiple of $s + 3$ and $s + 4$.

SOLUTION : Factor each expression completely.

$$s + 3 \qquad s + 4$$

cannot be factored.

The LCM will contain the factors $s + 3$ and $s + 4$ each with an exponent of 1. Thus,

$$\text{LCM} = (s + 3)(s + 4)$$

Note from this example that if the expressions do not contain a common factor, the LCM is the product of the given expressions. A common mistake for this problem is to answer that the LCM $= s + 12$. This answer is wrong because neither $(s + 4)$ or $(s + 3)$ is a divisor of $s + 12$.

EXERCISES 5–3

In Exercises 1–10 find the least common multiple of the collection of numbers.

1. 16 and 20   **2.** 15 and 24

3. 28 and 49   **4.** 24 and 32

5. 2, 3, and 19   **6.** 5, 7, and 11

7. 6, 15, and 21   **8.** 8, 18, and 30

9. 60, 80, and 108   **10.** 38, 66, and 209

11. Find 10 prime numbers that are greater than 29.

In Exercises 12–30 find the LCM of the expression.

12. a. $3s$ and $12s^2$ **b.** $3 \sin \theta$ and $12 \sin^2 \theta$

13. a. $4t$ and $4t^3$ **b.** $4 \tan \theta$ and $4 \tan^3 \theta$

14. a. $4s^2t$ and $6st^2$ **b.** $4 \sin^2 \theta \tan \theta$ and $6 \sin \theta \tan^2 \theta$

15. a. $6ct$ and $9ct$ **b.** $6 \cos \theta \tan \theta$ and $9 \cos \theta \tan \theta$

16. a. $2s^3tc$, $4st^2c$, and $5s^2t^2c$ **b.** $2 \sin^3 \theta \tan \theta \cos \theta$, $4 \sin \theta \tan^2 \theta \cos \theta$, and $5 \sin^2 \theta \tan^2 \theta \cos \theta$

17. a. $6s^4tc^3$, $15sc^2$, and $18t$ **b.** $6 \sin^4 \theta \tan \theta \cos^3 \theta$, $15 \sin \theta \cos^2 \theta$, and $18 \tan \theta$

18. a. $t(t-1)^2$ and $(t+1)(t-1)$ **b.** $\tan \theta(\tan \theta - 1)^2$ and $(\tan \theta + 1)(\tan \theta - 1)$

19. a. $c^2(c+1)^3$ and $c^3(c+1)^2$ **b.** $\cos^2 \theta(\cos \theta + 1)^3$ and $\cos^3 \theta(\cos \theta + 1)^2$

20. a. $t^2 - 9$ and $t + 3$ **b.** $\tan^2 \theta - 9$ and $\tan \theta + 3$

21. a. $s^2 + 4s + 4$ and $4s + 8$ **b.** $\sin^2 \theta + 4 \sin \theta + 4$ and $4 \sin \theta + 8$

22. a. $s - 2$ and $s - 4$ **b.** $\sin \theta - 2$ and $\sin \theta - 4$

23. a. $t + c$ and $t - c$ **b.** $\tan \theta + \cos \theta$ and $\tan \theta - \cos \theta$

24. a. s and $c + 1$ **b.** $\sin \theta$ and $\cos \theta + 1$

25. a. $t - 2$ and $2 - t$ **b.** $\tan \theta - 2$ and $2 - \tan \theta$

26. a. $t^2 + 4t - 5$ and $t^2 - 1$ **b.** $\tan^2 \theta + 4 \tan \theta - 5$ and $\tan^2 \theta - 1$

27. a. $2s - 1$ and $2s^2 + s - 1$ **b.** $2 \sin \theta - 1$ and $2 \sin^2 \theta + \sin \theta - 1$

28. a. $c^2 - c - 12$ and $c^2 + 2c - 3$ **b.** $\cos^2 \theta - \cos \theta - 12$ and $\cos^2 \theta + 2 \cos \theta - 3$

29. a. $4t^2 - 9$, $4t^2 - 12t + 9$, and $12t + 18$ **b.** $4 \tan^2 \theta - 9$, $4 \tan^2 \theta - 12 \tan \theta + 9$, and $12 \tan \theta + 18$

30. a. $5(s^2 - c^2)$, $s^2 + 3sc - 4c^2$, and $21(s - c)^2$

b. $5(\sin^2 \theta - \cos^2 \theta)$, $\sin^2 \theta + 3 \sin \theta \cos \theta - 4 \cos^2 \theta$, and $21(\sin \theta - \cos \theta)^2$

5–4 ADDITION AND SUBTRACTION

In arithmetic we know that the sum (or difference) of two or more fractions which have the same denominator is given by the sum (or difference) of the numerators divided by the common denominator. For example:

$$\frac{1}{5} + \frac{2}{5} = \frac{1+2}{5} = \frac{3}{5}$$

$$\frac{4}{7} - \frac{9}{7} = \frac{4-9}{7} = \frac{-5}{7}$$

$$\frac{8}{9} + \frac{2}{9} + \frac{5}{9} = \frac{8+2+5}{9} = \frac{15}{9} \quad \text{or} \quad \frac{5}{3}$$

The justification for this rule of fractions is the distributive property as illustrated in the following example:

$$\frac{8}{9} + \frac{2}{9} + \frac{5}{9} = 8 \cdot \frac{1}{9} + 2 \cdot \frac{1}{9} + 5 \cdot \frac{1}{9} \qquad \text{definition of division}$$

$$= (8 + 2 + 5)\frac{1}{9} \qquad \text{distributive property}$$

$$= \frac{(8 + 2 + 5)}{9} \qquad \text{definition of division}$$

$$= \frac{15}{9} = \frac{5}{3}$$

We can add or subtract fractions that contain algebraic or trigonometric expressions in a similar way.

EXAMPLE 1: Write as a single fraction in simplest form:

a. $\dfrac{2}{s} + \dfrac{5}{s}$

b. $\dfrac{2}{\sin \theta} + \dfrac{5}{\sin \theta}$

SOLUTION:

a. $\dfrac{2}{s} + \dfrac{5}{s} = \dfrac{2 + 5}{s}$

$\qquad = \dfrac{7}{s}$

b. $\dfrac{2}{\sin \theta} + \dfrac{5}{\sin \theta} = \dfrac{2 + 5}{\sin \theta}$

$\qquad = \dfrac{7}{\sin \theta}$

EXAMPLE 2: Write as a single fraction in simplest form:

a. $\dfrac{2t + 1}{3} - \dfrac{t - 2}{3}.$

b. $\dfrac{2 \tan \theta + 1}{3} - \dfrac{\tan \theta - 2}{3}.$

SOLUTION:

a. $\dfrac{2t + 1}{3} - \dfrac{t - 2}{3}$

$\qquad = \dfrac{(2t + 1) - (t - 2)}{3}$

$\qquad = \dfrac{2t + 1 - t + 2}{3}$

$\qquad = \dfrac{t + 3}{3}$

b. $\dfrac{2 \tan \theta + 1}{3} - \dfrac{\tan \theta - 2}{3}$

$\qquad = \dfrac{(2 \tan \theta + 1) - (\tan \theta - 2)}{3}$

$\qquad = \dfrac{2 \tan \theta + 1 - \tan \theta + 2}{3}$

$\qquad = \dfrac{\tan \theta + 3}{3}$

Note: Be careful in this and other subtraction problems to remove correctly the parentheses in an expression like $-(t - 2)$.

EXAMPLE 3: Write as a single fraction in simplest form:

a. $\dfrac{4}{s-2} + \dfrac{3}{2-s}$

b. $\dfrac{4}{\sin\theta - 2} + \dfrac{3}{2 - \sin\theta}$

SOLUTION:

a. $\dfrac{4}{s-2} + \dfrac{3}{2-s}$

$= \dfrac{4}{s-2} + \dfrac{3(-1)}{(2-s)(-1)}$

$= \dfrac{4}{s-2} + \dfrac{-3}{s-2}$

$= \dfrac{4 + (-3)}{s-2}$

$= \dfrac{1}{s-2}$

b. $\dfrac{4}{\sin\theta - 2} + \dfrac{3}{2 - \sin\theta}$

$= \dfrac{4}{\sin\theta - 2} + \dfrac{3(-1)}{(2 - \sin\theta)(-1)}$

$= \dfrac{4}{\sin\theta - 2} + \dfrac{-3}{\sin\theta - 2}$

$= \dfrac{4 + (-3)}{\sin\theta - 2}$

$= \dfrac{1}{\sin\theta - 2}$

In arithmetic, when fractions have different denominators we must change them into equivalent fractions with the same denominator before we can add them. For example, let us add the fractions $\frac{1}{4}$ and $\frac{1}{6}$.

$$\tfrac{1}{4} + \tfrac{1}{6}$$

The least common denominator (LCD) is the smallest integer that is exactly divisible by 4 and 6, or equivalently, the least common denominator is the least common multiple (12) of the denominators (4 and 6). Therefore, change each fraction into an equivalent fraction with a denominator of 12.

$$\frac{1}{4} + \frac{1}{6} = \frac{1\cdot 3}{4\cdot 3} + \frac{1\cdot 2}{6\cdot 2}$$

$$= \frac{3}{12} + \frac{2}{12} = \frac{5}{12}$$

We can add or subtract fractions that contain algebraic and trigonometric expressions with different denominators in a similar way.

EXAMPLE 4: Write as a single fraction in simplest form:

a. $\dfrac{5}{3s^2} + \dfrac{7}{6s}$

b. $\dfrac{5}{3\sin^2\theta} + \dfrac{7}{6\sin\theta}$

SOLUTION:

a. $\dfrac{5}{3s^2} + \dfrac{7}{6s}$ (LCD: $6s^2$)

$$= \dfrac{5(2)}{3s^2(2)} + \dfrac{7(s)}{6s(s)}$$

$$= \dfrac{10}{6s^2} + \dfrac{7s}{6s^2}$$

$$= \dfrac{10 + 7s}{6s^2}$$

b. $\dfrac{5}{3\sin^2\theta} + \dfrac{7}{6\sin\theta}$ (LCD: $6\sin^2\theta$)

$$= \dfrac{5(2)}{3\sin^2\theta(2)} + \dfrac{7(\sin\theta)}{6\sin\theta(\sin\theta)}$$

$$= \dfrac{10}{6\sin^2\theta} + \dfrac{7\sin\theta}{6\sin^2\theta}$$

$$= \dfrac{10 + 7\sin\theta}{6\sin^2\theta}$$

EXAMPLE 5: Write as a single fraction in simplest form:

a. $\dfrac{1}{s^2 - 4} - \dfrac{2}{s - 2}$

b. $\dfrac{1}{\sin^2\theta - 4} - \dfrac{2}{\sin\theta - 2}$

SOLUTION:

a. $\dfrac{1}{s^2 - 4} - \dfrac{2}{s - 2} = \dfrac{1}{(s + 2)(s - 2)} - \dfrac{2}{s - 2}$ $\big[$LCD: $(s + 2)(s - 2)\big]$

$$= \dfrac{1}{(s + 2)(s - 2)} - \dfrac{2(s + 2)}{(s - 2)(s + 2)}$$

$$= \dfrac{1 - 2(s + 2)}{(s + 2)(s - 2)}$$

$$= \dfrac{-2s - 3}{(s + 2)(s - 2)}$$

b. $\dfrac{1}{\sin^2\theta - 4} - \dfrac{2}{\sin\theta - 2}$

$$= \dfrac{1}{(\sin\theta + 2)(\sin\theta - 2)} - \dfrac{2}{\sin\theta - 2}$$ $\big[$LCD: $(\sin\theta + 2)(\sin\theta - 2)\big]$

$$= \dfrac{1}{(\sin\theta + 2)(\sin\theta - 2)} - \dfrac{2(\sin\theta + 2)}{(\sin\theta - 2)(\sin\theta + 2)}$$

$$= \dfrac{1 - 2(\sin\theta + 2)}{(\sin\theta + 2)(\sin\theta - 2)}$$

$$= \dfrac{-2\sin\theta - 3}{(\sin\theta + 2)(\sin\theta - 2)}$$

Note that the general procedure for adding or subtracting two or more fractions is to:

1. Completely factor each denominator.
2. Find the least common multiple of the denominators (LCD).
3. For each fraction, obtain an equivalent fraction by multiplying the top and bottom of the fraction by the factors of the LCD that are not contained in the denominator of that fraction.
4. Add or subtract the numerators and divide this result by the common denominator.

EXAMPLE 6: Write as a single fraction in simplest form:

a. $\dfrac{t}{t+1} + \dfrac{4}{t+2}$
b. $\dfrac{\tan\theta}{\tan\theta + 1} + \dfrac{4}{\tan\theta + 2}$

SOLUTION:

a. $\dfrac{t}{t+1} + \dfrac{4}{t+2}$ $[\text{LCD: } (t+1)(t+2)]$

$$= \frac{t(t+2)}{(t+1)(t+2)} + \frac{4(t+1)}{(t+2)(t+1)}$$

$$= \frac{t(t+2) + 4(t+1)}{(t+2)(t+1)}$$

$$= \frac{t^2 + 2t + 4t + 4}{(t+2)(t+1)}$$

$$= \frac{t^2 + 6t + 4}{(t+2)(t+1)}$$

b. $\dfrac{\tan\theta}{\tan\theta + 1} + \dfrac{4}{\tan\theta + 2}$ $[\text{LCD: } (\tan\theta + 1)(\tan\theta + 2)]$

$$= \frac{\tan\theta(\tan\theta + 2)}{(\tan\theta + 1)(\tan\theta + 2)} + \frac{4(\tan\theta + 1)}{(\tan\theta + 2)(\tan\theta + 1)}$$

$$= \frac{\tan\theta(\tan\theta + 2) + 4(\tan\theta + 1)}{(\tan\theta + 2)(\tan\theta + 1)}$$

$$= \frac{\tan^2\theta + 2\tan\theta + 4\tan\theta + 4}{(\tan\theta + 2)(\tan\theta + 1)}$$

$$= \frac{\tan^2\theta + 6\tan\theta + 4}{(\tan\theta + 2)(\tan\theta + 1)}$$

EXAMPLE 7: Write as a single fraction in simplest form:

a. $\dfrac{2c+3}{c^2+c} - \dfrac{c}{c^2-1}$
b. $\dfrac{2\cos\theta + 3}{\cos^2\theta + \cos\theta} - \dfrac{\cos\theta}{\cos^2\theta - 1}$

SOLUTION:

a. $\dfrac{2c + 3}{c^2 + c} - \dfrac{c}{c^2 - 1} = \dfrac{2c + 3}{c(c + 1)} - \dfrac{c}{(c + 1)(c - 1)}$ [LCD: $c(c + 1)(c - 1)$]

$$= \dfrac{(2c + 3)(c - 1)}{c(c + 1)(c - 1)} - \dfrac{c(c)}{(c + 1)(c - 1)(c)}$$

$$= \dfrac{(2c + 3)(c - 1) - c \cdot c}{c(c + 1)(c - 1)}$$

$$= \dfrac{2c^2 + c - 3 - c^2}{c(c + 1)(c - 1)}$$

$$= \dfrac{c^2 + c - 3}{c(c + 1)(c - 1)}$$

b. $\dfrac{2 \cos \theta + 3}{\cos^2 \theta + \cos \theta} - \dfrac{\cos \theta}{\cos^2 \theta - 1}$

$$= \dfrac{2 \cos \theta + 3}{\cos \theta(\cos \theta + 1)} - \dfrac{\cos \theta}{(\cos \theta + 1)(\cos \theta - 1)}$$

$$[\text{LCD: } \cos \theta(\cos \theta + 1)(\cos \theta - 1)]$$

$$= \dfrac{(2 \cos \theta + 3)(\cos \theta - 1)}{\cos \theta(\cos \theta + 1)(\cos \theta - 1)} - \dfrac{\cos \theta \cos \theta}{(\cos \theta + 1)(\cos \theta - 1)\cos \theta}$$

$$= \dfrac{(2 \cos \theta + 3)(\cos \theta - 1) - \cos \theta \cos \theta}{\cos \theta(\cos \theta + 1)(\cos \theta - 1)}$$

$$= \dfrac{2 \cos^2 \theta + \cos \theta - 3 - \cos^2 \theta}{\cos \theta(\cos \theta + 1)(\cos \theta - 1)}$$

$$= \dfrac{\cos^2 \theta + \cos \theta - 3}{\cos \theta(\cos \theta + 1)(\cos \theta - 1)}$$

EXERCISES 5–4

Combine each expression into a single fraction in simplest form.

1. $\frac{2}{11} + \frac{5}{11}$

2. $\frac{4}{9} - \frac{7}{9}$

3. a. $\dfrac{5}{s} - \dfrac{11}{s}$ b. $\dfrac{5}{\sin \theta} - \dfrac{11}{\sin \theta}$

4. a. $\dfrac{6}{7t} + \dfrac{1}{7t}$ b. $\dfrac{6}{7 \tan \theta} + \dfrac{1}{7 \tan \theta}$

5. a. $\dfrac{4s}{tc} + \dfrac{2s}{tc} - \dfrac{3s}{tc}$ b. $\dfrac{4 \sin \theta}{\tan \theta \cos \theta} + \dfrac{2 \sin \theta}{\tan \theta \cos \theta} - \dfrac{3 \sin \theta}{\tan \theta \cos \theta}$

6. a. $\dfrac{1}{s^2 c^2} - \dfrac{5}{s^2 c^2} + \dfrac{11}{s^2 c^2}$ **b.** $\dfrac{1}{\sin^2 \theta \cos^2 \theta} - \dfrac{5}{\sin^2 \theta \cos^2 \theta} + \dfrac{11}{\sin^2 \theta \cos^2 \theta}$

7. a. $\dfrac{t}{2} - \dfrac{t-1}{2}$ **b.** $\dfrac{\tan \theta}{2} - \dfrac{\tan \theta - 1}{2}$

8. a. $\dfrac{2s-3}{4} - \dfrac{3s+2}{4}$ **b.** $\dfrac{2\sin \theta - 3}{4} - \dfrac{3\sin \theta + 2}{4}$

9. a. $\dfrac{3c+4}{c+2} - \dfrac{2c+5}{c+2}$ **b.** $\dfrac{3\cos \theta + 4}{\cos \theta + 2} - \dfrac{2\cos \theta + 5}{\cos \theta + 2}$

10. a. $\dfrac{4c+1}{2c-6} - \dfrac{2c+7}{2c-6}$ **b.** $\dfrac{4\cos \theta + 1}{2\cos \theta - 6} - \dfrac{2\cos \theta + 7}{2\cos \theta - 6}$

11. a. $\dfrac{1}{t-1} + \dfrac{2}{1-t}$ **b.** $\dfrac{1}{\tan \theta - 1} + \dfrac{2}{1 - \tan \theta}$

12. a. $\dfrac{2-t}{3-2t} - \dfrac{4}{2t-3}$ **b.** $\dfrac{2 - \tan \theta}{3 - 2\tan \theta} - \dfrac{4}{2\tan \theta - 3}$

13. a. $\dfrac{2s-3}{s-6} - \dfrac{s+1}{6-s}$ **b.** $\dfrac{2\sin \theta - 3}{\sin \theta - 6} - \dfrac{\sin \theta + 1}{6 - \sin \theta}$

14. a. $\dfrac{2c-5}{4-3c} + \dfrac{2-c}{3c-4}$ **b.** $\dfrac{2\cos \theta - 5}{4 - 3\cos \theta} + \dfrac{2 - \cos \theta}{3\cos \theta - 4}$

15. a. $\dfrac{3t}{4} + \dfrac{7t}{8}$ **b.** $\dfrac{3\tan \theta}{4} + \dfrac{7\tan \theta}{8}$

16. a. $\dfrac{s}{12} - \dfrac{s}{15}$ **b.** $\dfrac{\sin \theta}{12} - \dfrac{\sin \theta}{15}$

17. a. $\dfrac{4t}{5} + t$ **b.** $\dfrac{4\tan \theta}{5} + \tan \theta$

18. a. $s + \dfrac{1}{s}$ **b.** $\sin \theta + \dfrac{1}{\sin \theta}$

19. a. $\dfrac{1}{t} - t$ **b.** $\dfrac{1}{\tan \theta} - \tan \theta$

20. a. $\dfrac{t}{s^2} + s^3$ **b.** $\dfrac{\tan \theta}{\sin^2 \theta} + \sin^3 \theta$

21. a. $\dfrac{1}{3s} + \dfrac{1}{4s}$ **b.** $\dfrac{1}{3\sin \theta} + \dfrac{1}{4\sin \theta}$

22. a. $\dfrac{5}{8t} - \dfrac{2}{12t}$ **b.** $\dfrac{5}{8\tan \theta} - \dfrac{2}{12\tan \theta}$

23. a. $\dfrac{5}{2t} - \dfrac{6}{5t^2}$ **b.** $\dfrac{5}{2 \tan \theta} - \dfrac{6}{5 \tan^2 \theta}$

24. a. $\dfrac{11}{3s} + \dfrac{2}{5}$ **b.** $\dfrac{11}{3 \sin \theta} + \dfrac{2}{5}$

25. a. $\dfrac{2}{c^3} - \dfrac{3}{c^2} + \dfrac{5}{c}$ **b.** $\dfrac{2}{\cos^3 \theta} - \dfrac{3}{\cos^2 \theta} + \dfrac{5}{\cos \theta}$

26. a. $\dfrac{16}{s^2 t} + \dfrac{1}{st} - \dfrac{6}{st^2}$ **b.** $\dfrac{16}{\sin^2 \theta \tan \theta} + \dfrac{1}{\sin \theta \tan \theta} - \dfrac{6}{\sin \theta \tan^2 \theta}$

27. a. $\dfrac{4c}{5} + \dfrac{2}{5c} + \dfrac{c}{2}$ **b.** $\dfrac{4 \cos \theta}{5} + \dfrac{2}{5 \cos \theta} + \dfrac{\cos \theta}{2}$

28. a. $5 + \dfrac{6}{c^2} + \dfrac{3}{c}$ **b.** $5 + \dfrac{6}{\cos^2 \theta} + \dfrac{3}{\cos \theta}$

29. a. $\dfrac{2s + 3}{5s} - \dfrac{2s - 1}{10s} + \dfrac{4}{s}$ **b.** $\dfrac{2 \sin \theta + 3}{5 \sin \theta} - \dfrac{2 \sin \theta - 1}{10 \sin \theta} + \dfrac{4}{\sin \theta}$

30. a. $\dfrac{7}{s^2} - \dfrac{5s - 2}{s} + 6$ **b.** $\dfrac{7}{\sin^2 \theta} - \dfrac{5 \sin \theta - 2}{\sin \theta} + 6$

31. a. $\dfrac{3c - 4}{10c} + \dfrac{5c}{3} - \dfrac{4 - c}{15}$ **b.** $\dfrac{3 \cos \theta - 4}{10 \cos \theta} + \dfrac{5 \cos \theta}{3} - \dfrac{4 - \cos \theta}{15}$

32. a. $\dfrac{t}{t^2 - 4} + \dfrac{1}{t + 2}$ **b.** $\dfrac{\tan \theta}{\tan^2 \theta - 4} + \dfrac{1}{\tan \theta + 2}$

33. a. $\dfrac{2 - s}{9s + 6} + \dfrac{s - 2}{6s + 4}$ **b.** $\dfrac{2 - \sin \theta}{9 \sin \theta + 6} + \dfrac{\sin \theta - 2}{6 \sin \theta + 4}$

34. a. $\dfrac{2c + 3}{2c^3 - 4c^2} - \dfrac{1}{c - 2}$ **b.** $\dfrac{2 \cos \theta + 3}{2 \cos^3 \theta - 4 \cos^2 \theta} - \dfrac{1}{\cos \theta - 2}$

35. a. $\dfrac{t}{2t - 6} + \dfrac{5t}{4t - 12}$ **b.** $\dfrac{\tan \theta}{2 \tan \theta - 6} + \dfrac{5 \tan \theta}{4 \tan \theta - 12}$

36. a. $\dfrac{1}{t + 6} + \dfrac{1}{t + 12}$ **b.** $\dfrac{1}{\tan \theta + 6} + \dfrac{1}{\tan \theta + 12}$

37. a. $\dfrac{3c}{c - 3} - \dfrac{3}{c}$ **b.** $\dfrac{3 \cos \theta}{\cos \theta - 3} - \dfrac{3}{\cos \theta}$

38. a. $\dfrac{t}{s + t} - \dfrac{s}{s - t}$ **b.** $\dfrac{\tan \theta}{\sin \theta + \tan \theta} - \dfrac{\sin \theta}{\sin \theta - \tan \theta}$

39. a. $\dfrac{4}{c - s} + \dfrac{4}{c + s}$ **b.** $\dfrac{4}{\cos \theta - \sin \theta} + \dfrac{4}{\cos \theta + \sin \theta}$

40. a. $\dfrac{1}{c^2 - 9} + \dfrac{1}{3 - c}$ **b.** $\dfrac{1}{\cos^2 \theta - 9} + \dfrac{1}{3 - \cos \theta}$

41. a. $\dfrac{3}{2t - 3} - \dfrac{2}{9 - 4t^2}$ **b.** $\dfrac{3}{2 \tan \theta - 3} - \dfrac{2}{9 - 4 \tan^2 \theta}$

42. a. $3t + 1 + \dfrac{5}{t + 2}$ **b.** $3 \tan \theta + 1 + \dfrac{5}{\tan \theta + 2}$

43. a. $\dfrac{2 - 5s}{4s - 1} - 6s + 7$ **b.** $\dfrac{2 - 5 \sin \theta}{4 \sin \theta - 1} - 6 \sin \theta + 7$

44. a. $s + t + \dfrac{t^2}{s - t}$ **b.** $\sin \theta + \tan \theta + \dfrac{\tan^2 \theta}{\sin \theta - \tan \theta}$

45. a. $\dfrac{c}{c - 1} - \dfrac{1}{c} + 1$ **b.** $\dfrac{\cos \theta}{\cos \theta - 1} - \dfrac{1}{\cos \theta} + 1$

46. a. $\dfrac{4t}{t^2 + 2t - 3} - \dfrac{5t}{t^2 + 5t + 6}$ **b.** $\dfrac{4 \tan \theta}{\tan^2 \theta + 2 \tan \theta - 3} - \dfrac{5 \tan \theta}{\tan^2 \theta + 5 \tan \theta + 6}$

47. a. $\dfrac{4s - 1}{s^2 - 4s - 5} + \dfrac{2s + 3}{s^2 - 3s - 10}$ **b.** $\dfrac{4 \sin \theta - 1}{\sin^2 \theta - 4 \sin \theta - 5} + \dfrac{2 \sin \theta + 3}{\sin^2 \theta - 3 \sin \theta - 10}$

48. a. $\dfrac{5c + 3}{c^2 - c - 6} - \dfrac{2 - c}{c^2 - 6c + 9}$ **b.** $\dfrac{5 \cos \theta + 3}{\cos^2 \theta - \cos \theta - 6} - \dfrac{2 - \cos \theta}{\cos^2 \theta - 6 \cos \theta + 9}$

49. a. $\dfrac{5}{t + 1} - \dfrac{1}{t^2 - 1} + \dfrac{2}{t - 1}$ **b.** $\dfrac{5}{\tan \theta + 1} - \dfrac{1}{\tan^2 \theta - 1} + \dfrac{2}{\tan \theta - 1}$

50. a. $\dfrac{2c}{c^2 - 1} - \dfrac{c + 1}{c - 1} + 7$ **b.** $\dfrac{2 \cos \theta}{\cos^2 \theta - 1} - \dfrac{\cos \theta + 1}{\cos \theta - 1} + 7$

51. a. $\dfrac{1 + (1/t)}{1 - (1/t)}$ **b.** $\dfrac{1 + (1/\tan \theta)}{1 - (1/\tan \theta)}$

52. a. $\dfrac{s + (1/c)}{(2/c) - s}$ **b.** $\dfrac{\sin \theta + (1/\cos \theta)}{(2/\cos \theta) - \sin \theta}$

5–5 COMPLEX FRACTIONS

A complex fraction is a fraction in which the numerator or denominator or both involve fractions. The procedure for simplifying complex fractions is as follows:

1. Find the least common multiple of all the denominators that appear in the top and bottom of the complex fraction.
2. Multiply the top and bottom of the complex fraction by this least common multiple.
3. Simplify your results.

EXAMPLE 1: Simplify:

$$\frac{4 + \frac{1}{3}}{5 - \frac{1}{2}}$$

SOLUTION: The LCM for 2 and 3 is 6. If we multiply the top and bottom of the complex fraction by 6, we obtain

$$\frac{6(4 + \frac{1}{3})}{6(5 - \frac{1}{2})} = \frac{24 + 2}{30 - 3} = \frac{26}{27}$$

EXAMPLE 2: Simplify:

a. $\dfrac{1 + (1/s)}{(1/s^2)}$

b. $\dfrac{1 + (1/\sin \theta)}{(1/\sin^2 \theta)}$

SOLUTION:

a. LCM of s and s^2 is s^2

$$= \frac{s^2 \left(1 + \dfrac{1}{s}\right)}{s^2 \left(\dfrac{1}{s^2}\right)}$$

$$= \frac{s^2 + s}{1} \quad \text{or} \quad s^2 + s$$

b. LCM of $\sin \theta$ and $\sin^2 \theta$ is $\sin^2 \theta$

$$= \frac{\sin^2 \theta \left(1 + \dfrac{1}{\sin \theta}\right)}{\sin^2 \theta \left(\dfrac{1}{\sin^2 \theta}\right)}$$

$$= \frac{\sin^2 \theta + \sin \theta}{1} \quad \text{or} \quad \sin^2 \theta + \sin \theta$$

EXAMPLE 3: Simplify:

a. $\dfrac{(1/c) + (1/s)}{(c/s) - (s/c)}$

b. $\dfrac{(1/\cos \theta) + (1/\sin \theta)}{(\cos \theta/\sin \theta) - (\sin \theta/\cos \theta)}$

SOLUTION:

a. LCM of c and s is cs

$$= \frac{cs \left(\dfrac{1}{c} + \dfrac{1}{s}\right)}{cs \left(\dfrac{c}{s} - \dfrac{s}{c}\right)}$$

$$= \frac{s + c}{c^2 - s^2}$$

$$= \frac{\overset{1}{\cancel{s + c}}}{\underset{1}{\cancel{(c + s)}(c - s)}}$$

$$= \frac{1}{c - s}$$

b. LCM of $\cos \theta$ and $\sin \theta$ is $\cos \theta \sin \theta$

$$= \frac{\cos \theta \sin \theta \left(\dfrac{1}{\cos \theta} + \dfrac{1}{\sin \theta}\right)}{\cos \theta \sin \theta \left(\dfrac{\cos \theta}{\sin \theta} - \dfrac{\sin \theta}{\cos \theta}\right)}$$

$$= \frac{\sin \theta + \cos \theta}{\cos^2 \theta - \sin^2 \theta}$$

$$= \frac{\overset{1}{\cancel{\sin \theta + \cos \theta}}}{\underset{1}{\cancel{(\cos \theta + \sin \theta)}(\cos \theta - \sin \theta)}}$$

$$= \frac{1}{\cos \theta - \sin \theta}$$

EXERCISES 5–5

In Exercises 1–20 change the complex fraction to a fraction in simplest form.

1. $\dfrac{3 + \frac{1}{4}}{4 - \frac{2}{5}}$

2. $\dfrac{11 - \frac{1}{10}}{\frac{2}{5} + 7}$

3. $\dfrac{\frac{1}{8} - \frac{3}{2}}{\frac{5}{4} + \frac{1}{6}}$

4. $\dfrac{\frac{5}{18} + \frac{11}{12}}{\frac{7}{6} - \frac{2}{9}}$

5. a. $\dfrac{1 + (1/t)}{1 - (1/t)}$

   b. $\dfrac{1 + (1/\tan\theta)}{1 - (1/\tan\theta)}$

6. a. $\dfrac{s + (1/c)}{(2/c) - s}$

   b. $\dfrac{\sin\theta + (1/\cos\theta)}{(2/\cos\theta) - \sin\theta}$

7. a. $\dfrac{s - (5/s)}{5}$

   b. $\dfrac{\sin\theta - (5/\sin\theta)}{5}$

8. a. $\dfrac{t}{(1/t^2) + (2/t)}$

   b. $\dfrac{\tan\theta}{(1/\tan^2\theta) + (2/\tan\theta)}$

9. a. $\dfrac{(s^2/c^2) - 1}{(s/c) - 1}$

   b. $\dfrac{(\sin^2\theta/\cos^2\theta) - 1}{(\sin\theta/\cos\theta) - 1}$

10. a. $\dfrac{t - (1/c^2)}{2 - (t/c)}$

    b. $\dfrac{\tan\theta - (1/\cos^2\theta)}{2 - (\tan\theta/\cos\theta)}$

11. a. $\dfrac{(s/c) - (c/s)}{(s/c) + (c/s)}$

    b. $\dfrac{(\sin\theta/\cos\theta) - (\cos\theta/\sin\theta)}{(\sin\theta/\cos\theta) + (\cos\theta/\sin\theta)}$

12. a. $\dfrac{(1/c^2) + (1/t^2)}{(t/c) - (c/t)}$

    b. $\dfrac{(1/\cos^2\theta) + (1/\tan^2\theta)}{(\tan\theta/\cos\theta) - (\cos\theta/\tan\theta)}$

13. a. $\dfrac{s + 1 + (2/s)}{s - 1 - (2/s)}$

    b. $\dfrac{(\sin\theta + 1) + (2/\sin\theta)}{(\sin\theta - 1) - (2/\sin\theta)}$

14. a. $\dfrac{t + 4 - (5/t)}{t + 1 - (2/t)}$

    b. $\dfrac{(\tan\theta + 4) - (5/\tan\theta)}{(\tan\theta + 1) - (2/\tan\theta)}$

15. a. $\dfrac{4 + \dfrac{1}{c - 1}}{\dfrac{2}{c - 1} + 3}$

    b. $\dfrac{4 + \dfrac{1}{\cos\theta - 1}}{\dfrac{2}{\cos\theta - 1} + 3}$

16. a. $\dfrac{5 - \dfrac{t}{t - 2}}{\dfrac{4}{2 - t} + 1}$

    b. $\dfrac{5 - \dfrac{\tan\theta}{\tan\theta - 2}}{\dfrac{4}{2 - \tan\theta} + 1}$

17. a. $\dfrac{1 + \dfrac{3}{t + 1}}{\dfrac{4}{t^2 - 1}}$

b. $\dfrac{1 + \dfrac{3}{\tan \theta + 1}}{\dfrac{4}{\tan^2 \theta - 1}}$

18. a. $\dfrac{\dfrac{s}{s - c} - \dfrac{c}{s + c}}{\dfrac{s^2 + c^2}{s^2 - c^2}}$

b. $\dfrac{\dfrac{\sin \theta}{\sin \theta - \cos \theta} - \dfrac{\cos \theta}{\sin \theta + \cos \theta}}{\dfrac{\sin^2 \theta + \cos^2 \theta}{\sin^2 \theta - \cos^2 \theta}}$

19. a. $\dfrac{\dfrac{1}{s + 1} + \dfrac{1}{s - 1}}{\dfrac{1}{s - 1} - \dfrac{1}{s + 1}}$

b. $\dfrac{\dfrac{1}{\sin \theta + 1} + \dfrac{1}{\sin \theta - 1}}{\dfrac{1}{\sin \theta - 1} - \dfrac{1}{\sin \theta + 1}}$

20. a. $\dfrac{\dfrac{c + 1}{c - 1} - \dfrac{c - 1}{c + 1}}{\dfrac{c - 1}{c + 1} + \dfrac{c + 1}{c - 1}}$

b. $\dfrac{\dfrac{\cos \theta + 1}{\cos \theta - 1} - \dfrac{\cos \theta - 1}{\cos \theta + 1}}{\dfrac{\cos \theta - 1}{\cos \theta + 1} + \dfrac{\cos \theta + 1}{\cos \theta - 1}}$

21. A person drives from his home to work at an average speed of 30 mi/hour. On the way home he takes the same route and averages 20 mi/hour. What is his average speed for the round trip? (Think!)

5–6 EQUATIONS THAT CONTAIN FRACTIONS

There are many important equations or formulas that contain fractions. The procedure for solving these equations is to remove the fractions by multiplying both sides of the equation by the *least common multiple of the denominators*, or *LCD*. The resulting equation will not contain fractions and can be solved by methods that we have already discussed.

EXAMPLE 1: For what value of s is the equation

$$\frac{3s}{8} - \frac{1}{4} = \frac{s + 5}{2}$$

a true statement?

SOLUTION: Multiplying both sides of the equation by the LCD of 8, we have

$$8\left(\frac{3s}{8} - \frac{1}{4}\right) = 8\left(\frac{s + 5}{2}\right)$$

$$3s - 2 = 4s + 20$$

$$-22 = s$$

Check: Replace s by 8 in the original equation.

$$\frac{3(-22)}{8} - \frac{1}{4} \stackrel{?}{=} \frac{(-22) + 5}{2}$$

$$\frac{-66}{8} - \frac{1}{4} \stackrel{?}{=} \frac{-17}{2}$$

$$\frac{-66}{8} - \frac{2}{8} \stackrel{?}{=} \frac{-68}{8}$$

$$\frac{-68}{8} \stackrel{\checkmark}{=} \frac{-68}{8}$$

EXAMPLE 2: Solve and check. In part (b) solve for the function of θ.

a. $\dfrac{4}{t} - \dfrac{1}{2t} = \dfrac{1}{2}$ **b.** $\dfrac{4}{\tan \theta} - \dfrac{1}{2 \tan \theta} = \dfrac{1}{2}$

SOLUTION:

a. LCD: $2t$

$$2t\left(\frac{4}{t} - \frac{1}{2t}\right) = 2t\left(\frac{1}{2}\right)$$

$$8 - 1 = t$$

$$7 = t$$

b. LCD: $2 \tan \theta$

$$2 \tan \theta \left(\frac{4}{\tan \theta} - \frac{1}{2 \tan \theta}\right) = 2 \tan \theta \left(\frac{1}{2}\right)$$

$$8 - 1 = \tan \theta$$

$$7 = \tan \theta$$

Check:

$$\frac{4}{7} - \frac{1}{2(7)} \stackrel{?}{=} \frac{1}{2}$$

$$\frac{8}{14} - \frac{1}{14} \stackrel{?}{=} \frac{7}{14}$$

$$\frac{7}{14} \stackrel{\checkmark}{=} \frac{7}{14}$$

EXAMPLE 3: Solve and check. In part (b) solve for the function of θ.

a. $\dfrac{2}{3c} = \dfrac{3}{2c - 1}$ **b.** $\dfrac{2}{3 \cos \theta} = \dfrac{3}{2 \cos \theta - 1}$

SOLUTION:

a. LCD: $3c(2c - 1)$

$$3c(2c - 1)\frac{2}{3c} = 3c(2c - 1)\frac{3}{2c - 1}$$

$$2(2c - 1) = 9c$$
$$4c - 2 = 9c$$
$$-2 = 5c$$
$$\frac{-2}{5} = c$$

b. LCD: $3 \cos \theta(2 \cos \theta - 1)$

$$3 \cos \theta(2 \cos \theta - 1)\frac{2}{3 \cos \theta} = 3 \cos \theta(2 \cos \theta - 1)\frac{3}{2 \cos \theta - 1}$$

$$2(2 \cos \theta - 1) = 9 \cos \theta$$
$$4 \cos \theta - 2 = 9 \cos \theta$$
$$-2 = 5 \cos \theta$$
$$\frac{-2}{5} = \cos \theta$$

Check:

$$\frac{2}{3(-2/5)} \overset{?}{=} \frac{3}{2(-2/5) - 1}$$

$$\frac{2}{(-6/5)} \overset{?}{=} \frac{3}{(-9/5)}$$

$$\frac{5}{-3} \overset{\checkmark}{=} \frac{5}{-3}$$

Note: If there is only one term on both sides of the equation, we can simplify by cross multiplying:

$$\frac{2}{3c} \diagdown \diagup \frac{3}{2c - 1}$$

EXAMPLE 4: For what value of x is the equation

$$\frac{2}{x + 1} + \frac{3}{x - 1} = \frac{-4}{x^2 - 1}$$

a true statement?

SOLUTION: LCD is $(x + 1)(x - 1)$ or $x^2 - 1$.

$$(x + 1)(x - 1)\left(\frac{2}{x + 1} + \frac{3}{x - 1}\right) = (x + 1)(x - 1)\frac{-4}{x^2 - 1}$$

$$2(x - 1) + 3(x + 1) = -4$$
$$2x - 2 + 3x + 3 = -4$$
$$5x + 1 = -4$$
$$5x = -5$$
$$x = -1$$

Check:

$$\frac{2}{(-1) + 1} + \frac{3}{(-1) - 1} \overset{?}{=} \frac{-4}{(-1)^2 - 1}$$

$$\frac{2}{0} + \frac{3}{-2} \overset{?}{=} \frac{-4}{0}$$

Since division by 0 is undefined, -1 is not a solution. This example illustrates the need to check solutions in the original equation, for there is no solution to this problem. Whenever we multiply both sides of an equation by an expression that contains a *variable*, we obtain an equation which, although true for the original solutions, might have *additional* solutions. Thus, the two equations are not equivalent. We obtain an equivalent (same solution) equation when we multiply both sides of the equation by a *constant* (except zero).

EXAMPLE 5: One pipe can fill a swimming pool in 4 hours; a second pipe can fill the pool in 12 hours. How long will it take to fill the pool if both pipes operate together?

SOLUTION: Let x = the number of hours required to fill the pool with both pipes operating. If the first pipe fills the pool in 4 hours, the pipe fills $\frac{1}{4}$ of the tank for each hour it operates. Thus, in 1 hour it fills $\frac{1}{4}$ of the tank; in 2 hours it fills $(\frac{1}{4})2$, or $\frac{2}{4}$, of the tank; and in x hours it fills $\frac{1}{4}x$ of the tank.

Similarly, if the second pipe fills the pool in 12 hours, the pipe fills $\frac{1}{12}$ of the tank for each hour it operates. Thus, in x hours it fills $\frac{1}{12}x$ of the tank.

In order to complete the job, the sum of the two fractions must be 1. For example, if the first pipe did $\frac{3}{5}$ of the job and the second pipe did $\frac{2}{5}$ of the job, the job would be completed (for $\frac{3}{5} + \frac{2}{5} = \frac{5}{5}$ of the job, or the whole job). Therefore,

$$\frac{1}{4}x + \frac{1}{12}x = 1 \qquad \text{LCD: } 12$$
$$12(\frac{1}{4}x + \frac{1}{12}x) = 12(1)$$
$$3x + x = 12$$
$$4x = 12$$
$$x = 3$$

It takes the two pipes 3 hours to fill the pool when they are in operation simultaneously.

EXERCISES 5-6

In Exercises 1–30, part a, solve and check each equation. In part b, solve for the various functions of θ.

1. a. $\dfrac{s}{4} + \dfrac{s}{6} = 5$ **b.** $\dfrac{\sec \theta}{4} + \dfrac{\sec \theta}{6} = 5$

2. a. $\dfrac{t}{5} + 1 = \dfrac{t + 2}{10}$ **b.** $\dfrac{\tan \theta}{5} + 1 = \dfrac{\tan \theta + 2}{10}$

3. a. $\dfrac{2c - 5}{3} - \dfrac{c}{4} = \dfrac{1}{2}$ **b.** $\dfrac{2 \cot \theta - 5}{3} - \dfrac{\cot \theta}{4} = \dfrac{1}{2}$

4. a. $\dfrac{3}{7} - \dfrac{s - 5}{14} = \dfrac{11s}{14}$ **b.** $\dfrac{3}{7} - \dfrac{\sin \theta - 5}{14} = \dfrac{11 \sin \theta}{14}$

5. a. $\dfrac{4}{t} - \dfrac{1}{t} = \dfrac{1}{4}$ **b.** $\dfrac{4}{\tan \theta} - \dfrac{1}{\tan \theta} = \dfrac{1}{4}$

6. a. $\dfrac{2}{c} + \dfrac{1}{2c} = \dfrac{1}{8}$ **b.** $\dfrac{2}{\cot \theta} + \dfrac{1}{2 \cot \theta} = \dfrac{1}{8}$

7. a. $\dfrac{5}{3s} - \dfrac{7}{4s} = \dfrac{5}{6}$ **b.** $\dfrac{5}{3 \sin \theta} - \dfrac{7}{4 \sin \theta} = \dfrac{5}{6}$

8. a. $\dfrac{2 + t}{6t} - 2 = \dfrac{3}{5t}$ **b.** $\dfrac{2 + \tan \theta}{6 \tan \theta} - 2 = \dfrac{3}{5 \tan \theta}$

9. a. $\dfrac{9}{2s + 1} = 3$ **b.** $\dfrac{9}{2 \sin \theta + 1} = 3$

10. a. $-3 = \dfrac{t + 4}{t - 8}$ **b.** $-3 = \dfrac{\tan \theta + 4}{\tan \theta - 8}$

11. a. $\dfrac{9c - 5}{7} = \dfrac{2c - 4}{3}$ **b.** $\dfrac{9 \cos \theta - 5}{7} = \dfrac{2 \cos \theta - 4}{3}$

12. a. $\dfrac{c}{c + 2} = \dfrac{2}{3}$ **b.** $\dfrac{\csc \theta}{\csc \theta + 2} = \dfrac{2}{3}$

13. a. $\dfrac{3}{t} = \dfrac{4}{t - 1}$ **b.** $\dfrac{3}{\tan \theta} = \dfrac{4}{\tan \theta - 1}$

14. a. $\dfrac{2}{5 - s} = \dfrac{-3}{s}$ **b.** $\dfrac{2}{5 - \sec \theta} = \dfrac{-3}{\sec \theta}$

15. a. $\dfrac{s + 4}{s + 1} = \dfrac{s + 2}{s + 3}$ **b.** $\dfrac{\sec \theta + 4}{\sec \theta + 1} = \dfrac{\sec \theta + 2}{\sec \theta + 3}$

16. a. $\dfrac{t-1}{t-2} = \dfrac{t+1}{t-3}$

    b. $\dfrac{\tan\theta - 1}{\tan\theta - 2} = \dfrac{\tan\theta + 1}{\tan\theta - 3}$

17. a. $\dfrac{5}{t-1} = \dfrac{7t-4}{t^2-t}$

    b. $\dfrac{5}{\tan\theta - 1} = \dfrac{7\tan\theta - 4}{\tan^2\theta - \tan\theta}$

18. a. $\dfrac{1}{c^2-c} = \dfrac{2}{c-1}$

    b. $\dfrac{1}{\cos^2\theta - \cos\theta} = \dfrac{2}{\cos\theta - 1}$

19. a. $\dfrac{1}{t-3} + \dfrac{2}{3-t} = \dfrac{1}{2}$

    b. $\dfrac{1}{\tan\theta - 3} + \dfrac{2}{3 - \tan\theta} = \dfrac{1}{2}$

20. a. $\dfrac{2}{2-c} + \dfrac{c}{c-2} = 1$

    b. $\dfrac{2}{2 - \cos\theta} + \dfrac{\cos\theta}{\cos\theta - 2} = 1$

21. a. $\dfrac{3}{t} + \dfrac{2}{t-1} = \dfrac{12}{t^2-t}$

    b. $\dfrac{3}{\tan\theta} + \dfrac{2}{\tan\theta - 1} = \dfrac{12}{\tan^2\theta - \tan\theta}$

22. a. $\dfrac{3}{s+3} - \dfrac{s}{3-s} = \dfrac{s^2+9}{s^2-9}$

    b. $\dfrac{3}{\sin\theta + 3} - \dfrac{\sin\theta}{3 - \sin\theta} = \dfrac{\sin^2\theta + 9}{\sin^2\theta - 9}$

23. a. $\dfrac{4}{t+1} - \dfrac{3}{t-1} = \dfrac{-6}{t^2-1}$

    b. $\dfrac{4}{\tan\theta + 1} - \dfrac{3}{\tan\theta - 1} = \dfrac{-6}{\tan^2\theta - 1}$

24. a. $\dfrac{2}{c^2+c} = \dfrac{1}{c} - \dfrac{2}{c+1}$

    b. $\dfrac{2}{\cos^2\theta + \cos\theta} = \dfrac{1}{\cos\theta} - \dfrac{2}{\cos\theta + 1}$

25. a. $\dfrac{1}{s-2} - \dfrac{3}{s+2} = \dfrac{4}{s^2-4}$

    b. $\dfrac{1}{\sin\theta - 2} - \dfrac{3}{\sin\theta + 2} = \dfrac{4}{\sin^2\theta - 4}$

26. a. $1 + \dfrac{12}{c^2-4} = \dfrac{3}{c-2}$

    b. $1 + \dfrac{12}{\cos^2\theta - 4} = \dfrac{3}{\cos\theta - 2}$

27. a. $\dfrac{3}{c+2} - 2 = \dfrac{-5}{2c+4}$

    b. $\dfrac{3}{\cos\theta + 2} - 2 = \dfrac{-5}{2\cos\theta + 4}$

28. a. $\dfrac{5s}{s-2} - \dfrac{4s}{2s-7} = 3$

    b. $\dfrac{5\sec\theta}{\sec\theta - 2} - \dfrac{4\sec\theta}{2\sec\theta - 7} = 3$

29. a. $\dfrac{3c}{c+2} - 2 = \dfrac{2c-3}{2c-1}$

    b. $\dfrac{3\cos\theta}{\cos\theta + 2} - 2 = \dfrac{2\cos\theta - 3}{2\cos\theta - 1}$

30. a. $3 - \dfrac{t-4}{t-3} = \dfrac{4t-1}{2t+3}$

    b. $3 - \dfrac{\tan\theta - 4}{\tan\theta - 3} = \dfrac{4\tan\theta - 1}{2\tan\theta + 3}$

In Exercises 31–40 solve each formula for the letter indicated.

31. $\dfrac{W_1}{W_2} = \dfrac{L_2}{L_1}$ for L_1

32. $\dfrac{d_1}{d_2} = \dfrac{F_2}{F_1}$ for d_2

33. $t = \dfrac{v - v_0}{a}$ for v

34. $I = \dfrac{E}{R+r}$ for r

35. $S = \dfrac{a - rL}{1 - r}$ for r

36. $I = \dfrac{nE}{R + nr}$ for n

37. $Z = \dfrac{Z_1 Z_2}{Z_1 + Z_2}$ for Z_2

38. $\dfrac{E}{e} = \dfrac{R + r}{r}$ for r

39. $\dfrac{1}{f} = \dfrac{1}{a} + \dfrac{1}{b}$ for a

40. $\dfrac{1}{R} = \dfrac{1}{R_1} + \dfrac{1}{R_2}$ for R_2

41. A man left $\frac{1}{2}$ of his property to his wife, $\frac{1}{3}$ to his son, and the remainder, which was $10,000, to his daughter. How much money did the man leave?

42. One card sorter can process a deck of punched cards in 1 hour; another can sort the deck in 2 hours. How long would it take the two sorters together to process the cards?

43. An oil storage tank is being filled through two pipes. One pipe working alone can fill the tank in 6 hours; the other pipe can fill the tank in 9 hours. How long will it take to fill the tank when both pipes are used?

44. A swimming pool can be filled through one pipe in 16 hours and through a second pipe in 12 hours. It can be emptied through a drain in 8 hours. If the drain is accidentally left open while both pipes are turned on, how long does it take to fill the pool?

45. A motorboat can travel downstream 20 mi in the same time that it can travel upstream 10 mi. If the current of the river is 4 mi/hour, what is the boat's rate in still water?

46. The sum of a number and its reciprocal is $\frac{29}{10}$. What are the numbers?

47. An organization rented a bus for a trip to New York City for $80. When four were unable to go, each member was charged an extra dollar. How many members made the trip?

48. Consider the following statement: "If two fractions are equal and have equal numerators, they also have equal denominators." Now consider the following equation, which we wish to solve for x:

$$\frac{x - 1}{x + 1} + 2 = \frac{3x + 1}{x + 5}$$

$$\frac{x - 1}{x + 1} + \frac{2(x + 1)}{x + 1} = \frac{3x + 1}{x + 5}$$

$$\frac{x - 1 + 2(x + 1)}{x + 1} = \frac{3x + 1}{x + 5}$$

$$\frac{3x + 1}{x + 1} = \frac{3x + 1}{x + 5}$$

By the above statement it follows that $x + 1 = x + 5$, or, upon subtracting x from both sides, that $1 = 5$. What is wrong? What value of x satisfies the equation?

49. A race car driver must average 60 mi/hour for two laps around the track if he is to qualify for the finals. Because of mechanical trouble he is only able to average

30 mi/hour for the first lap. What speed must he average for the second lap if he is to qualify for the finals?

50. The age at which Diophantus (the Greek father of algebra) died is preserved in the following riddle: "Diophantus's youth lasted $\frac{1}{6}$ of his life. He grew a beard after $\frac{1}{12}$ more. After $\frac{1}{7}$ more of his life Diophantus married; five years later he had a son. The son lived exactly $\frac{1}{2}$ as long as his father, and Diophantus died just four years after his son." At what age did Diophantus die?

Sample Test: Chapter 5

In questions 1–3 express each fraction in simplest form.

1. $\dfrac{(c-3)(c+3)(c-3)}{(3-c)(3+c)(-3+c)} = -1$

2. $\dfrac{2x+y}{x+y}$

3. $\dfrac{(\cos\theta-1)^2}{\cos^2\theta-1}$

4. For what value(s) of x is the fraction $2x/(6x^2-2x)$ undefined?

In questions 5–7 express each of the products or quotients in simplest form.

5. $\dfrac{10sc^2}{21t^4}\cdot\dfrac{7t}{5sc}$

6. $\dfrac{\sin^2\theta-5\sin\theta+6}{\sin\theta+6}\div(\sin\theta-2)$

7. $\dfrac{c^2+1}{4c+4}\cdot\dfrac{(c+1)^2}{c^2+c}$

In questions 8 and 9 determine the least common multiple.

8. $18s^2tc^3,\ 24st^2c$

9. $t^2-4,\ t^2-4t+4$

In questions 10–13 combine each expression into a single fraction in simplest form.

10. $\dfrac{2c+1}{c-1}-\dfrac{3c-4}{c-1}$

11. $\dfrac{2}{x}+\dfrac{3}{y}$

12. $\dfrac{2\sin\theta}{5}-\dfrac{7}{3\sin\theta}+\dfrac{2}{\sin\theta}$

13. $\dfrac{2t}{t^2+5t+6}+\dfrac{5t}{t^2+2t-3}$

In questions 14–16 change each complex fraction to a fraction in simplest form.

14. $\dfrac{1-(1/s)}{1+(1/s)}$

15. $\dfrac{\cos\theta}{(1/\cos^2\theta)+(2/\cos\theta)}$

16. $\dfrac{\dfrac{x}{1+x}-\dfrac{1-x}{x}}{\dfrac{x-1}{x}-\dfrac{x}{x+1}}$

17. Solve for $\tan \theta$: $\dfrac{3}{\tan \theta + 2} - 2 = \dfrac{1}{\tan \theta + 2}$

18. Solve for x: $\dfrac{x}{3} - \dfrac{5}{3} = \dfrac{2}{x}$

19. Solve for c: $1 - \dfrac{12}{c^2 - 4} = \dfrac{3}{c + 2}$

20. Solve for F_1: $\dfrac{d_1}{d_2} = \dfrac{F_2}{F_1}$

Chapter 6
Exponents and radicals

6–1 ZERO AND NEGATIVE EXPONENTS

In Section 4-1 we saw that the expression x^n, where n is a positive integer, means to use x as a factor n times. That is,

$$x^n \text{ is shorthand for } \underbrace{x \cdot x \cdot x \cdots x}_{n \text{ factors}}$$

We then used this definition to develop the following laws of exponents, where m and n denote positive integers.

1. $x^m \cdot x^n = x^{m+n}$.
2. $(x^m)^n = x^{mn}$.

3. $\dfrac{x^m}{x^n} = \begin{cases} x^{m-n} & \text{if } m > n \\ \dfrac{1}{x^{n-m}} & \text{if } n > m \end{cases} \qquad x \neq 0.$

We now wish to extend our definition of exponents to zero and negative integers. Note that it is meaningless to use x as a factor either zero times or a negative number of times, so we must define these exponents in a different manner. However, our guideline in these new definitions is to retain the laws of exponents developed for positive integers.

We start by considering the first law of exponents:

$$x^m \cdot x^n = x^{m+n}$$

If $n = 0$, we have

$$x^m \cdot x^0 = x^{m+0} = x^m$$

When we multiply x^m by x^0, our result is x^m. Thus, x^0 must equal one. In general, if we define

$$x^0 = 1 \qquad (x \neq 0)$$

our previous laws of exponents will hold for a zero exponent.

EXAMPLE 1: $3^0 = 1$

$$(a + b)^0 = 1$$
$$(2x)^0 = 1$$
$$2x^0 = 2(1) = 2$$

In order to obtain a definition for exponents that are negative integers, we will again consider the first law of exponents.

$$x^m \cdot x^n = x^{m+n}$$

If $m = 5$ and $n = -5$, we have

$$x^5 \cdot x^{-5} = x^{5+(-5)} = x^0 = 1$$

When we multiply x^5 by x^{-5} the result is 1. Thus x^{-5} is the reciprocal of x^5 or $x^{-5} = 1/x^5$. In·general, if we define

$$x^{-n} = \frac{1}{x^n} \qquad (x \neq 0)$$

our previous laws of exponents will hold for exponents that are negative integers.

EXAMPLE 2: Simplify: $2^{-5} \cdot 2^3$.

SOLUTION:

Method 1: $2^{-5} \cdot 2^3 = 2^{(-5)+3} = 2^{-2} = \dfrac{1}{2^2} \quad \text{or} \quad \dfrac{1}{4}$

Method 2: $2^{-5} \cdot 2^3 = \dfrac{1}{2^5} \cdot 2^3 = \dfrac{2^3}{2^5} = \dfrac{1}{2^{5-3}} = \dfrac{1}{2^2} \quad \text{or} \quad \dfrac{1}{4}$

EXAMPLE 3: Simplify: $(2x)^{-3}$.

SOLUTION:

Method 1: $(2x)^{-3} = \dfrac{1}{(2x)^3} = \dfrac{1}{2^3 \cdot x^3} = \dfrac{1}{8x^3}$

Method 2: $(2x)^{-3} = 2^{-3} \cdot x^{-3} = \dfrac{1}{8} \cdot \dfrac{1}{x^3} = \dfrac{1}{8x^3}$

Note from Examples 2 and 3 that there are different ways of simplifying these expressions. An important principle to keep in mind is that any *factor* of the numerator may be made a factor of the denominator (and vice versa) by changing the sign of the exponent. For example:

$$\frac{3}{5x^{-2}} = \frac{3x^2}{5} \qquad \frac{2^{-3}}{5^{-4}} = \frac{5^4}{2^3} \qquad \frac{3^{-2}x^2}{y+1} = \frac{x^2}{3^2(y+1)} \qquad \left(\frac{2}{3}\right)^{-1} = \frac{2^{-1}}{3^{-1}} = \frac{3}{2}$$

It is important to remember that this principle applies only to factors.

EXAMPLE 4: Simplify: $\dfrac{3^{-2}}{2^{-2} + 4^{-1}}$

SOLUTION:

$$\frac{3^{-2}}{2^{-2} + 4^{-1}} = \frac{1/3^2}{(1/2^2) + (1/4)} = \frac{1/9}{(1/4) + (1/4)} = \frac{1/9}{1/2} = \frac{(1/9) \cdot 18}{(1/2) \cdot 18} = \frac{2}{9}$$

Note that 2^{-2} and 4^{-1} could not be moved to the numerator with positive exponents since they are not factors in the denominator.

EXAMPLE 5: Simplify

$$\frac{3^{-1}x^2y^{-3}}{2x^{-5}y}$$

and write the result using only positive exponents.

SOLUTION:

$$\frac{3^{-1}x^2y^{-3}}{2x^{-5}y} = \frac{x^{2-(-5)}}{3 \cdot 2y^{1-(-3)}} = \frac{x^7}{6y^4}$$

EXAMPLE 6: Simplify

$$\left(\frac{x^2 - 9}{x + 2}\right)^{-1} \div \frac{2}{x + 3}$$

SOLUTION:

$$\left(\frac{x^2 - 9}{x + 2}\right)^{-1} \div \frac{2}{x + 3} = \left(\frac{x^2 - 9}{x + 2}\right)^{-1} \cdot \frac{x + 3}{2}$$

$$= \frac{x + 2}{x^2 - 9} \cdot \frac{x + 3}{2}$$

$$= \frac{x + 2}{\underset{1}{(\cancel{x + 3})}(x - 3)} \cdot \frac{\overset{1}{\cancel{x + 3}}}{2}$$

$$= \frac{x + 2}{2(x - 3)}$$

EXERCISES 6–1

In Exercises 1–30 express each number in simplest form.

1. 3^{-2} **2.** 1^{-5}

3. 7^0 **4.** -7^0

5. $\dfrac{1}{2^{-3}}$ **6.** $\dfrac{1}{4^{-1}}$

7. $(3-5)^{-2}$ **8.** $(1-4)^{-3}$

9. $\left(\frac{4}{3}\right)^{-1}$ **10.** $\left(\frac{2}{5}\right)^{-3}$

11. $\left(\dfrac{4}{3-5}\right)^{-3}$ **12.** $\left(\dfrac{1}{1+6}\right)^{-2}$

13. $4^5 \cdot 4^{-2}$ **14.** $5^{-6} \cdot 5^5$

15. $(-2)^{-1} \cdot (-2)^{-3}$ **16.** $(-6)^4 \cdot (-6)^{-6}$

17. 4.12×10^2 **18.** 3.65×10^{-1}

19. 2.04×10^{-3} **20.** 6.71×10^0

21. $\dfrac{1}{4^{-1}} + 1^0$ **22.** $\dfrac{1}{2^{-3}} - \dfrac{1}{3^{-2}}$

23. $\dfrac{2}{2^{-2}}$ **24.** $\dfrac{3^{-1}}{3}$

25. $\dfrac{4^{-5}}{4^{-3}}$ **26.** $\dfrac{5^{-2}}{5^{-4}}$

27. $\dfrac{2^{-3} \cdot 7^0}{4^{-1} \cdot 5^0}$ **28.** $\dfrac{2^{-3} + 7^0}{4^{-1} + 5^0}$

29. $\dfrac{6^{-1} \cdot 2^{-4}}{3^{-2} \cdot 2^{-3}}$ **30.** $\dfrac{6^{-1} - 2^{-4}}{3^{-2} - 2^{-3}}$

In Exercises 31–50 write each expression with only positive exponents.

31. x^{-3} **32.** $-x^{-3}$

33. $4x^{-2}$ **34.** $(4x)^{-2}$

35. $(2x)^0$ **36.** $2x^0$

37. $x^{-4}y^{-1}$ **38.** $5x^0y^{-3}$

39. $(x+y)^0y^{-2}$ **40.** $2^{-1}x^{-3}y$

41. $\dfrac{1}{x^{-2}}$ **42.** $\dfrac{1}{2x^{-3}}$

43. $\dfrac{5}{x^{-1}}$ **44.** $\dfrac{4^{-1}}{y^{-3}}$

45. $\dfrac{3^{-2}x^3}{5y^{-1}}$ **46.** $\dfrac{-4^{-1}y^{-2}}{2^0x^{-4}}$

47. $(x^{-1}y^{-3})^{-2}$

48. $(xy^{-2})^{-1}$

49. $\dfrac{x^{-1}+y^{-1}}{(x+y)^{-1}}$

50. $\dfrac{2x^{-1}}{x^{-1}-3y^{-1}}$

In Exercises 51–70 use the laws of exponents to perform the operation and write the result in the simplest form that contains only positive exponents.

51. $x^5 \cdot x^{-3}$

52. $3y \cdot y^{-4}$

53. $(2x^{-3})(-3x^{-2})$

54. $(4x^0)(-2x^{-3})$

55. $(-x)^{-1}(4x^{-2})^2$

56. $(3x^{-2}y)^{-3}(-4x^2y^{-3})^{-4}$

57. $\dfrac{3^{-1}x^2y^{-3}}{9x^{-2}y^{-3}}$

58. $\dfrac{2^2x^{-3}y^4}{2^{-1}x^3y^{-2}}$

59. $\dfrac{(xyz)^{-1}}{x^{-2}yz^{-3}}$

60. $\dfrac{x^0y^{-2}z^3}{(xy^{-1}z^{-3})^{-1}}$

61. $\dfrac{5x-5}{2}\left(\dfrac{x-1}{4}\right)^{-1}$

62. $\dfrac{3x+6}{x^2+4x} \div \left(\dfrac{x^2+3x-4}{x^2-4}\right)^{-1}$

63. $2+3y^{-1}-4y^{-2}$

64. $(x-4)^{-2}-5(x-4)^{-1}$

65. $4(3x+1)^{-1}-(4x-1)(3x+1)^{-2}$

66. $(2x^{-3}-3x^{-2}+x^{-1})(x^{-2}-x^0)$

67. $\dfrac{x^{-1}+y^{-1}}{(xy)^{-1}}$

68. $\dfrac{xy^{-1}+x^{-1}y}{x^{-1}y-xy^{-1}}$

69. $(x^{-1}+y^{-1})^{-1}$

70. $1+(x+x^{-1})^{-1}$

6–2 SCIENTIFIC NOTATION

Many numbers that appear in scientific work are either very large or very small. For example, the average distance from the earth to the sun is approximately 93,000,000 mi, and the mass of an atom of hydrogen is approximately

0.0000000000000000000000017 g

In order to work conveniently with these numbers, we often write them in a form called *scientific notation*.

A positive number is expressed in scientific notation when it is written in the following form:

$N = m \times 10^k$ where $1 \le m < 10$ and k is some integer

For example,

$93,000,000 = 9.3(10,000,000) = 9.3 \times 10^7$

$0.00103 = 1.03\left(\dfrac{1}{1000}\right) = 1.03 \times 10^{-3}$

To convert a number from regular notation to scientific notation, use the following procedure:

1. Immediately after the first nonzero digit of the number, place an apostrophe (').
2. Starting at the apostrophe, count the number of places to the decimal point. If you move to the right, your count is expressed as a positive number; if you move to the left, the count is negative.
3. The apostrophe indicates the position of the decimal in the factor between 1 and 10; the count represents the exponent to be used in the factor, which is a power of 10.

The following examples illustrate how this procedure is used. (*Note:* The arrow indicates the direction of the counting.)

Number	=	number from 1 to 10	×	power of 10
$\overrightarrow{9'3000000.}$	=	9.3	×	10^7
$\overrightarrow{1'36.}$	=	1.36	×	10^2
$\overleftarrow{0.0001'36}$	=	1.36	×	10^{-4}
$\overleftarrow{0.6'2}$	=	6.2	×	10^{-1}
$6'.2$	=	6.2	×	10^0

Note from the above examples that a number can be changed from scientific notation to regular notation by adding zeros and moving the decimal point the appropriate number of spaces to the right if the exponent is positive, to the left if the exponent is negative.

EXERCISES 6–2

In Exercises 1–10 write the number in scientific notation by replacing the question marks in the following table with the appropriate number.

Number (N)	=	m	×	10^k
1. 740000	=	?	×	10^5
2. 0.0005	=	?	×	10^{-4}
3. 0.024	=	2.4	×	$10^?$
4. 240	=	2.4	×	$10^?$
5. 12300000	=	?	×	?
6. 0.00000123	=	?	×	?
7. ?	=	4.5	×	10^1
8. ?	=	4.5	×	10^{-1}
9. ?	=	9.08	×	10^{-7}
10. ?	=	9.08	×	10^7

In Exercises 11–20 express each number in scientific notation.

11. 42

12. 0.6

13. 34251

14. 7.21

15. A light year (that is, the distance light travels in 1 year) is about 5,900,000,000,000 mi.

16. A single red cell of human blood contains about 270,000,000 hemoglobin molecules.

17. A certain radio station broadcasts at a frequency of about 1,260,000 hertz (cycles per second).

18. A certain computer can perform an addition in about 0.000014 second.

19. The weight of an oxygen molecule is approximately 0.000000000000000000000053 g.

20. The length of a wave of yellow light is about 0.000023 in.

In Exercises 21–30 express each number in regular notation.

21. 9.2×10^4

22. 3×10^{-1}

23. 4.21×10^1

24. 6.3×10^0

25. The earth travels about 5.8×10^8 mi in its trip around the sun each year.

26. The number of atoms in 1 oz of gold is approximately 8.65×10^{21}.

27. One coulomb is equal to about 6.28×10^{18} electrons.

28. An atom is about 5×10^{-9} in. in diameter.

29. The mass of a molecule of water is about 3×10^{-23} g.

30. The wavelength of red light is approximately 6.6×10^{-5} cm.

31. If light travels about 186,000 mi/second, about how far will it travel in 5 minutes?

32. If one red blood cell contains about 270,000,000 hemoglobin molecules, about how many molecules of hemoglobin are there in 2 million red cells?

33. If the mass of one electron is about 0.0000000000000000000000000009 g, what is approximately the mass of 400 electrons?

34. If a certain computer can perform an addition in about 1.4×10^{-5} second, about how long will it take the computer to perform 50 additions?

6–3 FRACTIONAL EXPONENTS

In Section 6-1 we extended our definition of exponents to include zero and negative integers by defining them so that they satisfied the laws of exponents that we developed for positive integers. We will now follow the same procedure to develop a consistent definition for fractional exponents.

If the laws of exponents are to hold for fractions, consider the result of the following product:

$$9^{1/2} \cdot 9^{1/2} = 9^{1/2 + 1/2} = 9^1$$

Here we have two equal factors whose product is 9. Since the square root of a number is one of its two equal factors, $9^{1/2}$ must mean the square root of 9. That is,

$$9^{1/2} = \sqrt{9}$$

Similarly,

$$8^{1/3} \cdot 8^{1/3} \cdot 8^{1/3} = 8^{1/3+1/3+1/3} = 8$$

The factor $8^{1/3}$ is one of three equal factors whose product is 8. Since the cube root of a number is one of its three equal factors, $8^{1/3}$ must mean the cube root of 8.

$$8^{1/3} = \sqrt[3]{8}$$

In general, the nth root of a number is one of its n equal factors, and if we define

$$x^{1/n} = \sqrt[n]{x}$$

our previous laws of exponents will hold for exponents that are fractions.

When we take the root of a number it is important to remember the following:

1. There are two square roots for a positive number such as 4 since $(2)(2) = 4$ and $(-2)(-2) = 4$. To avoid ambiguity we define the *principal square root* of a positive number to be its *positive* square root. Thus, the principal square root of 4 is 2 and not -2. The radical sign, $\sqrt{}$, is used to symbolize the principal square root, so $\sqrt{4} = 2$. To symbolize the negative square root of 4 (which is -2) we use $-\sqrt{4}$. In general when n is even, $\sqrt[n]{}$ means the positive nth root.

2. No real number is the square root of a negative number such as -4 since the product of two equal factors is never negative. Thus, $\sqrt{-4}$ is not a real number. In general, when n is even, the nth root of a negative number does not exist among the real numbers.

3. The square root of any number that is not a perfect square is an irrational number. Consequently, if we wish to express the square root as a decimal, we can only *approximate* the number to some desired number of significant digits. In most cases we will leave these numbers in radical form.

EXAMPLE 1: Simplify:
  a. $9^{1/2}$ **b.** $8^{1/3}$ **c.** $(-8)^{1/3}$ **d.** $-8^{1/3}$ **e.** $16^{-1/4}$

SOLUTION:
a. $9^{1/2} = \sqrt{9} = 3$
b. $8^{1/3} = \sqrt[3]{8} = 2$
c. $(-8)^{1/3} = \sqrt[3]{-8} = -2$
d. $-8^{1/3} = -\sqrt[3]{8} = -2$
e. $16^{-1/4} = \dfrac{1}{16^{1/4}} = \dfrac{1}{\sqrt[4]{16}} = \dfrac{1}{2}$

When the numerator in the fractional exponent is not 1, we can evaluate the expression by considering the following law of exponents:

$$(x^m)^n = x^{mn}$$

For example, consider

$9^{3/2}$

Rewrite the expression in either of the following forms and simplify:

$(9^{1/2})^3$ or $(9^3)^{1/2}$

$(\sqrt{9})^3$ or $\sqrt{9^3}$

$(3)^3$ or $\sqrt{729}$

27   or   27

Thus, $9^{3/2} = 27$. (*Note:* Although either form may be used, you should notice that if the root results in a rational value, it is easier to find the root first.) In general, if m/n represents a reduced fraction such that $x^{1/n}$ represents a real number, then

$$x^{m/n} = (x^{1/n})^m = (x^m)^{1/n}$$

or equivalently,

$$x^{m/n} = (\sqrt[n]{x})^m = \sqrt[n]{x^m}$$

You should notice that in this definition we avoid the situation where the base is negative and the exponent is a reduced fraction in which n is an even number. This is done because such a condition deals with numbers that are not real numbers and, in such a case, the laws of exponents do not necessarily hold.

EXAMPLE 2: Simplify:

a. $8^{2/3}$ **b.** $49^{3/2}$ **c.** $(-32)^{4/5}$ **d.** $125^{-2/3}$

SOLUTION:

a. $8^{2/3} = (8^{1/3})^2 = (\sqrt[3]{8})^2 = (2)^2 = 4$

b. $49^{3/2} = (49^{1/2})^3 = (\sqrt{49})^3 = (7)^3 = 343$

c. $(-32)^{4/5} = [(-32)^{1/5}]^4 = (\sqrt[5]{-32})^4 = (-2)^4 = 16$

d. $125^{-2/3} = \dfrac{1}{125^{2/3}} = \dfrac{1}{(125^{1/3})^2} = \dfrac{1}{(\sqrt[3]{125})^2} = \dfrac{1}{(5)^2} = \dfrac{1}{25}$

EXAMPLE 3: Use the laws of exponents to simplify:

a. $6^{7/5} \cdot 6^{8/5}$ **b.** $(4^8)^{-3/16}$

SOLUTION:

a. $6^{7/5} \cdot 6^{8/5} = 6^{7/5 + 8/5}$

                $= 6^{15/5} = 6^3$

                $= 216$

b. $(4^8)^{-3/16} = 4^{8(-3/16)}$

            $= 4^{-3/2} = \dfrac{1}{4^{3/2}} = \dfrac{1}{(4^{1/2})^3}$

            $= \dfrac{1}{(2)^3} = \dfrac{1}{8}$

EXAMPLE 4: Simplify each of the following by performing the indicated operations and writing your results with only positive exponents. Also, express the result in radical form.

a. $\dfrac{2^{-1/2} \cdot 3x^{1/3}}{2 \cdot 3^{1/4}x}$

b. $\left(\dfrac{4x^{-1}y^{1/5}}{32x^5y^0}\right)^{1/3}$

SOLUTION:

a. $\dfrac{2^{-1/2} \cdot 3x^{1/3}}{2 \cdot 3^{1/4}x} = \dfrac{3^{1-1/4}}{2^{1-(-1/2)}x^{1-(1/3)}}$

$\quad = \dfrac{3^{3/4}}{2^{3/2}x^{2/3}}$ or $\dfrac{\sqrt[4]{3^3}}{\sqrt{2^3}\sqrt[3]{x^2}}$

b. $\left(\dfrac{4x^{-1}y^{1/5}}{32x^5y^0}\right)^{1/3} = \left(\dfrac{y^{1/5}}{8x^{5-(-1)} \cdot 1}\right)^{1/3}$

$\quad = \left(\dfrac{y^{1/5}}{8x^6}\right)^{1/3}$

$\quad = \dfrac{y^{(1/5)(1/3)}}{8^{1(1/3)}x^{6(1/3)}}$

$\quad = \dfrac{y^{1/15}}{2x^2}$ or $\dfrac{\sqrt[15]{y}}{2x^2}$

EXERCISES 6–3

In Exercises 1–40 simplify the expression.

1. $25^{1/2}$
2. $81^{1/2}$
3. $4^{-1/2}$
4. $49^{-1/2}$
5. $1^{-1/3}$
6. $27^{-1/3}$
7. $8^{1/3}$
8. $64^{1/3}$
9. $(-27)^{1/3}$
10. $(-64)^{1/3}$
11. $16^{1/4}$
12. $81^{1/4}$
13. $(\frac{1}{4})^{1/2}$
14. $(\frac{1}{16})^{1/2}$
15. $(\frac{9}{25})^{-1/2}$
16. $(\frac{49}{4})^{-1/2}$
17. $(-8)^{-1/3}$
18. $(-32)^{-1/5}$
19. $(0.04)^{-1/2}$
20. $(0.027)^{-1/3}$
21. $-4^{1/2}$
22. $-16^{1/4}$
23. $9^{3/2}$
24. $4^{5/2}$
25. $8^{5/3}$
26. $16^{3/4}$
27. $(-8)^{5/3}$
28. $(-32)^{3/5}$
29. $(-27)^{2/3}$
30. $(-27)^{-2/3}$
31. $(-32)^{-4/5}$
32. $(-1)^{-7/3}$
33. $(\frac{4}{9})^{-3/2}$
34. $(\frac{27}{8})^{-2/3}$
35. $9^{-3/2} + 7^0$
36. $4(4)^0 + 2(4)^{-1/2}$
37. $8^{2/3} - 2^{-1} + 4^0$
38. $(\frac{1}{9})^{-2} + 2(3)^0 - 27^{2/3}$
39. $4(16)^{-1/2} + (\frac{1}{64})^{-2/3} - 16^0$
40. $(3^2 + 4^2)^{-1/2}$

In Exercises 41–50 use the laws of exponents to simplify the expression.

41. $2^{1/2} \cdot 2^{3/2}$
42. $5^{1/2} \cdot 5^{1/2}$
43. $(-3)^{5/3} \cdot (-3)^{4/3}$
44. $(-4)^{7/5} \cdot (-4)^{8/5}$
45. $(4^2)^{1/2}$
46. $[(-4)^{10}]^{2/5}$

47. $[(-8)^6]^{-1/3}$

48. $(9^6)^{-2/3}$

49. $\dfrac{5^{1/2}}{5^{3/2}}$

50. $\dfrac{6^{1/5} \cdot 6^{3/5}}{6^{-1/5}}$

In Exercises 51–70 simplify each expression by performing the indicated operations and writing your result with only positive exponents. Also, show your result in radical form. (*Note:* When the denominator in a fractional exponent is even, assume that the base is a positive number.)

51. $x^{4/3} \cdot x^{-1}$

52. $x^{1/2} \cdot x^{2/3}$

53. $\dfrac{x^{-1/3}}{x^{5/6}}$

54. $\dfrac{x}{x^{1/2}}$

55. $(8x^3y^6)^{1/3}$

56. $(3^{1/2}x^{3/4})^{-2}$

57. $(4a^{-2}b^6)^{-1/2}$

58. $(2x^{1/3}y^{5/6})^6$

59. $\dfrac{x^{1/2}yz^{1/3}}{x^0y^{-1}z}$

60. $\dfrac{2x^{4/5}y^{-2/3}}{6^0x^{-1}y}$

61. $\left(\dfrac{x^{2/3}y^{-3/2}}{x^{-1/3}y}\right)^{-1}$

62. $\left(\dfrac{27x^{-3}}{64y^{-3}}\right)^{-1/3}$

63. $\left(\dfrac{49x^{\ 4}y^6}{25x^{-2}y^{-10}}\right)^{-3/2}$

64. $\left(\dfrac{x^{1/4}y^{-2/3}z^0}{x^{3/4}y^{-2/3}z^{1/2}}\right)^{-3/2}$

65. $\dfrac{3x^{1/2}y^{-2/3}}{5y^{-1}} \cdot \dfrac{20y^{1/4}}{27x^{-2}}$

66. $\dfrac{7^{-1}x^{1/4}}{3^{-2}y^{-1}} \div \dfrac{2(3x^{-1/2})^2}{(49y)^0}$

67. $\sqrt[3]{2} \cdot \sqrt[4]{2}$

68. $\sqrt[3]{3} \cdot \sqrt{3}$

69. $\dfrac{16}{\sqrt[3]{16}}$

70. $\dfrac{\sqrt[4]{2}}{\sqrt[8]{16}}$

71. Explain the following paradox:

$$[(-3)^2]^{1/2} = (-3)^{2(1/2)} = (-3)^1 = -3$$

But

$$[(-3)^2]^{1/2} = (9)^{1/2} = \sqrt{9} = 3$$

Therefore,

$$-3 = 3$$

72. Explain the following paradox:

$$(-1)^{1/2} \cdot (-1)^{1/2} = [(-1)(-1)]^{1/2} = (1)^{1/2} = \sqrt{1} = 1$$

But

$$(-1)^{1/2}(-1)^{1/2} = [(-1)^{1/2}]^2 = -1$$

Therefore,

$$-1 = 1$$

6–4 SIMPLIFYING RADICALS

Although it is possible to use fractional exponents for any operation required with radicals, computations are often performed by use of radical symbolism. For this reason we now define the properties of radicals in such a way that they are consistent with the laws of exponents.

PROPERTIES OF RADICALS: (x, y, $\sqrt[n]{x}$, and $\sqrt[n]{y}$ denote real numbers)

1. $(\sqrt[n]{x})^n = x$

2. $\sqrt[n]{x} \cdot \sqrt[n]{y} = x^{1/n} \cdot y^{1/n} = (x \cdot y)^{1/n} = \sqrt[n]{x \cdot y}$

3. $\dfrac{\sqrt[n]{x}}{\sqrt[n]{y}} = \dfrac{x^{1/n}}{y^{1/n}} = \left(\dfrac{x}{y}\right)^{1/n} = \sqrt[n]{\dfrac{x}{y}} \quad (y \neq 0)$

With the aid of these properties we can now use either radicals or fractional exponents whichever is more convenient.

When working with radicals we must be able to change the radical to simplest form. The first step that should be taken to simplify a radical is illustrated in the following examples.

EXAMPLE 1: Simplify $\sqrt{50}$.

SOLUTION: To simplify $\sqrt{50}$ we rewrite the 50 as the product of a perfect square and another factor.

$$\sqrt{50} = \sqrt{25 \cdot 2} = \sqrt{25} \cdot \sqrt{2} = 5\sqrt{2}$$

Thus, $\sqrt{50} = 5\sqrt{2}$.

EXAMPLE 2: Simplify $\sqrt[3]{16}$.

SOLUTION: To simplify $\sqrt[3]{16}$ we rewrite 16 as the product of a perfect cube and another factor.

$$\sqrt[3]{16} = \sqrt[3]{8 \cdot 2} = \sqrt[3]{8} \cdot \sqrt[3]{2} = 2\sqrt[3]{2}$$

Thus, $\sqrt[3]{16} = 2\sqrt[3]{2}$.

EXAMPLE 3: Simplify $3\sqrt{98}$.

SOLUTION: Rewrite 98 as the product of a perfect square and another factor.

$$3\sqrt{98} = 3\sqrt{49 \cdot 2} = 3\sqrt{49}\sqrt{2} = 3 \cdot 7\sqrt{2} = 21\sqrt{2}$$

Thus, $3\sqrt{98} = 21\sqrt{2}$.

EXAMPLE 4: Simplify $\sqrt{8x^9y^6}$ (assume that x and y denote positive numbers).

SOLUTION: Rewrite $8x^9y^6$ as the product of a perfect square and another factor. Remember that any expression with an even exponent is a perfect square.

$$\sqrt{8x^9y^6} = \sqrt{(4x^8y^6)(2x)} = \sqrt{4x^8y^6}\sqrt{2x} = 2x^4y^3\sqrt{2x}$$

Thus, $\sqrt{8x^9y^6} = 2x^4y^3\sqrt{2x}$.

In general, the first procedure for simplifying a radical is to remove any factor of the expression under the radical whose indicated root can be taken exactly.

The second procedure for simplifying a radical is to eliminate any fractions that appear under the radical.

EXAMPLE 5: Simplify $\sqrt{\frac{3}{2}}$.

SOLUTION: Rewrite $\frac{3}{2}$ as an equivalent fraction whose denominator is a perfect square.

$$\sqrt{\frac{3}{2}} = \sqrt{\frac{3\cdot 2}{2\cdot 2}} = \sqrt{\frac{6}{4}} = \frac{\sqrt{6}}{\sqrt{4}} = \frac{\sqrt{6}}{2}$$

Thus,

$$\sqrt{\frac{3}{2}} = \frac{\sqrt{6}}{2} \quad\text{or}\quad \frac{1}{2}\sqrt{6}$$

EXAMPLE 6: Simplify $\sqrt{\frac{5}{12}}$.

SOLUTION: Rewrite $\frac{5}{12}$ as an equivalent fraction whose denominator is a perfect square.

$$\sqrt{\frac{5}{12}} = \sqrt{\frac{5\cdot 3}{12\cdot 3}} = \sqrt{\frac{15}{36}} = \frac{\sqrt{15}}{\sqrt{36}} = \frac{\sqrt{15}}{6}$$

Thus,

$$\sqrt{\frac{5}{12}} = \frac{\sqrt{15}}{6} \quad\text{or}\quad \frac{1}{6}\sqrt{15}$$

(*Note:* We attain the same result if we simplify $\sqrt{\frac{5}{12}}$ as follows:

$$\sqrt{\frac{5}{12}} = \sqrt{\frac{5\cdot 12}{12\cdot 12}} = \sqrt{\frac{60}{144}} = \frac{\sqrt{60}}{\sqrt{144}} = \frac{\sqrt{60}}{12} = \frac{\sqrt{4\cdot 15}}{12} = \frac{2\sqrt{15}}{12} = \frac{\sqrt{15}}{6}$$

Notice that it is more convenient to change the denominator to the smallest possible perfect square.)

EXAMPLE 7: Simplify $\sqrt[3]{\frac{7}{9}}$.

SOLUTION: Rewrite $\frac{7}{9}$ as an equivalent fraction whose denominator is a perfect cube:

$$\sqrt[3]{\frac{7}{9}} = \sqrt[3]{\frac{7\cdot 3}{9\cdot 3}} = \sqrt[3]{\frac{21}{27}} = \frac{\sqrt[3]{21}}{\sqrt[3]{27}} = \frac{\sqrt[3]{21}}{3}$$

EXAMPLE 8: Simplify $\sqrt{2x/y}$ (assume that x and y denote positive numbers).

SOLUTION: Rewrite $2x/y$ as an equivalent fraction whose denominator is a perfect square.

$$\sqrt{\frac{2x}{y}} = \sqrt{\frac{2x \cdot y}{y \cdot y}} = \sqrt{\frac{2xy}{y^2}} = \frac{\sqrt{2xy}}{\sqrt{y^2}} = \frac{\sqrt{2xy}}{y}$$

Thus, $\sqrt{2x/y} = \sqrt{2xy}/y$.

EXERCISES 6–4

Express each radical in simplest form. (Assume that x and y represent positive numbers.)

1. $\sqrt{45}$	**2.** $\sqrt{48}$
3. $\sqrt{150}$	**4.** $\sqrt{162}$
5. $\sqrt{108}$	**6.** $-\sqrt{200}$
7. $-\sqrt{72}$	**8.** $-\sqrt{28}$
9. $-\sqrt{98}$	**10.** $-\sqrt{80}$
11. $\sqrt[3]{54}$	**12.** $\sqrt[3]{-32}$
13. $\sqrt[4]{48}$	**14.** $\sqrt[5]{64}$
15. $5\sqrt{8}$	**16.** $\frac{1}{3}\sqrt{50}$
17. $-\frac{1}{4}\sqrt{96}$	**18.** $-2\sqrt{300}$
19. $\sqrt{xy^2}$	**20.** $\sqrt{x^3y^5}$
21. $\sqrt{12x^7y^4}$	**22.** $\sqrt{54x^{11}y^5}$
23. $\sqrt{250x^3y^{15}}$	**24.** $\sqrt[3]{5x^2y^4}$
25. $\sqrt[4]{9x^6y^7}$	**26.** $\sqrt[5]{16x^3y^5}$
27. $\sqrt{\frac{5}{4}}$	**28.** $\sqrt{\frac{2}{9}}$
29. $\sqrt{\frac{1}{2}}$	**30.** $\sqrt{\frac{1}{6}}$
31. $\sqrt{\frac{8}{7}}$	**32.** $\sqrt{\frac{3}{8}}$
33. $5\sqrt{\frac{2}{5}}$	**34.** $9\sqrt{\frac{4}{3}}$
35. $-12\sqrt{\frac{20}{27}}$	**36.** $-3\sqrt{\frac{5}{32}}$
37. $\frac{1}{2}\sqrt{\frac{11}{2}}$	**38.** $\frac{3}{5}\sqrt{\frac{13}{18}}$
39. $\sqrt[3]{\frac{1}{2}}$	**40.** $\sqrt[3]{\frac{-3}{4}}$
41. $\sqrt[4]{\frac{5}{8}}$	**42.** $\sqrt[5]{\frac{2}{9}}$
43. $\sqrt{x/y}$	**44.** $\sqrt{x^2/y}$
45. $\sqrt{9/x}$	**46.** $\sqrt{5/(3x^2y)}$
47. $\sqrt[3]{x/y}$	**48.** $\sqrt[4]{2x/y^5}$
49. $\sqrt{\frac{1}{2} + \frac{1}{3}}$	**50.** $\sqrt{\frac{5}{4} - \frac{4}{5}}$

6–5 ADDITION AND SUBTRACTION OF RADICALS

We can simplify an algebraic expression such as $7x + 2x - x$ by combining similar terms. That is,

$$7x + 2x - x = (7 + 2 - 1)x = 8x$$

Similarly, if x is replaced by $\sqrt{2}$, we have

$$7\sqrt{2} + 2\sqrt{2} - \sqrt{2} = (7 + 2 - 1)\sqrt{2} = 8\sqrt{2}$$

or if x is replaced by $\sqrt[3]{5x}$, we have

$$7\sqrt[3]{5x} + 2\sqrt[3]{5x} - \sqrt[3]{5x} = (7 + 2 - 1)\sqrt[3]{5x} = 8\sqrt[3]{5x}$$

Thus, we can combine similar radicals by adding or subtracting their coefficients. By similar radicals we mean radicals that have the same expression under the radical sign and the same indicated root. In general, in order to combine radicals, we first express each radical in simplest form and then we combine those which are similar.

EXAMPLE 1: Combine $\sqrt{45} + \sqrt{20}$.

SOLUTION:

$$\sqrt{45} = \sqrt{9 \cdot 5} = \sqrt{9}\sqrt{5} = 3\sqrt{5}$$
$$\sqrt{20} = \sqrt{4 \cdot 5} = \sqrt{4}\sqrt{5} = 2\sqrt{5}$$
$$\sqrt{45} + \sqrt{20} = 3\sqrt{5} + 2\sqrt{5} = (3 + 2)\sqrt{5} = 5\sqrt{5}$$

Thus, $\sqrt{45} + \sqrt{20} = 5\sqrt{5}$.

EXAMPLE 2: Combine $2\sqrt{32} - 4\sqrt{\frac{1}{2}}$.

SOLUTION:

$$2\sqrt{32} = 2\sqrt{16 \cdot 2} = 2\sqrt{16}\sqrt{2} = 2 \cdot 4\sqrt{2} = 8\sqrt{2}$$

$$4\sqrt{\frac{1}{2}} = 4\sqrt{\frac{1 \cdot 2}{2 \cdot 2}} = 4\sqrt{\frac{2}{4}} = \frac{4\sqrt{2}}{\sqrt{4}} = \frac{4\sqrt{2}}{2} = 2\sqrt{2}$$

$$2\sqrt{32} - 4\sqrt{\frac{1}{2}} = 8\sqrt{2} - 2\sqrt{2} = (8 - 2)\sqrt{2} = 6\sqrt{2}$$

Thus, $2\sqrt{32} - 4\sqrt{\frac{1}{2}} = 6\sqrt{2}$.

EXAMPLE 3: Combine $\sqrt[3]{-16} + \sqrt[3]{54}$.

SOLUTION:

$$\sqrt[3]{-16} = \sqrt[3]{-8 \cdot 2} = \sqrt[3]{-8}\sqrt[3]{2} = -2\sqrt[3]{2}$$
$$\sqrt[3]{54} = \sqrt[3]{27 \cdot 2} = \sqrt[3]{27}\sqrt[3]{2} = 3\sqrt[3]{2}$$
$$\sqrt[3]{-16} + \sqrt[3]{54} = -2\sqrt[3]{2} + 3\sqrt[3]{2} = (-2 + 3)\sqrt[3]{2} = 1\sqrt[3]{2}$$

Thus, $\sqrt[3]{-16} + \sqrt[3]{54} = \sqrt[3]{2}$.

EXAMPLE 4: Combine $\sqrt{50x^3y} - \sqrt{8xy^3}$. (Assume that x and y denote positive numbers.)

SOLUTION:

$$\sqrt{50x^3y} = \sqrt{(25x^2)(2xy)} = \sqrt{25x^2}\sqrt{2xy} = 5x\sqrt{2xy}$$
$$\sqrt{8xy^3} = \sqrt{(4y^2)(2xy)} = \sqrt{4y^2}\sqrt{2xy} = 2y\sqrt{2xy}$$
$$\sqrt{50x^3y} - \sqrt{8xy^3} = 5x\sqrt{2xy} - 2y\sqrt{2xy} = (5x - 2y)\sqrt{2xy}$$

Thus, $\sqrt{50x^3y} - \sqrt{8xy^3} = (5x - 2y)\sqrt{2xy}$.

EXERCISES 6–5

Perform the indicated operation and express the result in simplest form. Assume that x and y denote positive real numbers.

1. $4\sqrt{2} + 7\sqrt{2}$

2. $9\sqrt{5} - 15\sqrt{5}$

3. $\sqrt{3} - 4\sqrt{3}$

4. $-2\sqrt{7} - \sqrt{7}$

5. $\sqrt{27} + \sqrt{12}$

6. $\sqrt{8} - \sqrt{32}$

7. $3\sqrt{72} - 5\sqrt{50}$

8. $3\sqrt{28} - 10\sqrt{63}$

9. $-2\sqrt{176} - 3\sqrt{44}$

10. $9\sqrt{10} - 2\sqrt{1000}$

11. $\sqrt{18} - \sqrt{12}$

12. $3\sqrt{48} - \sqrt{20}$

13. $4\sqrt{12} - \sqrt{27} + 2\sqrt{48}$

14. $-4\sqrt{45} + 2\sqrt{125} - 6\sqrt{20}$

15. $2\sqrt{50} - 6\sqrt{\frac{1}{2}}$

16. $4\sqrt{\frac{1}{5}} + 2\sqrt{125}$

17. $\sqrt{56} - \frac{1}{2}\sqrt{\frac{2}{7}}$

18. $10\sqrt{\frac{2}{5}} - 2\sqrt{\frac{1}{10}}$

19. $6\sqrt{\frac{1}{6}} - \sqrt{\frac{2}{3}}$

20. $5\sqrt{\frac{1}{3}} - 3\sqrt{\frac{1}{12}}$

21. $3\sqrt{\frac{3}{2}} - \sqrt{\frac{2}{3}} + 5\sqrt{6}$

22. $\frac{1}{3}\sqrt{\frac{5}{3}} - \frac{1}{2}\sqrt{60} + \frac{1}{6}\sqrt{15}$

23. $\sqrt[3]{24} + 7\sqrt[3]{3}$

24. $2\sqrt[3]{-54} - \sqrt[3]{128}$

25. $\sqrt[3]{\frac{1}{9}} + 2\sqrt[3]{-375}$

26. $3\sqrt[4]{32} - 5\sqrt[4]{2}$

27. $\sqrt{8x} + \sqrt{72x}$

28. $9\sqrt{3y} - 4\sqrt{12y}$

29. $\sqrt{20xy^2} - 2\sqrt{45x^3}$

30. $\sqrt{x^3y/2} + \frac{1}{2}\sqrt{32xy^7}$

31. $\sqrt{50x^3y} + x\sqrt{xy/2} - 2\sqrt{9x^3y/2}$

32. $2\sqrt{x/y} - \sqrt{y/x} + 3\sqrt{1/xy}$

33. $2\sqrt{2x/y} - 4\sqrt{y/2x^3} + 5\sqrt{\frac{1}{8}x^3y}$

34. $\sqrt[3]{54x^4y} - \sqrt[3]{xy^4/4}$

35. $\sqrt[4]{x/y} - \sqrt[4]{xy^3}$

36. $\sqrt{4 + 4x} + \sqrt{16 + 16x}$

37. $\dfrac{-(-4) + \sqrt{(-4)^2 - 4(1)(-3)}}{2(1)}$

38. $\dfrac{-b + \sqrt{b^2 - 4ac}}{2a} + \dfrac{-b - \sqrt{b^2 - 4ac}}{2a}$

6–6 MULTIPLICATION OF RADICALS

When multiplying expressions containing radicals, we use the second property of radicals that we developed in Section 6-4. That is,

$$\sqrt[n]{x} \cdot \sqrt[n]{y} = \sqrt[n]{xy}$$

This property, together with the normal rules of algebraic multiplication, enables us to multiply radicals. It should be noted that both radicals must have the same indicated root if the second property of radicals is to be used.

EXAMPLE 1: Multiply and express the result in simplest form:
a. $\sqrt{7} \cdot \sqrt{2}$ **b.** $\sqrt[3]{6x} \cdot \sqrt[3]{3x}$ **c.** $\sqrt{2} \cdot \sqrt{6}$

SOLUTION:
a. $\sqrt{7} \cdot \sqrt{2} = \sqrt{7 \cdot 2} = \sqrt{14}$
b. $\sqrt[3]{6x} \cdot \sqrt[3]{3x} = \sqrt[3]{6x \cdot 3x} = \sqrt[3]{18x^2}$
c. $\sqrt{2} \cdot \sqrt{6} = \sqrt{2 \cdot 6} = \sqrt{12}$
Simplify:
$$\sqrt{12} = \sqrt{4 \cdot 3} = \sqrt{4} \cdot \sqrt{3} = 2\sqrt{3}$$
Thus, $\sqrt{2} \cdot \sqrt{6} = 2\sqrt{3}$.

EXAMPLE 2: Multiply and express the result in simplest form:
a. $2\sqrt{6} \cdot 3\sqrt{3}$ **b.** $(3\sqrt{5})^2$ **c.** $4\sqrt[3]{2} \cdot 3\sqrt[3]{4}$

SOLUTION:
a. $2\sqrt{6} \cdot 3\sqrt{3} = 2 \cdot 3\sqrt{6 \cdot 3} = 6\sqrt{18}$
Simplify:
$$6\sqrt{18} = 6\sqrt{9 \cdot 2} = 6\sqrt{9} \cdot \sqrt{2} = 6 \cdot 3\sqrt{2} = 18\sqrt{2}$$
Thus, $2\sqrt{6} \cdot 3\sqrt{3} = 18\sqrt{2}$.
b. $(3\sqrt{5})^2 = 3\sqrt{5} \cdot 3\sqrt{5} = 3 \cdot 3 \cdot \sqrt{5} \cdot \sqrt{5} = 9 \cdot 5 = 45$

(*Note:* By definition $\sqrt{x}\sqrt{x} = x$ is always true.)
c. $4\sqrt[3]{2} \cdot 3\sqrt[3]{4} = 4 \cdot 3\sqrt[3]{2 \cdot 4} = 12\sqrt[3]{8}$
Simplify:
$$12\sqrt[3]{8} = 12 \cdot 2 = 24$$
Thus, $4\sqrt[3]{2} \cdot 3\sqrt[3]{4} = 24$.

EXAMPLE 3: Multiply and express the result in simplest form: $\sqrt{3}(2\sqrt{6} - 3\sqrt{3})$.

SOLUTION:
$$\sqrt{3}(2\sqrt{6} - 3\sqrt{3}) = 2\sqrt{3}\sqrt{6} - 3\sqrt{3} \cdot \sqrt{3} = 2\sqrt{18} - 3 \cdot 3$$
Simplify:
$$2\sqrt{18} - 9 = 2\sqrt{9 \cdot 2} - 9 = 2\sqrt{9} \cdot \sqrt{2} - 9 = 2 \cdot 3\sqrt{2} - 9 = 6\sqrt{2} - 9$$
Thus, $\sqrt{3}(2\sqrt{6} - 3\sqrt{3}) = 6\sqrt{2} - 9$.

EXAMPLE 4: Multiply and express the result in simplest form: $(2 + 3\sqrt{5})^2$.

SOLUTION:

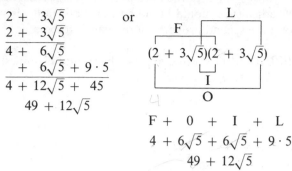

$$\begin{array}{r} 2 + 3\sqrt{5} \\ 2 + 3\sqrt{5} \\ \hline 4 + 6\sqrt{5} \\ + 6\sqrt{5} + 9 \cdot 5 \\ \hline 4 + 12\sqrt{5} + 45 \\ 49 + 12\sqrt{5} \end{array}$$

or

$$(2 + 3\sqrt{5})(2 + 3\sqrt{5})$$

$$F + 0 + I + L$$
$$4 + 6\sqrt{5} + 6\sqrt{5} + 9 \cdot 5$$
$$49 + 12\sqrt{5}$$

EXAMPLE 5: Multiply and express the result in simplest form:
$(3\sqrt{2} + 2\sqrt{3})(3\sqrt{2} - 2\sqrt{3})$

SOLUTION:

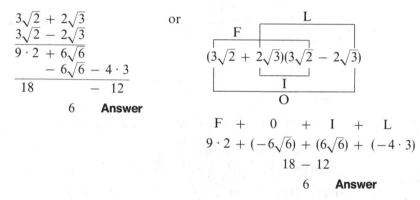

$$\begin{array}{r} 3\sqrt{2} + 2\sqrt{3} \\ 3\sqrt{2} - 2\sqrt{3} \\ \hline 9 \cdot 2 + 6\sqrt{6} \\ - 6\sqrt{6} - 4 \cdot 3 \\ \hline 18 \phantom{+ 6\sqrt{6}} - 12 \\ 6 \quad \textbf{Answer} \end{array}$$

or

$$(3\sqrt{2} + 2\sqrt{3})(3\sqrt{2} - 2\sqrt{3})$$

$$F + 0 + I + L$$
$$9 \cdot 2 + (-6\sqrt{6}) + (6\sqrt{6}) + (-4 \cdot 3)$$
$$18 - 12$$
$$6 \quad \textbf{Answer}$$

EXERCISES 6–6

Perform the indicated operation and express the result in simplest form. Assume that x and y denote positive real numbers.

1. $\sqrt{3} \cdot \sqrt{5}$
2. $\sqrt{2} \cdot \sqrt{11}$
3. $\sqrt{6} \cdot \sqrt{2}$
4. $\sqrt{15} \cdot \sqrt{5}$
5. $\sqrt{27} \cdot \sqrt{3}$
6. $\sqrt{2} \cdot \sqrt{50}$
7. $3\sqrt{10} \cdot 2\sqrt{6}$
8. $\frac{1}{2}\sqrt{24} \cdot 8\sqrt{3}$
9. $-3\sqrt{12} \cdot \frac{2}{3}\sqrt{3}$
10. $4\sqrt{6} \cdot -3\sqrt{8}$
11. $(\sqrt{2})^2$
12. $(4\sqrt{2})^2$
13. $(\frac{1}{2}\sqrt{6})^2$
14. $(-3\sqrt{5})^2$
15. $(\sqrt[3]{4})^3$
16. $\sqrt[3]{6} \cdot \sqrt[3]{4}$
17. $\sqrt[3]{-4} \cdot \sqrt[3]{16}$
18. $\sqrt[4]{8} \cdot \sqrt[4]{2}$
19. $\sqrt{15x} \cdot \sqrt{6x}$
20. $2\sqrt{x} \cdot 5\sqrt{9x^3}$
21. $-3\sqrt{8xy} \cdot 4\sqrt{3x^3y^2}$
22. $\sqrt{5x^3y} \cdot \sqrt{10y}$

23. $\sqrt[3]{4x^2y} \cdot \sqrt[3]{6x^2y^3}$

24. $4\sqrt[4]{27x^3} \cdot 5\sqrt[4]{3x^5}$

25. $\sqrt{2}(3\sqrt{2} + 2\sqrt{8})$

26. $\sqrt{3}(4\sqrt{6} - \sqrt{30})$

27. $2\sqrt{5}(4\sqrt{10} - 3\sqrt{5})$

28. $-3\sqrt{3}(4 - 2\sqrt{27})$

29. $(1 + \sqrt{2})(1 - \sqrt{2})$

30. $(3 + \sqrt{5})(3 - \sqrt{5})$

31. $(3 + \sqrt{2})(3 - \sqrt{2})$

32. $(5 + \sqrt{7})(5 - \sqrt{7})$

33. $(2\sqrt{3} - 4\sqrt{5})(2\sqrt{3} + 4\sqrt{5})$

34. $(4\sqrt{2} - 3\sqrt{5})(4\sqrt{2} + 3\sqrt{5})$

35. $(1 + \sqrt{2})^2$

36. $(2 - \sqrt{5})^2$

37. $(2\sqrt{3} + 2)^2$

38. $(\sqrt{5} - 2\sqrt{6})^2$

39. $(-5\sqrt{11} - 7\sqrt{2})^2$

40. $(3\sqrt{5} + 2\sqrt{6})^2$

41. $(1 + \sqrt{2})(2 - \sqrt{3})$

42. $(3 - \sqrt{2})(5 - \sqrt{6})$

43. $(2\sqrt{3} + 4\sqrt{5})(5\sqrt{3} + 4\sqrt{2})$

44. $(3\sqrt{3} - 2\sqrt{6})(2\sqrt{6} - 3\sqrt{3})$

45. $(5\sqrt{7} - 3\sqrt{8})(2\sqrt{7} + 4\sqrt{2})$

46. $(3\sqrt{5} - 6\sqrt{10})(\sqrt{2} - 2\sqrt{5})$

47. $(\sqrt{x} - \sqrt{y})^2$

48. $(\sqrt{x} + \sqrt{y})^2$

49. $(\sqrt{x + y})^2$

50. $\left(\dfrac{-b + \sqrt{b^2 - 4ac}}{2a}\right)\left(\dfrac{-b - \sqrt{b^2 - 4ac}}{2a}\right)$

51. If $f(x) = x^2 - 5$. find $f(\sqrt{5})$.

52. If $f(x) = x^2 - 5$, find $f(-\sqrt{5})$.

53. If $f(x) = x^2 - 4x - 3$, find $f(2 + \sqrt{7})$.

54. If $f(x) = x^2 - 4x - 3$, find $f(2 - \sqrt{7})$.

55. If $f(x) = x^2 - 2x + 5$, find $f(1 + \sqrt{2})$.

56. If $f(x) = x^2 - 2x + 5$, find $f(1 - \sqrt{2})$.

57. If $f(x) = 3x^2 + 4x - 2$, find $f\left(\dfrac{-2 + \sqrt{10}}{3}\right)$.

58. If $f(x) = 3x^2 + 4x - 2$, find $f\left(\dfrac{-2 - \sqrt{10}}{3}\right)$.

59. If $f(x) = 2x^2 - x + 1$, find $f(\sqrt{2} + \sqrt{3})$.

60. If $f(x) = 2x^2 - x + 1$, find $f(\sqrt{2} - \sqrt{3})$.

6–7 DIVISION OF RADICALS

When dividing expressions containing radicals, we use the third property of radicals that we developed in Section 6-4. That is,

$$\frac{\sqrt[n]{x}}{\sqrt[n]{y}} = \sqrt[n]{\frac{x}{y}} \qquad (y \neq 0)$$

It should be noted that both radicals must have the same indicated root if the third property of radicals is to be used.

EXAMPLE 1: Divide and express the result in simplest form:

a. $\dfrac{\sqrt{7}}{\sqrt{2}}$ **b.** $\dfrac{50\sqrt{32}}{5\sqrt{2}}$ **c.** $\dfrac{\sqrt[3]{48}}{\sqrt[3]{2}}$

SOLUTION:

a. $\dfrac{\sqrt{7}}{\sqrt{2}} = \sqrt{\dfrac{7}{2}}$

Simplify:

$$\sqrt{\dfrac{7}{2}} = \sqrt{\dfrac{7\cdot 2}{2\cdot 2}} = \dfrac{\sqrt{14}}{\sqrt{4}} = \dfrac{\sqrt{14}}{2}$$

Thus,

$$\dfrac{\sqrt{7}}{\sqrt{2}} = \dfrac{\sqrt{14}}{2} \quad \text{or} \quad \tfrac{1}{2}\sqrt{14}$$

b. $\dfrac{50\sqrt{32}}{5\sqrt{2}} = \dfrac{50}{5}\sqrt{\dfrac{32}{2}} = 10\sqrt{16} = 10\cdot 4 = 40$

c. $\dfrac{\sqrt[3]{48}}{\sqrt[3]{2}} = \sqrt[3]{\dfrac{48}{2}} = \sqrt[3]{24}$

Simplify:

$$\sqrt[3]{24} = \sqrt[3]{8\cdot 3} = \sqrt[3]{8}\cdot\sqrt[3]{3} = 2\sqrt[3]{3}$$

Thus:

$$\dfrac{\sqrt[3]{48}}{\sqrt[3]{2}} = 2\sqrt[3]{3}$$

One main objective in dividing radicals is to eliminate any radical in the denominator of the fraction. For example, consider the fraction

$$\dfrac{1}{\sqrt{2}}$$

If we wish to find an approximate value for this fraction in its present form, we would approximate $\sqrt{2}$, which is an irrational number, and then divide:

$$\dfrac{1}{\sqrt{2}} \approx \dfrac{1}{1.414}$$

We now have a difficult division to perform. Instead of dividing, let's change the form of $1/\sqrt{2}$ so that we eliminate the radical in the denominator:

$$\dfrac{1}{\sqrt{2}} = \dfrac{1\sqrt{2}}{\sqrt{2}\sqrt{2}} = \dfrac{\sqrt{2}}{2} \approx \dfrac{1.414}{2}$$

Note that the division is now easier if we wish to approximate $1/\sqrt{2}$ because we made the denominator a rational number. Thus, when dividing radicals we should always *rationalize the denominator* by eliminating any radicals in the denominator of the fraction.

EXAMPLE 2: Rationalize the denominator:

a. $\dfrac{2}{\sqrt{5}}$ b. $\dfrac{7}{4\sqrt{3}}$ c. $\dfrac{9}{2\sqrt{18}}$ d. $\dfrac{9}{\sqrt{3x}}$

SOLUTION:

a. $\dfrac{2}{\sqrt{5}} = \dfrac{2\sqrt{5}}{\sqrt{5}\sqrt{5}} = \dfrac{2\sqrt{5}}{5}$

b. $\dfrac{7}{4\sqrt{3}} = \dfrac{7\sqrt{3}}{4\sqrt{3}\sqrt{3}} = \dfrac{7\sqrt{3}}{4\cdot 3} = \dfrac{7\sqrt{3}}{12}$

c. $\dfrac{9}{2\sqrt{18}} = \dfrac{9\sqrt{2}}{2\sqrt{18}\sqrt{2}} = \dfrac{9\sqrt{2}}{2\sqrt{36}} = \dfrac{9\sqrt{2}}{2\cdot 6} = \dfrac{3\sqrt{2}}{4}$

d. $\dfrac{9}{\sqrt{3x}} = \dfrac{9\sqrt{3x}}{\sqrt{3x}\sqrt{3x}} = \dfrac{9\sqrt{3x}}{3x} = \dfrac{3\sqrt{3x}}{x}$

If the denominator contains square roots and is the *sum* of two terms, we rationalize the denominator by multiplying the top and bottom of the fraction by the *difference* of the same two terms, and vice versa.

EXAMPLE 3a: Rationalize the denominator:

a. $\dfrac{5}{1 + \sqrt{2}}$

SOLUTION:

a. $\dfrac{5}{1 + \sqrt{2}} = \dfrac{5(1 - \sqrt{2})}{(1 + \sqrt{2})(1 - \sqrt{2})}$

$(1 + \sqrt{2})(1 - \sqrt{2})$

F + O + I + L

$1 + (-\sqrt{2}) + \sqrt{2} + (-2) = -1$

Thus,

$$\dfrac{5}{1 + \sqrt{2}} = \dfrac{5 - 5\sqrt{2}}{-1} \quad \text{or} \quad -5 + 5\sqrt{2}$$

EXAMPLE 3b:

b. $\dfrac{2 + \sqrt{3}}{4\sqrt{3} - \sqrt{2}}$

SOLUTION:

b. $\dfrac{2 + \sqrt{3}}{4\sqrt{3} - \sqrt{2}} = \dfrac{(2 + \sqrt{3})(4\sqrt{3} + \sqrt{2})}{(4\sqrt{3} - \sqrt{2})(4\sqrt{3} + \sqrt{2})}$

$(2 + \sqrt{3})(4\sqrt{3} + \sqrt{2})$

$(4\sqrt{3} - \sqrt{2})(4\sqrt{3} + \sqrt{2})$

$\begin{array}{cccc} F & + \ O & + \ I & + \ L \\ 8\sqrt{3} & + \ 2\sqrt{2} & + \ 4\cdot 3 & + \ \sqrt{6} \end{array}$

$\begin{array}{cccc} F & + \ O & + \ I & + \ L \\ 16\cdot 3 & + \ 4\sqrt{6} & + \ (-4\sqrt{6}) & + \ (-2) \\ 48\ + & & 0 & - \ 2 \\ & & 46 & \end{array}$

Thus,

$$\frac{2 + \sqrt{3}}{4\sqrt{3} - \sqrt{2}} = \frac{8\sqrt{3} + 2\sqrt{2} + 12 + \sqrt{6}}{46}$$

EXERCISES 6–7

In Exercises 1–20 perform the indicated operation and express your result in simplest form. Assume that x and y denote positive numbers.

1. $\dfrac{\sqrt{20}}{\sqrt{5}}$ 2. $\dfrac{\sqrt{50}}{\sqrt{2}}$

3. $\dfrac{\sqrt{24}}{\sqrt{3}}$ 4. $\dfrac{\sqrt{24}}{\sqrt{2}}$

5. $8\sqrt{7} \div 3\sqrt{7}$ 6. $\sqrt{2} \div 2\sqrt{2}$

7. $\dfrac{\sqrt{10}}{\sqrt{3}}$ 8. $\dfrac{\sqrt{14}}{\sqrt{5}}$

9. $\dfrac{\sqrt{18}}{\sqrt{3x}}$ 10. $\dfrac{\sqrt{4x^3}}{\sqrt{5x}}$

11. $\dfrac{\sqrt{18} - \sqrt{50}}{\sqrt{2}}$

12. $\dfrac{\sqrt{15} + \sqrt{60}}{\sqrt{3}}$

13. $\dfrac{6\sqrt{2x}}{5\sqrt{5x}}$

14. $\dfrac{4\sqrt{5x}}{\sqrt{2x^3}}$

15. $\dfrac{-2\sqrt{54x^3y}}{4\sqrt{3xy}}$

16. $\dfrac{5\sqrt{6xy^2}}{-12\sqrt{54x^3y^5}}$

17. $\dfrac{\sqrt[3]{-3y}}{\sqrt[3]{48y}}$

18. $\dfrac{\sqrt[3]{2x^2y^2}}{\sqrt[3]{54x^4y}}$

19. $\dfrac{\sqrt[4]{7x^2}}{\sqrt[4]{2x}}$

20. $\dfrac{\sqrt[5]{5y^4}}{\sqrt[5]{3y^2}}$

In Exercises 21–42 rationalize the denominator.

21. $\dfrac{2}{\sqrt{3}}$

22. $\dfrac{3}{\sqrt{5}}$

23. $\dfrac{7}{2\sqrt{18}}$

24. $\dfrac{-4}{3\sqrt{12}}$

25. $\dfrac{\sqrt{18} + 1}{\sqrt{2}}$

26. $\dfrac{1 - \sqrt{3}}{\sqrt{3}}$

27. $\dfrac{-10}{3\sqrt{20x}}$

28. $\dfrac{15x}{2\sqrt{5x^3}}$

29. $\dfrac{1}{\sqrt[3]{2}}$

30. $\dfrac{6}{-2\sqrt[3]{-3xy}}$

31. $\dfrac{3}{1 + \sqrt{2}}$

32. $\dfrac{9}{3 - \sqrt{5}}$

33. $\dfrac{1}{\sqrt{3} - \sqrt{5}}$

34. $\dfrac{-6}{\sqrt{2} + \sqrt{7}}$

35. $\dfrac{2\sqrt{3}}{4 - 2\sqrt{3}}$

36. $\dfrac{4\sqrt{7}}{2\sqrt{2} + 1}$

37. $\dfrac{2 - \sqrt{2}}{2 + \sqrt{2}}$

38. $\dfrac{5 + 2\sqrt{7}}{2 - 2\sqrt{7}}$

39. $\dfrac{\sqrt{3} + \sqrt{5}}{2\sqrt{3} - 7\sqrt{5}}$

40. $\dfrac{3\sqrt{6} - 2\sqrt{11}}{\sqrt{6} - 6\sqrt{11}}$

41. $\dfrac{\sqrt{x}}{\sqrt{x} + \sqrt{y}}$

42. $\dfrac{2\sqrt{y}}{x - 2\sqrt{y}}$

6-8 RADICAL EQUATIONS

An equation in which the unknown appears under a radical is called a *radical equation*. To solve these equations we raise both sides of the equation to the same power. However, when we do this, the equation that we form may have more solutions than the original equation. To eliminate such extraneous solutions it is necessary to check our solutions in the original equation. The following examples illustrate the procedure for solving radical equations.

EXAMPLE 1: For what value of x is the equation $\sqrt{x-1} = 4$ a true statement?

SOLUTION:

$$\sqrt{x-1} = 4$$
$$(\sqrt{x-1})^2 = (4)^2 \qquad \text{square both sides of the equation}$$
$$x - 1 = 16$$
$$x = 17$$

Check:

$$\sqrt{17-1} \overset{?}{=} 4$$
$$\sqrt{16} \overset{?}{=} 4$$
$$4 \overset{\checkmark}{=} 4$$

Therefore, 17 is the solution of the equation.

EXAMPLE 2: Solve the equation: $\sqrt{2x+1} + 3 = 0$.

SOLUTION: To solve the equation we must first isolate the radical.

$$\sqrt{2x+1} + 3 = 0$$
$$\sqrt{2x+1} = -3$$
$$(\sqrt{2x+1})^2 = (-3)^2 \qquad \text{square both sides of the equation}$$
$$2x + 1 = 9$$
$$2x = 8$$
$$x = 4$$

Check:

$$\sqrt{2(4)+1} + 3 \overset{?}{=} 0$$
$$\sqrt{9} + 3 \overset{?}{=} 0$$
$$3 + 3 \overset{?}{=} 0$$
$$6 \neq 0$$

Therefore, the equation $\sqrt{2x + 1} + 3 = 0$ has no solution. [*Note:* Remember that by definition the square root of a number is either positive or zero. Thus, $\sqrt{9} \neq -3$, even though $(-3)(-3) = 9$.]

EXAMPLE 3: Solve the equation: $\sqrt[3]{3x - 1} = 2$.

SOLUTION:

$$\sqrt[3]{3x - 1} = 2$$
$$(\sqrt[3]{3x - 1})^3 = 2^3 \qquad \text{raise both sides of the equation to the third power}$$
$$3x - 1 = 8$$
$$3x = 9$$
$$x = 3$$

Check:

$$\sqrt[3]{3(3) - 1} \overset{?}{=} 2$$
$$\sqrt[3]{8} \overset{?}{=} 2$$
$$2 = 2$$

Therefore, 3 is the solution of the equation.

EXAMPLE 4: Solve the equation: $x + \sqrt{x} - 2 = 0$.

SOLUTION: To solve the equation, we must first isolate the radical.

$$x + \sqrt{x} - 2 = 0$$
$$x - 2 = -\sqrt{x}$$
$$(x - 2)^2 = (-\sqrt{x})^2 \qquad \text{square both sides of the equation}$$
$$x^2 - 4x + 4 = x$$
$$x^2 - 5x + 4 = 0$$
$$(x - 4)(x - 1) = 0$$
$$x = 4 \qquad x = 1$$

Check:

$$4 + \sqrt{4} - 2 \overset{?}{=} 0 \qquad 1 + \sqrt{1} - 2 \overset{?}{=} 0$$
$$4 + 2 - 2 \overset{?}{=} 0 \qquad 1 + 1 - 2 \overset{?}{=} 0$$
$$4 \neq 0 \qquad\qquad 0 \overset{\checkmark}{=} 0$$

Therefore, 1 is the solution of the equation.

EXAMPLE 5: Solve the equation: $\sqrt{x} + \sqrt{x + 5} = 5$.

SOLUTION: To solve the equation we first isolate one of the radicals.

$$\sqrt{x} + \sqrt{x + 5} = 5$$
$$\sqrt{x + 5} = 5 - \sqrt{x}$$
$$(\sqrt{x + 5})^2 = (5 - \sqrt{x})^2 \quad \text{square both sides of the equation}$$
$$x + 5 = 25 - 10\sqrt{x} + x$$
$$-20 = -10\sqrt{x}$$
$$2 = \sqrt{x}$$
$$(2)^2 = (\sqrt{x})^2 \quad \text{square both sides of the equation}$$
$$4 = x$$

Check:

$$\sqrt{4} + \sqrt{4 + 5} \overset{?}{=} 5$$
$$2 + 3 \overset{?}{=} 5$$
$$5 \overset{\checkmark}{=} 5$$

Thus, 4 is the solution of the equation.

Note that it is sometimes necessary to raise both sides of the equation to some power more than once.

EXERCISES 6–8

Solve each equation and check the solution.

1. $\sqrt{x + 5} = 2$

2. $\sqrt{2x + 1} = 7$

3. $\sqrt{x - 5} = -3$

4. $\sqrt{x - 2} + 3 = 0$

5. $2\sqrt{3x + 3} = 13$

6. $3\sqrt{2x} - 4 = 5$

7. $\sqrt{2x + 1} = \sqrt{3x - 5}$

8. $\sqrt{x^2 + x + 6} = \sqrt{x^2 + 3x + 2}$

9. $\sqrt[3]{x + 1} = -2$

10. $\sqrt[3]{x - 2} = 4$

11. $\sqrt{x^2 + 9} = x + 1$

12. $\sqrt{x^2 - x + 1} = x - 1$

13. $\sqrt{8 - x^2} + x = 0$

14. $\sqrt{x + 1} + 1 = x$

15. $\sqrt{x - 2} + x = 4$

16. $\sqrt{x + 4} - 4 = x$

17. $2\sqrt{x + 1} = 2 - x$

18. $2\sqrt{x - 1} = x - 1$

19. $\sqrt{2x^2 - 2} + x - 1 = 0$

20. $\sqrt{x - 4} = 5 - \sqrt{x + 1}$

21. $\sqrt{x - 8} + \sqrt{x} = 2$

22. $\sqrt{x + 5} - 9 = \sqrt{x - 4}$

23. $\sqrt{x} - \sqrt{2x + 1} = -1$

24. $\sqrt{4x + 2} + \sqrt{2x} - \sqrt{2} = 0$

25. $\sqrt{2x} + \sqrt{2x + 3} = 3$

26. $\dfrac{10}{\sqrt{x - 5}} = \sqrt{x - 5} - 3$

27. $\sqrt{x^2} = x$

6–9 COMPLEX NUMBERS

In the last few sections we have been careful to ensure that we were not taking the square root of a negative number. Such numbers as $\sqrt{-1}$ are not real numbers, for the product of two equal real factors never results in a negative number. So we must extend our number system to include a new kind of number, called an *imaginary number*.

> IMAGINARY NUMBER: An imaginary number is an even root of a negative number.

For example, $\sqrt{-4}$, $\sqrt[4]{-2}$, and $\sqrt[6]{-5}$ are imaginary numbers. In this section we will only consider the square root of a negative number.

The basic unit in imaginary numbers is $\sqrt{-1}$ and it is designated by i. Thus, by definition we have

$$i = \sqrt{-1} \quad \text{and} \quad i^2 = -1$$

When working with the square roots of negative numbers, the first step is to express these imaginary numbers as the product of a real number and i. For example:

$$\sqrt{-4} = \sqrt{(-1)(4)} = \sqrt{-1}\sqrt{4} = 2i$$
$$\sqrt{-5} = \sqrt{(-1)(5)} = \sqrt{-1}\sqrt{5} = i\sqrt{5}$$
$$\sqrt{-12} = \sqrt{(-1)(4)(3)} = \sqrt{-1}\sqrt{4}\sqrt{3} = 2i\sqrt{3}$$

Using real and imaginary numbers we may define a new type of number. A number of the form $a + bi$, where a and b are real numbers and $i = \sqrt{-1}$, is called a *complex number*. The first term, a, is the real part and bi is the imaginary part. Since a and/or b may be zero, the complex numbers include the real numbers and the imaginary numbers. Figure 6.1 illustrates the relationships among the various types of numbers. Thus, every number that we have discussed in this book is a complex number.

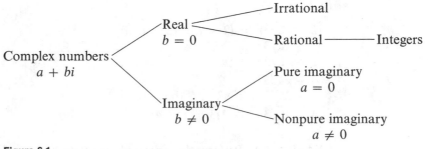

Figure 6.1

In many ways, computations with complex numbers are similar to computations with real numbers that contain square roots. The important difference is that we must always express the imaginary part in terms of i, remembering that $i^2 = -1$. The following examples illustrate how to perform the various operations with complex numbers.

EXAMPLE 1: Express in terms of i and combine:

a. $\sqrt{-100} + \sqrt{-64}$ 　　　　　　　　**b.** $3\sqrt{-3} - 2\sqrt{-48}$

SOLUTION:

a. $\quad \sqrt{-100} + \sqrt{-64}$ 　　　　　　**b.** $\quad 3\sqrt{-3} - 2\sqrt{-48}$

$\quad\quad \sqrt{-1} \cdot \sqrt{100} + \sqrt{-1} \cdot \sqrt{64}$ 　　$\quad 3\sqrt{-1} \cdot \sqrt{3} - 2\sqrt{-1} \cdot \sqrt{16} \cdot \sqrt{3}$

$\quad\quad\quad\quad 10i + 8i$ 　　　　　　　　　　$\quad\quad 3i\sqrt{3} - 8i\sqrt{3}$

$\quad\quad\quad\quad\quad 18i$ 　　　　　　　　　　　$\quad\quad\quad -5i\sqrt{3}$

EXAMPLE 2: Perform the indicated operation and simplify: $\sqrt{-2} \cdot \sqrt{-32}$

SOLUTION:

$$\sqrt{-2} \cdot \sqrt{-32} = (\sqrt{-1} \cdot \sqrt{2})(\sqrt{-1} \cdot \sqrt{16} \cdot \sqrt{2}) = (i\sqrt{2})(4i\sqrt{2}) = 4 \cdot i^2 \cdot 2$$
$$= 4(-1)2 \quad \text{(since } i^2 = -1)$$
$$= -8$$

Note that the property of radicals $\sqrt{x}\sqrt{y} = \sqrt{xy}$ does not hold in this case. That is, it is *not* true that

$$\sqrt{-2}\sqrt{-32} = \sqrt{(-2)(-32)} = \sqrt{64} = 8$$

The correct result is -8. This is why we must be careful when we work with complex numbers. Always remember to express the complex numbers in terms of i before performing computations.

EXAMPLE 3: Perform the indicated operation and simplify: $(2 - i\sqrt{3})^2$.

SOLUTION:

$$(2 - i\sqrt{3})^2 = (2 - i\sqrt{3})(2 - i\sqrt{3})$$

$\quad\quad 2 - i\sqrt{3}$

$\quad\quad \underline{2 - i\sqrt{3}}$

$\quad\quad 4 - 2i\sqrt{3}$

$\quad\quad\quad \underline{- 2i\sqrt{3} + i^2 \cdot 3} \quad\quad \text{(but } i^2 = -1)$

$\quad\quad 4 - 4i\sqrt{3} + (-1)3 = 1 - 4i\sqrt{3}$

EXAMPLE 4: Perform the indicated operation and simplify: $(4 + 3i) \div (1 - 2i)$

SOLUTION: As in our work with radicals, we rationalize the denominator. The *conjugate* of $a + bi$ is $a - bi$ and conversely. The product of two conjugate complex numbers is always a real number. Thus, two complex numbers are divided by multiplying the numerator and denominator by the conjugate of the denominator.

$$\frac{4 + 3i}{1 - 2i} = \frac{4 + 3i}{1 - 2i} \cdot \frac{1 + 2i}{1 + 2i}$$

$$\begin{array}{cc}
\begin{array}{r}
4 + 3i \\
1 + 2i \\
\hline
4 + 3i \\
+\ 8i + 6i^2 \\
\hline
4 + 11i + 6(-1) = -2 + 11i
\end{array}
&
\begin{array}{r}
1 + 2i \\
1 - 2i \\
\hline
1 + 2i \\
-\ 2i - 4i^2 \\
\hline
1 \qquad - 4(-1) = 5
\end{array}
\end{array}$$

Thus,

$$\frac{4 + 3i}{1 - 2i} = \frac{-2 + 11i}{5} \quad \text{or} \quad \frac{-2}{5} + \frac{11}{5}i$$

To summarize, the following rules are used to combine complex numbers:

1. Two complex numbers are added by adding separately their real parts and their imaginary parts.

$$(a + bi) + (c + di) = (a + c) + (b + d)i$$

2. Two complex numbers are subtracted by subtracting separately their real parts and their imaginary parts.

$$(a + bi) - (c + di) = (a - c) + (b - d)i$$

3. Two complex numbers are multiplied in the usual FOIL method with i^2 being replaced by -1.

$$\text{F} + \text{O} + \text{I} + \text{L}$$
$$\downarrow \quad \downarrow \quad \downarrow \quad \downarrow$$
$$(a + bi)(c + di) = ac + adi + bci + bdi^2 = (ac - bd) + (ad + bc)i$$

4. Two complex numbers are divided by multiplying the numerator and denominator by the conjugate of the denominator.

$$\frac{a + bi}{c + di} = \frac{(a + bi)(c - di)}{(c + di)(c - di)} = \frac{(ac + bd) + (bc - ad)i}{c^2 + d^2}$$

EXERCISES 6-9

In Exercises 1–20 express in terms of i and simplify when possible.

1. $\sqrt{-25}$ 2. $\sqrt{-49}$

3. $2\sqrt{-16}$ 4. $5\sqrt{-121}$

5. $\sqrt{-22}$ 6. $-3\sqrt{-7}$

7. $\frac{1}{2}\sqrt{-12}$ 8. $-3\sqrt{-98}$

9. $2 - \sqrt{-128}$ 10. $-3 + \frac{1}{2}\sqrt{-162}$

11. $2\sqrt{-9/16}$

12. $10\sqrt{-49/4}$

13. $3\sqrt{-1/2}$

14. $6\sqrt{-2/3}$

15. $\frac{1}{2}\sqrt{-63x^2}$

16. $\frac{1}{4}\sqrt{-48x^3}$

17. $\sqrt{(-4)^2 - 4(1)(5)}$

18. $\sqrt{(12)^2 - 4(9)(8)}$

19. $\sqrt{(2)^2 - 4(-1)(-3)}$

20. $\sqrt{(-5)^2 - 4(2)(7)}$

In Exercises 21–30 simplify in terms of i and combine.

21. $2\sqrt{-9} - 3\sqrt{-9} + 4\sqrt{-9}$

22. $4\sqrt{-1} - 2\sqrt{-4} + 5\sqrt{-16}$

23. $\sqrt{-8} + \sqrt{-18} - \sqrt{-50}$

24. $4\sqrt{-27} - 2\sqrt{-48} + 3\sqrt{-75}$

25. $\sqrt{-72} - \frac{1}{2}\sqrt{-32} + 4\sqrt{-1/2}$

26. $-2\sqrt{-12} + \sqrt{-48} - 9\sqrt{-1/3}$

27. $(3 - i) - (-2 + 5i)$

28. $(6 + 2i) + (2 - 4i)$

29. $(1 - \sqrt{-4}) + (5 + 3\sqrt{-4})$

30. $(2 + 3\sqrt{-5}) - (-1 - 4\sqrt{-5})$

In Exercises 31–50 perform the indicated operation and simplify.

31. $\sqrt{-3} \cdot \sqrt{-12}$

32. $\sqrt{-5} \cdot \sqrt{-18}$

33. $(2i)^2$

34. $(-i)^2$

35. $\sqrt{-4} \div \sqrt{-25}$

36. $4\sqrt{-2} \div \sqrt{-32}$

37. $(4 + 2i)(1 - 3i)$

38. $(2 - 5i)(3 - i)$

39. $(2 + 3i)(2 - 3i)$

40. $(6 - 5i)(6 + 5i)$

41. $(1 - 2i)^2$

42. $(5 + i)^2$

43. $(-3 + \frac{1}{3}i)^2$

44. $(-1 - \frac{1}{2}i)^2$

45. $1 \div (1 + 2i)$

46. $-3 \div (2 - 3i)$

47. $i \div (-1 - 4i)$

48. $-2i \div (6 - 2i)$

49. $(2 + i) \div (3 + i)$

50. $(2 + 5i) \div (4 + 2i)$

51. If $f(x) = x^2 + 4$, find $f(2i)$.

52. If $f(x) = x^2 + 2$, find $f(i\sqrt{2})$.

53. If $f(x) = x^2 - 4x + 5$, find $f(2 + i)$.

54. If $f(x) = x^2 - 4x + 5$, find $f(2 - i)$.

55. If $f(x) = 3x^2 - 2x + 1$, find $f\left(\dfrac{1 + i\sqrt{2}}{3}\right)$.

56. If $f(x) = 3x^2 - 2x + 1$, find $f\left(\dfrac{1 - i\sqrt{2}}{3}\right)$.

57. If $f(x) = 2x^2 + x + 1$, find $f(5 + 2i)$.

58. If $f(x) = 2x^2 + x + 1$, find $f(5 - 2i)$.

59. Write each expression in a simpler form:

    a. i^3

    b. i^4

    c. i^5

    d. i^6

    e. i^7

    f. i^9

    g. i^{93}

    h. i^{100}

60. Complex numbers are used extensively in electrical engineering to designate voltage (E), current (I), and resistance (R). However, since i usually symbolizes current in this discipline, complex numbers are written in the form $a + bj$ with

$j = \sqrt{-1}$. In the following exercises use Ohm's law, $E = IR$, to determine the value of the missing variable.

    a. $I = 1 - 2j$ amps, $R = 2 + 3j$ ohms.
    b. $E = 2$ volts, $I = 1 + 4j$ amps.
    c. $R = 7 - 5j$ ohms, $E = 1 - j$ volts.

6-10 TRIGONOMETRIC FORM AND DEMOIVRE'S THEOREM

Each complex number $a + bi$ involves a pair of real numbers, a and b. Graphically, this means that we may represent a complex number as a point in the Cartesian coordinate system. The values for a are plotted on the x axis and the values for b are plotted on the y axis. Thus, the complex number $a + bi$ is represented by the point (a, b) with x value a and y value b. Consider Figure 6.2, which indicates the points corresponding to several complex numbers.

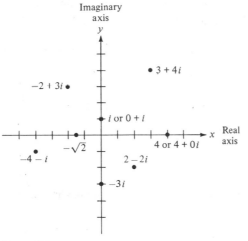

Figure 6.2

A useful new way of writing a complex number is to interpret the position of its geometric representation in terms of the trigonometric functions. Consider the point corresponding to the complex number $a + bi$ in Figure 6.3. When we use

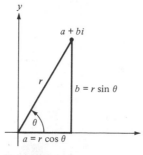

Figure 6.3

trigonometry, we obtain the following relationships:

$$\sin \theta = \frac{b}{r} \quad \text{therefore} \quad b = r \sin \theta$$

$$\cos \theta = \frac{a}{r} \quad \text{therefore} \quad a = r \cos \theta$$

$$r = \sqrt{a^2 + b^2}$$

$$\tan \theta = \frac{b}{a}$$

Thus, we may change the form of the complex number as follows:

$$a + bi = r \cos \theta + (r \sin \theta)i = r(\cos \theta + i \sin \theta)$$

In this trigonometric form, θ is called the *amplitude* of the complex number and r is called the *absolute value*. Note that the absolute value of a complex number $a + bi$ is interpreted geometrically as the distance from the origin to (a, b).

EXAMPLE 1: Write the number $3 - 3i$ in trigonometric form.

SOLUTION: First determine the absolute value, r. Since $a = 3$ and $b = -3$, we have

$$r = \sqrt{a^2 + b^2} = \sqrt{(3)^2 + (-3)^2} = \sqrt{18} \quad \text{or} \quad 3\sqrt{2}$$

Now determine the amplitude,

$$\tan \theta = \frac{b}{a} = \frac{-3}{3} = -1$$

Since $(3, -3)$ is in Q_4 and the reference angle is $45°$, we conclude that

$$\theta = 315°$$

The number is written in trigonometric form as

$$3\sqrt{2}(\cos 315° + i \sin 315°)$$

See Figure 6.4. (*Note:* Any angle coterminal with $315°$ may also be used. For example, we may write $3\sqrt{2}[\cos(-45°) + i \sin(-45°)]$.)

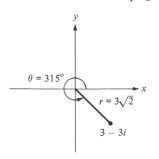

Figure 6.4

EXAMPLE 2: Write the number -2 in trigonometric form.

SOLUTION: First determine the absolute value, r. Since $a = -2$ and $b = 0$, we have

$$r = \sqrt{a^2 + b^2} = \sqrt{(-2)^2 + 0^2} = \sqrt{4} = 2$$

The amplitude is $180°$ since -2 is a point on the negative portion of the x axis. The number in trigonometric form is

$$2(\cos 180° + i \sin 180°) \qquad \text{(See Figure 6.5.)}$$

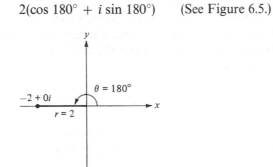

Figure 6.5

EXAMPLE 3: Write the number $2(\cos 150° + i \sin 150°)$ in the form $a + bi$.

SOLUTION: Since $\cos 150° = -\sqrt{3}/2$ and $\sin 150° = 1/2$, we have

$$2(\cos 150° + i \sin 150°) = 2\left[\frac{-\sqrt{3}}{2} + i\left(\frac{1}{2}\right)\right]$$
$$= -\sqrt{3} + i$$

The trigonometric form is very useful when working with powers and roots of complex numbers. Consider the following statement, which is known as *Demoivre's theorem.*

DEMOIVRE'S THEOREM: If $r(\cos \theta + i \sin \theta)$ is any complex number and n is any real number, then

$$[r(\cos \theta + i \sin \theta)]^n = r^n(\cos n\theta + i \sin n\theta)$$

We illustrate this theorem in the following example.

EXAMPLE 4: Find $(1 + i)^{12}$.

SOLUTION: First write the number in trigonometric form.

$$r = \sqrt{a^2 + b^2} = \sqrt{(1)^2 + (1)^2} = \sqrt{2}$$

$$\tan \theta = \frac{b}{a} = \frac{1}{1} = 1 \qquad \theta = 45°$$

Thus, $1 + i = \sqrt{2}(\cos 45° + i \sin 45°)$. Now Demoivre's theorem tells us that

$$[\sqrt{2}(\cos 45° + i \sin 45°)]^{12} = (\sqrt{2})^{12}[\cos(12 \cdot 45°) + i \sin(12 \cdot 45°)]$$
$$= (2^{1/2})^{12}[\cos 540° + i \sin 540°]$$
$$= 2^6[(-1) + i(0)]$$
$$= -64$$

Any complex number has two square roots, three cube roots, four fourth roots, and so on. To find these roots we use Demoivre's theorem. However, to find all of them you must remember that there are many trigonometric representations for the same complex number. For example, since the angle $\theta + k \cdot 360°$ (k any integer) is coterminal to θ, we have

$$1 + i = \sqrt{2}(\cos 45° + i \sin 45°) = \sqrt{2}(\cos 405° + i \sin 405°)$$
$$= \sqrt{2}(\cos 765° + i \sin 765°) \quad \text{etc.}$$

Thus, the method is to find one root, add 360° to θ, use the new trigonometric representation to find another root, and repeat this procedure until all n roots are found.

EXAMPLE 5: Find the five fifth roots of 32.

SOLUTION: First we write 32 in trigonometric form as

$$32(\cos 0° + i \sin 0°)$$

Now applying Demoivre's theorem, we have

$$[32(\cos 0° + i \sin 0°)]^{1/5} = 32^{1/5}\left(\cos \frac{0°}{5} + i \sin \frac{0°}{5}\right)$$

$$= 2(\cos 0° + i \sin 0°) = 2$$

The first root is 2. To find another root, first add 360° to θ to obtain a different trigonometric representation for 32.

$$32 = 32[\cos(0° + 360°) + i \sin(0° + 360°)]$$

Then, by Demoivre's theorem,

$$\text{2nd root is } 32^{1/5}\left(\cos \frac{0° + 360°}{5} + i \sin \frac{0° + 360°}{5}\right) = 2(\cos 72° + i \sin 72°)$$

Repeating this procedure, we have

$$\text{3rd root is } 32^{1/5}\left(\cos \frac{0° + 2 \cdot 360°}{5} + i \sin \frac{0° + 2 \cdot 360°}{5}\right) = 2(\cos 144° + i \sin 144°)$$

$$\text{4th root is } 32^{1/5}\left(\cos \frac{0° + 3 \cdot 360°}{5} + i \sin \frac{0° + 3 \cdot 360°}{5}\right) = 2(\cos 216° + i \sin 216°)$$

$$\text{5th root is } 32^{1/5}\left(\cos \frac{0° + 4 \cdot 360°}{5} + i \sin \frac{0° + 4 \cdot 360°}{5}\right) = 2(\cos 288° + i \sin 288°)$$

Note that except for the first root of 2, these numbers are expressed more simply in trigonometric form.

EXERCISES 6-10

In Exercises 1–14 graph each complex number and write the number in trigonometric form.

1. 3

2. i

3. $-2i$

4. -4

5. $1 - i$

6. $-1 + i$

7. $4 + 3i$

8. $-5 - 12i$

9. $-3 - 2i$

10. $7 - 4i$

11. $-1 + \sqrt{3}i$

12. $\sqrt{3} - i$

13. $\sqrt{2} - \sqrt{2}i$

14. $-2 - 2\sqrt{3}i$

In Exercises 15–24 write the number in the form $a + bi$.

15. $3(\cos 90° + i \sin 90°)$

16. $5(\cos 0° + i \sin 0°)$

17. $4(\cos 180° + i \sin 180°)$

18. $\sqrt{2}(\cos 270° + i \sin 270°)$

19. $\sqrt{3}(\cos 120° + i \sin 120°)$

20. $2(\cos 210° + i \sin 210°)$

21. $2(\cos 225° + i \sin 225°)$

22. $\sqrt{6}(\cos 315° + i \sin 315°)$

23. $\cos 52° + i \sin 52°$

24. $10(\cos 115° + i \sin 115°)$

In Exercises 25–30 use Demoivre's theorem and express the result in the form $a + bi$.

25. $(1 + i)^8$

26. $(-1 + i)^6$

27. $(1 - \sqrt{3}i)^5$

28. $(-\sqrt{3} - i)^7$

29. $\left(\dfrac{\sqrt{2}}{2} + \dfrac{\sqrt{2}}{2}i\right)^{16}$

30. $\left(\dfrac{\sqrt{2}}{2} - \dfrac{\sqrt{2}}{2}i\right)^{14}$

In Exercises 31–36 use Demoivre's theorem to find all roots. Leave the answers in trigonometric form when Table 1 in the Appendix is required.

31. $\sqrt[3]{1}$

32. $\sqrt[4]{-1}$

33. $\sqrt[4]{-16i}$

34. $\sqrt[3]{8i}$

35. $\sqrt[5]{1 + i\sqrt{3}}$

36. $\sqrt[5]{1 + i}$

In Exercises 37–40 find all the solutions of the equation. Leave the answers in trigonometric form when Table 1 in the Appendix is required.

37. $x^3 + 8 = 0$

38. $x^3 - 27i = 0$

39. $x^4 - 1 = 0$

40. $x^5 + 32 = 0$

Sample Test: Chapter 6

In questions 1–12 perform the indicated operations and express your result in the simplest form that contains only positive exponents.

1. $\dfrac{2}{2^{-1}}$

2. $(-2)^{-1} \cdot (-2)^{-3}$

3. $2^0 - \left(\dfrac{4}{9}\right)^{-3/2}$

4. $\dfrac{2^{-3} - 7^0}{4^{-1} + 5^0}$

5. $5\sqrt{12} - 2\sqrt{27}$

6. $\sqrt{\frac{4}{3}} + \sqrt{\frac{3}{4}}$

7. $\dfrac{\sqrt{3} + \sqrt{2}}{\sqrt{2}}$

8. $\dfrac{1}{3 + i}$

9. $(3\sqrt{-5})^2$

10. $(2 + 3i^2) - (3 - 2i)$

11. $\dfrac{(x^{-2/3}x^{1/6})^{-1}}{x^{1/3}}$

12. $\dfrac{a^{-2/3}}{\sqrt[3]{a}}$

13. If $f(x) = x^{3/2} + 2x^0$, find $f(4)$.

14. If $x = 3 + \sqrt{2}$, determine the value of x^2.

15. The total surface area of the earth is about 1.97×10^8 mi^2. Express this number in regular notation.

16. The wavelength of violet light is about 0.000040 cm. Express this number in scientific notation.

17. Solve for x: $\sqrt{2x + 1} - 3 = 0$.

18. Solve for x: $2\sqrt{x + 1} = x - 2$.

19. Write the complex number $-1 - i$ in trigonometric form.

20. Write the number $4(\cos 300° + i \sin 300°)$ in the form $a + bi$.

PART
TWO

Elementary functions

Chapter 7
Functions revisited

7–1 FUNCTIONS

In Chapter 2 we introduced the idea of a function. For the next few chapters this concept is the central theme as we investigate the behavior of the "elementary functions" studied in calculus. These important functions are called the polynomial, rational, exponential, logarithmic and trigonometric functions. We begin by restating some of the key ideas in Chapter 2.

In mathematics one of the most important considerations is determining how two variables are related. For example:

The sales tax on a purchase is related to the price charged for the item(s).

The exposure time for photographing an object is related to the diameter of the camera lens.

The weight of an object is related to its distance from the center of the earth.

The circumference of a circle is related to the length of the diameter of the circle.

We can analyze each relationship if we find a rule that gives the correspondences between values of each variable. This rule can be given in different ways. Formulas or equations are frequently used to define relationships. For example the formula $C = \pi d$ establishes a correspondence between the length of the diameter and the circumference of a circle. Tables are another common way to state a rule. In many college catalogs you find the table in Figure 7.1, which idealizes the relationship between final averages and final grades.

$$\text{Final grade} = \begin{cases} A & \text{if} \quad 90 \le a \le 100 \\ B & \text{if} \quad 80 \le a < 90 \\ C & \text{if} \quad 70 \le a < 80 \\ D & \text{if} \quad 60 \le a < 70 \\ F & \text{if} \quad 0 \le a < 60 \end{cases}$$

Figure 7.1

Since formulas and tables are not always applicable, it is sometimes best to give the rule verbally or to make a list of the correspondences. The relationship between students and their social security numbers is such a case.

Whether the rule is given by formula or table, verbally, or by a list, the rule is most useful if we obtain *exactly one* answer whenever we use it. For example, the rule in Figure 7.1 assigns to each final average exactly one final grade. Once we compute the final average, the rule tells us exactly what grade to assign for the course. Some typical assignments are shown in Figure 7.2.

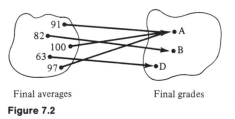

Final averages   Final grades

Figure 7.2

However some correspondences do not always give us exactly one answer. For example, we cannot determine a final average if we know the final grade; we can determine only that the average lies in some interval. An A grade may correspond to any average between 90 and 100 (inclusive). This concept is shown in Figure 7.3.

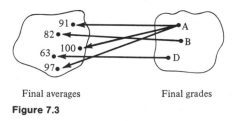

Final averages   Final grades

Figure 7.3

Thus, in defining the relationship between two variables, we wish to pick an arbitrary value for a variable (which is then called the independent variable) and have some rule assign exactly one value for the other variable (called the dependent variable). This observation leads us to the following definition of a function.

DEFINITION OF A FUNCTION: A function consists of three features:
1. A collection of values that may be substituted for the independent variable, called the *domain* of the function.
2. A collection of assigned values for the dependent variable, called the *range* of the function.
3. A rule (usually represented by f) that assigns to each member of the domain *exactly one* member of the range.

The analogy between a function and a computing machine may help to clarify the concept. Consider Figure 7.4, which shows a machine processing domain elements (x values) into range elements (y values). In goes an x value, out comes exactly one y value. With this in mind, you should have a clear image of the three features of a function: (1) the domain (the input), (2) the range (the output), and (3) the rule f (the machine).

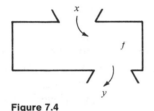

Figure 7.4

It is common practice to refer to a function by stating only the rule; no domain or range is specified. Thus we say, "Consider the circumference function $C = \pi d$." The domain is then assumed to be the collection of values for d that are interpretable in the problem. Since the length of the diameter of a circle must be positive in this case, the domain is $d > 0$. For functions defined by algebraic equations like $y = 1/\sqrt{x - 1}$, the domain is the collection of all real numbers for which a real number exists in the range. Thus we exclude from the domain values for the independent variable (x) that result in the square root of a negative number or in division by zero. For $y = 1/\sqrt{x - 1}$ we want $x - 1 > 0$, so that the domain is $x > 1$. The range of a function is usually more difficult to determine, and we will gradually consider this topic as we progress in the text.

EXAMPLE 1: The formula $A = s^2$ defines a function since to each value of the side (s) of a square there corresponds exactly one area (A). The domain of the function is all the positive real numbers since the side of a square must be positive.

EXAMPLE 2: The equation $y = 1/(x - 2)$ defines a function since to each value of x there corresponds exactly one y value. The domain of the function is all the real numbers except two. We exclude two since when $x = 2$, $y = 1/0$, which is undefined.

EXAMPLE 3: The equation $y = \sqrt{1 - x^2}$ defines a function. Since the square root of a negative number is not a real number, $1 - x^2$ cannot be less than 0. Thus, the domain is given by $-1 \le x \le 1$.

EXAMPLE 4: The equation $y = 4$ (or $y = 4 + 0 \cdot x$) defines a function since each x value is assigned exactly one y value. In this case the y value is 4 regardless of the x chosen. This type of function, in which the y value does not change, is called a *constant function*. The domain is all the real numbers.

EXAMPLE 5: The deduction that single persons in New York State may take on their federal income tax based on state sales tax expended on items purchased during the year depends on their income. The following formula table shows the amount that may be deducted for incomes (i) up to $5000:

$$\text{Deduction} = \begin{cases} \$37 & \text{if} \quad \$0 \quad \le i < \$3000 \\ \$49 & \text{if} \quad \$3000 \le i < \$4000 \\ \$60 & \text{if} \quad \$4000 \le i < \$5000 \end{cases}$$

This table defines a function, since to each income there corresponds exactly one deduction. The domain of this function would be: $\$0 \le i < \$5000$. Remember that one does not think of this function as three separate functions but rather as one function that describes the relationship between an individual's income and the allowable sales tax deduction. Notice that if we reverse the assignments, we do not have a function. That is, we cannot determine an individual's income if we know his deduction.

EXAMPLE 6: Consider the following rule:

$$y = \begin{cases} x - 1 & \text{if} \quad 0 \le x \le 3 \\ 2 & \text{if} \quad x > 3 \end{cases}$$

To any value of x between 0 and 3 (inclusive) this rule assigns the number one less than the x value. For example, when $x = 2$, $y = 1$ and when $x = \frac{1}{2}$, $y = -\frac{1}{2}$. This rule also assigns the number 2 to each value of x greater than 3. The rule defines a function, since each x value is assigned exactly one y value. The domain is $x \ge 0$.

EXAMPLE 7: Consider the rule

$$y = \begin{cases} x & \text{if} \quad x \ge 0 \\ -x & \text{if} \quad x < 0 \end{cases}$$

This rule assigns to any nonnegative value of x the number itself. For example, when $x = 3$, $y = 3$, and when $x = 0$, $y = 0$. The rule also assigns to each negative value of x the negative of that number. For example, when $x = -3$, $y = -(-3) = 3$, and when $x = -\frac{1}{2}$, $y = -(-\frac{1}{2}) = \frac{1}{2}$. This rule defines the *absolute value function* $y = |x|$. The domain is all the real numbers.

EXAMPLE 8: A piece of wire 46 in. long is bent into a rectangle of sides x and y units. Express y as a function of x. Express the area (A) of the rectangle as a function of x. Find the domain of the function.

SOLUTION: First express y as a function of x by using the perimeter formula:

$$P = 2x + 2y$$
$$46 = 2x + 2y$$
$$23 - x = y$$

To express the area (A) as a function of x we now replace y by $23 - x$ in the area formula:

$$A = x \cdot y$$
$$= x(23 - x)$$

Since x and A must both be positive, the domain is

$$0 < x < 23$$

In mathematical notation we represent the correspondences in a function by using *ordered pairs*. For example, consider the equation $y = x^2 - 1$.

If x equals	then $y = x^2 - 1$	thus the ordered pairs
2	$(2)^2 - 1 = 3$	$(2, 3)$
0	$(0)^2 - 1 = -1$	$(0, -1)$
$\sqrt{2}$	$(\sqrt{2})^2 - 1 = 1$	$(\sqrt{2}, 1)$
-3	$(-3)^2 - 1 = 8$	$(-3, 8)$

In the pairs that represent the correspondences in a function, the value for the independent variable is listed first and the value for the dependent variable comes second. Thus, the order of the numbers in the pair is significant. The pairing $(2, 3)$ indicates that when $x = 2$, $y = 3$; $(3, 2)$ means when $x = 3$, $y = 2$. In the equation $y = x^2 - 1$, $(2, 3)$ is an ordered pair that makes the equation a true statement, so $(2, 3)$ is said to be a solution of the equation; $(3, 2)$ is not a solution of this equation.

EXAMPLE 9: Which of the following ordered pairs are solutions of the equation $y = 2x^3 - 1$?

a. $(-1, -3)$ **b.** $(-3, -1)$

SOLUTION:

a. $(-1, -3)$ means that when

$$x = -1 \qquad y = -3$$
$$y = 2x^3 - 1$$
$$(-3) \overset{?}{=} 2(-1)^3 - 1$$
$$-3 \overset{?}{=} 2(-1) - 1$$
$$-3 \overset{\checkmark}{=} -3$$

Thus, $(-1, -3)$ is a solution.

b. $(-3, -1)$ means that when

$$x = -3 \qquad y = -1$$
$$y = 2x^3 - 1$$
$$(-1) \overset{?}{=} 2(-3)^3 - 1$$
$$-1 \overset{?}{=} 2(-27) - 1$$
$$-1 \neq -55$$

Thus, $(-3, -1)$ is not a solution.

The representation of the correspondences in a function as ordered pairs gives us a different perspective of the function concept. In higher mathematics the following definition of a function is very popular.

DEFINITION OF A FUNCTION: A function is a collection of ordered pairs in which no two different ordered pairs have the same first component. The collection of all first components (of the ordered pairs) is called the *domain* of the function. The collection of all second components is called the *range* of the function.

EXAMPLE 10: Consider the ordered pairs $(-1, 0)(0, 0)$ and $(1, 0)$. This collection is a function because the first component in the ordered pairs is always different. Domain: $-1, 0, 1$; range: 0 (see Figure 7.5).

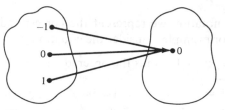

Figure 7.5 Function

EXAMPLE 11: Consider the ordered pairs $(0, -1)(0, 0)$ and $(0, 1)$. This collection is not a function because the number 0 is the first component in more than one ordered pair (see Figure 7.6).

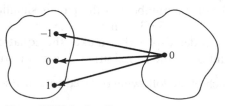

Figure 7.6 Not a function

One of the sources of information and insight about a relationship is a picture that describes the particular situation. We can pictorially represent a function by using the Cartesian coordinate system. The *graph* of a function is the collection of all the points in this coordinate system that correspond to ordered pairs in the function. A detailed explanation of the Cartesian coordinate system and graphing techniques may be found in Section 2-4.

Consider Figures 7.7–7.9. They represent the graphs of the functions in Examples 3, 5, and 7. Note that the domain and range of each function may be easily read from the appropriate graph.

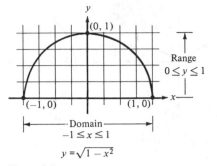

Figure 7.7

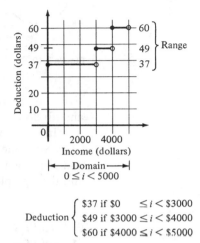

$$\text{Deduction} \begin{cases} \$37 \text{ if } \$0 \quad \leq i < \$3000 \\ \$49 \text{ if } \$3000 \leq i < \$4000 \\ \$60 \text{ if } \$4000 \leq i < \$5000 \end{cases}$$

Figure 7.8

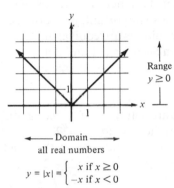

Figure 7.9

At present we graph functions by plotting enough points to determine the basic behavior of the function. In succeeding chapters we develop the more efficient method of determining the essential characteristics of the graph from the form of the equation. For example, the graph of a function of the form $y = mx + b$, where m and b are real number constants, is a straight line.

In a function no two ordered pairs may have the same first component. Graphically, this means that a function cannot contain two points that lie on the same vertical line. This observation leads to the following simple test for determining if a graph represents a function:

VERTICAL LINE TEST: Imagine a vertical line sweeping across the graph. If the vertical line at any position intersects the graph in more than one point, the graph is not the graph of a function.

EXAMPLE 12: Use the vertical line test to determine which graphs in Figure 7.10 represent the graph of a function.

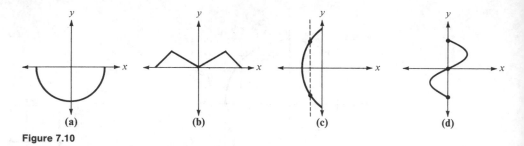

Figure 7.10

SOLUTION: By the vertical line test, graphs (a) and (b) represent functions, but (c) and (d) do not.

In this section we have defined a function in terms of (1) a rule and (2) ordered pairs. Since the function concept is so important, you should consider both definitions and satisfy yourself that these definitions are equivalent. Other examples of functions and their graphs may be found in Sections 2-1 and 2-4.

EXERCISES 7–1

1. Find five ordered pairs that are solutions of the equation $y = 2x^2 - 3$.
2. Find five correspondences in the function $y = 2$.
3. Which of the following are correspondences in the function $y = |x|$?
   a. $(-2, 2)$ b. $(2, -2)$
   c. $(-2, -2)$ d. $(2, 2)$
4. Which of the following ordered pairs are solutions of the equation, $y = 1/x$?
   a. $(\frac{1}{3}, 3)$ b. $(3, \frac{1}{3})$
   c. $(-1, -1)$ d. $(0, 0)$

In Exercises 5–10 state which collections of ordered pairs are functions.
5. $(1, -1)(2, 5)(3, 1)$ 6. $(-1, 1)(5, 2)(1, 3)$
7. $(-2, 2)(2, -2)(2, 0)$ 8. $(0, 2)(-2, 2)(2, -2)$
9. $(1, 5)(1, 6)(1, 7)$ 10. $(5, 1)(6, 1)(7, 1)$

In Exercises 11–23 find the domain of the function.
11. $V = \frac{4}{3}\pi r^3$ (volume of a sphere)   12. $y = x^3$
13. $y = \sqrt[3]{x}$ 14. $y = 1/(x + 1)$
15. $y = 1/(x^2 + 1)$ 16. $y = \sqrt{x}$
17. $y = 1/\sqrt{x}$ 18. $y = \sqrt{9 - x}$
19. $y = \sqrt{9 - x^2}$ 20. $y = \sqrt{x^2 - 9}$
21. $y = 5$ 22. $(3, 6)(4, 7)(5, 8)$
23. $(-2, 2)(-1, 1)(0, 0)$
24. Write five *different* functions that have the same domain and the same range as the function in Exercise 22.

In Exercises 25–32 graph the functions. In each case find the domain of the function.
25. $y = 2$ 26. $y = -1$

27. $y = \begin{cases} 2 & \text{if } x > 1 \\ -2 & \text{if } x \leq -1 \end{cases}$

28. $y = \begin{cases} 1 & \text{if } x > 0 \\ 0 & \text{if } x = 0 \\ -1 & \text{if } x < 0 \end{cases}$

29. $y = \begin{cases} x & \text{if } x \leq 0 \\ -x & \text{if } x > 0 \end{cases}$

30. $y = \begin{cases} x - 1 & \text{if } 0 \leq x \leq 3 \\ 2 & \text{if } x > 3 \end{cases}$

31. The deduction that motorists in New York may take on their federal income tax as a result of paying gasoline taxes is as follows for mileages (m) up to 4000 miles:

$$\text{Deduction} = \begin{cases} \$12 & \text{if } 0 \leq m < 3000 \\ \$19 & \text{if } 3000 \leq m < 3500 \\ \$22 & \text{if } 3500 \leq m < 4000 \end{cases}$$

Graph this function.

32. The New York Automobile Club measures the intensity (i) of bumps resulting from potholes in city streets through an instrument called a joltmeter (an electrocardiograph wired to the rear wheels of a car). The following formula table gives the club's rating scale:

$$\text{Rating} = \begin{cases} 1 \text{ (smooth)} & \text{if } 0 \leq i \leq 20 \\ 2 \text{ (wavy)} & \text{if } 20 < i \leq 25 \\ 3 \text{ (bumpy)} & \text{if } 25 < i \leq 30 \\ 4 \text{ (jarring)} & \text{if } 30 < i \leq 35 \\ 5 \text{ (teeth-rattling)} & \text{if } 35 < i \leq 40 \\ 6 \text{ (intolerable)} & \text{if } i > 40 \end{cases}$$

Graph this function.

33. Which of the graphs in the figure represent the graph of a function?

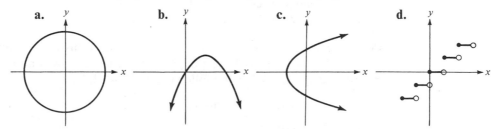

34. Find the domain and range of the functions shown.

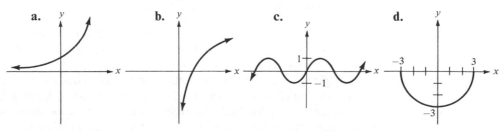

In Exercises 35–38 each table gives a few of the correspondences in some function. Write a formula (rule) to define the relationship between the variables.

35.

If x is	1	2	3	4	5
y is	0	1	2	3	4

36.

If x is	−2	−1	0	1	2
y is	2	1	0	1	2

37.

If x is	1	2	3	4	5
y is	2	6	10	14	18

38.

If x is	1	2	3	4	5
y is	0	3	8	15	24

In Exercises 39–50 find a formula or table that defines the functional relationship between the two variables; in each case indicate the domain of the function.

39. Express the area (A) of a circle as a function of its diameter (d).

40. The total cost of producing a certain product consists of $600 per month in rent plus $7 per unit for material and labor. Express the company's monthly total costs (c) as a function of the number of units (x) they produce that month.

41. Express the area (A) of a square as a function of its perimeter (p).

42. Express the area (A) of a square as a function of the length of its diagonal (d).

43. Express the federal income tax (t) for single persons as a function of their taxable income (i) if their taxable income is between $4000 and $6000 (inclusive) and the tax rate is $690 plus 21 percent of the excess over $4000.

44. Express the monthly earnings (e) of a salesman in terms of the cash amount (a) of merchandise sold if the salesman earns $400 per month plus 10 percent commission on sales above $5000.

45. Express the monthly cost (c) of a customer's water bill in terms of the number (n) of units purchased (1 unit equals 1000 gal) if no more than 100 units were used and the water company has the following rate schedule:

	Amount	Charge
First	5 units or less	$2.00
Next	95 units at	31 cents/unit

46. Express the monthly cost (c) of a gas bill in terms of the number (n) of units purchased (1 unit equals 100 ft^3 of gas) if the customer used no more than 14 units and the gas company has the following rate schedule:

	Amount	Charge
First	2 units or less	$2.06
Next	6 units at	40 cents/unit
Next	6 units at	26.5 cents/unit

47. A farmer encloses an area by connecting 2000 ft of fencing in the shape of a rectangle of sides x and y units. Express y as a function of x. Find a formula for the enclosed area (A) as a function of x.

48. A parking lot is adjacent to a building and is to have fencing on three sides, the side on the building requiring no fencing. If 100 yd of fencing is used to form a rectangular parking lot, find a formula expressing the area (A) of the lot as a function of the length (x) of the side of the fencing that is perpendicular to the building.

49. A rectangular sheet of tin 8 in. by 15 in. is to be used to make an open top box by cutting a small square of tin from each corner and bending up the sides. Find an equation for the volume (v) of the box in terms of the length (x) of the side of the square that is to be cut from each corner (see the figure).

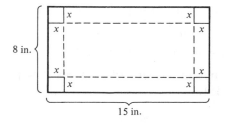

50. A rectangular billboard is to display 100 ft^2 of advertising material with borders of 3 ft top and bottom and 5 ft on the sides. Express the total area (A) of the billboard as a function of the width (w) of the advertised material (see the figure).

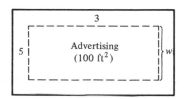

7–2 FUNCTIONAL NOTATION

Functional notation is used extensively for the remainder of this text and in calculus. It is useful because it allows us to represent more conveniently the value of the independent variable for a particular value of the dependent variable. In this notation we write an equation like $y = 3x^2 - 2$ as $f(x) = 3x^2 - 2$. The dependent variable y is replaced by $f(x)$, with the independent variable x appearing in the parentheses. The term $f(x)$ is read "f at x" and means the value of the function (the y value) corresponding to some x value. The term $f(x)$ does *not* mean f times x. Similarly $f(-2)$ is read "f at -2" and means the value of the function when $x = -2$. If $f(x) = 3x^2 - 2$ we find $f(-2)$ by substituting -2 for x in the equation. Thus $f(-2) = 3(-2)^2 - 2 = 10$.

EXAMPLE 1: If $y = f(x) = 7 + 1/x$ find $f(1)$, $f(\frac{1}{3})$, $f(0)$ and $f(a)$.

SOLUTION:

$$y_{\text{when } x=1} = f(1) = 7 + \frac{1}{1} = 8$$

$$y_{\text{when } x=\frac{1}{3}} = f(\tfrac{1}{3}) = 7 + \frac{1}{\frac{1}{3}} = 10$$

$$y_{\text{when } x=0} = f(0) = 7 + \frac{1}{0} \quad \text{undefined}$$

$$y_{\text{when } x=a} = f(a) = 7 + \frac{1}{a}$$

EXAMPLE 2: If $f(x) = 2x - 1$ and $g(x) = x^2$, find $4f(-3) + 2g(-2)$.

SOLUTION:

$$f(x) = 2x - 1 \qquad\qquad g(x) = x^2$$
$$f(-3) = 2(-3) - 1 \qquad g(-2) = (-2)^2$$
$$f(-3) = -7 \qquad\qquad g(-2) = 4$$

Thus $4f(-3) + 2g(-2) = 4(-7) + 2(4) = -20$

(Note that a function is not always labeled f. In this example we also used g; other symbols are also possible.)

EXAMPLE 3: If $h(x) = \begin{cases} 4 & \text{if } x < 0 \\ -2x & \text{if } x \geq 0 \end{cases}$ find $h(-2)$, $h(0)$ and $h(3)$.

SOLUTION:

Since $-2 < 0$, $h(-2) = 4$

Since $\quad 0 \geq 0$, $h(0) = -2(0) = 0$

Since $\quad 3 \geq 0$, $h(3) = -2(3) = -6$

EXAMPLE 4: The difference quotient of a function $y = f(x)$ is defined as

$$\frac{f(x_1 + h) - f(x_1)}{h} \qquad h \neq 0$$

Computing this ratio is an important consideration in calculus. Find the difference quotient for $f(x) = 3x^2 + 1$.

SOLUTION: If $f(x) = 3x^2 + 1$, we have

$$f(x_1 + h) = 3(x_1 + h)^2 + 1 = 3x_1{}^2 + 6x_1 h + 3h^2 + 1$$

Then

$$\frac{f(x_1 + h) - f(x_1)}{h} = \frac{(3x_1{}^2 + 6x_1 h + 3h^2 + 1) - (3x_1{}^2 + 1)}{h}$$

$$= \frac{3x_1{}^2 + 6x_1 h + 3h^2 + 1 - 3x_1{}^2 - 1}{h}$$

$$= \frac{6x_1 h + 3h^2}{h}$$

$$= 6x_1 + 3h \qquad h \neq 0$$

When working with functions it is important to recognize that we frequently use the symbols f and x more out of custom than necessity, and that other symbols work just as well. The notations $f(x) = 2x$, $f(t) = 2t$, $g(y) = 2y$, and $h(z) = 2z$ all define exactly the same function if x, t, y, and z may be replaced by the same numbers. Other examples of functional notation may be found in Section 2-3.

EXERCISES 7-2

1. If $f(x) = x + 1$, find $f(1)$, $f(0)$, and $f(-1)$.
2. If $g(y) = y + 1$, find $g(1)$, $g(0)$, and $g(-1)$.
3. If $h(x) = -3x + 2$, find $h(0)$, $h(-2)$, and $h(4)$.
4. If $f(t) = t^2 - 1$, find $f(2)$, $f(1)$, and $f(t_0 + h)$.
5. If $g(x) = (1/x) - 5$, find $g(0)$, $g(-1)$, and $g(x_0 + h)$.
6. If $f(x) = -x^2$, find $f(3)$, $f(-3)$, and $f(h)$.

7. If $h(t) = \dfrac{t + 2}{t - 1}$, find $h(0)$, $h(1)$, and $h(a)$.

8. If $f(y) = \dfrac{2y - 5}{3y + 1}$, find $f(0)$, $f(-\frac{1}{3})$, and $f(0.5)$.

9. If $f(x) = x + 2$, for what value of x will:
   a. $f(x) = 6$? **b.** $f(x) = 0$? **c.** $f(x) = -3$?
10. If $g(x) = 1/(x - 1)$, find (where possible) the value of x for which:
   a. $g(x) = 5$ **b.** $g(x) = 0$ **c.** $g(x) = 0.5$
11. If $f(x) = 2x - 1$ and $g(x) = x^2$, find:
   a. $f(0) + g(2)$
   c. $4g(-3) + 3f(1)$
   b. $f(-1) + g(-4)$
   d. $2f(5) - 5g(-2)$

   e. $g(-1) \cdot f(2)$
   f. $\dfrac{g(5)}{f(0.5)}$

   g. $[f(-3)]^2$
   i. $g[f(2)]$
   h. $f[g(2)]$
   j. $g[f(x)]$
12. If $f(x) = \sqrt{x}$:
   a. Does $f(9 + 16) = f(9) + f(16)$?
   b. Does $f(a + b) = f(a) + f(b)$?
13. If $f(x) = cx$, where c is a constant, show that $f(a + b) = f(a) + f(b)$.
14. If $g(x) = |x|$, find:
   a. $g(5)$ **b.** $g(-5)$ **c.** Does $g(x) = g(-x)$ for all values of x?
15. A function in which $f(-x) = f(x)$ for all x in the domain is called an *even function*; a function in which $f(-x) = -f(x)$ for all x in the domain is called an *odd function*. Find a function $y = f(x)$ that is:
   a. Even. **b.** Not even. **c.** Odd. **d.** Not odd.
16. Classify each of the following functions as even, odd, or neither:
   a. $f(x) = x^3$ **b.** $f(x) = x^4$ **c.** $f(x) = -|x|$
   d. $f(x) = x - 1$ **e.** $f(x) = 1$
17. Postage rates for first class mail were once as follows for weights (w) of up to 3 oz:

$$\text{Charge} = f(w) = \begin{cases} 10 \text{ cents} & \text{if } 0 \text{ oz} < w \leq 1 \text{ oz} \\ 20 \text{ cents} & \text{if } 1 \text{ oz} < w \leq 2 \text{ oz} \\ 30 \text{ cents} & \text{if } 2 \text{ oz} < w \leq 3 \text{ oz} \end{cases}$$

Find:
   a. $f(1.6)$ **b.** $f(\frac{1}{4})$ **c.** $f(3)$

18. The deduction that motorists in New York may take on their federal income tax due to gasoline taxes is given in the following formula table for mileages (m) up to 4000 miles:

$$\text{Deduction} = f(m) = \begin{cases} \$11 & \text{if} \quad 0 \quad \le m < 3000 \\ \$19 & \text{if} \quad 3000 \le m < 3500 \\ \$21 & \text{if} \quad 3500 \le m < 4000 \end{cases}$$

Find:
   a. $f(3798)$ **b.** $f(3000)$ **c.** $f(1818)$

19. If $f(x) = \begin{cases} 2 & \text{if} \quad x \ge 0 \\ 0 & \text{if} \quad x < 0 \end{cases}$

Find:
   a. $f(-2)$ **b.** $f(0)$ **c.** $f(100)$

20. If $f(x) = \begin{cases} 1 & \text{if} \quad x > 0 \\ 0 & \text{if} \quad x = 0 \\ -1 & \text{if} \quad x < 0 \end{cases}$

Find:
   a. $f(\pi)$ **b.** $f(0)$ **c.** $f(-\sqrt{2})$ **d.** $f(-0.0001)$

21. If $g(x) = \begin{cases} x & \text{if} \quad x \ge 0 \\ -x & \text{if} \quad x < 0 \end{cases}$

Find:
   a. $g(\frac{1}{2})$ **b.** $g(0)$ **c.** $g(-\pi)$

22. If $h(x) = \begin{cases} 2x & \text{if} \quad 0 \le x < 3 \\ x^2 & \text{if} \quad x \ge 3 \end{cases}$

Find:
   a. $h(3)$ **b.** $h(\frac{1}{4})$ **c.** $h(4)$ **d.** $h(-1)$

23. If $f(x) = \begin{cases} 1 & \text{if} \quad x \text{ is rational} \\ -1 & \text{if} \quad x \text{ is irrational} \end{cases}$

Find:
   a. $f(\pi)$ **b.** $f(\frac{22}{7})$ **c.** $f(0)$ **d.** $f(\sqrt{2})$

24. For each of the following functions write the difference quotient in simplest form (see Example 4):
   a. $f(x) = 4x - 1$ **b.** $f(x) = 2x^2 + 3$
   c. $f(x) = 1$ **d.** $f(x) = x^3$

25. The bracket notation $[x]$ means the greatest integer less than or equal to x. For example, $[4.3] = 4$, $[\pi] = 3$, $[-\frac{1}{2}] = -1$, and $[2] = 2$. If $f(x) = [x]$, find:
   a. $f(\frac{1}{2})$ **b.** $f(\sqrt{2})$ **c.** $f(-1.4)$
   d. $f(-\pi)$ **e.** Graph this function for $-2 \le x < 3$.

7–3 OPERATIONS WITH FUNCTIONS

We may add, subtract, multiply, and divide functions to form new functions. For example, suppose that an item which cost \$5 to manufacture sells for \$7. The total cost of x items is given by the function

$$C(x) = 5x$$

The total sales revenue from x items is given by

$$S(x) = 7x$$

Since profit is the difference between sales and cost, we form the profit function as the difference of the two other functions:

$$P(x) = S(x) - C(x)$$
$$= 7x - 5x$$
$$= 2x$$

In general, for two functions f and g, the functions $f + g, f - g, f \cdot g, f/g$ are defined as follows:

$$(f + g)(x) = f(x) + g(x)$$
$$(f - g)(x) = f(x) - g(x)$$
$$(f \cdot g)(x) = f(x) \cdot g(x)$$

$$\left(\frac{f}{g}\right)(x) = \frac{f(x)}{g(x)} \qquad g(x) \neq 0$$

The domain of the resulting function is the common domain of f and g. The domain of the quotient function excludes any x for which $g(x) = 0$. In effect, the definitions say that for any x at which both functions are defined, we combine the y values.

EXAMPLE 1: If $f(x) = x$ and $g(x) = \sqrt{x}$, find $(f + g)(x)$, $(f - g)(x)$, $(f \cdot g)(x)$, and $(f/g)x$. In each case give the domain of the resulting function.

SOLUTION: Using the definitions, we have

$$(f + g)(x) = f(x) + g(x) = x + \sqrt{x}$$
$$(f - g)(x) = f(x) - g(x) = x - \sqrt{x}$$
$$(f \cdot g)(x) = f(x) \cdot g(x) = x\sqrt{x} \quad \text{or} \quad \sqrt{x^3}$$

$$\left(\frac{f}{g}\right)(x) = \frac{f(x)}{g(x)} = \frac{x}{\sqrt{x}} \quad \text{or} \quad \sqrt{x}$$

The domain of f is all real numbers and the domain of g is $x \geq 0$. The domain of $f + g$, $f - g$, and $f \cdot g$ is the collection of numbers common to both domains, which is $x \geq 0$. In the quotient function $x \neq 0$. Thus, the domain of f/g is $x > 0$.

EXAMPLE 2: If $f = (0, 1)(1, 2)(2, 3)(3, 4)$ and $g = (2, 0)(3, 1)(4, -5)(5, -3)$, find $f + g, f - g, f \cdot g$, and f/g.

SOLUTION: Both functions are defined at $x = 2$ and $x = 3$. Thus,

$$f + g = (2, 3 + 0)(3, 4 + 1) = (2, 3)(3, 5)$$
$$f - g = (2, 3 - 0)(3, 4 - 1) = (2, 3)(3, 3)$$
$$f \cdot g = (2, 3 \cdot 0)(3, 4 \cdot 1) = (2, 0)(3, 4)$$
$$f/g = (2, 3/0)(3, 4/1) = (3, 4)$$

There is another important way of combining functions that has no counterpart in arithmetic. The operation is called *composition*. Basically, composition is a substitution that causes a "chain reaction" in which two functions are applied in succession.

Consider the problem of determining the area of a square whose perimeter is 20 in. You would probably reason as follows: If the perimeter is 20 in., the side is 5 in.; the area is then 25 in.² Given a specific value for the perimeter (P), we first determine the side (S) by using the function

$$S = \frac{P}{4}$$

The result of this step is then substituted in the function

$$A = S^2$$

to determine the area. Let us put these formulas in functional notation and use the analogy of a function as a machine. Consider Figure 7.11, in which $S = f(P) = P/4$ and $A = g(S) = S^2$. The output of the f machine feeds the input of the g machine. Thus, we first apply the f function to determine S and then apply the g function to determine A. The area is given by the function

$$A = g[f(P)]$$

This function is said to be a *composition* of f and g. By substitution we have

$$A = g[f(P)] = g\left(\frac{P}{4}\right) = \left(\frac{P}{4}\right)^2$$

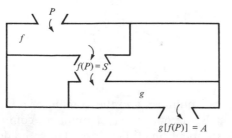

Figure 7.11

The symbol "∘" is also used for composition. We could write

$$A = g \circ f$$

In general, the composite functions of functions f and g are defined as follows:

$$(g \circ f)(x) = g[f(x)]$$
$$(f \circ g)(x) = f[g(x)]$$

EXAMPLE 3: If $f(x) = x - 1$ and $g(x) = x^2$, find $(g \circ f)(x)$ and $(f \circ g)(x)$.

SOLUTION:

$$(g \circ f)(x) = g[f(x)] = g(x - 1)$$
$$= (x - 1)^2 = x^2 - 2x + 1$$
$$(f \circ g)(x) = f[g(x)] = f(x^2)$$
$$= x^2 - 1$$

Note that the order in which the functions are applied is important. In general, $(g \circ f)(x) \neq (f \circ g)(x)$.

A number x must satisfy two requirements to belong to the domain of $g \circ f$.

1. Since we first apply the f function, x must be in the domain of f.
2. The output from step 1, $f(x)$, must be in the domain of g.

Thus the domain of $g \circ f$ is the collection of all real numbers x in the domain of f for which $f(x)$ is in the domain of g. This collection consists of all the values in the domain of f for which $g \circ f$ is defined.

EXAMPLE 4: If $f(x) = \sqrt{x}$ and $g(x) = x^2$, find $(g \circ f)(x)$ and $(f \circ g)(x)$. Indicate the domain of each function.

SOLUTION:

$$(g \circ f)(x) = g[f(x)] = g(\sqrt{x}) = (\sqrt{x})^2 = x$$
Domain of f: $x \geq 0$
$$(g \circ f)(x) = x \text{ is defined for all real numbers}$$

Thus, the domain of $g \circ f$ is $x \geq 0$.

$$(f \circ g)(x) = f[g(x)] = f(x^2) = \sqrt{x^2} \quad \text{or} \quad |x|$$
Domain of g: all real numbers
$$(f \circ g)(x) = |x| \text{ is defined for all real numbers}$$

Thus, the domain of $f \circ g$ is all the real numbers.

EXAMPLE 5: If $f = (0, 1)(1, 2)(2, 3)(3, 4)$ and $g = (2, 0)(3, 1)(4, -5)(5, -3)$, find $f \circ g$ and $g \circ f$. Indicate the domain of each composite function.

SOLUTION: To determine $f \circ g$ we first apply the g function. As illustrated below, the range elements 0 and 1 are in the domain of f. Thus, $f \circ g = (2, 1)(3, 2)$ and the domain of $f \circ g$ is 2, 3.

$2 \xrightarrow{g} \quad 0 \xrightarrow{f} 1$

$3 \rightarrow \quad 1 \rightarrow 2$

$4 \rightarrow -5$

$5 \rightarrow -3$

To find $g \circ f$ we first apply the f function. As illustrated below, the range elements 2, 3, and 4 are in the domain of g. Thus $g \circ f = (1, 0)(2, 1)(3, -5)$ and the domain of $g \circ f$ is 1, 2, 3.

$0 \xrightarrow{f} 1$

$1 \rightarrow 2 \xrightarrow{g} \quad 0$

$2 \rightarrow 3 \rightarrow \quad 1$

$3 \rightarrow 4 \rightarrow -5$

EXERCISES 7–3

In Exercises 1–10 find $(f + g)(x), (f - g)(x), (f \cdot g)(x), (f/g)(x), (f \circ g)(x)$, and $(g \circ f)(x)$. Indicate the domain of each function.

1. $f(x) = 2x; g(x) = x - 1$ **2.** $f(x) = x - 2; g(x) = x + 2$

3. $f(x) = 4x - 5; g(x) = -x + 3$ **4.** $f(x) = 2; g(x) = 3x + 1$

5. $f(x) = x^2; g(x) = 1$ **6.** $f(x) = x; g(x) = 1/x$

7. $f(x) = x^3; g(x) = \sqrt[3]{x}$ **8.** $f(x) = \sqrt{x}; g(x) = -x^2$

9. $f = (0, 2)(1, 3)(2, 4)(3, 5)$ **10.** $f = (-3, 5)(0, 1)(2, 6)(4, 11)$

     $g = (2, -1)(3, 0)(4, 1)(5, 2)$ $g = (0, 0)(2, 2)(5, 5)(-3, -3)$

11. If $f(x) = \sqrt{x}$ and $g(x) = x$, find:

     a. $(f - g)(4)$ **b.** $[f(x) - g(x)](4)$ **c.** $(g - f)(4)$ **d.** $[g(x) - f(x)](4)$

12. If $f(x) = -x$ and $g(x) = x^2$, find:

     a. $\left(\dfrac{f}{g}\right)(-2)$ **b.** $\left[\dfrac{f(x)}{g(x)}\right](-2)$ **c.** $\left(\dfrac{g}{f}\right)(-2)$ **d.** $\left[\dfrac{g(x)}{f(x)}\right](-2)$

13. If $f = (1, 3)(2, -2)(3, 5)$ and $g = (2, -1)(3, 4)(4, 0)$, find:

     a. $(f + g)(2)$ **b.** $(f \cdot g)(3)$ **c.** $(g \cdot f)(1)$ **d.** $(g \circ f)(1)$

14. If $f = (1, 5)(-2, 4)(-3, 0)$ and $g = (0, 2)(-3, 1)(-5, -1)$, find:

     a. $\left(\dfrac{f}{g}\right)(-3)$ **b.** $\left(\dfrac{g}{f}\right)(-3)$ **c.** $(f \circ g)(-3)$ **d.** $(g \circ f)(-3)$

In calculus it is often necessary to recognize a function as the composition of simpler functions. For example, $h(x) = (2x + 3)^2$ may be thought of as $(f \circ g)(x)$, where $g(x) = 2x + 3$ and $f(x) = x^2$. In Exercises 15–20 find simpler functions $f(x)$ and $g(x)$ so that $h(x) = (f \circ g)(x)$.

15. $h(x) = (4x - 1)^3$

16. $h(x) = (x^2 + 1)^2$

17. $h(x) = 2(3 - x)^4$

18. $h(x) = \sqrt{1 - x^2}$

19. $h(x) = \sqrt[3]{2x + 1}$

20. $h(x) = \dfrac{1}{3x - 5}$

21. If $f(x) = \sqrt{x - 7}$ and $g(x) = \sqrt{2 - x}$, consider the domains of the functions and explain why it does not make sense to form $(f + g)(x)$.

22. If $f(x) = \sqrt{x - 7}$ and $g(x) = 2 - x^2$, explain why it does not make sense to form $(f \circ g)(x)$.

Sample Test: Chapter 7

1. Give three correspondences in the function $f(x) = 7$.

2. Is the collection of ordered pairs $(-3, 0)(0, 0)(1, 0)$ a function?

3. What is the domain of the function $y = 1/\sqrt{4 - x^2}$?

4. Write a different function that has the same domain and the same range as the function $(1, 4)(2, 5)$.

In questions 5–7 consider the function

$$f(x) = \begin{cases} 1 & \text{if } x > 2 \\ -x & \text{if } x \le -2 \end{cases}$$

5. Find $f(-3)$.

6. Graph the function.

7. What is the domain of the function?

8. Is the graph in the figure the graph of a function?

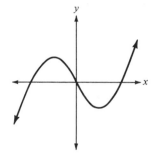

9. What is the domain and range of the function in the figure?

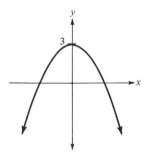

10. Consider the following table, which gives a few of the correspondences in some function:

If x is	1	2	3	4	5
y is	6	6	6	6	6

Write a formula to define the relationship.

11. Express the monthly earnings (e) of a salesman in terms of the cash amount (a) of merchandise sold if the salesman earns \$500 per month plus 9 percent commission on sales above \$2000.

12. A rectangle of sides x and y units is inscribed in a circle of radius 5 units. Express the area of the rectangle as a function of x. What is the domain of the function?

In questions 13 and 14 use $f(x) = 2x + 1$ and $g(x) = \sqrt{x}$.

13. Find $3f(-2) - 4g(9)$.

14. Find $(g - f)(4)$.

15. If $g(x) = 3x - 4$, for what value of x does $g(x) = 0$?

16. Is the function $f(x) = x^2 - 4$ even, odd, or neither?

17. For the function $f(x) = 3x^2 - 1$, write the difference quotient

$$\frac{f(x_1 + h) - f(x_1)}{h}$$

in simplest form.

18. If $f(x) = x^2$ and $g(x) = 1/x$, find $(f \cdot g)(x)$. What is the domain of $f \cdot g$?

19. If $f = (0, -3)(-1, -2)(2, -1)$ and $g = (-1, 2)(-2, 1)(-3, 3)$, find $(g \circ f)(-1)$.

20. If $h(x) = 3(1 - x)^2$, find simpler functions $f(x)$ and $g(x)$ so that $h(x) = (f \circ g)(x)$.

Chapter 8
Topics concerning polynomials

8–1 SLOPE AND MATHEMATICAL CHANGE

An understanding of the process of change is vital to the investigation of the laws of nature, for everything is constantly changing in such significant respects as size, position, and temperature. Since almost nothing in the world is immune from change, an important topic in mathematics is finding the rate at which one variable changes with respect to another. For example:

An investor is interested in the rate of change (interest rate) of his bank deposit so that his investment will yield a maximum return.

An economist is interested in knowing the rate at which the demand for a product will change with respect to the price charged for the product.

A motorist is interested in knowing his speed, the rate of change of the distance he is traveling with respect to time.

A businessman needs to know the rate at which the total cost of a production is changing with respect to each additional unit he produces.

Social scientists are groping for a way to analyze the accelerating rate of social processes, which crowds an increasing number of events into an arbitrary interval of time. For example, in *Future Shock* (p. 2) Alvin Toffler argues that "unless man quickly learns to control the rate of change in his personal affairs as well as in society at large, we are doomed to a massive adaptational breakdown." This disease of change is labeled "future shock."

A health official helps prevent an epidemic by analyzing the rate at which an infectious disease (such as flu) might spread in a certain community.

These examples illustrate the necessity of analyzing change. The most important tool in this analysis is calculus. In fact, it is this mathematical ability of calculus to analyze movement and change that makes it the principal link between practical science and mathematical thought. At this time we avoid more complicated functions and study only functions that always change at the same rate, These functions are called *linear functions* because their graphs are straight lines. To find the rate of change of a linear function, pick *any* two points on the line and calculate the change in the dependent variable and divide it by the change in the independent variable. This ratio is called the *slope of the line.*

DEFINITION OF SLOPE: If (x_1, y_1) and (x_2, y_2) are any two points on a nonvertical line, as shown in Figure 8.1, then

$$\text{slope} = m = \frac{\Delta y}{\Delta x} = \frac{y_2 - y_1}{x_2 - x_1}$$

(*Note:* The symbol Δ is the Greek capital letter delta. The symbol is used to indicate a change.)

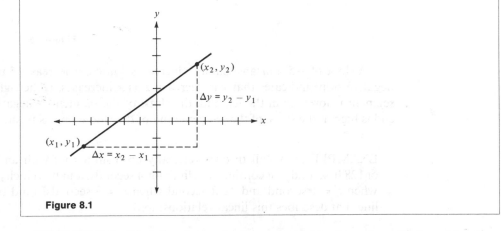

Figure 8.1

EXAMPLE 1: Find the slope of a line containing the points (2, 1) and (5, 4).

SOLUTION: If we label (2, 1) as point 1 (Figure 8.2),

$$x_1 = 2 \quad y_1 = 1 \quad x_2 = 5 \quad y_2 = 4$$

$$m = \frac{\Delta y}{\Delta x} = \frac{y_2 - y_1}{x_2 - x_1} = \frac{4 - 1}{5 - 2} = \frac{3}{3} = 1$$

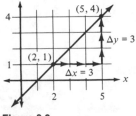

Figure 8.2

A slope of 1 means that as x increases 1 unit, y increases 1 unit. Notice that the slope is unaffected by the way in which we label the points. If we label (5, 4) as point 1,

$$x_1 = 5 \quad y_1 = 4 \qquad x_2 = 2 \quad y_2 = 1$$

$$m = \frac{\Delta y}{\Delta x} = \frac{y_2 - y_1}{x_2 - x_1} = \frac{1 - 4}{2 - 5} = \frac{-3}{-3} = 1$$

EXAMPLE 2: Find the slope of the line containing the points $(-1, -1)$ and $(3, -6)$.

SOLUTION: (See Figure 8.3.)

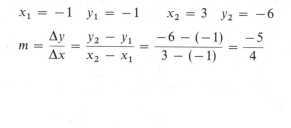

$$x_1 = -1 \quad y_1 = -1 \qquad x_2 = 3 \quad y_2 = -6$$

$$m = \frac{\Delta y}{\Delta x} = \frac{y_2 - y_1}{x_2 - x_1} = \frac{-6 - (-1)}{3 - (-1)} = \frac{-5}{4}$$

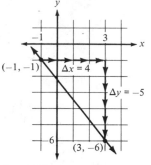

Figure 8.3

A slope of $-5/4$ means that as x increases 4 units, y decreases 5 units. [*Note:* A negative slope indicates that y is decreasing as x increases. If the right end of a line segment is lower than the left end, the slope of the segment is negative. If the right end is higher, the slope of the segment (and of the whole line) is positive.]

EXAMPLE 3: A ball thrown vertically up from a roof with an initial velocity of 128 ft/second will continue to climb for 4 seconds, and its velocity is 96 ft/second when $t = 1$ second and 32 ft/second when $t = 3$ seconds. Find the slope of the line that describes this linear relationship.

SOLUTION: (See Figure 8.4.)

$$t_1 = 1 \quad v_1 = 96 \qquad t_2 = 3 \quad v_2 = 32$$

$$m = \frac{\Delta v}{\Delta t} = \frac{v_2 - v_1}{t_2 - t_1} = \frac{32 - 96}{3 - 1} = \frac{-64}{2} = -32$$

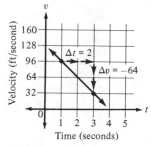

Figure 8.4

A slope of -32 means that for every additional second (up to 4 seconds) the velocity decreases by 32 ft/second.

EXAMPLE 4: Find the slope of the line containing the points $(-1, 4)$ and $(5, 4)$.

SOLUTION: (See Figure 8.5.)

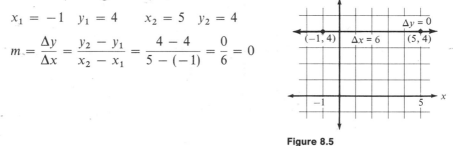

$$x_1 = -1 \quad y_1 = 4 \qquad x_2 = 5 \quad y_2 = 4$$

$$m = \frac{\Delta y}{\Delta x} = \frac{y_2 - y_1}{x_2 - x_1} = \frac{4 - 4}{5 - (-1)} = \frac{0}{6} = 0$$

Figure 8.5

The slope of every horizontal line is zero since the numerator of the slope ratio (Δy) is always zero.

EXAMPLE 5: Find the slope of the line containing the points $(3, 2)$ and $(3, -4)$.

SOLUTION: (See Figure 8.6.)

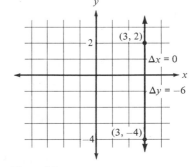

$$x_1 = 3 \quad y_1 = 2 \qquad x_2 = 3 \quad y_2 = -4$$

$$m = \frac{\Delta y}{\Delta x} = \frac{y_2 - y_1}{x_2 - x_1} = \frac{-4 - 2}{3 - 3} = \frac{-6}{0}$$

m is undefined

Figure 8.6

The slope of every vertical line is undefined since the denominator of the slope ratio (Δx) is always zero.

EXERCISES 8–1

In Exercises 1–10 draw the line determined by each pair of points and compute its slope.

1. $(1, 2)$ and $(3, 4)$ **2.** $(3, -1)$ and $(5, -7)$
3. $(5, 1)$ and $(-2, -3)$ **4.** $(-1, 2)$ and $(0, 0)$
5. $(4, -4)$ and $(2, 7)$ **6.** $(-5, -1)$ and $(-2, -4)$
7. $(-1, -3)$ and $(2, -3)$ **8.** $(-2, 0)$ and $(5, 0)$
9. $(1, 3)$ and $(1, -1)$ **10.** $(-3, 2)$ and $(-3, -2)$

In Exercises 11–14 find two ordered pairs that satisfy the equation and use them to find the slope of the line.

11. $y = -x + 1$

12. $y = 4x - 5$

13. $y = \frac{1}{2}x + 3$

14. $y = -\frac{5}{4}x - 2$

15. Approximate the slope of each line in the figure.

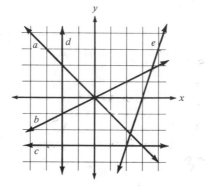

16. Approximate the slope of each line in the figure.

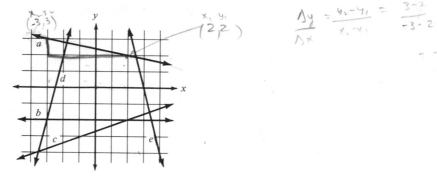

17. On a piece of graph paper draw lines through the point (1, 2) with the following slopes:

    a. 2   **b.** −3   **c.** $-\frac{1}{4}$ **d.** $\frac{2}{3}$ **e.** 0

18. On a piece of graph paper draw lines through the point $(-2, -1)$ with the following slopes:

    a. 1   **b.** −4   **c.** $-\frac{1}{3}$ **d.** $\frac{1}{3}$ **e.** $-\frac{3}{2}$

In Exercises 19–26 find the rate of change of each linear relationship by calculating the slope. In each case interpret the physical meaning of the slope.

19. An object that weighs 24 lb on earth weighs 4 lb on the moon; a different object, which weighs 42 lb on earth, weighs 7 lb on the moon.

20. The freezing point is 32° Fahrenheit compared to 0° Celsius; the boiling point is 212° Fahrenheit compared to 100° Celsius.

21. A ball thrown vertically up from the ground with an initial velocity of 320 ft/second will continue to climb for 10 seconds; its velocity is 256 ft/second when $t = 2$ seconds and 64 ft/second when $t = 8$ seconds.

22. A projectile fired vertically up from the ground with an initial velocity of 224 ft/second will continue to climb for 7 seconds; its velocity is 96 ft/second when $t = 4$ seconds.

23. The cost of 50 brochures is \$11; the cost of 100 brochures is \$17.

24. The total cost of manufacturing 100 units of certain product is \$450; the cost of manufacturing 200 units is \$600.

25. A moving company charges \$50 to move a certain machine 5 mi and \$71 to move the same machine 8 mi.

26. The monthly cost of renting a computer is \$400 if it is used for 5 hours and \$500 if it is used for 7 hours.

8–2 LINEAR FUNCTIONS

In Section 8-1 we discussed functions that always change at the same rate and we called these functions linear functions, since their graph is a straight line. We will now take up the problem of finding an equation that defines a linear function.

Since the graph of a linear function is always a straight line, we will start by considering any nonvertical line (see Figure 8.7). (We eliminate vertical lines since they are not functions and their slope is undefined.) This line must cross the y axis at some point. The x coordinate of this point is zero and we will label the y coordinate b. This point $(0, b)$, where the line crosses the y axis, is called the y intercept. Now consider any other point (x, y) on the line. The slope of this line would be

$$m = \frac{y - b}{x - 0} = \frac{y - b}{x}$$

Solving this equation for y, we have

$$m = \frac{y - b}{x}$$

$$mx = y - b$$

$$mx + b = y$$

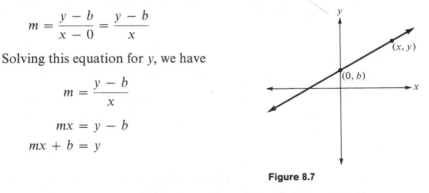

Figure 8.7

Thus, any linear function can be written in the form $y = f(x) = mx + b$, where m is the slope of the line and b is the y coordinate of the y intercept.

EXAMPLE 1: Find an equation for the line whose slope is 4 and whose y intercept is $(0, -7)$.

SOLUTION: Substituting $m = 4$ and $b = -7$ in the equation $y = mx + b$, we have

$$y = 4x - 7$$

EXAMPLE 2: Find the slope and the y intercept of the line defined by the equation $2x + 3y = 6$.

SOLUTION: First, transform the equation to the form $y = mx + b$:

$$2x + 3y = 6$$
$$3y = -2x + 6$$
$$y = \frac{-2}{3}x + 2$$

Matching this equation to the form $y = mx + b$, we conclude that

$$m = \frac{-2}{3} \qquad b = 2$$

Thus, the slope is $-\frac{2}{3}$ and the y intercept is $(0, 2)$.

EXAMPLE 3: Find the equation of the line that contains the point $(1, 3)$ and whose slope is 2.

SOLUTION: Substituting $m = 2$ in the equation $y = mx + b$, we have

$$y = 2x + b$$

To find b we utilize the principle that every point on the line corresponds to an ordered pair that satisfies the equation; conversely, every ordered pair that satisfies the equation corresponds to a point on the line. Thus, substitute the coordinates of the point $(1, 3)$ in the equation.

$$3 = 2(1) + b$$
$$1 = b$$

The equation of the line is $y = 2x + 1$.

Note: There are many ways of writing the equation of a line. The form $y = mx + b$ is called the *slope-intercept form*. If the slope and one point of the line are given, the *point-slope form*,

$$y - y_1 = m(x - x_1)$$

is useful. For instance, in Example 3, $x_1 = 1$, $y_1 = 3$, and $m = 2$. Substituting these numbers in the point-slope form, we have

$$y - 3 = 2(x - 1)$$

which simplifies to

$$y = 2x + 1$$

EXAMPLE 4: Find the equation that defines the linear function f if $f(1) = 6$ and $f(-1) = 2$.

SOLUTION: If $f(1) = 6$, then when $x = 1$, $y = 6$. Similarly, if $f(-1) = 2$, then when $x = -1$, $y = 2$. Thus, $(1, 6)$ and $(-1, 2)$ are points on the graph. First calculate the slope:

$$m = \frac{y_2 - y_1}{x_2 - x_1} = \frac{6 - 2}{1 - (-1)} = 2$$

Thus,

$$y = 2x + b \quad \text{or} \quad f(x) = 2x + b$$

To find b, substitute the coordinates of either point since they are both on the line.

$$y = 2x + b \qquad \text{or} \qquad y = 2x + b$$
$$6 = 2(1) + b \qquad\qquad 2 = 2(-1) + b$$
$$4 = b \qquad\qquad\qquad 4 = b$$

The equation that defines the function f is

$$f(x) = 2x + 4$$

EXAMPLE 5: Find the equation that defines the linear relationship between Fahrenheit and Celsius temperatures if the freezing point is 32° Fahrenheit and 0° Celsius and the boiling point is 212° Fahrenheit and 100° Celsius.

SOLUTION: First calculate the slope:

$$m = \frac{\Delta F}{\Delta C} = \frac{212 - 32}{100 - 0} = \frac{180}{100} = \frac{9}{5}$$

Thus,

$$F = \frac{9}{5}C + b$$

To find b, substitute the coordinates of the point $(0, 32)$ since this choice makes the arithmetic easier.

$$F = \frac{9}{5}C + b$$

$$32 = \frac{9}{5}(0) + b$$

$$32 = b$$

The equation that defines the linear relationship is

$$F = \frac{9}{5}C + 32$$

Note: If we calculated the slope by finding $\Delta C / \Delta F$, the resulting equation would be $C = \frac{5}{9}F - \frac{160}{9}$ or $C = \frac{5}{9}(F - 32)$.

EXERCISES 8–2

In Exercises 1–6 write the equation of the straight line whose slope and y intercept are given.

1. $m = 2; (0, 3)$

2. $m = 5; (0, 1)$

3. $m = -3; (0, -1)$

4. $m = -\frac{3}{4}; (0, -2)$

5. $m = 0; (0, -5)$

6. $m = 0; (0, 0)$

In Exercises 7–16 find the slope and the y intercept of the line defined by the following equations.

7. $y = x + 7$

8. $y = \dfrac{-2}{3} x - 5$

9. $y = 5x$

10. $y = -x$

11. $y = -2$

12. $y = 0$

13. $3y + 2x = -2$

14. $3x + 5y = 2$

15. $6x - y = -7$

16. $2x - 7y = -4$

In Exercises 17–22 find the equation of the line that has the given slope, m, and passes through the given point.

17. $m = 5; (4, 3)$

18. $m = -4; (2, -1)$

19. $m = \frac{1}{2}; (-2, 0)$

20. $m = -\frac{3}{4}; (-5, 2)$

21. $m = -\frac{1}{2}; (0, 0)$

22. $m = 0; (1, 2)$

In Exercises 23–26 find the equation of the line that contains the points.

23. $(3, 2)$ and $(4, 1)$

24. $(-4, 3)$ and $(-2, 5)$

25. $(-3, -3)$ and $(-4, -2)$

26. $(2, -3)$ and $(3, -3)$

In Exercises 27–32 find the equation that defines the linear function f.

27. $f(3) = 0$ and $f(0) = -2$

28. $f(0) = -1$ and $f(2) = 0$

29. $f(-1) = -3$ and $f(2) = -7$

30. $f(-6) = -1$ and $f(-2) = -5$

31. $f(-3) = 1$ and $f(0) = 1$

32. $f(4) = -6$ and $f(-1) = -6$

In Exercises 33–36 determine the given function value; f is a linear function.

33. $f(3) = 1$ and $f(5) = -3$; find $f(4)$.

34. $f(-2) = 0$ and $f(1) = 6$; find $f(0)$.

35. $f(-4) = -1$ and $f(-1) = -3$; find $f(0)$.

36. $f(-5) = 2$ and $f(1) = 2$; find $f(5)$.

37. The cost of 50 brochures is $9 and the cost of 200 brochures is $15. Find the equation that defines this linear relationship. How much would 150 brochures cost?

38. A moving company charges $70 to move a certain machine 5 mi and $106 to move the same machine 14 mi. Find the equation that defines this linear relationship. How much would the company charge for moving the machine 10 mi?

39. A mechanic estimates a job to be $76 if he can complete the job in 2 hours and $103 if it takes him 5 hours.

    a. Find the equation that defines this linear relationship.

    b. What would he charge if he took 4 hours to do the job?

c. How much did he charge for parts?

d. How much did he charge per hour for labor?

40. A certain fish tank weighs 51 lb when it contains 5 gal of water and 78 lb when it contains 8 gal of water.

   a. Find the equation that defines this linear relationship.

   b. What would the tank weigh if it contained 10 gal of water?

   c. How much did the tank weigh before the water was added?

   d. How much did the water weigh per gallon?

41. A moving company charges $46 to move a certain machine 10 mi and $58 to move the same machine 30 mi.

   a. Find the equation that defines this linear equation.

   b. What will it cost to move the machine 25 mi?

   c. What is the minimum charge for moving the machine?

   d. What is the rate for each mile the machine is moved?

42. The total cost of producing a certain item consists of paying rent for the building and paying a fixed amount per unit for material. The total cost is $250 if 10 units are produced and $330 if 30 units are produced.

   a. Find the equation that defines the linear relationship.

   b. What will it cost to produce 100 units?

   c. How much is paid in rent?

   d. What is the cost of the material for each unit?

43. A salesman receives a monthly salary plus a commission on sales. He earned $1050 during a month in which his sales totaled $5000; the next month he sold $6000 worth of merchandise and earned $1150.

   a. Find the equation that defines this linear relationship.

   b. How much will he earn if he sells $9000 worth of merchandise for the month?

   c. What is his monthly salary?

   d. What is his rate of commission?

44. A projectile shot vertically up has a velocity of 192 ft/second when $t = 4$ seconds and 96 ft/second when $t = 7$ seconds.

   a. Find the equation that defines this linear relationship.

   b. Find the velocity of the projectile when $t = 9$ seconds.

   c. What is the initial velocity of the projectile (velocity when $t = 0$)?

   d. When is the velocity 0? What is the physical significance of this?

45. A ball thrown vertically down from a building has a velocity of -252 ft/second when $t = 1$ second and -316 ft/second when $t = 3$ seconds.

   a. Find the equation that defines this linear relationship.

   b. Find the velocity of the ball when $t = 4$ seconds.

   c. What is the initial velocity of the ball?

   d. What is the significance of whether the velocity is positive or negative?

46. A spring that is 24 in. long is compressed to 20 in. by a force of 16 lb and to 15 in. by a force of 36 lb.

   a. Find the equation that defines this linear relationship.

   b. What is the length of the spring if a force of 28 lb is applied?

   c. How much force is needed to compress the spring to 10 in.?

47. For the function $f(x) = mx + b$, find the difference quotient $\dfrac{f(x_1 + h) - f(x_1)}{h}$

in simplest form.

48. An important application of linear functions is found in statistics, where it is often desirable to approximate the relationship between two variables in terms of a linear function. For example, suppose that the guidance counselor at a two-year college wishes to develop a technique for estimating the performance of his students when they transfer to a four-year college. From his files he randomly picks 10 students who previously transferred to a four-year college and obtains the following comparison between their cumulative averages:

Student	Two-Year College Cumulative Average	Four-Year College Cumulative Average
1	2.1	2.5
2	3.5	3.2
3	2.9	3.0
4	2.6	2.3
5	2.5	2.7
6	2.9	2.8
7	2.3	2.3
8	3.1	2.9
9	3.4	3.6
10	2.5	2.4

The guidance counselor now graphs the information (as shown in the figure) and notices that the points basically form a linear pattern. From this information he uses

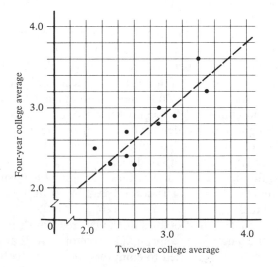

statistical techniques to calculate the equation ($y = 0.79x + 0.57$) of the dashed line in the figure, which comes closest to all the points. This "best fitting" line can now be used to make predictions about the performance of his two-year college students when they transfer. (*Note:* We have left many important questions concerning this application unanswered. For example: What criterion is used to decide which line "best fits" the data? How good are predictions based upon this equation? At this time we only wish to give an indication of how linear functions are used. For a more detailed account of this application, consult the topic "regression analysis" in an elementary statistics book.)

Use the equation $y = 0.79x + 0.57$ to predict the four-year college average of the two-year college transfer students with the following averages:

  a. 2.0

  b. 3.0

  c. 2.7

  d. 3.3

State in each case whether you would expect the slope of the line that best describes the relationship between the given variables to be positive or negative.

  e. An individual's height and weight.

  f. The number of hours that a runner practices and his time for a 1-mi race.

  g. A student's cumulative average and the number of hours per week that she works at a job.

  h. The number of automobile accidents and the amount of insurance premiums that the driver has to pay.

  i. The percentage of nitrogen in a fertilizer and the height to which a treated plant will grow.

  j. The amount of alcohol consumed by an individual and the length of time in which he responds to a given stimulus.

8–3 PARALLEL AND PERPENDICULAR LINES

Two lines that are always the same distance apart are called *parallel lines*. The slopes of two parallel lines are related as follows:

Case 1: Two vertical lines are parallel to each other.

Case 2: If two lines with slopes m_1 and m_2 are parallel, their slopes are equal ($m_1 = m_2$); conversely, if two lines have the same slope ($m_1 = m_2$), the lines are parallel.

  EXAMPLE 1: Find the equation of the line passing through the point $(1, -2)$ and parallel to $y = 2x - 5$.

SOLUTION: By inspection, the slope of the line $y = 2x - 5$ is 2. Since the slopes of parallel lines are equal, the desired equation is of the form

$$y = 2x + b$$

To find b, substitute the coordinates of the given point, here $(1, -2)$, in the equation

$$-2 = 2(1) + b$$
$$-4 = b$$

The equation of the line is $y = 2x - 4$ (see Figure 8.8).

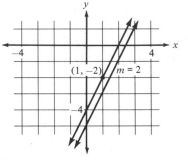

Figure 8.8

Two lines that intersect at right angles are called *perpendicular lines*. The slopes of two perpendicular lines are related as follows:

Case 1: When one line is vertical, the other is horizontal.

Case 2: If two lines with slopes m_1 and m_2 are perpendicular, their slopes are negative reciprocals of each other ($m_1m_2 = -1$); conversely, if two lines have slopes that are negative reciprocals of each other, the lines are perpendicular.

EXAMPLE 2: Find the equation of the line passing through the point $(-4, 1)$ and perpendicular to $5x - 2y = 3$.

SOLUTION: First, find the slope of the given line by changing the equation to the form $y = mx + b$.

$$5x - 2y = 3$$
$$-2y = -5x + 3$$
$$y = \tfrac{5}{2}x - \tfrac{3}{2}$$

The slope of the given line is $\tfrac{5}{2}$. Since the slopes of perpendicular lines are negative reciprocals of each other, the slope of the desired line is $-\tfrac{2}{5}$ and the equation is of the form

$$y = \frac{-2}{5}x + b$$

To find b, substitute the coordinates of the given point, here $(-4, 1)$, in the equation.

$$1 = \frac{-2}{5}(-4) + b$$

$$1 = \frac{8}{5} + b$$

$$\frac{-3}{5} = b$$

The equation of the line is $y = -\frac{2}{5}x - \frac{3}{5}$ (see Figure 8.9).

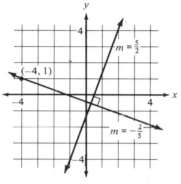

Figure 8.9

EXAMPLE 3: Find the equation of the line that is the perpendicular bisector of the line segment from $(3, 5)$ to $(11, -7)$.

SOLUTION:

a. The perpendicular bisector of a line segment passes through the midpoint of the segment. If (x, y) is the midpoint of the segment joining (x_1, y_1) and (x_2, y_2), then

$$x = \frac{x_1 + x_2}{2} \qquad y = \frac{y_1 + y_2}{2}$$

Substituting the coordinates of the points $(3, 5)$ and $(11, -7)$, we have

$$x = \frac{3 + 11}{2} = 7 \qquad y = \frac{5 + (-7)}{2} = -1$$

The coordinates of the midpoint are $(7, -1)$.

b. The slope of the line segment is

$$m = \frac{y_2 - y_1}{x_2 - x_1} = \frac{-7 - 5}{11 - 3} = \frac{-12}{8} = \frac{-3}{2}$$

The slope of the perpendicular bisector, which is the negative reciprocal of the slope of the line segment, is $\frac{2}{3}$.

c. Since the slope of the perpendicular bisector is $\frac{2}{3}$, the line is of the form

$$y = \tfrac{2}{3}x + b$$

Since the line goes through the midpoint $(7, -1)$, we have

$$-1 = \frac{2}{3}(7) + b$$

$$-1 = \frac{14}{3} + b$$

$$\frac{-17}{3} = b$$

The equation of the perpendicular bisector is $y = \frac{2}{3}x - \frac{17}{3}$ (see Figure 8.10).

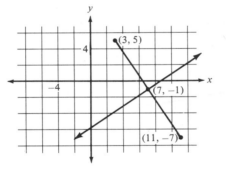

Figure 8.10

EXERCISES 8–3

In Exercises 1–12 find the equation of the line passing through the given point that is:

a. Parallel to the given line.
b. Perpendicular to the given line.

1. $y = x - 2; (1, 2)$ **2.** $y = 3x + 1; (-2, -1)$
3. $y = -2x + 7; (3, 0)$ **4.** $y = -x; (-4, 2)$
5. $x + 7y = 2; (0, 0)$ **6.** $3x - 2y = 5; (0, -5)$
7. $5x - 3y = 2; (1, -3)$ **8.** $4x - y = -8; (-1, -4)$
9. $3x + 12y = 10; (5, 2)$ **10.** $x + 5y = -3; (-2, 3)$
11. $2y - 3x + 1 = 0; (\frac{1}{2}, -\frac{1}{3})$ **12.** $4x - y - 1 = 0; (-\frac{3}{4}, -\frac{1}{2})$

In Exercises 13–20 find the equation of the line that is the perpendicular bisector of the line segment whose end points are given.

13. $(1, 2)$ and $(7, 6)$ **14.** $(1, 3)$ and $(-3, -1)$
15. $(-4, -9)$ and $(0, -5)$ **16.** $(-3, 5)$ and $(1, -1)$
17. $(2, -3)$ and $(-3, 4)$ **18.** $(1, 0)$ and $(-2, 7)$
19. $(6, -1)$ and $(-1, 2)$ **20.** $(-4, 4)$ and $(3, -5)$

8–4 SYSTEMS OF LINEAR FUNCTIONS

More than one linear function is often needed to describe a situation adequately. For example, suppose that you have to decide between two positions as a salesman. The first company offers a straight 20 percent commission, whereas the second company offers a salary of $70 per week plus a 10 percent commission. Before you choose your job, it is important to determine how much you would have to sell per week before the two job offers would pay the same salary. For the purpose of comparison, we have graphed the two job offers in Figure 8.11.

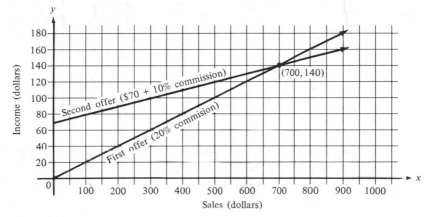

Figure 8.11

First offer (20% commission): $y = 0.2x$

sales (in dollars) x	100	200	300	400	500	600	700	800	900	1000
income (in dollars) y	20	40	60	80	100	120	140	160	180	200

Second offer ($70 + 10% commission): $y = 70 + 0.1x$

sales (in dollars) x	100	200	300	400	500	600	700	800	900	1000
income (in dollars) y	80	90	100	110	120	130	140	150	160	170

By examining the graph we can see that the solution is the point (700, 140) where the two lines intersect. That is, you earn a salary of $140 from either job when you sell $700 worth of merchandise. Above $700 the larger commission is the better offer. Notice in our example that there are many ordered pairs that satisfy the first job offer and many ordered pairs that satisfy the second offer, but our solution is the only ordered pair common to the two functions. In general, the solution of a system of linear functions is the collection of all the ordered pairs that satisfy both equations. Graphically, this corresponds to the collection of points where the lines intersect. Usually, as in our example, the two lines will intersect at only one point and, consequently, our solution is only one ordered pair. However, the next two examples illustrate two other possibilities.

EXAMPLE 1: Find all the ordered pairs that satisfy the following pair of equations:

$$y = 2x + 1$$
$$y = 2x + 3$$

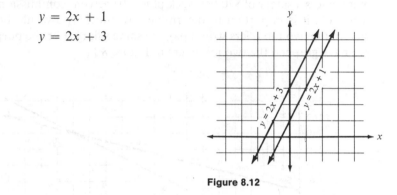

Figure 8.12

SOLUTION: These two lines are parallel and do not intersect, as is shown in Figure 8.12. Thus, there is no ordered pair that satisfies both equations. This situation will arise whenever two lines have the same slope and different y intercepts. Such a system of equations is called *inconsistent*.

EXAMPLE 2: Find all the ordered pairs that satisfy the following pair of equations:

$$y = 2x + 3$$
$$2y = 4x + 6$$

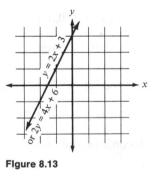

Figure 8.13

SOLUTION: When we attempt to graph both functions (see Figure 8.13), we find that any ordered pair that satisfies the first equation also satisfies the second equation. Thus, the two equations are equivalent. This means that the same line is the graph of both functions and that the solution is all the points on that line. This situation occurred because the second equation is obtained if we multiply both sides of the first equation by 2. In more complicated situations it is often difficult to determine if two equations are equivalent. A system containing equivalent equations is called *dependent*.

By far the most useful situation is when the graphs of the two linear functions intersect at one point so that there is only one ordered pair that satisfies both equations. We have been finding this solution graphically. This technique is good for illustrating the principle involved and for obtaining an approximate solution. However, if an exact solution is required, we must use algebraic methods. The first method we discuss makes use of a substitution, as in the following example.

EXAMPLE 3: Find all the ordered pairs that satisfy the following pair of equations:

$$y = x + 2$$
$$y = 2x - 3$$

SOLUTION: We wish to find the value for x that makes both y values the same, or equivalently, we want to know the value for x that makes $x + 2$ equal to $2x - 3$.

$$x + 2 = 2x - 3$$
$$2 = x - 3$$
$$5 = x$$

Thus, the x coordinate of the solution is 5. To find the y coordinate, substitute 5 for x in either of the given equations:

$$y = x + 2 \quad \text{or} \quad y = 2x - 3$$
$$= (5) + 2 \qquad\qquad = 2(5) - 3$$
$$= 7 \qquad\qquad\qquad = 7$$

Thus the ordered pair (5, 7) satisfies both equations. (*Note:* A good check of your result is to find the y coordinate by substituting the x coordinate in both equations.)

EXAMPLE 4: The total cost of producing gadgets consists of $300 per month for rent plus $4 per unit for material. If the selling price for a gadget is $9, how many units must be made and sold per month for the company to break even?

SOLUTION: If x represents the number of gadgets made and sold, then

$$\text{Total cost} = 300 + 4x$$
$$\text{Total revenue} = 9x$$

We wish to find the value for x that makes the total cost equal to the total revenue, or equivalently, we want to know the value for x that makes $300 + 4x$ equal to $9x$.

$$300 + 4x = 9x$$
$$300 = 5x$$
$$60 = x$$

Thus, the company will break even when it makes and sells 60 gadgets. To check the result, substitute 60 in the original equations (see Figure 8.14).

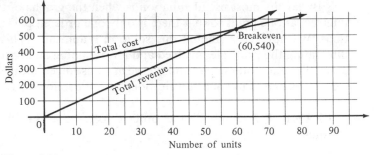

Figure 8.14

$$\text{Total cost} = 300 + 4x \qquad \text{or} \qquad \text{Total revenue} = 9x$$
$$= 300 + 4(60) \qquad\qquad\qquad = 9(60)$$
$$= 540 \qquad\qquad\qquad\qquad = 540$$

In these examples the substitution method has been appropriate since all the linear functions have been given in the form $y = mx + b$. However, if one of the linear functions is given as $3x - 5y - 4 = 0$, we would have to change the equation to $y = \frac{3}{5}x - \frac{4}{5}$ and continue with this equation. This procedure is awkward algebraically. Thus, a different method is used. The following example illustrates this new method. It is very important because it can be used to solve complicated systems of equations with more variables.

EXAMPLE 5: Find all the ordered pairs that satisfy the following pair of equations:

$$x + y = 9$$
$$x - y = 5$$

SOLUTION: This method attempts to eliminate one of the variables by adding the two equations together. In this example the coefficients for y are $+1$ and -1. If we add the equations, the result will be an equation that contains only x.

$$
\begin{array}{ll}
x + y = & 9 \\
x - y = & 5 \\
\hline
2x \quad = & 14 \qquad \text{add the equations} \\
x \quad = & 7
\end{array}
$$

Thus, the x coordinate of the solution is 7.

To find the y coordinate, substitute 7 for x in either of the given equations.

$$
\begin{array}{ll}
x + y = 9 & \text{or} \quad x - y = 5 \\
(7) + y = 9 & \qquad 7 - y = 5 \\
\quad y = 2 & \qquad\quad y = 2
\end{array}
$$

Thus, the ordered pair (7, 2) satisfies both equations.

EXAMPLE 6: Find all the ordered pairs that satisfy the following pair of equations:

$$3x - 2y = 27$$
$$2x + 5y = -1$$

SOLUTION: If we form equivalent equations by multiplying the top equation by -2 and the bottom equation by 3, we can eliminate the x variable.

$$
\begin{array}{rl}
-6x + 4y = & -54 \\
6x + 15y = & -3 \\
\hline
19y = & -57 \qquad \text{add the equations} \\
y = & -3
\end{array}
$$

Thus, the y coordinate of the solution is -3.

To find the x coordinate, substitute -3 for y in either of the given equations.

$$
\begin{array}{lll}
3x - \quad 2y = 27 & \text{or} & 2x + \quad 5y = -1 \\
3x - 2(-3) = 27 & & 2x + 5(-3) = -1 \\
\quad\quad x = 7 & & \quad\quad x = 7
\end{array}
$$

Thus, the ordered pair $(7, -3)$ satisfies both equations.

EXAMPLE 7: Find the forces F_1 and F_2 that achieve equilibrium for the beam in Figure 8.15.

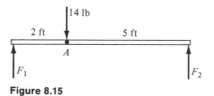

Figure 8.15

SOLUTION: If we consider two of the laws of physics, we can determine two equations that contain F_1 and F_2. First, since the system is in equilibrium, the sum of the forces pointing down must be equal to the sum of the forces pointing up.

$$F_1 + F_2 = 14$$

Second, the tendency of the lumber to rotate about point A in a clockwise direction is the product of F_1 and the distance (2 ft) between F_1 and the turning point A; the tendency of the lumber to rotate in a counterclockwise direction is the product of F_2 and the distance (5 ft) between F_2 and the turning point A. Since the system is in equilibrium, we have

$$\text{clockwise turning effect} = \text{counterclockwise turning effect}$$
$$2 \cdot F_1 = 5 \cdot F_2$$

Thus, we can determine F_1 and F_2 by finding the ordered pair that satisfies these two equations.

$$F_1 + F_2 = 14$$
$$2F_1 - 5F_2 = 0 \quad \text{or} \quad (2F_1 = 5F_2)$$

We can eliminate F_2 by multiplying the top equation by 5 and then adding the two equations.

$$
\begin{array}{rl}
5F_1 + 5F_2 &= 70 \\
2F_1 - 5F_2 &= 0 \\
\hline
7F_1 &= 70 \\
F_1 &= 10
\end{array}
$$

Thus, F_1 is 10 lb. To find F_2 substitute 10 for F_1 in either of the given equations.

$$
\begin{array}{ll}
F_1 + F_2 = 14 \quad \text{or} & 2F_1 = 5F_2 \\
(10) + F_2 = 14 & 2(10) = 5F_2 \\
\quad\quad F_2 = 4 & \quad\; 4 = F_2
\end{array}
$$

Thus, $F_1 = 10$ lb and $F_2 = 4$ lb.

In economics an analysis of the law of supply and demand involves the intersection of two curves. Basically as the price for an item increases, the quantity of the product that is supplied increases while the quantity that is demanded decreases. The point at which the supply and demand curves intersect is called the *point of market equilibrium*. This principle is shown in Figure 8.16, where, for illustrative purposes, we assume the supply and demand functions to be linear.

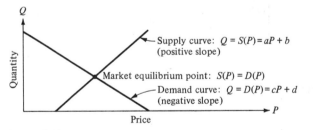

Figure 8.16

The equilibrium price is the value for p at which supply equals demand. The law of supply and demand states that in a system of free enterprise (a big assumption!) a product will sell near its equilibrium price. The theory asserts that if the price is above the equilibrium price, more of the product will be put on the market to capitalize on the higher price. A surplus results, causing the price of the product to lower. When the price is below the equilibrium price, the demand for the product will exceed the supply, so the sellers can raise their prices.

EXERCISES 8–4

In Exercises 1–20 find all the ordered pairs that satisfy the pair of equations. In 1–10 first estimate the solution graphically.

1. $y = x - 2$
   $y = 5x + 6$
2. $y = -x - 3$
   $y = 2x + 3$
3. $y = 3x + 4$
   $y = -x - 2$
4. $y = 4x + 7$
   $y = 3x + 5$
5. $y = x + 4$
   $y = 2x + 4$
6. $y = -5x + 2$
   $y = 4x - 7$
7. $y = x$
   $x - 2y = 6$
8. $y = -3x$
   $2x + 3y = -21$
9. $y = \frac{1}{3}x - 5$
   $x - 3y - 15 = 0$
10. $y = -\frac{5}{2}x + 3$
    $5x + 2y + 6 = 0$
11. $x + y = 25$
    $6x - y = 3$
12. $2x + 3y = 8$
    $2x - 7y = -32$
13. $5x + 3y = -2$
    $x - 2y = -3$
14. $-2x - y = -5$
    $5x + 2y = -17$
15. $3x - 2y = 1$
    $6x - 4y = 5$
16. $2x - 3y = 1$
    $6x - 9y = 3$
17. $2x - 5y = 5$
    $4x + 3y = 23$
18. $4x + 2y = 2$
    $6x - 5y = 27$
19. $7x - 2y - 19 = 0$
    $3x + 5y + 14 = 0$
20. $6x + 10y - 7 = 0$
    $15x - 4y - 3 = 0$

In Exercises 21–39 set up a system of linear functions and solve by an appropriate method illustrated in this section.

21. The sum of two numbers is 70, their difference is 22. What are the numbers?
22. Find two complementary angles whose difference is 20°.
23. Find two supplementary angles whose difference is 100°.
24. A piece of lumber is 120 in. long. Where must it be cut for one piece to be four times longer than the other piece?
25. A container with a liquid weighs 500 g. If one-half the liquid is poured out, the weight is 350 g. What is the weight of the empty container?
26. If four black metal balls and one red metal ball are placed on a scale, they balance a weight of 100 g. A weight of 90 g will balance two black and three red balls. Find the weight of each kind of metal ball.
27. The velocity of a particle that accelerates at a uniform rate is linearly related to the elapsed time by the equation $v = v_0 + at$, where v_0 is the initial velocity and a the acceleration. If $v = 36$ ft/second when $t = 2$ seconds and $v = 4$ ft/second when $t = 3$ seconds, find values for v_0 and a.
28. In Exercise 27 find the values of v_0 and a if $v = 45$ ft/second when $t = 2$ seconds and $v = 72$ ft/second when $t = 5$ seconds.
29. Find the forces F_1 and F_2 that achieve equilibrium for the beam in the figure.

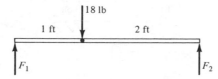

30. Find the forces F_1 and F_2 that achieve equilibrium for the beam in the figure.

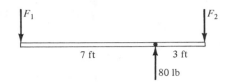

31. It takes a boat 1 hour to go 6 mi upstream (against the current); the return trip downstream (with the current) takes $\frac{1}{2}$ hour. What is the rate of the current? What would be the speed of the boat if there was no current?

32. A chemist has a 10 percent solution and a 25 percent solution of alcohol. How much of each should be mixed together to obtain 100 gal of a 20 percent solution?

33. A college wants to invest $10,000 in order to have an annual income of $700 to be used for a scholarship. They plan to invest part of the money in a bank which yields 6 percent and the remainder in a speculative fund which promises to yield 11 percent. How much should be invested in each to obtain the desired income?

34. You are trying to decide between two positions as a salesman. The first offer is for a straight 15 percent commission while the second offer pays a salary of $60 per week plus a 10 percent commission. How much must you sell each week for the two jobs to pay the same?

35. A manufacturer wants to know whether it will pay him to buy and install a special machine to turn out widgets that he has been purchasing from an outside supplier for $1 each. The machine will cost him $500 per year and will be able to produce widgets for $0.50 each. How many widgets would he have to make and use each year to justify purchasing this new machine?

36. A company is trying to decide between two machines for packaging its new product. Machine A will cost $5000 per year plus $2 to package each unit; machine B will cost $8000 per year plus $1 to package each unit.

    a. How many units must be produced for the cost of the two machines to be the same? If the company plans to produce more units, which machine should they purchase?

    b. If packaging can be subcontracted to another firm at a cost of $5 per unit, how many units would have to be made before the purchase of machine A would be worthwhile?

37. A company is planning to purchase and store two kinds of items, gaskets and widgets. Each gasket costs $3 and occupies 1 ft^2 of floor space; each widget cost $1 and occupies 2 ft^2 of floor space. If the company plans to purchase exactly $1000 worth of items and store them in exactly 700 ft^2 of floor space, how many units of each item may be purchased?

38. Consider the figure. How many nails are needed to balance the cube?

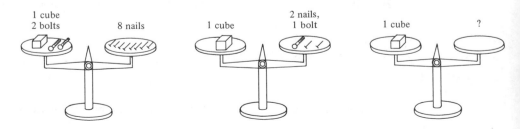

39. One of the classic stories from the history of mathematics concerns Archimedes (250 B.C.), the best mathematician and scientist of ancient times. Archimedes lived in the Greek city-state of Syracuse, where the king, King Hieron, suspected a goldsmith of giving him a gold crown that contained hidden silver. The king referred the problem to Archimedes and asked him to determine, without destroying the crown, the percent of pure gold in the crown. Archimedes found the principle needed to solve this problem while taking a bath. It is reported that Archimedes, who usually became very engrossed in his work, was so excited by his discovery that he forgot to put on his clothes and ran naked through the streets yelling "Eureka!" (I have found it!) What Archimedes noticed in his bath was the obvious fact that when a body is immersed in water, it displaces a volume of water that is equal to the volume of the body. He also knew that bodies of the same weight do not necessarily have the same volume. Using these principles, Archimedes then filled a bucket of water to the brim. Suppose the crown weighed 10 lb and displaced 18 cubic in. of water and that 10 lb of pure gold and 10 lb of pure silver displaced 15 and 30 in.^{3} of water, respectively. What percent of the crown would you tell King Hieron is made of gold?

8-5 DETERMINANTS

A system of linear equations may be solved by formula if we find the solution of the general system.

$$a_1 x + b_1 y = c_1$$
$$a_2 x + b_2 y = c_2$$

Once we do the hard work of solving this system we need only substitute values for the constants to solve a specific problem. Also, a problem that may be solved by formula is done easily on a computer.

To solve for x, we multiply each member of the first equation by b_2 and each member of the second equation by b_1, to obtain

$$a_1 b_2 x + b_1 b_2 y = c_1 b_2$$
$$b_1 a_2 x + b_1 b_2 y = b_1 c_2$$

Now we subtract the second equation from the first, to obtain

$$a_1 b_2 x - b_1 a_2 x = c_1 b_2 - b_1 c_2$$

Factoring the left side of the equation, we have

$$(a_1 b_2 - b_1 a_2)x = c_1 b_2 - b_1 c_2$$

Thus,

$$x = \frac{c_1 b_2 - b_1 c_2}{a_1 b_2 - b_1 a_2}$$

In a similar manner, if we multiply each member of the first equation by a_2 and each member of the second equation by a_1 and then subtract, we have

$$y = \frac{a_1 c_2 - c_1 a_2}{a_1 b_2 - b_1 a_2}$$

These formulas may now be used to find x and y whenever $a_1b_2 - b_1a_2 \neq 0$. If $a_1b_2 - b_1a_2 = 0$, the system is either dependent (in which case the numerators of x and y are zero) or inconsistent (in which case the numerators of x and y are not zero).

We do not memorize the formulas in this form since they may be obtained by defining what is called a *determinant*. Consider the expression $a_1b_2 - b_1a_2$, which is the denominator in the formulas for both x and y. We define a determinant to be a square array of numbers enclosed by vertical bars, such as

$$\begin{vmatrix} a_1 & b_1 \\ a_2 & b_2 \end{vmatrix}$$

and we define the value of this determinant to be $a_1b_2 - b_1a_2$.

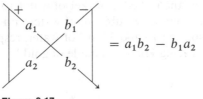

$$= a_1b_2 - b_1a_2$$

Figure 8.17

The numbers a_1 and b_2 are elements of the principal diagonal and the numbers b_1 and a_2 are elements of the secondary diagonal. Note in Figure 8.17 that the value of the determinant is the product of the elements of the principal diagonal minus the product of the elements of the secondary diagonal.

EXAMPLE 1:

$$\begin{vmatrix} 1 & 2 \\ 3 & 4 \end{vmatrix} = 1(4) - 2(3) = 4 - 6 = -2$$

EXAMPLE 2:

$$\begin{vmatrix} 1 & -5 \\ 4 & -7 \end{vmatrix} = 1(-7) - (-5)(4) = -7 + 20 = 13$$

Using determinants a system of the form

$$a_1x + b_1y = c_1$$
$$a_2x + b_2y = c_2$$

may be solved by the following formulas:

$$x = \frac{c_1b_2 - b_1c_2}{a_1b_2 - b_1a_2} = \frac{\begin{vmatrix} c_1 & b_1 \\ c_2 & b_2 \end{vmatrix}}{\begin{vmatrix} a_1 & b_1 \\ a_2 & b_2 \end{vmatrix}}$$

$$y = \frac{a_1c_2 - c_1a_2}{a_1b_2 - b_1a_2} = \frac{\begin{vmatrix} a_1 & c_1 \\ a_2 & c_2 \end{vmatrix}}{\begin{vmatrix} a_1 & b_1 \\ a_2 & b_2 \end{vmatrix}}$$

In determinant form the formulas for x and y are not difficult to remember. In both cases the determinant in the denominator is formed from the coefficients of x and y. Different determinants are used in the numerators. When solving for x, the coefficients of x, a_1 and a_2, are replaced by the constants c_1 and c_2. Similarly, when solving for y, the coefficients of y, b_1 and b_2, are replaced by the constants c_1 and c_2. These formulas written in determinant form are called *Cramer's rule*.

EXAMPLE 3: Use determinants to solve the following system of equations:

$$3x - 2y = 27$$
$$2x + 5y = -1$$

SOLUTION: First, evaluate the determinant in the denominator, which consists of the coefficients of x and y.

$$\begin{vmatrix} 3 & -2 \\ 2 & 5 \end{vmatrix} = 3(5) - (-2)(2) = 15 + 4 = 19$$

Second, evaluate the determinant in the numerator of the formula for x. Replace the column containing the coefficients of x by the column with the constants on the right side of the equations.

$$\begin{vmatrix} 27 & -2 \\ -1 & 5 \end{vmatrix} = 27(5) - (-2)(-1) = 135 - 2 = 133$$

Third, evaluate the determinant in the numerator of the formula for y. We replace the column containing the coefficients of y by the column with the constants on the right side of the equations.

$$\begin{vmatrix} 3 & 27 \\ 2 & -1 \end{vmatrix} = 3(-1) - 27(2) = -3 - 54 = -57$$

Fourth, substitute the values of the determinants in the formulas

$$x = \frac{\begin{vmatrix} 27 & -2 \\ -1 & 5 \end{vmatrix}}{\begin{vmatrix} 3 & -2 \\ 2 & 5 \end{vmatrix}} = \frac{133}{19} = 7$$

$$y = \frac{\begin{vmatrix} 3 & 27 \\ 2 & -1 \end{vmatrix}}{\begin{vmatrix} 3 & -2 \\ 2 & 5 \end{vmatrix}} = \frac{-57}{19} = -3$$

Thus, the ordered pair $(7, -3)$ satisfies both equations.

EXAMPLE 4: Use determinants to solve the following system of equations:

$$5x + 2y - 1 = 0$$
$$x - 3y - 7 = 0$$

SOLUTION: First, change the equations to the form from which the formulas were derived

$$5x + 2y = 1$$
$$x - 3y = 7$$

$$x = \frac{\begin{vmatrix} 1 & 2 \\ 7 & -3 \end{vmatrix}}{\begin{vmatrix} 5 & 2 \\ 1 & -3 \end{vmatrix}} = \frac{1(-3) - 2(7)}{5(-3) - 2(1)} = \frac{-17}{-17} = 1$$

$$y = \frac{\begin{vmatrix} 5 & 1 \\ 1 & 7 \end{vmatrix}}{\begin{vmatrix} 5 & 2 \\ 1 & -3 \end{vmatrix}} = \frac{5(7) - 1(1)}{-17} = \frac{34}{-17} = -2$$

Thus, the solution is $(1, -2)$.

Determinants may be used to solve more complicated systems, such as

$$a_1x + b_1y + c_1z = d_1$$
$$a_2x + b_2y + c_2z = d_2$$
$$a_3x + b_3y + c_3z = d_3$$

This system involves three equations and three variables. Its solution is the collection of all values for x, y, and z that satisfy all three equations simultaneously. To solve this system we must first learn to evaluate determinants with three rows and three columns.

EXAMPLE 5: Evaluate the determinant:

$$\begin{vmatrix} -2 & 3 & 1 \\ 2 & -1 & 5 \\ 4 & 7 & -3 \end{vmatrix}$$

SOLUTION: We evaluate a determinant with three rows and three columns by re-writing the first and second columns to the right of the third column (see Figure 8.18). The value of the determinant is the sum of the products of the elements along the three principal diagonals minus the sum of the products of the elements along the three secondary diagonals.

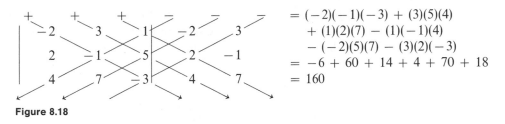

$$= (-2)(-1)(-3) + (3)(5)(4)$$
$$+ (1)(2)(7) - (1)(-1)(4)$$
$$- (-2)(5)(7) - (3)(2)(-3)$$
$$= -6 + 60 + 14 + 4 + 70 + 18$$
$$= 160$$

Figure 8.18

In determinant form the formulas for x, y, and z follow an arrangement similar to that of the system with two equations and two variables. In each case, the determinant in the denominator is formed from the coefficients of x, y, and z. Different determinants are used in the numerators. When solving for x, the coefficients of x are replaced by the constants d_1, d_2, and d_3. Similarly, when solving for y or z, the constants d_1, d_2, and d_3 replace the coefficients of the desired variable.

EXAMPLE 6: Use determinants to solve the following system of equations:

$$3x - y + 6z = 1$$
$$x + 2y - 3z = 0$$
$$2x - 3y - z = -9$$

SOLUTION: The determinant in the denominator is formed from the coefficients of x, y, and z. To find the determinant in the numerator for x, we replace the column containing the coefficients of x by the column containing the constants on the right side of the equations.

$$x = \frac{\begin{vmatrix} 1 & -1 & 6 \\ 0 & 2 & -3 \\ -9 & -3 & -1 \end{vmatrix} \begin{matrix} 1 & -1 \\ 0 & 2 \\ -9 & -3 \end{matrix}}{\begin{vmatrix} 3 & -1 & 6 \\ 1 & 2 & -3 \\ 2 & -3 & -1 \end{vmatrix} \begin{matrix} 3 & -1 \\ 1 & 2 \\ 2 & -3 \end{matrix}} = \frac{-2 - 27 + 0 + 108 - 9 - 0}{-6 + 6 - 18 - 24 - 27 - 1} = \frac{70}{-70} = -1$$

The value of the determinant in the denominator for both y and z is -70. The column containing the constants on the right side of the equations first replaces the coefficients of y (to find y) and then the coefficients of z (to find z).

$$y = \frac{\begin{vmatrix} 3 & 1 & 6 \\ 1 & 0 & -3 \\ 2 & -9 & -1 \end{vmatrix} \begin{matrix} 3 & 1 \\ 1 & 0 \\ 2 & -9 \end{matrix}}{-70} = \frac{0 - 6 - 54 - 0 - 81 + 1}{-70} = \frac{-140}{-70} = 2$$

$$z = \frac{\begin{vmatrix} 3 & -1 & 1 \\ 1 & 2 & 0 \\ 2 & -3 & -9 \end{vmatrix} \begin{matrix} 3 & -1 \\ 1 & 2 \\ 2 & -3 \end{matrix}}{-70} = \frac{-54 + 0 - 3 - 4 - 0 - 9}{-70} = \frac{-70}{-70} = 1$$

The solution is $x = -1$, $y = 2$, $z = 1$. (*Note:* If the determinant in the denominator is zero, merely write that there is no unique solution to the problem. You will learn how to handle this case in a future course in linear algebra.)

EXERCISES 8–5

In Exercises 1–20 evaluate the determinant.

1. $\begin{vmatrix} 1 & 3 \\ 4 & 2 \end{vmatrix}$

2. $\begin{vmatrix} -2 & 5 \\ 1 & 3 \end{vmatrix}$

3. $\begin{vmatrix} 1 & -5 \\ 2 & -8 \end{vmatrix}$

4. $\begin{vmatrix} 2 & -3 \\ 2 & -5 \end{vmatrix}$

5. $\begin{vmatrix} 5 & -3 \\ 2 & 4 \end{vmatrix}$

6. $\begin{vmatrix} -2 & 14 \\ 3 & -3 \end{vmatrix}$

7. $\begin{vmatrix} -3 & 0 \\ 4 & 1 \end{vmatrix}$

8. $\begin{vmatrix} 5 & -28 \\ 4 & -10 \end{vmatrix}$

9. $\begin{vmatrix} 3 & 8 \\ -8 & 11 \end{vmatrix}$

10. $\begin{vmatrix} -7 & -9 \\ -10 & -6 \end{vmatrix}$

11. $\begin{vmatrix} 2 & 5 \\ -4 & -10 \end{vmatrix}$

12. $\begin{vmatrix} 4 & -8 \\ 6 & -12 \end{vmatrix}$

13. $\begin{vmatrix} -2 & 3 & 2 \\ 0 & 7 & 4 \\ 1 & -1 & 3 \end{vmatrix}$

14. $\begin{vmatrix} 0 & 2 & -1 \\ 5 & 1 & 0 \\ -3 & 0 & -4 \end{vmatrix}$

15. $\begin{vmatrix} 2 & 1 & 1 \\ 1 & -2 & -1 \\ 3 & 3 & -1 \end{vmatrix}$

16. $\begin{vmatrix} 2 & 3 & 2 \\ 1 & -3 & 3 \\ 5 & 1 & -4 \end{vmatrix}$

17. $\begin{vmatrix} 2 & -7 & 1 \\ 4 & 1 & 2 \\ 6 & -3 & 3 \end{vmatrix}$

18. $\begin{vmatrix} 9 & 10 & -2 \\ 4 & 1 & 8 \\ -1 & -7 & 3 \end{vmatrix}$

19. $\begin{vmatrix} 6 & 3 & 11 \\ -3 & 4 & 5 \\ -1 & -2 & 3 \end{vmatrix}$

20. $\begin{vmatrix} -1 & 2 & -1 \\ 5 & -2 & -3 \\ 2 & -4 & 2 \end{vmatrix}$

In Exercises 21–30 use determinants to solve the systems of equations in Exercises 11–20 of Section 8-4.

In Exercises 31–40 use determinants to solve the given systems of equations. If a variable is missing from an equation, enter zero in the determinant as its coefficient.

31. $\begin{aligned} x - y + z &= -1 \\ x + y - 3z &= 3 \\ 2x + y - 2z &= 4 \end{aligned}$

32. $\begin{aligned} x + y + z &= 1 \\ x + y - z &= 3 \\ x - y - z &= 5 \end{aligned}$

33. $-2x + y - z = 3$
$\quad 2x + 3y + 2z = 5$
$\quad 3x + 2y + z = -3$

34. $x + y - 3z = -4$
$\quad -x - y + 5z = 8$
$\quad 2x + 3y - z = 3$

35. $x - y + z = 3$
$\quad 3x + y + z = 1$
$\quad 2x + 2y - z = -3$

36. $x + y + 3z = 1$
$\quad 2x + 5y + 2z = 0$
$\quad 3x - 2y - z = 3$

37. $2x + 3y = 10$
$\quad 3x + 2z = 2$
$\quad 4y + z = 6$

38. $x + y + z = 0$
$\quad 2x - 5y = 3$
$\quad -5y + 3z = 7$

39. $2x - y + 3z = 1$
$\quad -x + y - z = -1$
$\quad -4x + 2y - 6z = -2$

40. $x + y + z = 1$
$\quad 2x - 3y - 2z = -4$
$\quad 3x - 2y - z = -2$

In Exercises 41–44 solve the equations simultaneously to determine the value of the currents I_1, I_2, and I_3 shown in the figure.

$$I_1 + I_2 + I_3 = 0$$
$$R_1 I_1 - R_3 I_3 = E_1$$
$$R_2 I_2 - R_3 I_3 = E_2$$

	E_1 (volts)	E_2 (volts)	R_1 (ohms)	R_2 (ohms)	R_3 (ohms)
41.	3	10	2	9	5
42.	8	5	6	10	3
43.	7	2	5	4	9
44.	10	20	50	12	100

8-6 QUADRATIC FUNCTIONS

A function of the form $y = ax^2 + bx + c$ (where a, b, and c are constants with $a \neq 0$) is called a *quadratic function*. To see the essential properties of this function, consider the graphs of the quadratic functions $y = x^2 - 4x + 3$ (Figure 8.19) and $y = 4x - 2x^2$ (Figure 8.20). We obtained these graphs by finding some ordered pairs that satisfy the relationships.

x	$x^2 - 4x + 3 = y$	Solutions	x	$4x - 2x^2 = y$	Solutions
-1	$(-1)^2 - 4(-1) + 3 = 8$	$(-1, 8)$	-1	$4(-1) - 2(-1)^2 = -6$	$(-1, -6)$
0	$(0)^2 - 4(0) + 3 = 3$	$(0, 3)$	0	$4(0) - 2(0)^2 = 0$	$(0, 0)$
1	$(1)^2 - 4(1) + 3 = 0$	$(1, 0)$	1	$4(1) - 2(1)^2 = 2$	$(1, 2)$
2	$(2)^2 - 4(2) + 3 = -1$	$(2, -1)$	2	$4(2) - 2(2)^2 = 0$	$(2, 0)$
3	$(3)^2 - 4(3) + 3 = 0$	$(3, 0)$	3	$4(3) - 2(3)^2 = -6$	$(3, -6)$
4	$(4)^2 - 4(4) + 3 = 3$	$(4, 3)$			
5	$(5)^2 - 4(5) + 3 = 8$	$(5, 8)$			

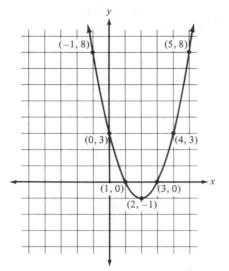

Figure 8.19

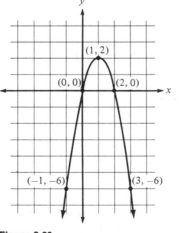

Figure 8.20

The graph in both instances is a curve that is called a *parabola*. For $y = x^2 - 4x + 3$ the parabola has a minimum turning point and opens upward like a cup. This occurs because the coefficient of the dominant term (x^2) is a positive number. Similarly, the graph of $y = 4x - 2x^2$ has a turning point at the maximum y value. The parabola opens down since the coefficient of x^2 is a negative number. In general:

1. If a is positive $(a > 0)$, the parabola opens upward and turns at the lowest point on the graph.

2. If a is negative $(a < 0)$, the parabola opens downward and turns at the highest point on the graph.

Another essential property of this graph is that a vertical line which passes through the turning point divides the parabola into two segments such that if the curve were folded over on this line, its left half would coincide with its right half (see Figure 8.21). This line is called the *axis of symmetry* for the parabola and it is useful to be able to determine the equation of this line. To find the equation of the axis of symmetry, note that this line is halfway between any pair of points on the parabola that have the same y coordinate. The simplest pair of points to work with is the pair of points on the graph of $y = ax^2 + bx + c$ whose y coordinate is c. Thus,

$$y = ax^2 + bx + c$$

Let $y = c$; then

$$c = ax^2 + bx + c$$
$$0 = ax^2 + bx$$
$$0 = (ax + b)x$$

$$ax + b = 0 \qquad x = 0$$
$$ax = -b$$
$$x = \frac{-b}{a}$$

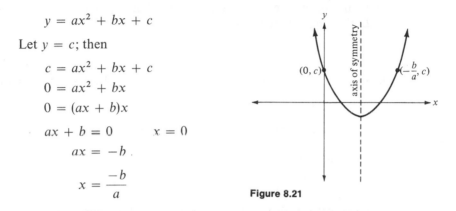

Figure 8.21

The x coordinate halfway between these two x values is

$$x = \frac{0 + (-b/a)}{2} \quad \text{or} \quad x = \frac{-b}{2a}$$

Since the maximum or minimum point lies on the axis of symmetry, the x coordinate of this turning point is $-b/2a$. We determine the y coordinate of this point by finding the y value when $x = -b/2a$. Thus, the coordinates of the turning point of the parabola (which is also called the *vertex*) are

$$\left(\frac{-b}{2a}, \ f\left(\frac{-b}{2a} \right) \right)$$

The last important property of the graph of a quadratic function that we consider is the coordinates of the points where the curve crosses the axes. The point where the parabola crosses the y axis (that is, the y intercept) can be found by substituting 0 for x in the equation $y = ax^2 + bx + c$.

$$y = a(0)^2 + b(0) + c$$
$$= c$$

Thus, the y intercept is always $(0, c)$.

To find the points where the parabola crosses the x axis (that is, the x intercepts) we substitute 0 for y in the equation $y = ax^2 + bx + c$. Thus, the x coordinates of the x intercepts are found by solving the equation $ax^2 + bx + c = 0$. In the next section we will develop a formula that can be used to solve this equation for all values of a, b, and c; for now we restrict ourselves to equations that can be solved by the methods of Section 4-5.

EXAMPLE 1: Graph the function defined by $y = x^2 - 7x + 6$ and indicate
a. the coordinates of the x and y intercepts
b. the equation of the axis of symmetry
c. the coordinates of maximum or minimum point
d. the range of the function.

SOLUTION:
a. To find the x intercepts, set $y = 0$ and solve the resulting equation.

$$x^2 - 7x + 6 = 0$$
$$(x - 6)(x - 1) = 0$$

$$x - 6 = 0 \qquad x - 1 = 0$$
$$x = 6 \qquad\qquad x = 1$$

x intercepts: $(1, 0)$ and $(6, 0)$; y intercept: $(0, 6)$ since $c = 6$.
b. Axis of symmetry:

$$x = \frac{-b}{2a} = \frac{-(-7)}{2(1)} = \frac{7}{2}$$

c. Since $a > 0$, we have a minimum point. The x coordinate of the minimum point is $\frac{7}{2}$ since the minimum point lies on the axis of symmetry. We find the y coordinate of the minimum point by finding the y value when $x = \frac{7}{2}$.

$$y = f(x) = x^2 - 7x + 6$$
$$f(\tfrac{7}{2}) = (\tfrac{7}{2})^2 - 7(\tfrac{7}{2}) + 6$$
$$= \tfrac{49}{4} - \tfrac{49}{2} + 6$$
$$= \tfrac{49}{4} - \tfrac{98}{4} + \tfrac{24}{4}$$
$$= \frac{-25}{4}$$

Minimum point: $(7/2, -25/4)$.
d. As illustrated in Figure 8.22, the range is $y \geq -25/4$.

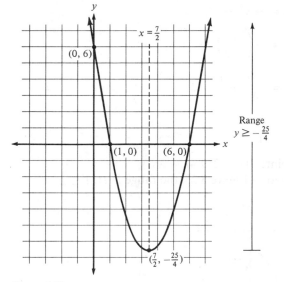

Figure 8.22

EXAMPLE 2: Graph the function defined by $f(x) = -x^2 - x + 2$ and indicate
a. the coordinates of x and y intercepts
b. the equation of the axis of symmetry
c. the coordinates of maximum or minimum point
d. the range of the function.

SOLUTION:
a. To find the x intercepts, set $y = 0$ and solve the resulting equation.

$$-x^2 - x + 2 = 0$$
$$(-x - 2)(x - 1) = 0$$

$$-x - 2 = 0 \qquad x - 1 = 0$$
$$x = -2 \qquad x = 1$$

x intercepts: $(-2, 0)$ and $(1, 0)$; y intercept: $(0, 2)$ since $c = 2$.
b. Axis of symmetry:

$$x = \frac{-b}{2a} = \frac{-(-1)}{2(-1)} = \frac{1}{-2}$$

c. Since $a < 0$, we have a maximum point. The x coordinate of the maximum point is $1/-2$, since the maximum point lies on the axis of symmetry. We find the y coordinate of the maximum point by finding $f(1/-2)$, the value of the function when $x = 1/-2$.

$$f(x) = -x^2 - x + 2$$

$$f\left(\frac{1}{-2}\right) = -\left(\frac{1}{-2}\right)^2 - \left(\frac{1}{-2}\right) + 2$$

$$= \frac{-1}{4} + \frac{1}{2} + 2$$

$$= \frac{9}{4}$$

Maximum point: $(1/-2, 9/4)$.

d. As illustrated in Figure 8.23, the range is $f(x) \leq \frac{9}{4}$.

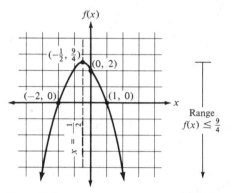

Figure 8.23

EXAMPLE 3: Graph the function defined by $y = x^2 - 2$ and indicate

a. the coordinates of x and y intercepts

b. the equation of the axis of symmetry

c. the coordinates of maximum or minimum point

d. the range of the function.

SOLUTION:

a. To find the x intercepts, set $y = 0$ and solve the resulting equation.

$$x^2 - 2 = 0$$

$$x^2 = 2$$

$$\sqrt{x^2} = \sqrt{2} \qquad \text{take the square root of both sides of the equation}$$

$$|x| = \sqrt{2}$$

$$x = \pm\sqrt{2}$$

x intercepts: $(\sqrt{2}, 0)$ and $(-\sqrt{2}, 0)$; y intercept: $(0, -2)$ since $c = -2$.

b. Axis of symmetry:

$$x = \frac{-b}{2a} = \frac{-(0)}{2(1)} = 0$$

c. Since $a > 0$, we have a minimum point. The x coordinate of the minimum point is 0 since the minimum point lies on the axis of symmetry. Thus, in this problem the minimum point is the y intercept. Minimum point: $(0, -2)$.

d. As illustrated in Figure 8.24, the range is $y \geq -2$.

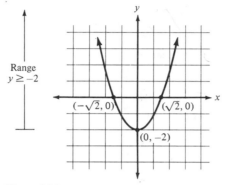

Figure 8.24

The following examples illustrate how quadratic functions can be applied in practical situations. Notice, in particular, the significance of the maximum or minimum value, which is an important topic in applied mathematics.

EXAMPLE 4: The height (y) of a projectile shot vertically up from the ground with an initial velocity of 144 ft/second is given by the formula $y = 144t - 16t^2$. Graph the function defined by this formula and indicate:

a. when the projectile will strike the ground

b. when the projectile attains its maximum height

c. the maximum height.

SOLUTION:

a. When the projectile hits the ground, the height (y) of the projectile is 0. Therefore, we want to find the values of t for which $144t - 16t^2$ equals 0.

$$144t - 16t^2 = 0$$
$$16t(9 - t) = 0$$
$$16t = 0 \qquad 9 - t = 0$$
$$t = 0 \qquad\quad t = 9$$

The projectile will hit the ground 9 seconds later.

b. Since $a < 0$, we have a maximum point. Axis of symmetry:

$$t = \frac{-b}{2a} = \frac{-(144)}{2(-16)} = \frac{-144}{-32} = 4.5$$

The projectile reaches its highest point when $t = 4.5$ seconds.

c. We can find the maximum height by finding the value of y when $t = 4.5$ (or $t = \frac{9}{2}$) seconds.

$$y = f(t) = 144t - 16t^2$$
$$f(\tfrac{9}{2}) = 144(\tfrac{9}{2}) - 16(\tfrac{9}{2})^2$$
$$= 72 \cdot 9 - 16(\tfrac{81}{4})$$
$$= 324$$

As shown in Figure 8.25, the projectile reaches a maximum height of 324 ft.

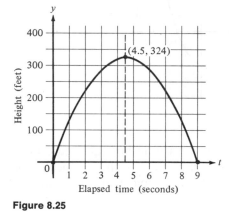

Figure 8.25

EXAMPLE 5: A businessman wants to fence in a parking lot. One side of the lot is bounded by a building 200 ft in length. If a total of 300 ft of fencing is avail--able, what are the dimensions of the largest rectangular parking lot that can be enclosed with the available fencing?

SOLUTION: First, draw a sketch of the situation (Figure 8.26). If we let x represent the length of one side of the lot, the opposite side also has length x. The length of the side opposite the building is then $300 - 2x$. We wish to maximize the area of this rectangular region, which is given by the following quadratic function:

$$\text{Area} = x(300 - 2x)$$
$$A = 300x - 2x^2$$

Building

————200 ft————

x x

$300 - 2x$

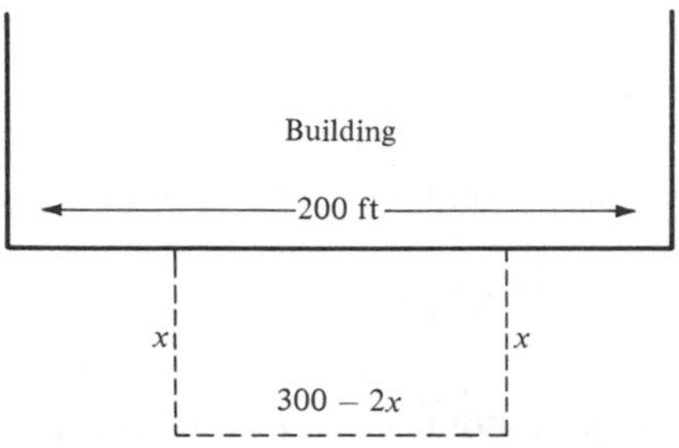

Figure 8.26

Since $a < 0$, we have a maximum point. Axis of symmetry:

$$x = \frac{-b}{2a} = \frac{-300}{2(-2)} = \frac{-300}{-4} = 75$$

Length of the side opposite building:

$$300 - 2x = 300 - 2(75) = 300 - 150 = 150$$

The area is a maximum when the dimensions are 75 ft by 150 ft.

Finally, in Section 8-4 we showed how to determine any points at which two straight lines intersect. In calculus it is frequently necessary to find the intersection points of a parabola with a line or another parabola. Example 6 illustrates the procedure for solving such a system of equations.

EXAMPLE 6: Find all the ordered pairs that satisfy the following pair of equations:

$$y = x^2 - 6x + 8$$
$$y = x + 2$$

SOLUTION: We wish to find the value(s) for x that make both y values the same, or equivalently, we want to know the value(s) for x that make $x^2 - 6x + 8$ equal to $x + 2$.

$$x^2 - 6x + 8 = x + 2$$
$$x^2 - 7x + 6 = 0$$
$$(x - 6)(x - 1) = 0$$

$$x - 6 = 0 \qquad x - 1 = 0$$
$$x = 6 \qquad\quad x = 1$$

Thus, the parabola and the line intersect at $x = 1$ and $x = 6$. To find the y coordinates, substitute these numbers in the simpler equation $y = x + 2$.

$$f(x) = x + 2$$
$$f(1) = 1 + 2 = 3$$
$$f(6) = 6 + 2 = 8$$

Thus, the ordered pairs $(1, 3)$ and $(6, 8)$ satisfy both equations. Consider Figure 8.27, which shows the parabola and the line intersecting at these points.

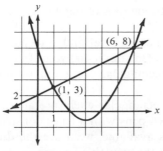

Figure 8.27

EXERCISES 8–6

In Exercises 1–20 graph the quadratic function defined by the equation. On the graph indicate:

    a. the coordinates of x and y intercepts
    b. the equation of the axis of symmetry
    c. the coordinates of maximum or minimum point
    d. the range of the function.

1. $y = x^2 - 3x + 2$ **2.** $y = x^2 + 6x + 5$
3. $f(x) = x^2 - 4$ **4.** $g(x) = 3x^2 - 3$
5. $y = x^2 - 2x$ **6.** $f(x) = 2x^2 - 3x$
7. $g(x) = 4x - x^2$ **8.** $y = 1 - x^2$
9. $y = x^2 + 2x + 1$ **10.** $y = x^2 - 4x + 4$
11. $f(x) = -x^2 + x + 2$ **12.** $f(x) = -x^2 + 5x - 4$
13. $y = x^2 - 5$ **14.** $y = 2 - x^2$
15. $y = 5 - 4x^2$ **16.** $y = 2x^2 - 3$
17. $f(x) = 2x^2 - x - 1$ **18.** $y = 3x^2 - 4x + 1$
19. $y = -2x^2 + 7x - 3$ **20.** $g(x) = -3x^2 + 5x + 2$

21. What is the domain of all the functions in Exercises 1–20?

22. The height of a ball thrown up from a roof 144 ft high with an initial velocity of 128 ft/second is given by the formula $y = 144 + 128t - 16t^2$. What is the maximum height attained by the ball? When does the ball hit the ground?

23. The height of a projectile shot vertically up from the ground with an initial velocity of 96 ft/second is given by the formula $y = 96t - 16t^2$. What is the maximum height attained by the projectile? When does the projectile hit the ground?

24. The total sales revenue for a company is estimated by the formula $S = 8x - x^2$, where x is the unit selling price (in dollars) and S is the total sales revenue (1 unit equals $10,000). What should be the unit selling price if the company wishes to maximize the total revenue?

25. In a 120-volt line having a resistance of 10 ohms, the power W in watts when the current I is flowing is given by the formula $W = 120I - 10I^2$. What is the maximum power that can be delivered in this circuit?

26. Find two positive numbers whose sum is 8 and whose product is a maximum.

27. Find two positive numbers whose sum is 20 and whose product is a maximum.

28. What positive number exceeds its square by the largest amount?

29. Find two positive numbers whose sum is 20 such that the sum of their squares is a minimum.

30. The sum of the base and altitude of a triangle is 12 in. Find each dimension if the area is to be a maximum.

31. A rectangular field is adjacent to a river and is to have fencing on three sides, the side on the river requiring no fencing. If 400 yd of fencing is available, what are the dimensions of the largest rectangular section that can be enclosed with the available fencing?

32. A farmer wants to make a rectangular enclosure along the side of a barn and then divide the enclosure into two pens with a fence constructed at a right angle to the barn. If 300 yd of fencing is available, what are the dimensions of the largest section that can be enclosed with the available fencing?

33. A sheet of metal 12 in. wide and 20 ft long is to be made into a gutter by turning up the same amount of material on each edge at right angles to the base. Determine the amount of material that should be turned up to maximize the volume that the gutter can carry.

34. Find the maximum possible area of a rectangle with a perimeter of 100 ft.

35. Show that the largest rectangular field that can be enclosed by a given length of fence is a square.

36. A businessman owns a parking lot that he wants to enclose with a fence. One side of the lot is bounded by a building, so no fencing is required. There is to be a 16-ft opening in the front of the fence which is opposite the building. If side fencing cost $1 per foot and front fencing cost $1.50 per foot, what are the dimensions of the largest rectangular lot that can be fenced for $300?

37. A bus tour charges a fare of $10 per person and carries 200 people per day. The manager estimates that he will lose 10 passengers for each $1 increase in fare. Find the most profitable fare for him to charge.

38. An airline offers a charter flight at a fare of $100 per person if at least 100 passengers sign up. For each passenger beyond 100, the fare for each passenger is reduced by 50 cents. What is the maximum total revenue that the airline can obtain if the plane has 200 seats?

39. A farmer finds that if he plants 20 peach trees per acre, on the average each tree will yield 300 peaches. He also finds that for each tree in excess of 20 per acre the average yield is reduced by 10 peaches per tree. How many trees per acre should he plant to obtain the highest total yield?

In Exercises 40–45 find all the ordered pairs that satisfy the pairs of equations. In each case sketch the functions and indicate the solution graphically.

40. $y = x^2$
$y = x$

41. $f(x) = 2 - x^2$
$g(x) = -x$

42. $g(x) = 10 - x^2$
$f(x) = 1$

43. $y = x^2 - 2x - 3$
$y = 2x + 2$

44. $y = x^2 - 4$
$y = 4 - x^2$

45. $f(x) = x^2$
$g(x) = 2 - x^2$

8-7 QUADRATIC FORMULA

As we have seen from Sections 8-6 and 4-5, it is important to be able to solve second-degree (or quadratic) equations. Remember that these equations may always be written in the form $ax^2 + bx + c = 0$ (where a, b, and c represent constants with $a \neq 0$). We have developed methods for solving these equations in the case when $ax^2 + bx + c$ can be factored and in the case when $b = 0$. We now develop a technique that can be used to solve all quadratic equations regardless of the values of a, b, and c.

The procedure for solving any quadratic equation is to substitute the values of a, b, and c into a formula and then simplify. To obtain an idea of the origin of this formula, let us solve two equations using a method which works for every quadratic equation; the particular equation $2x^2 - 5x + 1 = 0$ on the left side of the page, and the general equation $ax^2 + bx + c = 0$ on the right.

Particular equation: **General equation:**

$$2x^2 - 5x + 1 = 0 \qquad\qquad ax^2 + bx + c = 0 \quad (\text{with } a \neq 0)$$

We shall attempt to make the left side of the equation a perfect square so that we can take the square root of both sides of the equation.

First, we will divide each term by the coefficient of x^2.

$$x^2 - \frac{5}{2}x + \frac{1}{2} = 0 \qquad\qquad x^2 + \frac{b}{a}x + \frac{c}{a} = 0$$

Subtract the constant term from both sides of the equation.

$$x^2 - \frac{5}{2}x = \frac{-1}{2} \qquad\qquad x^2 + \frac{b}{a}x = \frac{-c}{a}$$

We now add a number to both sides of the equation so that the left side becomes a perfect square. The desired number can always be found by taking half the coefficient of the x term and squaring the result.

Half of $\dfrac{-5}{2}$ is $\dfrac{-5}{4}$ and $\left(\dfrac{-5}{4}\right)^2$ is $\dfrac{25}{16}$ Half of $\dfrac{b}{a}$ is $\dfrac{b}{2a}$ and $\left(\dfrac{b}{2a}\right)^2$ is $\dfrac{b^2}{4a^2}$

$$x^2 - \frac{5}{2}x + \frac{25}{16} = \frac{-1}{2} + \frac{25}{16} \qquad x^2 + \frac{b}{a}x + \frac{b^2}{4a^2} = \frac{-c}{a} + \frac{b^2}{4a^2}$$

Write the left side as a perfect square and simplify the right side.

$$\left(x - \frac{5}{4}\right)^2 = \frac{17}{16} \qquad\qquad \left(x + \frac{b}{2a}\right)^2 = \frac{b^2 - 4ac}{4a^2}$$

Take the square root of both sides of the equation.

$$\left|x - \frac{5}{4}\right| = \sqrt{\frac{17}{16}} \qquad\qquad \left|x + \frac{b}{2a}\right| = \sqrt{\frac{b^2 - 4ac}{4a^2}}$$

$$x - \frac{5}{4} = \pm\sqrt{\frac{17}{16}} \qquad\qquad x + \frac{b}{2a} = \pm\sqrt{\frac{b^2 - 4ac}{4a^2}}$$

Simplify the result.

$$x - \frac{5}{4} = \frac{\pm\sqrt{17}}{4} \qquad\qquad x + \frac{b}{2a} = \frac{\pm\sqrt{b^2 - 4ac}}{2a}$$

$$x = \frac{5}{4} + \frac{\pm\sqrt{17}}{4} \qquad\qquad x = \frac{-b}{2a} + \frac{\pm\sqrt{b^2 - 4ac}}{2a}$$

$$= \frac{5 \pm \sqrt{17}}{4} \qquad\qquad = \frac{-b \pm \sqrt{b^2 - 4ac}}{2a}$$

Thus, our two solutions x_1 and x_2 are

$$x_1 = \frac{5 + \sqrt{17}}{4} \qquad\qquad x_1 = \frac{-b + \sqrt{b^2 - 4ac}}{2a}$$

$$x_2 = \frac{5 - \sqrt{17}}{4} \qquad\qquad x_2 = \frac{-b - \sqrt{b^2 - 4ac}}{2a}$$

The end result of applying this method (which is called *completing the square*) to the general equation $ax^2 + bx + c = 0$ is the *quadratic formula.*

QUADRATIC FORMULA: If $ax^2 + bx + c = 0$ and $a \neq 0$, then

$$x = \frac{-b \pm \sqrt{b^2 - 4ac}}{2a}$$

Note that this formula results in two solutions.

$$x_1 = \frac{-b + \sqrt{b^2 - 4ac}}{2a} \quad \text{and} \quad x_2 = \frac{-b - \sqrt{b^2 - 4ac}}{2a}$$

We can solve *any* quadratic equation with this formula by identifying the values of a, b, and c and substituting these numbers into the formula.

EXAMPLE 1: Solve the equation $3x^2 + 5x + 2 = 0$ by using the quadratic formula.

SOLUTION: In this equation $a = 3$, $b = 5$, and $c = 2$. Therefore,

$$x = \frac{-b \pm \sqrt{b^2 - 4ac}}{2a} = \frac{-(5) \pm \sqrt{(5)^2 - 4(3)(2)}}{2(3)}$$

$$= \frac{-5 \pm \sqrt{25 - 24}}{6}$$

$$= \frac{-5 \pm \sqrt{1}}{6}$$

$$= \frac{-5 \pm 1}{6}$$

$$x_1 = \frac{-5 + 1}{6} = \frac{-4}{6} = \frac{-2}{3} \qquad x_2 = \frac{-5 - 1}{6} = \frac{-6}{6} = -1$$

EXAMPLE 2: Solve the equation $x^2 - 4x - 3 = 0$ by using the quadratic formula.

SOLUTION: In this equation $a = 1$, $b = -4$, and $c = -3$; therefore,

$$x = \frac{-b \pm \sqrt{b^2 - 4ac}}{2a} = \frac{-(-4) \pm \sqrt{(-4)^2 - 4(1)(-3)}}{2(1)}$$

$$= \frac{4 \pm \sqrt{16 + 12}}{2}$$

$$= \frac{4 \pm \sqrt{28}}{2}$$

$$= \frac{4 \pm 2\sqrt{7}}{2}$$

$$= 2 \pm \sqrt{7}$$

$$x_1 = 2 + \sqrt{7} \qquad x_2 = 2 - \sqrt{7}$$

EXAMPLE 3: Solve the equation $4x^2 = 20x - 25$ by using the quadratic formula.

SOLUTION: First, express the equation in the form $ax^2 + bx + c = 0$.

$$4x^2 - 20x + 25 = 0$$

In this equation $a = 4$, $b = -20$, and $c = 25$. Therefore,

$$x = \frac{-b \pm \sqrt{b^2 - 4ac}}{2a} = \frac{-(-20) \pm \sqrt{(-20)^2 - 4(4)(25)}}{2(4)}$$

$$= \frac{20 \pm \sqrt{400 - 400}}{8}$$

$$= \frac{20 \pm \sqrt{0}}{8}$$

$$x_1 = \frac{20 + 0}{8} = \frac{20}{8} = \frac{5}{2} \qquad x_2 = \frac{20 - 0}{8} = \frac{20}{8} = \frac{5}{2}$$

It is common practice to think of this equation as having two solutions that are equal.

EXAMPLE 4: Solve the equation $2x^2 - x + 3 = 0$ by using the quadratic formula.

SOLUTION: In this equation $a = 2$, $b = -1$, and $c = 3$. Therefore,

$$x = \frac{-b \pm \sqrt{b^2 - 4ac}}{2a} = \frac{-(-1) \pm \sqrt{(-1)^2 - 4(2)(3)}}{2(2)}$$

$$= \frac{1 \pm \sqrt{1 - 24}}{4}$$

$$= \frac{1 \pm \sqrt{-23}}{4}$$

$$= \frac{1 \pm i\sqrt{23}}{4}$$

$$x_1 = \frac{1 + i\sqrt{23}}{4} \qquad x_2 = \frac{1 - i\sqrt{23}}{4}$$

The solution of this equation contains imaginary numbers.

Notice from the examples that the nature of the solutions to a quadratic equation depends upon the value of $b^2 - 4ac$ (called the *discriminant*), which appears under the radical in the quadratic formula. That is,

When a, b, c are rational and	the solutions are
$b^2 - 4ac < 0$	imaginary
$b^2 - 4ac = 0$	real, rational, equal
$b^2 - 4ac > 0$ and a perfect square	real, rational, unequal
$b^2 - 4ac > 0$, not a perfect square	real, irrational, unequal

For example, in the equation $3x^2 + 5x + 2 = 0$ (Example 1), $b^2 - 4ac = (5)^2 - 4(3)(2) = 1$. Since 1 is greater than 0 and a perfect square, we know that there are two different solutions that are rational numbers. Similarly, in the equation $2x^2 - x + 3 = 0$ (Example 4), $b^2 - 4ac = (-1)^2 - 4(2)(3) = -23$. Since -23 is less than 0, our solutions are imaginary numbers. Thus, without solving the equation, we can determine the nature of the solutions by finding the value of $b^2 - 4ac$.

Finally, you should notice the close connection between the solutions of the quadratic equation ($ax^2 + bx + c = 0$) and the x intercepts of the quadratic function ($y = ax^2 + bx + c$). That is, in order to find the x coordinates where the function crosses the x axis, we set $y = 0$ and solve the equation $ax^2 + bx + c = 0$. Thus, the nature of the solutions of the resulting quadratic equation will indicate the manner in which the graph crosses the x axis.

EXAMPLE 5: The graph of $y = 3x^2 + 5x + 2$ intersects the x axis at two points that correspond to rational numbers since $b^2 - 4ac = 1$ (see Figure 8.28).

EXAMPLE 6: The graph of $y = x^2 - 4x - 3$ intersects the x axis at two points that correspond to irrational numbers since $b^2 - 4ac = 28$ (see Figure 8.29).

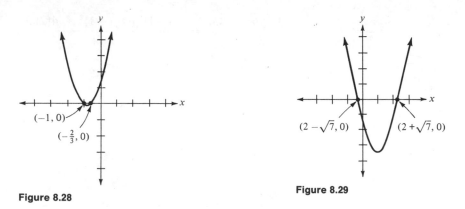

Figure 8.28

Figure 8.29

EXAMPLE 7: The graph of $y = 4x^2 - 20x + 25$ intersects the x axis at one point that corresponds to a rational number since $b^2 - 4ac = 0$ (see Figure 8.30).

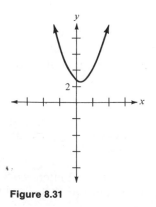

Figure 8.30

EXAMPLE 8: The graph of $y = 2x^2 - x + 3$ does not intersect the x axis since $b^2 - 4ac = -23$ (see Figure 8.31).

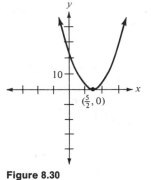

Figure 8.31

We can summarize this result in the following chart:

When a, b, c are rational and	the graph intersects the x axis at
$b^2 - 4ac < 0$	no points
$b^2 - 4ac = 0$	one point (corresponding to rational numbers)
$b^2 - 4ac > 0$ and a perfect square	two points (corresponding to rational numbers)
$b^2 - 4ac > 0$ and not a perfect square	two points (corresponding to irrational numbers)

EXERCISES 8–7

In Exercises 1–20 solve each equation by using the quadratic formula.

1. $x^2 + 5x + 4 = 0$
2. $x^2 - 3x + 2 = 0$
3. $x^2 + x - 3 = 0$
4. $x^2 + 2x - 1 = 0$
5. $x^2 - 2x + 2 = 0$
6. $x^2 + 4x + 5 = 0$
7. $x^2 - 6x + 9 = 0$
8. $x^2 + 4x + 4 = 0$
9. $x^2 - 4 = 0$
10. $x^2 + 2 = 0$
11. $3x^2 - 2x = 0$
12. $2x^2 + x = 0$
13. $3x^2 - 4x - 2 = 0$
14. $5x^2 - 3x - 4 = 0$
15. $-5x^2 = 2x + 3$
16. $-2x^2 = 6x + 5$
17. $3x^2 = 2x - 1$
18. $4x^2 = 12x - 9$
19. $4x^2 + x = 3$
20. $5x^2 - 28x = 12$

In Exercises 21–30 find the x intercepts for the function.

21. $y = x^2 - 4x - 5$
22. $y = 6x^2 - x - 2$
23. $f(x) = x^2 - 4x + 5$
24. $f(x) = 3x^2 + x + 1$
25. $g(x) = x^2 - 4x + 4$
26. $g(x) = 25x^2 - 10x + 1$
27. $y = x^2 - 4x - 3$
28. $f(x) = 2x^2 + 3x - 2$
29. $f(x) = 6x^2 + 5x - 6$
30. $y = 11x^2 - x + 2$

In Exercises 31–40 use the discriminant to determine the nature of the solutions of the equation.

31. $2x^2 - x + 2 = 0$
32. $-3x^2 + x + 1 = 0$
33. $x^2 + 2x + 1 = 0$
34. $5x^2 - 7x + 2 = 0$
35. $x^2 - 10x - 9 = 0$
36. $-2x^2 + 4x + 9 = 0$
37. $4x^2 = 2x + 5$
38. $3x^2 = -4x - 1$
39. $x^2 - 5x = -7$
40. $-2x^2 - x = -2$

In Exercises 41–50 use the discriminant to determine the manner in which the graph of the function intersects the x axis.

41. $y = x^2 + x + 1$
42. $y = 2x^2 + x - 6$
43. $f(x) = 3x^2 - 5x + 1$
44. $f(x) = -x^2 + x + 1$
45. $g(x) = x^2 - 1$
46. $g(x) = x^2 + 3$
47. $y = 9x^2 - 6x + 1$
48. $f(x) = -4x^2 + 7x + 2$
49. $f(x) = -2x^2 - x + 5$
50. $y = x^2 + 8x + 16$

In Exercises 51–62 solve each problem by use of the quadratic formula.

51. The sum of a positive number and its square is 56. What is the number?

52. The perimeter of a rectangle is 100 ft and the area is 400 ft². Find the dimensions of the rectangle.

53. A rectangular plate is to have an area of 30 in.² and the sides of the plate are to differ in length by 4 in. To the nearest tenth of an inch, find the dimensions of the plate.

54. The total cost of manufacturing x units of a certain product is estimated by the formula $C = 0.1x^2 + 0.7x + 3$. How many units can be made for a total cost of $20?

55. The height (y) above water of a diver t seconds after he steps off a platform 10 ft high is given by the formula $y = 10 - 16t^2$. When will the diver hit the water?

56. The height (y) of a ball that is thrown down from a roof 200 ft high with an initial velocity of 50 ft/second is given by the formula $y = 200 - 50t - 16t^2$. To the nearest tenth of a second, when will the ball hit the ground?

57. The height (y) of a projectile that is shot directly up from the ground with an initial velocity of 100 ft/second is given by the formula $y = 100t - 16t^2$. To the nearest tenth of a second, when will the projectile initially attain a height of 50 ft?

58. A manufacturer sells his product for $10 per unit. The cost of manufacturing x units is estimated by the formula $C = 200 + x + 0.01x^2$. How many units must he manufacture and sell to earn a profit of $1200?

59. From a square sheet of metal, an open box is made by cutting 2-in. squares from the four corners and folding up the ends. To the nearest tenth of an inch, how large a piece of metal should be used if the box is to have a volume of 100 in.³?

60. When two resistors R_1 and R_2 are connected in series, their combined resistance is given by $R = R_1 + R_2$; if the resistors are connected in parallel, their combined resistance is given by $1/R = (1/R_1) + (1/R_2)$. Find the value of resistors R_1 and R_2 if their combined resistance is 32 ohms when connected in series and 6 ohms when connected in parallel.

61. Show that the sum of the solutions of the equation $ax^2 + bx + c = 0$ (with $a \neq 0$) is $-b/a$ and that the product of the solutions is c/a.

62. Determine the value of x in the following expression. The solution requires a clever substitution.

$$x = \cfrac{2}{2 + \cfrac{2}{2 + \text{etc.}}}$$

8–8 QUADRATIC INEQUALITIES

In Section 1-8 we discussed the meaning and importance of inequalities. We now extend our ability to work with these statements by considering quadratic inequalities. They are expressed (with $a \neq 0$) in the form $ax^2 + bx + c > 0$ or in the form $ax^2 + bx + c < 0$. For example, let us find the values of x for which the inequality $x^2 - x - 6 > 0$ is a true statement.

1. Sketch the graph of $y = x^2 - x - 6$ with emphasis on the x intercepts and on whether the curve has a maximum or minimum turning point.

a. To find the x intercepts, set $y = 0$ and solve the resulting equation.

$$0 = x^2 - x - 6$$
$$0 = (x - 3)(x + 2)$$
$$x = 3 \qquad x = -2$$

x intercepts: $(3, 0)$ and $(-2, 0)$

b. Since $a > 0$, the graph has a minimum turning point and opens up (see Figure 8.32).

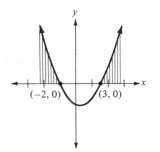

Figure 8.32

2. To solve the inequality $x^2 - x - 6 > 0$, we want only those values of x for which $x^2 - x - 6$ (or y) is positive. Observing the graph, we see that values of x which are to the right of 3 (that is, $x > 3$) or values of x that are to the left of -2 (that is, $x < -2$) make $x^2 - x - 6$ (or y) positive. Therefore, the solution is that $x < -2$ or $x > 3$ make $x^2 - x - 6 > 0$ a true statement.

EXAMPLE 1: For what values of x is the inequality $4 + 3x - x^2 > 0$ a true statement?

SOLUTION:

a. Sketch the graph of $y = 4 + 3x - x^2$.

(1) To find the x intercepts, set $y = 0$ and solve the resulting equation, $0 = 4 + 3x - x^2$.

$$x = \frac{-b \pm \sqrt{b^2 - 4ac}}{2a} = \frac{-(3) \pm \sqrt{(3)^2 - 4(-1)(4)}}{2(-1)}$$

$$= \frac{-3 \pm \sqrt{9 + 16}}{-2}$$

$$= \frac{-3 \pm \sqrt{25}}{-2}$$

$$x_1 = \frac{-3 + 5}{-2} = \frac{2}{-2} = -1 \qquad x_2 = \frac{-3 - 5}{-2} = \frac{-8}{-2} = 4$$

x intercepts: $(-1, 0)$ and $(4, 0)$.

(2) Since $a < 0$, the graph has a maximum turning point and opens down (see Figure 8.33).

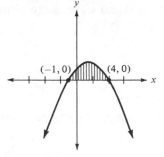

Figure 8.33

b. To solve the inequality $4 + 3x - x^2 > 0$ we want only those values of x for which $4 + 3x - x^2$ (or y) is positive. Observing the graph we see that values of x between -1 and 4 (that is, $-1 < x < 4$) make $4 + 3x - x^2$ (or y) positive. Thus, the solution is $-1 < x < 4$.

EXAMPLE 2: Solve the inequality: $2x^2 + 5x \leq 2$.

SOLUTION:
a. $2x^2 + 5x \leq 2$ in the form $ax^2 + bx + c \leq 0$ is $2x^2 + 5x - 2 \leq 0$. Therefore, we want to sketch the graph of $y = 2x^2 + 5x - 2$.
(1) To find the x intercepts, set $y = 0$ and solve the resulting equation, $0 = 2x^2 + 5x - 2$.

$$x = \frac{-b \pm \sqrt{b^2 - 4ac}}{2a} = \frac{-(5) \pm \sqrt{(5)^2 - 4(2)(-2)}}{2(2)}$$

$$= \frac{-5 \pm \sqrt{25 + 16}}{4}$$

$$= \frac{-5 \pm \sqrt{41}}{4}$$

$$x_1 = \frac{-5 + \sqrt{41}}{4} \qquad x_2 = \frac{-5 - \sqrt{41}}{4}$$

x intercepts: $\left(\dfrac{-5 + \sqrt{41}}{4}, 0\right)$ and $\left(\dfrac{-5 - \sqrt{41}}{4}, 0\right)$

(2) Since $a > 0$, the graph has a minimum point and opens up (see Figure 8.34).

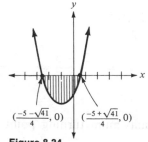

Figure 8.34

b. To solve the inequality $2x^2 + 5x - 2 \le 0$ we want only those values of x for which $2x^2 + 5x - 2$ (or y) either equals 0 or is negative. Observing the graph we see that $2x^2 + 5x - 2$ is less than or equal to zero when

$$\frac{-5 - \sqrt{41}}{4} \le x \le \frac{-5 + \sqrt{41}}{4}$$

EXAMPLE 3: Solve the inequality: $x^2 + 2 < 0$.

SOLUTION:
a. Sketch the graph of $y = x^2 + 2$.
(1) To find the x intercepts, set $y = 0$ and solve the resulting equation.

$$0 = x^2 + 2$$
$$-2 = x^2$$
$$\pm\sqrt{-2} = x$$

The solutions are imaginary numbers. Therefore, there are no x intercepts.
(2) Since $a > 0$, the graph has a minimum turning point and opens up (see Figure 8.35).

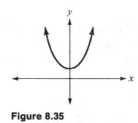

Figure 8.35

b. To solve the inequality $x^2 + 2 < 0$ we want only those values of x for which $x^2 + 2$ (or y) is negative. Observing the graph we see that $x^2 + 2$ (or y) is never negative. Therefore, there are no values of x for which the inequality will be a true statement.

Note from the examples that only a very rough sketch of the graph, which includes the x intercepts and the direction that the curve opens, is needed. In particular, it is not necessary to find the coordinates of the turning point of the parabola.

EXERCISES 8–8
Find the value of x that makes the inequality true.

1. $x^2 - 4x - 5 > 0$ 2. $x^2 + x - 6 < 0$
3. $x^2 < 4x + 5$ 4. $x^2 + 5 < 6x$
5. $-x^2 + x + 2 \le 0$ 6. $-10 \ge 3x - x^2$
7. $x^2 \ge 4$ 8. $x^2 - 5 \le 0$
9. $x^2 < 3x$ 10. $0 > 5x - 10x^2$
11. $2x^2 + x \le 6$ 12. $4 + x > x^2$

13. $3x^2 - 2x < 5$

14. $5x^2 \geq 4x - 3$

15. $x^2 + 2x > -1$

16. $4x^2 \leq 12x - 9$

17. $x^2 - 5x + 7 < 0$

18. $x^2 + 1 \geq 0$

19. $1 - x^2 \geq 0$

20. $-2x^2 + x > 4$

8–9 POLYNOMIALS

A function of the form

$$y = P(x) = a_n x^n + a_{n-1} x^{n-1} + \cdots + a_1 x + a_0 \qquad a_n \neq 0$$

where n is a nonnegative integer and $a_n, a_{n-1}, \ldots, a_1$ and a_0 are real number constants, is called a *polynomial function in x of degree n*. For example, $P(x) = 5x^3 + 2x^2 - 1$ is a polynomial function of degree 3. Since n must be a nonnegative integer, functions with terms such as $x^{1/2}$ (or $\sqrt{x}$) and x^{-2} (or $1/x^2$) are not polynomial functions.

We have already discussed the following polynomial functions:

1. The constant function $y = P(x) = a_0$ is a polynomial function of degree 0.

2. The linear function $y = P(x) = a_1 x + a_0$ is a polynomial function of degree 1.

3. The quadratic function $y = P(x) = a_2 x^2 + a_1 x + a_0$ is a polynomial function of degree 2.

To discuss polynomial functions of degree greater than 2, we must first consider the division of polynomials. We divide one polynomial by another in a manner similar to the division of two integers. Consider the following arrangement for dividing $3x^3 + x - 4$ by $x + 4$:

$$
\begin{array}{r}
3x^2 - 12x + 49 \quad \text{(quotient)} \\
\text{(divisor)} \quad x + 4 \overline{)\, 3x^3 + 0x^2 + x - 4} \quad \text{(dividend)} \\
3x^3 + 12x^2 \\
\overline{-12x^2 + x - 4} \\
-12x^2 - 48x \\
\overline{49x - 4} \\
49x + 196 \\
\overline{-200 \quad \text{(remainder)}}
\end{array}
$$

First, we arrange the terms of the dividend and the divisor in descending powers of x. If a term is missing, write 0 as its coefficient. Then divide the first term of the dividend by the first term of the divisor to obtain the first term of the quotient. Next, we multiply the entire divisor by the first term of the quotient and subtract this product from the dividend. Use the remainder as the new dividend and repeat the above procedure until the remainder is of lower degree than the divisor.

As with the division of numbers,

dividend = (divisor)(quotient) + remainder

For example, the division of 7 by 3 may be expressed as

$$7 = 3 \cdot 2 + 1$$

In the case of the polynomial division above we have

$$3x^3 + x - 4 = (x + 4)(3x^2 - 12x + 49) - 200$$

The division of any polynomial by a polynomial of the form $x - b$ is of theoretical and practical importance. This division may be performed by a shorthand method called *synthetic division*. Consider the arrangement for dividing $2x^3 + 5x^2 - 1$ by $x - 2$.

$$
\begin{array}{r}
2x^2 + 9x + 18 \\
x - 2\,\overline{)\,2x^3 + 5x^2 + 0x - 1} \\
\underline{2x^3 - 4x^2} \\
9x^2 + 0x - 1 \\
\underline{9x^2 - 18x} \\
18x - 1 \\
\underline{18x - 36} \\
35
\end{array}
$$

When the polynomials are written with terms in descending powers of x, there is no need to write all the x's. Only the coefficients are needed. Also notice that the encircled coefficients entailed needless writing. Using only the necessary coefficients we may abbreviate this division as follows:

$$
\begin{array}{r}
2 \quad 9 \quad 18 \\
-2\,\overline{)\,2 \quad 5 \quad 0 \quad -1} \\
-4 \quad -18 \quad -36 \\
\hline
9 \quad 18 \quad 35
\end{array}
$$

If we bring down 2 as the first entry in the bottom row, all the coefficients of the quotient appear. The arrangement may then be shortened to

$$
\begin{array}{llll}
-2 \;\Big| & 2 \quad 5 \quad 0 \quad -1 & \leftarrow \text{ coefficients of dividend} & \text{(row 1)} \\
& \quad -4 \quad -18 \quad -36 & & \text{(row 2)} \\
\text{coefficients of quotient} \rightarrow & 2 \quad 9 \quad 18 \quad\; 35 \leftarrow \text{ remainder} & & \text{(row 3)}
\end{array}
$$

Finally if we change -2 to 2, which is the value of b, we may change the sign of each number in row 2 and add at each step instead of subtracting. The final arrangement for synthetic division is then as follows:

$$
\begin{array}{llll}
2\,\Big|\; 2 \quad 5 \quad 0 \quad -1 & & & \text{(row 1)} \\
\quad\quad 4 \quad 18 \quad 36 & & & \text{(row 2)} \\
\hline
\quad 2 \quad 9 \quad 18 \quad 35 \;\rightarrow\; \text{remainder} & & & \text{(row 3)} \\
\quad\; \downarrow \quad\; \downarrow \quad\; \downarrow \\
\text{quotient} \;\rightarrow\; 2x^2 + 9x + 18
\end{array}
$$

Let's summarize the procedure for synthetic division. First, write the coefficients of the terms in the dividend. Be sure the powers are in descending order and enter a 0 as the coefficient of any missing term. Write the value of b (in this case 2) to the left of the dividend. Bring down the first dividend entry (2). Multiply this number by $b(2)$ and place the result (4) under the next number (5) in row 1. Add, then multiply the resulting 9 by $b(2)$. Place the result (18) under the next number (0) in row 1. Add, then multiply the resulting 18 by $b(2)$. Place the result (36) under the next number (-1) in row 1. Add and obtain 35. The last number 35 is the remainder. The other numbers in row 3 are the coefficients of the quotient. The degree of the polynomial in the quotient is always one less than the degree of the polynomial in the dividend.

A summary cannot do justice to the simple procedure in synthetic division. Careful consideration of Examples 1 and 2 should help to clarify the process. Remember, synthetic division applies only when we divide by a polynomial of the form $x - b$.

EXAMPLE 1: Use synthetic division to divide $x^5 - 1$ by $x - 1$. Express the result in the form: dividend = (divisor)(quotient) + remainder.

SOLUTION: We use 0's as the coefficients of the missing x^4, x^3, x^2, and x terms. We divide by $x - 1$ so that $b = 1$ (not -1). The arrangement is as follows:

$$\underline{1\,|}\quad 1 \quad\quad 0 \quad\quad 0 \quad\quad 0 \quad\quad 0 \quad -1$$
$$\quad\quad\quad\quad\quad 1 \quad\quad 1 \quad\quad 1 \quad\quad 1 \quad\quad 1$$

$$\quad\quad 1 \quad\quad 1 \quad\quad 1 \quad\quad 1 \quad\quad 1 \quad\quad 0 \;\leftarrow\; \text{remainder}$$

quotient $\rightarrow$ $x^4 + x^3 + x^2 + x + 1$

The answer in the form requested is

$$x^5 - 1 = (x - 1)(x^4 + x^3 + x^2 + x + 1) + 0$$

EXAMPLE 2: Use synthetic division to divide $4x^3 - x^2 + 2$ by $x + 3$. Express the result in the form

$$\text{dividend} = (\text{divisor})(\text{quotient}) + \text{remainder}$$

SOLUTION: We use 0 as the coefficient of the missing x term. We divide by $x + 3$ so that $b = -3$ (not 3). The arrangement is as follows:

$$\underline{-3\,|}\quad 4 \quad\quad -1 \quad\quad 0 \quad\quad 2$$
$$\quad\quad\quad\quad -12 \quad\quad 39 \quad -117$$

$$\quad\quad 4 \quad\quad -13 \quad\quad 39 \quad -115 \;\leftarrow\; \text{remainder}$$

quotient $\rightarrow$ $4x^2 - 13x + 39$

The answer in the form requested is

$$4x^3 - x^2 + 2 = (x + 3)(4x^2 - 13x + 39) - 115$$

In Example 2 we found that the function

$$P(x) = 4x^3 - x^2 + 2$$

may be written as

$$P(x) = (x + 3)(4x^2 - 13x + 39) - 115$$

When $x = -3$, the factor $x + 3 = 0$; thus

$$P(-3) = 0 - 115$$
$$= -115$$

The value of the function when $x = -3$ is the same as the remainder obtained when $P(x)$ is divided by $x - (-3)$. This discussion suggests the following theorem:

> **REMAINDER THEOREM**: If a polynomial function $P(x)$ is divided by $x - b$, the remainder is $P(b)$.

We now prove this theorem. Let $Q(x)$ and r represent the quotient and remainder when $P(x)$ is divided by $x - b$. Then,

$$\text{dividend} = (\text{divisor})(\text{quotient}) + \text{remainder}$$
$$P(x) = (x - b)Q(x) + r$$

This statement is true for all values of x. If $x = b$, then

$$P(b) = (b - b)Q(b) + r$$
$$= 0 \cdot Q(b) + r$$

Thus, $P(b) = r$.

We know that $P(b)$ may be found by substituting b for x in the function. The remainder theorem provides an alternative method. That is, we find $P(b)$ by determining the remainder when $P(x)$ is divided by $x - b$. Since this remainder may be obtained by synthetic division, this approach is often simpler than direct substitution. In the next section other advantages of the remainder theorem method will be discussed.

EXAMPLE 3: If $P(x) = 3x^5 + 5x^4 + 7x^3 - 4x^2 + x - 24$, find $P(-2)$ by (a) direct substitution; (b) the remainder theorem.

SOLUTION:
a. By direct substitution, we have

$$P(-2) = 3(-2)^5 + 5(-2)^4 + 7(-2)^3 - 4(-2)^2 + (-2) - 24$$
$$= -96 + 80 - 56 - 16 - 2 - 24$$
$$= -114$$

b. By the remainder theorem, we have

$$
\begin{array}{r|rrrrrr}
-2 & 3 & 5 & 7 & -4 & 1 & -24 \\
   &   & -6 & 2 & -18 & 44 & -90 \\
\hline
   & 3 & -1 & 9 & -22 & 45 & -114 \\
\end{array}
$$

Since the remainder is -114, $P(-2) = -114$.

EXAMPLE 4: If $P(x) = 2x^4 - 5x^3 + 11x^2 - 3x - 5$, find $P(-\frac{1}{2})$ by (a) direct substitution; (b) the remainder theorem.

SOLUTION:

a. By direct substitution we have

$$
\begin{aligned}
P(-\tfrac{1}{2}) &= 2(-\tfrac{1}{2})^4 - 5(-\tfrac{1}{2})^3 + 11(-\tfrac{1}{2})^2 - 3(-\tfrac{1}{2}) - 5 \\
&= \tfrac{1}{8} + \tfrac{5}{8} + \tfrac{11}{4} + \tfrac{3}{2} - 5 \\
&= 0
\end{aligned}
$$

b. By the remainder theorem we have

$$
\begin{array}{r|rrrrr}
-\tfrac{1}{2} & 2 & -5 & 11 & -3 & -5 \\
   &   & -1 & 3 & -7 & 5 \\
\hline
   & 2 & -6 & 14 & -10 & 0 \\
\end{array}
$$

Since the remainder is 0, $P(-\frac{1}{2}) = 0$.

Notice in Examples 3 and 4 that the solution is obtained more easily by the remainder theorem method.

EXERCISES 8–9

In Exercises 1–10 is the function a polynomial function?

1. $f(x) = x^3 + \sqrt{5x} - 3$

2. $f(x) = x^3 + 5\sqrt{x} - 3$

3. $f(x) = 4x - 3$

4. $f(x) = 7 + 4x - x^2$

5. $f(x) = x^2 + x^{1/2} - 1$

6. $f(x) = x^{-3} + 2x^2 - x + 3$

7. $f(x) = \dfrac{1}{x}$

8. $f(x) = 3^{-1}$

9. $f(x) = \pi$

10. $f(x) = x^{100} - 1$

In Exercises 11–16 perform the indicated divisions by using long division. Express the result in the form: dividend = (divisor)(quotient) + remainder.

11. $(x^2 + 7x - 2) \div (x + 5)$

12. $(x^2 - 4) \div (x - 1)$

13. $(6x^3 - 3x^2 + 14x - 7) \div (2x - 1)$

14. $(4x^3 + 5x^2 - 10x + 4) \div (4x - 3)$

15. $(3x^4 + x - 2) \div (x^2 - 1)$

16. $(2x^3 - 3x^2 + 10x - 5) \div (x^2 + 5)$

In Exercises 17–22 perform the indicated division by using synthetic division. Express the result in the form: dividend = (divisor)(quotient) + remainder.

17. $(x^3 - 5x^2 + 2x - 3) \div (x - 1)$ **18.** $(4 - x + 3x^2 - x^3) \div (x - 4)$
19. $(2x^3 + x - 5) \div (x + 1)$ **20.** $(3x^4 - x^2 + 7) \div (x + 3)$
21. $(x^4 - 16) \div (x - 2)$ **22.** $(x^3 + 27) \div (x - 3)$

In Exercises 23–30 find the given function value by (a) direct substitution; (b) the remainder theorem.

23. $P(x) = x^3 - 4x^2 + x + 6$; find $P(2)$ and $P(3)$.
24. $P(x) = 2x^3 + 5x^2 - x - 7$; find $P(4)$ and $P(-3)$.
25. $P(x) = 2x^4 - 7x^3 - x^2 + 4x + 11$; find $P(2)$ and $P(-2)$.
26. $P(x) = x^4 - 5x^3 - 4x^2 + 17x + 15$; find $P(-1)$ and $P(5)$.
27. $P(x) = 2x^4 + 5x^3 - 20x - 32$; find $P(-2)$ and $P(2)$.
28. $P(x) = 2x^4 - x^3 + 2x - 1$; find $P(\frac{1}{2})$ and $P(-\frac{1}{2})$.
29. $P(x) = 6x^4 + 2x^3 - 5x - 5$; find $P(\frac{1}{3})$ and $P(\frac{-1}{3})$.
30. $P(x) = 3x^4 + x^2 - 7x + 2$; find $P(0.1)$ and $P(-0.1)$.

8–10 RATIONAL ZEROS OF POLYNOMIAL FUNCTIONS

A zero of a function f is a value of x for which $f(x) = 0$. For example, the zeros of the function

$$f(x) = x^2 - 5x + 4$$

are 1 and 4 since

$$f(1) = (1)^2 - 5(1) + 4 = 0$$

and

$$f(4) = (4)^2 - 5(4) + 4 = 0$$

Zeros is merely a new name applied to a familiar concept. Depending upon the frame of reference, consider three closely related names applied to the preceding example:

1. The *solutions* or *roots* of the *equation* $x^2 - 5x + 4 = 0$ are 1 and 4.
2. The *zeros* of the *function* $f(x) = x^2 - 5x + 4$ are 1 and 4.
3. The *x intercepts* of the *graph* of the function $y = x^2 - 5x + 4$ are $(1, 0)$ and $(4, 0)$.

We limit our discussion of polynomial functions of degree greater than 2 to finding rational zeros. The following important topics are deferred to the course indicated:

1. Irrational zeros: Unfortunately, irrational zeros of higher-degree polynomials are usually very difficult to find. There is no formula (like the quadratic formula) that can be used to produce them. Numerical methods that require a computer are used to approximate irrational zeros. We defer this topic to an introductory course in computer programming.
2. Graphing: We could graph higher degree polynomial functions by using synthetic division and plotting a lot of points. However, superior techniques are developed

in calculus. In Examples 2 and 3 we graph the function under discussion. These graphs are only for reference purposes. Note that each graph is a smooth unbroken curve. The number of turning points is at most one less than the degree of the polynomial.

The following theorem is used to obtain a list of *possible* rational zeros of a polynomial function with integer coefficients. We emphasize the word "possible." The zeros may all be either irrational or imaginary. This theorem states only that if there are rational zeros, they satisfy the requirement indicated.

RATIONAL ZERO THEOREM: If p/q, a rational number in lowest terms, is a zero of the polynomial function with integer coefficients,

$$P(x) = a_n x^n + a_{n-1} x^{n-1} + \cdots + a_1 x + a_0 \qquad (a_n \neq 0)$$

p is an integral factor of the constant term a_0 and q is an integral factor of the leading coefficient a_n.

EXAMPLE 1: List the possible rational zeros of the function

$$P(x) = 2x^3 + 7x^2 - x + 3.$$

SOLUTION: The constant term a_0 is 3. The possibilities for p are the integers that are factors of 3. Thus,

$$p = \pm 3, \pm 1$$

The leading coefficient a_n is 2. The possibilities for q are the integers that are factors of 2. Thus,

$$q = \pm 2, \pm 1$$

The possible rational zeros p/q are then

$$3, \tfrac{3}{2}, 1, \tfrac{1}{2}, -\tfrac{1}{2}, -1, -\tfrac{3}{2}, -3$$

Since we are often faced with a large list of possible rational zeros, we need some help in narrowing down the possibilities. Consider the two theorems that follow. Each theorem helps us chip away at the problem. Note that they are statements about real zeros. Thus, they apply to zeros that are rational or irrational.

1. *Factor theorem:* If b is a zero of the polynomial function $P(x)$, then $x - b$ is a factor of $P(x)$.
2. *Descartes' rule of signs:* The maximum number of positive real zeros of the polynomial function $P(x)$ is the number of changes in sign of the coefficients in $P(x)$. The number of changes in sign of the coefficients in $P(-x)$ is the maximum number of negative real zeros.

We illustrate these theorems in the following example.

EXAMPLE 2: Find the zeros of the polynomial function $P(x) = 2x^3 + 5x^2 + 4x + 1$ (Figure 8.36).

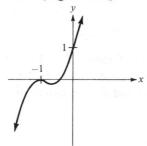

Figure 8.36

SOLUTION:

(1) Use Descartes' rule of signs to determine the maximum number of positive and negative real zeros.

$$P(x) = 2x^3 + 5x^2 + 4x + 1$$

All the coefficients in $P(x)$ are positive. There are no positive real zeros since there are no changes in sign.

$$P(-x) = 2(-x)^3 + 5(-x)^2 + 4(-x) + 1$$
$$= -2x^3 + 5x^2 - 4x + 1$$

There are three sign changes in $P(-x)$. At most, there may be three negative real zeros.

(2) Determine the possible rational zeros. The constant term a_0 is 1. The possibilities for p are the integers that are factors of 1. Thus,

$$p = \pm 1$$

The leading coefficient a_n is 2. The integers that are factors of 2 give the possibilities for q. Thus,

$$q = \pm 2, \pm 1$$

The possible rational zeros p/q are then

$$1, \tfrac{1}{2}, -\tfrac{1}{2}, -1$$

From step 1 we eliminate the positive possibilities, leaving

$$-\tfrac{1}{2}, -1$$

(3) Use synthetic division to test whether one of the possibilities, say -1, is a zero.

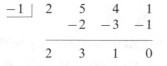

Since the remainder is zero, $P(-1) = 0$. Thus, -1 is a zero of the function.

(4) The factor theorem states that if -1 is a zero, $x - (-1)$ or $x + 1$ is a factor. The coefficients of the other factor are given in the bottom row of the synthetic division. That is,

$$2x^3 + 5x^2 + 4x + 1 = (x + 1)(2x^2 + 3x + 1)$$

We now find other zeros by setting the factor $2x^2 + 3x + 1$ equal to zero. The big advantage is that after finding a zero we may lower by 1 the degree of the equation we are solving. When we reach a second-degree equation, as in this case, we may apply the quadratic formula.

$$2x^2 + 3x + 1 = 0$$

$$x = \frac{-(3) \pm \sqrt{(3)^2 - 4(2)(1)}}{2(2)} = \frac{-3 \pm 1}{4}$$

$$x = -1 \qquad x = -\tfrac{1}{2}$$

Thus, the zeros are -1, -1, and $-\tfrac{1}{2}$. Notice that it is possible for a number to be counted as a zero more than once. Such zeros are called *multiple zeros*.

For the next example we add another useful rule, which permits us to establish upper and lower bounds on the zeros of a polynomial function.

1. If we test a positive possibility (say b) and all the numbers in the bottom row in the synthetic division have the same sign, there is no zero greater than b.
2. If we test a negative possibility (say c) and the numbers in the bottom row alternate in sign, there is no zero less than c.

For the purposes of this rule, zero may be denoted as either $+0$ or -0.

EXAMPLE 3: Find all the zeros of the function $P(x) = 2x^4 - 5x^3 + 11x^2 - 3x - 5$ (Figure 8.37).

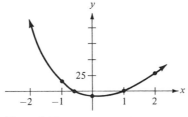

Figure 8.37

SOLUTION:
(1) Use Descartes' rule of signs to determine the maximum number of positive and negative real zeros.

$$P(x) = 2x^4 - 5x^3 + 11x^2 - 3x - 5$$

There are three sign changes in $P(x)$. At most, there may be three positive real zeros.

$$P(-x) = 2(-x)^4 - 5(-x)^3 + 11(-x)^2 - 3(-x) - 5$$
$$= 2x^4 + 5x^3 + 11x^2 + 3x - 5$$

There is one sign change in $P(-x)$. At most, there may be one negative real zero. (*Note:* A more thorough statement of Descartes' rule of signs guarantees exactly one negative *real* zero in this case.)

(2) Determine the possible rational zeros. The constant term a_0 is -5. The possibilities for p are the integers that are factors of -5. Thus,

$$p = \pm 5, \pm 1$$

The leading coefficient a_n is 2. The integral factors of 2 give the possibilities for q. Thus,

$$q = \pm 2, \pm 1$$

The possible rational zeros are then

$$5, \tfrac{5}{2}, 1, \tfrac{1}{2}, -\tfrac{1}{2}, -1, -\tfrac{5}{2}, -5$$

(3) We test negative zeros first, because step 1 indicated that there is, at most, one negative zero, If we find a negative zero, we then switch to the positive possibilities. Pick -1, a number in the middle of the negative choices. If it's not a zero, it may be a lower bound that eliminates other possibilities.

$$
\begin{array}{r|rrrrr}
-1 & 2 & -5 & 11 & -3 & -5 \\
   &   & -2 & 7 & -18 & 21 \\
\hline
   & 2 & -7 & 18 & -21 & 16
\end{array}
$$

The remainder is 16, so -1 is not a zero. We tested a negative possibility and the numbers in the bottom row alternate in sign. Thus, there is no zero less than -1. This eliminates $-\tfrac{5}{2}$ and -5, so try $-\tfrac{1}{2}$.

$$
\begin{array}{r|rrrrr}
-\tfrac{1}{2} & 2 & -5 & 11 & -3 & -5 \\
   &   & -1 & 3 & -7 & 5 \\
\hline
   & 2 & -6 & 14 & -10 & 0
\end{array}
$$

Since the remainder is zero, $P(-\tfrac{1}{2}) = 0$. Thus, $-\tfrac{1}{2}$ is a zero of the function.

(4) Since $-\tfrac{1}{2}$ is a zero, $x - (-\tfrac{1}{2})$ is a factor and we write

$$2x^4 - 5x^3 + 11x^2 - 3x - 5 = (x + \tfrac{1}{2})(2x^3 - 6x^2 + 14x - 10)$$

We now try to find the values of x for which $2x^3 - 6x^2 + 14x - 10$ equals zero. We found the one negative zero, so we switch to the positive possibilities. Pick 1. If it is not a zero, the numbers in the bottom row in the synthetic division may have the same sign, thus eliminating $\tfrac{5}{2}$ and 5.

$$
\begin{array}{r|rrrr}
1 & 2 & -6 & 14 & -10 \\
  &   & 2 & -4 & 10 \\
\hline
  & 2 & -4 & 10 & 0
\end{array}
$$

Therefore, 1 is a zero of the function.

(5) We now use the quadratic formula to complete the solution.

$$2x^2 - 4x + 10 = 0$$
$$x^2 - 2x + 5 = 0$$

$$x = \frac{-(-2) \pm \sqrt{(-2)^2 - 4(1)(5)}}{2(1)} = \frac{2 \pm \sqrt{-16}}{2}$$

$$= \frac{2 \pm 4i}{2} = 1 \pm 2i$$

The zeros are $-\frac{1}{2}$, 1, $1 + 2i$, and $1 - 2i$.

EXERCISES 8–10

In Exercises 1–4 list the possible rational zeros of the function.

1. $P(x) = x^3 + 8x^2 - 10x - 20$ **2.** $P(x) = x^4 - 3x^3 + x^2 - x - 6$

3. $P(x) = 4x^4 - 3x^3 + x^2 - x - 6$ **4.** $P(x) = 3x^3 + 7x^2 + 8$

In Exercises 5–8 use Descartes' rule of signs to determine the maximum number of positive and negative real zeros.

5. $P(x) = x^3 - 2x^2 + x - 3$ **6.** $P(x) = 5x^3 + x^2 + 4x + 1$

7. $P(x) = 2x^4 + 5x^3 - x^2 - 3x - 7$ **8.** $P(x) = x^8 + 1$

In Exercises 9–18 find the zeros of the function.

9. $P(x) = x^3 - x^2 - 10x - 8$ **10.** $P(x) = x^3 + 6x^2 + 11x + 6$

11. $P(x) = 3x^3 + x^2 + 15x + 5$ **12.** $P(x) = 4x^3 - x^2 - 28x + 7$

13. $P(x) = 8x^3 - 12x^2 + 6x - 1$ **14.** $P(x) = 3x^3 + 16x^2 - 8$

15. $P(x) = 4x^4 - 5x^3 - 2x^2 - 3x - 10$

16. $P(x) = x^4 - 6x^3 + 7x^2 + 12x - 18$

17. $P(x) = 2x^4 + 7x^3 + 25x^2 + 47x + 18$

18. $P(x) = 6x^4 + 11x^3 - 63x^2 - 7x + 5$

19. Prove the factor theorem. (*Hint:* Consider the proof of the remainder theorem in Section 8-9.)

20. Use the rational zero theorem and the function $P(x) = x^2 - 2$ to show that $\sqrt{2}$ is not a rational number.

8–11 RATIONAL FUNCTIONS

When we add, subtract, or multiply two polynomials, the result is another polynomial. However, a function of the form

$$y = \frac{P(x)}{Q(x)} \qquad Q(x) \neq 0$$

where $P(x)$ and $Q(x)$ are polynomials, is a new type of function called a *rational function*. For example,

$$y = \frac{1}{x} \qquad y = \frac{x - 4}{x^2 - 1} \qquad y = \frac{3x^5 + 2}{x^3}$$

are rational functions. *In our statements about rational functions we will assume that P(x) and Q(x) have no common factors, so the rational function is in lowest terms.*

The behavior of a rational function differs dramatically from that of a polynomial function. This difference may be easily seen by comparing the graph of a rational function to the smooth unbroken curves that characterize a polynomial.

Consider the rational function $y = 1/x$. Since division by zero is undefined, x cannot equal zero. Thus, the graph of this function does not intersect the line $x = 0$ (or the y axis). We can, however, let x approach zero and consider x values as close to zero as we wish. From the following tables note that as the x values squeeze in on zero, $|y|$ becomes larger and larger.

x	1	0.5	0.1	0.01	0.001
$y = \dfrac{1}{x}$	1	2	10	100	1000

x	-1	-0.5	-0.1	-0.01	-0.001
$y = \dfrac{1}{x}$	-1	-2	-10	-100	-1000

Figure 8.38 shows the behavior of $y = 1/x$ in the interval $-1 \le x \le 1$. The vertical line $x = 0$ that the curve approaches, but never touches, is called a *vertical asymptote*. We may use the following rule to determine if a rational function has any vertical asymptotes.

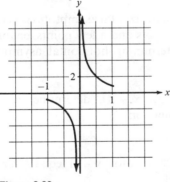

Figure 8.38

The rational function $y = P(x)/Q(x)$ has a vertical asymptote at $x = a$ for each value a at which $Q(a) = 0$.

EXAMPLE 1: Find any vertical asymptotes of the function

$$y = \frac{x + 5}{(x - 3)(x + 2)}$$

SOLUTION: The polynomial in the denominator

$$Q(x) = (x - 3)(x + 2)$$

equals 0 when x is 3 or -2. Thus, there are two vertical asymptotes: $x = 3$ and $x = -2$.

To complete the graph of $y = 1/x$ we must consider the behavior of the function as $|x|$ becomes larger. It is not difficult to see that $1/x$ squeezes in on zero from the positive side when x takes on larger and larger positive values. Similarly, $1/x$ squeezes in on zero from the negative side when x becomes more and more negative. Thus, the curve gets closer and closer to the line $y = 0$ and the x axis is a *horizontal asymptote*. Figure 8.39 shows the graph of $y = 1/x$.

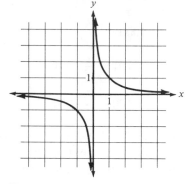

Figure 8.39

Identifying any vertical and horizontal asymptotes is very important. These lines, together with plotting the intercepts and a few points, enable us to graph a rational function. A technique for determining horizontal asymptotes is included in the following examples.

EXAMPLE 2: If $y = (x - 1)/(x + 1)$, determine any vertical or horizontal asymptotes and graph the function.

SOLUTION:

(1) The polynomial in the denominator $x + 1$ equals zero when x is -1. Thus, $x = -1$ is a vertical asymptote.

(2) To determine any horizontal asymptotes, we change the form of the function *by dividing each term in the numerator and denominator by the highest power of x in the expression.*

$$y = \frac{x - 1}{x + 1} = \frac{(x/x) - (1/x)}{(x/x) + (1/x)}$$

$$= \frac{1 - (1/x)}{1 + (1/x)} \qquad \text{(assuming that } x \neq 0)$$

Now as $|x|$ gets larger, $1/x$ approaches zero. Thus, y tends to become $(1 - 0)/(1 + 0)$ and $y = 1$ is a horizontal asymptote.

(3) By setting $x = 0$ we may determine $(0, -1)$ is the y intercept. Similarly, by setting $y = 0$ we may determine that $(1, 0)$ is the x intercept.

(4) The vertical asymptote divides the x axes into two regions. The intercepts are two points to the right of $x = -1$. If we plot a couple of points to the left of the vertical asymptote, say $(-2, 3)$ and $(-3, 2)$, we may complete the graph (see Figure 8.40).

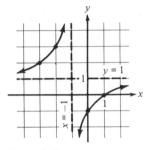

Figure 8.40

In Example 2 an important question is unanswered. We know that as $|x|$ becomes *large*, y approaches, but never quite reaches, 1. How do we know that y is not 1 when $|x|$ is small? We answer this question by determining if there is any value of x for which

$$\frac{x - 1}{x + 1} = 1$$

Solving this equation we have

$$x - 1 = x + 1$$
$$-1 = 1 \quad \text{false}$$

The equation has no solution, and we may conclude the curve never crosses the horizontal asymptote. Example 3 illustrates why this possibility must be considered.

EXAMPLE 3: If

$$y = \frac{2x^2 + x - 3}{x^2}$$

determine any vertical and horizontal asymptotes and graph the function.

SOLUTION:
(1) The polynomial in the denominator equals zero when x is zero. Thus, $x = 0$ is a vertical asymptote.

(2) To determine any horizontal asymptotes, we use the procedure from Example 2. Dividing each term in the numerator and denominator by x^2 gives

$$y = \frac{(2x^2/x^2) + (x/x^2) - (3/x^2)}{x^2/x^2} = \frac{2 + (1/x) - (3/x^2)}{1} \qquad (x \neq 0)$$

As $|x|$ gets larger, $1/x$ and $-3/x^2$ approach 0. Thus, y tends to become 2 and $y = 2$ is a horizontal asymptote.

(3) To determine if the curve ever crosses the horizontal asymptote, find if there are any values of x for which

$$\frac{2x^2 + x - 3}{x^2} = 2$$

Solving this equation for x, we have

$$2x^2 + x - 3 = 2x^2$$
$$x - 3 = 0$$
$$x = 3$$

The curve crosses the asymptote at $(3, 2)$.

(4) Since x cannot be zero there is no y intercept. After setting $y = 0$, we solve the equation $2x^2 + x - 3 = 0$ to find the x intercepts $(1, 0)$ and $(-\frac{3}{2}, 0)$.

(5) To determine the behavior of the curve before it drops and starts approaching 2, we plot a couple of points to the right of $x = 3$, say $(4, \frac{33}{16})$ and $(5, \frac{52}{25})$. Additional points may always be plotted. Figure 8.41 shows the graph of the function.

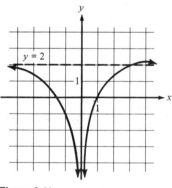

Figure 8.41

Finally, you may wish to determine horizontal asymptotes by using the following rule: The rational function

$$y = \frac{P(x)}{Q(x)} = \frac{a_n x^n + a_{n-1} x^{n-1} + \cdots + a_0}{b_m x^m + b_{m-1} x^{m-1} + \cdots + b_0}$$

where $a_n, b_m \neq 0$, has:

1. A horizontal asymptote at $y = 0$ (the x axis) if $n < m$.
2. A horizontal asymptote at $y = a_n/b_m$ if $n = m$.
3. No horizontal asymptote if $n > m$.

Consideration of the previous examples should clarify this theorem. We restrict this section to functions in cases 1 and 2. When there are no horizontal asymptotes, other techniques may be employed.

EXERCISES 8–11

In Exercises 1–4 find any vertical asymptotes of the function.

1. $y = \dfrac{x + 7}{2x - 3}$

2. $y = \dfrac{x^2 + 1}{x(x + 4)}$

3. $y = \dfrac{2}{x^2 - 5x - 6}$

4. $y = \dfrac{x}{x^2 + 1}$

In Exercises 5–20 determine any vertical or horizontal asymptotes and graph the function.

5. $y = \dfrac{-1}{x}$

6. $y = \dfrac{3}{x^2}$

7. $y = \dfrac{2}{x + 1}$

8. $y = \dfrac{-5}{3x + 2}$

9. $y = \dfrac{x - 2}{x + 2}$

10. $y = \dfrac{x + 3}{x - 5}$

11. $y = \dfrac{2x - 5}{3x + 5}$

12. $y = \dfrac{4x + 1}{2x + 7}$

13. $y = \dfrac{2}{(x - 1)^2}$

14. $y = \dfrac{2}{x^2 - 1}$

15. $y = \dfrac{1}{(x + 1)(x - 4)}$

16. $y = \dfrac{x - 1}{x^2 - 4}$

17. $y = \dfrac{x^2 - 1}{x^3}$

18. $y = \dfrac{x}{x^2 + 1}$

19. $y = \dfrac{x^2 - x - 6}{x^2}$

20. $y = \dfrac{3x^2 + x - 2}{2x^2}$

In our statements about rational functions we assumed that $P(x)$ and $Q(x)$ had no common factors. In Exercises 21 and 22 this assumption is not true. In each case graph the function with particular emphasis on any value of x for which $P(x) = Q(x) = 0$.

21. $y = \dfrac{x^2 - 1}{x + 1}$

22. $y = \dfrac{x}{x^2 + 2x}$

Sample Test: Chapter 8

1. Indicate if the given function is a polynomial.

   a. $P(x) = \sqrt{2}x^3 + \dfrac{3}{x} - 1$

   b. $P(x) = x^4 - \dfrac{3}{2}x^{3/2} + 3$

   c. $P(x) = x^2 + x + 2^{-1}$
   d. $P(x) = x^2 + x^{-1} + 2$
   e. $P(x) = x^4 + \sqrt{2}x + \pi$

2. Find the equation that defines the linear function f if $f(-1) = 6$ and $f(2) = -3$.

3. Find the slope and y intercept of the function $3x - 2y = 6$.

4. Find the equation of the line that passes through the point $(-3, 1)$ and is perpendicular to the line $y = \frac{1}{2}x + 4$.

5 and 6. Solve the given system of equations by the following methods:
   a. Addition–subtraction.
   b. Substitution.
   c. Determinants.
   d. Graphing.

   $2x + 3y = -17$
   $3x - y = \;\;2$

7. Find the x intercepts of the function $f(x) = x^2 - x - 5$. Plot the approximate position of these points on the x axis.

8. Find the value of the discriminant $b^2 - 4ac$ for the function $f(x) = 2x^2 - x + 6$. In what manner (if any) does the graph of this function cross the x axis?

9. What is the range of the function $y = 4 - x^2$?

10. An automobile manufacturer sells 6000 cars per week of his $3000 model. Because of a backlog of cars a rebate program is introduced. It is estimated that each $100 decrease in price will result in the sale of 300 more cars. The weekly revenue R is then given by the formula

    $R = (6000 + 300x)(3000 - 100x)$

    where x is the number of $100 discounts offered. What rebate should be given to produce maximum revenue?

11. Solve: $x^2 + x \geq 6$.

12. Perform the indicated operation. Express the result in the form: dividend = (divisor)(quotient) + remainder.

$$(2x^4 - x + 7) \div (x^2 + 2)$$

13. If $P(x) = 2x^4 - x + 7$, find $P(-2)$ by:
   a. Direct substitution.
   b. Synthetic division.

14. List the possible rational zeros of the function $P(x) = 3x^3 - 7x^2 + x - 2$.

15. Use Descartes' rule of signs to determine the maximum number of (a) positive and (b) negative real zeros of the function $P(x) = x^4 - 3x^3 + x^2 - x + 1$.

16. If 2 is a zero of the function $P(x) = 2x^3 - 4x^2 - 6x + 12$, find the other zeros.

17. Consider the following synthetic division:

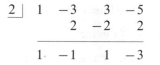

On the basis of this division can the function $P(x) = x^3 - 3x^2 + 3x - 5$ have any real zeros greater than 2? Justify your answer.

18. Determine the vertical asymptotes for the function

$$y = \frac{3x + 1}{x^2 + 2x}$$

19. Determine the horizontal asymptote for the function

$$y = \frac{2x^2 + 3}{3x^2 + x - 2}$$

Does the graph cross the horizontal asymptote? If yes, find the point at which it crosses.

20. Graph the function $y = -2/3x^2$.

Chapter 9
Exponential and logarithmic functions

9–1 EXPONENTIAL FUNCTIONS

In engineering, biology, economics, psychology, and other fields there are important phenomena whose behavior is described by functions in which the independent variable appears as the exponent of a fixed base. These functions are called *exponential functions* and the simplest form of an exponential function is

$$y = b^x \qquad \text{(where } b \text{ is any positive real number except 1)}$$

This equation is a rule that assigns to each exponent exactly one number. We eliminate a nonpositive base in the equation because such expressions as $(-1)^{1/2}$ and 0^{-2} are not real numbers. Since $1^x = 1$ for all values of x, we also eliminate 1.

EXAMPLE 1: In the biological reproduction of cells by division, one cell divides into two cells after a certain period of time. Each of these cells repeats the process and divides into two more, and so on. To be specific, assume that on a given day we start with one cell and the number doubles each day. Then, there are 2 cells present after 1 day; $2 \cdot 2$, or 2^2, after 2 days; $2 \cdot 2^2$, or 2^3, after 3 days, and so on. Therefore, there are 2^x cells present after x days. This phenomenon of cell reproduction is described by the function

$$y = 2^x$$

where x represents the number of complete days from the start of the experiment.

At present we are able to interpret exponents that are rational numbers. For example, if $f(x) = 2^x$, then

$$f(3) = 2^3 = 8$$
$$f(0) = 2^0 = 1$$

$$f(-2) = 2^{-2} = \frac{1}{2^2} = \frac{1}{4}$$

$$f(\tfrac{3}{2}) = 2^{3/2} = \sqrt{2^3} = \sqrt{8} \approx 2.83$$

(*Note:* A complete discussion of rational exponents may be found in Sections 6-1 and 6-3.)

For now, we assume that numbers with irrational exponents, such as

$$2^\pi \qquad 2^{\sqrt{3}} \qquad 2^{-\sqrt{2}}$$

are real numbers. In Section 9-6 we will find approximate values for these numbers. The domain of an exponential function is then the real numbers (that is, any real number may be substituted for x).

Let us look at the behavior of the exponential functions by considering the graphs of two typical cases:

$$y = (\tfrac{1}{2})^x \qquad \text{in which } 0 < b < 1$$
$$y = 2^x \qquad \text{in which } b > 1$$

Some of the ordered pairs of the functions and their graphs are shown in Figure 9.1.

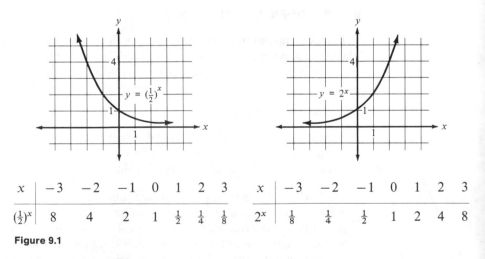

x	-3	-2	-1	0	1	2	3
$(\tfrac{1}{2})^x$	8	4	2	1	$\frac{1}{2}$	$\frac{1}{4}$	$\frac{1}{8}$

x	-3	-2	-1	0	1	2	3
2^x	$\frac{1}{8}$	$\frac{1}{4}$	$\frac{1}{2}$	1	2	4	8

Figure 9.1

Notice in the graph of $y = (\tfrac{1}{2})^x$ that as x increases, y decreases. This type of exponential function is useful in applications where a quantity is decaying or depreciating. The graph of $y = 2^x$ shows that as x increases, y increases. This type of exponential function is useful in applications in which a quantity is growing. In both graphs the curve approaches, but never touches, the x axis. Since the y value is always greater than zero, the range of an exponential function is the positive real numbers.

EXAMPLE 2: A company purchases a new machine for $3000. The value of the machine depreciates at a rate of 10 percent each year.
a. Find a formula (rule) showing the value (y) of the machine at the end of x years.
b. What is the machine worth after 4 years?

SOLUTION:
a. If the machine depreciates 10 percent each year, at the end of a year the machine is worth 90 percent of the value at which it began the year. Thus, the value is 3000(0.9) after 1 year; 3000(0.9)(0.9) or 3000(0.9)^2 after 2 years; and

$$y = 3000(0.9)^x \qquad \text{after } x \text{ years}$$

b. Evaluating the function when $x = 4$, we have

$$y = 3000(0.9)^4$$
$$= \$1968.30$$

EXAMPLE 3: $1000 is invested at 6 percent interest compounded annually.
a. Describe in functional notation the value of the investment at the end of t years.
b. How much is the investment worth after 3 years?

SOLUTION:
a. The $1000 amounts to $1000(1 + 0.06) or $1060 at the end of 1 year. The principal during the second year is 1000(1 + 0.06). Thus, the amount at the end of 2 years is 1000(1 + 0.06)(1 + 0.06) or 1000(1 + 0.06)^2. In general, the value of the investment after t years is

$$f(t) = 1000(1 + 0.06)^t$$

b. Evaluating the function when $t = 3$ we have

$$f(3) = 1000(1 + 0.06)^3$$
$$= \$1191.02$$

In this section the applications describe situations in which no change is considered to have occurred until an entire time period has elapsed. Thus, the $1000 investment in Example 3 is compounded annually and does not change until the end of the year. For this reason the domain of these functions is the nonnegative integers. In the next section we discuss phenomena that change continuously. The domain of the functions in these applications is usually the nonnegative real numbers.

Finally in this chapter we will frequently solve equations containing exponential expressions by using the principle that $b^x = b^y$ implies $x = y$. For example, what value of x satisfies the equation $9^x = \frac{1}{27}$? In most cases it is hoped that you can determine the solution by inspection. If not, then try to rewrite the expressions in terms of a common base. In this case since $9 = 3^2$ and $27 = 3^3$ we solve the equation as follows:

$$9^x = \frac{1}{27}$$

$$(3^2)^x = \frac{1}{3^3}$$

$$3^{2x} = 3^{-3}$$

Now since $b^x = b^y$ implies $x = y$ we have

$$2x = -3$$

$$x = -\frac{3}{2}$$

In Section 9-6 we will consider equations like $3^x = 8$ in which it is difficult to rewrite the expressions in terms of a common base.

EXERCISES 9–1

In Exercises 1–20 fill in the missing component of each of the ordered pairs that make the pair a solution of the given equation. Also, choose a convenient scale for the vertical axis and graph each function.

1. $y = 2^x$ $\qquad$ $(-3, \square)$ $(\square, 1)$ $(\square, \frac{1}{2})$ $(2, \square)$
2. $f(x) = 3^x$ $\qquad$ $(\square, 9)$ $(\square, \frac{1}{9})$ $(0, \square)$ $(1, \square)$
3. $f(x) = (\frac{1}{3})^x$ $\qquad$ $(1, \square)$ $(\square, 1)$ $(\square, \frac{1}{9})$ $(\square, 9)$
4. $y = (\frac{1}{2})^x$ $\qquad$ $(0, \square)$ $(2, \square)$ $(\square, 2)$ $(\square, \frac{1}{2})$
5. $y = (0.1)^x$ $\qquad$ $(\square, 0.01)$ $(\square, 1)$ $(-1, \square)$ $(-2, \square)$
6. $y = (0.8)^x$ $\qquad$ $(2, \square)$ $(-1, \square)$ $(\square, 1)$ $(1, \square)$
7. $g(x) = 9^{-x}$ $\qquad$ $(1, \square)$ $(-1, \square)$ $(-\frac{1}{2}, \square)$ $(\square, \frac{1}{3})$
8. $g(x) = 4^{-x}$ $\qquad$ $(2, \square)$ $(\square, 0.5)$ $(-0.5, \square)$ $(\square, 4)$
9. $y = (0.25)^{-x}$ $\qquad$ $(\frac{1}{2}, \square)$ $(\square, 1)$ $(1, \square)$ $(\square, \frac{1}{2})$
10. $y = (\frac{1}{8})^{-x}$ $\qquad$ $(0, \square)$ $(1, \square)$ $(\square, 2)$ $(\square, \frac{1}{2})$
11. $h(x) = -2^x$ $\qquad$ $(0, \square)$ $(1, \square)$ $(-1, \square)$ $(\square, -4)$
12. $y = -5^x$ $\qquad$ $(0, \square)$ $(1, \square)$ $(-1, \square)$ $(\square, -0.04)$
13. $h(x) = (\sqrt{2})^x$ $\qquad$ $(0, \square)$ $(-2, \square)$ $(1, \square)$ $(\square, 2)$
14. $y = (\sqrt{3})^x$ $\qquad$ $(2, \square)$ $(\square, 1)$ $(\square, 9)$ $(-2, \square)$
15. $y = 2(3)^x$ $\qquad$ $(-1, \square)$ $(0, \square)$ $(1, \square)$ $(2, \square)$
16. $f(x) = 3(2)^x$ $\qquad$ $(-1, \square)$ $(0, \square)$ $(1, \square)$ $(2, \square)$
17. $f(x) = 100(1.06)^x$ $\qquad$ $(-1, \square)$ $(0, \square)$ $(1, \square)$ $(2, \square)$
18. $f(x) = 1000(0.9)^x$ $\qquad$ $(-1, \square)$ $(0, \square)$ $(1, \square)$ $(2, \square)$
19. $y = 2^x + 2^{-x}$ $\qquad$ $(-1, \square)$ $(0, \square)$ $(1, \square)$ $(2, \square)$
20. $y = 3^x + 3^{-x}$ $\qquad$ $(-1, \square)$ $(0, \square)$ $(1, \square)$ $(2, \square)$
21. Find the base of the exponential function that contains the given point.
    a. $(1, 4)$ $\qquad\qquad\qquad\qquad$ b. $(2, 9)$
    c. $(-1, 3)$ $\qquad\qquad\qquad\quad$ d. $(-2, \frac{1}{25})$
    e. $(\frac{1}{2}, 4)$ $\qquad\qquad\qquad\quad$ f. $(0.5, 10)$

g. $(\frac{2}{3}, \frac{1}{4})$ **h.** $(\frac{3}{2}, 27)$

i. $(-\frac{1}{3}, 2)$ **j.** $(-\frac{2}{3}, \frac{1}{9})$

22. Determine an integer n such that $n < x < n + 1$.

   a. $10^x = 451$ **b.** $10^x = 45,100$

   c. $10^x = 0.451$ **d.** $10^x = 0.00451$

23. Show that if $f(x) = b^x$ is an exponential function with base b, then $f(x_1 + x_2) = f(x_1) \cdot f(x_2)$.

24. A biologist has 500 cells in a culture at the start of an experiment. Hourly readings indicate that the number of cells is doubling every hour.

   a. Find a formula (rule) showing the number of cells present at the end of t hours.

   b. How many cells are present at the end of 4 hours?

   c. At the end of how many hours are there 32,000 cells present?

25. $100 is invested at 5 percent compounded annually.

   a. Find a formula showing the value of the investment at the end of t years.

   b. How much is the investment worth at the end of 3 years?

26. The market value of a company's investment is $100,000. This value increases at a rate of 10 percent each year.

   a. Describe in functional notation the market value of the investment t years from now.

   b. What should be the market value of the investment 2 years from now?

27. A company purchases a machine for $10,000. The value of the machine depreciates at a rate of 20 percent each year.

   a. Find a formula to show the value of the machine after t years.

   b. What will be the value of the machine 3 years from now?

28. A biologist grows a colony of a certain kind of bacteria. It is found experimentally that $N = N_0 3^t$, where N represents the number of bacteria present at the end of t days. N_0 is the number of bacteria present at the start of the experiment. Suppose that there are 153,000 bacteria present at the end of 2 days.

   a. How many bacteria were present at the start of the experiment?

   b. How many bacteria are present at the end of 4 days?

   c. At the end of how many days are there 459,000 bacteria present?

29. A new element has a half-life of 30 minutes; that is, if x oz of the element exist at a given time, $x/2$ oz exist 30 minutes later. The other half disintegrates into another element. Suppose that there is 1 g of the element present 2 hours after the start of an experiment.

   a. Find a formula showing the amount of the element present t *hours* after the start of the experiment.

   b. How much of the element was present 1 hour after the start of the experiment?

30. At what interest rate (compounded annually) must a sum of money be invested if it is to double in 2 years?

9-2 MORE APPLICATIONS AND THE NUMBER e

In Section 9-1 the applications describe situations in which no change is considered to have occurred until an entire time period has elapsed. We now wish to discuss phenomena that change continuously. To illustrate this type of change we study more extensively the concept of compound interest.

EXAMPLE 1: $2000 is invested at 6 percent compounded semiannually.
a. Find a formula showing the value of the investment at the end of t years.
b. How much is the investment worth after 2 years?

SOLUTION:
a. Since the amount is compounded semiannually, there are two conversions each year at a 3 percent interest rate. The $2000 amounts to $2000(1 + 0.06/2)$ after $\frac{1}{2}$ year; $2000(1 + 0.06/2)^2$ after 1 year; and $2000(1 + 0.06/2)^{2t}$ after t years. Thus,

$$A = 2000(1.03)^{2t}$$

b. Evaluating the function when $t = 2$, we have

$$A = 2000(1.03)^{2(2)}$$
$$= 2000(1.03)^4$$
$$= \$2251.02$$

We can generalize the procedure from Example 1 to obtain the following formula for the amount (including principal and interest) of an investment in terms of four items: (1) the original principal, P; (2) the rate of interest, r; (3) the number of conversions per year, n; and (4) the number of years, t.

$$A = P\left(1 + \frac{r}{n}\right)^{nt}$$

EXAMPLE 2: $100 is invested at 5 percent interest compounded quarterly. How much will it amount to in 3 years?

SOLUTION: Substituting $P = 100$, $r = 0.05$, $n = 4$, and $t = 3$ in the formula above, we have

$$A = 100\left(1 + \frac{0.05}{4}\right)^{4(3)}$$
$$= 100(1.0125)^{12}$$
$$= \$116.08$$

You should note that if the interest is computed frequently and added to the principal, the amount grows at a faster rate. Thus, it is more profitable to an investor for the conversion periods to be shorter. For example, $10,000 amounts to $10,600 when compounded annually at 6 percent; $10,609 when compounded semiannually; and $10,613.64 when compounded quarterly. An important consideration is to determine the effect of compounding more frequently, such as every day, or even every minute. Does the compounded amount become astronomical or is the growth limited?

To determine the result of continuous growth, let us examine the return on a $1 investment for 1 year at an interest rate of 100 percent. Progressively, we will compound our investment more frequently and observe the result. (*Note:* A $1 investment and a 100 percent interest rate are very unlikely, but we choose them to illustrate the situation with the easiest numbers.)

$$P = \$1 \qquad r = 100\% \quad \text{or} \quad 1 \qquad t = 1$$

Substituting in the formula

$$A = P\left(1 + \frac{r}{n}\right)^{nt}$$

we have

$$A = 1\left(1 + \frac{1}{n}\right)^{n(1)}$$

which simplifies to

$$A = \left(1 + \frac{1}{n}\right)^{n}$$

If compounded

annually $(n = 1)$:	then $A = (1 + \frac{1}{1})^1 = 2.00$
semiannually $(n = 2)$:	then $A = (1 + \frac{1}{2})^2 = (1.5)^2 = 2.25$
quarterly $(n = 4)$:	then $A = (1 + \frac{1}{4})^4 = (1.25)^4 \approx 2.441\ldots$
monthly $(n = 12)$:	then $A = (1 + \frac{1}{12})^{12} \approx (1.08333)^{12} \approx 2.613\ldots$
daily $(n = 365)$:	then $A = (1 + \frac{1}{365})^{365} \approx (1.00274)^{365} \approx 2.7145\ldots$
hourly $(n = 8760)$:	then $A = (1 + \frac{1}{8760})^{8760} \approx (1.000114)^{8760} \approx 2.71828\ldots$

Note the small difference between compounding daily and compounding hourly. More frequent conversions result in even smaller changes in the amount. Thus, for large values of *n*,

$$\left(1 + \frac{1}{n}\right)^{n} \approx 2.71828$$

This number is extremely important in mathematics and is symbolized by the letter *e*.

$$e \approx 2.71828\ldots$$

The number *e* is an irrational number and may be approximated to any desired accuracy. In fact, computers have written the number to 250,000 significant digits.

Thus, a $1 investment compounded every instant (or continuously) for 1 year at an interest rate of 100 percent amounts to $2.71828\ldots, or $*e*. In the next year each dollar again grows to $*e*, so at the end of 2 years the amount is $*e*². We can therefore find the amount to which the $1 has grown in *t* years by the formula

$$A = e^t$$

where the base in the exponential function is the irrational number *e*.

This formula has to be generalized to handle more practical principals and interest rates as well as other physical situations that change in a similar manner. The generalized formula is

$$A = A_0 e^{kt}$$

where

A = amount

A_0 = initial amount

k = rate of growth or decay

t = number of time intervals

The important idea is that an exponential function with base e is used in a situation where a quantity is *continuously* either growing or decaying. The rate of increase or decrease is proportional to the amount present at any instant.

EXAMPLE 3: $2000 is invested at 6 percent compounded continuously. How much is the investment worth in 10 years?

SOLUTION: $A_0 = \$2000$, $k = 0.06$, and $t = 10$. Substituting in the formula,

$$A = A_0 e^{kt}$$

we have

$$A = 2000e^{(0.06)(10)}$$
$$= 2000e^{0.6}$$

Referring to Table 2 in the Appendix we find that $e^{0.6} \approx 1.8221$. Thus,

$$A \approx 2000(1.8221)$$
$$\approx \$3644.20$$

EXAMPLE 4: An amount is invested at 7 percent compounded continuously. In how many years does the amount double?

SOLUTION: $A = 2A_0$ and $k = 0.07$. Substituting in the formula,

$$A = A_0 e^{kt}$$

we have

$$2A_0 = A_0 e^{0.07t}$$
$$2 = e^{0.07t}$$

Referring to Table 2 we find that $e^{0.70} \approx 2$. Thus,

$$0.07t \approx 0.70$$
$$t \approx 10 \text{ years}$$

EXAMPLE 5: Physicists tell us that the half-life of radium is 1620 years; that is, a given amount (A) of radium decomposes to the amount $A/2$ in 1620 years. Radium (like all radioactive substances) decays at a rate proportional to the amount present. Thus, the formula $A = A_0 e^{kt}$ is appropriate. If a research center owns 1 g of radium, how much radium will it have in 100 years?

SOLUTION: We know that $A_0 = 1$ g; thus our formula

$$A = A_0 e^{kt}$$

becomes

$$A = 1 e^{kt}$$

We can find k since we know that $A = 0.5$ when $t = 1620$.

$$0.5 = e^{k(1620)}$$

Referring to Table 2 we find that $e^{-0.70} \approx 0.5$. Thus,

$$1620k = -0.70$$
$$k = -0.00043$$

The formula is then

$$A = e^{-0.00043t}$$

Substituting $t = 100$, we have

$$A = e^{-0.00043(100)}$$
$$= e^{-0.043}$$
$$A \approx 0.96 \text{ g}$$

EXAMPLE 6: If there is no restriction on food and living space, the rate of growth of a population of living organisms is proportional to the size of the population. Thus, the formula $A = A_0 e^{kt}$ is appropriate. Assume that in the absence of hunters, the deer population in New York State triples every 11 years.
a. Find the size of a herd of 500 deer after 5 years.
b. In how many years will the herd number 5000?

SOLUTION:
a. We know that $A_0 = 500$. Thus, our formula

$$A = A_0 e^{kt}$$

becomes

$$A = 500 e^{kt}$$

We can find k since we know that $A = 1500$ when $t = 11$:

$$1500 = 500 e^{k(11)}$$
$$3 = e^{11k}$$

Referring to Table 2 we find that $e^{1.1} \approx 3$. Thus,

$$11k = 1.1$$
$$k = 0.1$$

The formula is then

$$A = 500e^{0.1t}$$

Substituting $t = 5$, we have

$$A = 500e^{0.1(5)}$$
$$= 500e^{0.5}$$
$$= 500(1.6487)$$
$$A \approx 824 \text{ deer}$$

b. Substituting $A = 5000$ in our formula, we have

$$5000 = 500e^{0.1t}$$
$$10 = e^{0.1t}$$

Referring to Table 2 we find that $e^{2.3} \approx 10$. Thus,

$$0.1t = 2.3$$
$$t \approx 23 \text{ years}$$

EXAMPLE 7: Newton's law of cooling states that when a warm body is placed in colder surroundings at temperature t_α, the temperature (T) of the body at time t is given by

$$T - t_\alpha = D_0 e^{kt}$$

where D_0 is the initial difference in temperature and k is a constant. Because of an ice storm, there is a loss of power in a home heated to 68°F. If the outside temperature is 28°F and the temperature in the house drops from 68 to 64°F in 1 hour, when will the temperature in the house be down to 50°F?

SOLUTION: We know that $t_\alpha = 28$ and $D_0 = 68 - 28 = 40$. Thus, our formula becomes

$$T - 28 = 40e^{kt}$$

We can find k since we know that $T = 64$ when $t = 1$.

$$64 - 28 = 40e^{k(1)}$$
$$0.9 = e^k$$

Referring to Table 2 we find that $e^{-0.11} \approx 0.9$. Thus,

$$k = -0.11$$

(*Note:* k depends upon the insulation in the house.) The formula is then

$$T - 28 = 40e^{-0.11t}$$

Substituting $T = 50$, we have

$$50 - 28 = 40e^{-0.11t}$$
$$0.55 = e^{-0.11t}$$

Referring to Table 2 we find $e^{-0.60} \approx 0.55$. Thus,

$$-0.11t = -0.60$$
$$t \approx 5.45 \text{ hours}$$

(*Note:* This example also suggests why the amount of oil consumed to maintain a temperature in a home increases exponentially as the thermostat setting is increased.)

Note from the examples that k is positive if the quantity is increasing and negative if the quantity is decreasing. In this section we discuss in detail only a few of the basic applications of the number e. However, keep in mind that the scope and power of the number e is immense. For example, consider the following relationships:

1. When the power is turned off, the electric current in a circuit does not vanish instantly but rather decreases exponentially.
2. The intensity of sunlight decreases exponentially as a diver descends to further ocean depths.
3. After a drug reaches maximum concentration in an organ, the body dilutes the solution exponentially.
4. Atmospheric pressure decreases exponentially as altitude above sea level increases.

The list could easily be continued. In calculus you will probably consider some of these applications.

EXERCISES 9–2

1. Find the compounded amount for the following investments.
   a. $1000 compounded semiannually for 3 years at 6 percent.
   b. $5000 compounded quarterly for 2 years at 5 percent.
   c. $3000 compounded monthly for 1 year at 5.4 percent.
2. Use Table 2 to approximate the value of x in each of the following.
   a. $e^{0.16} = x$ b. $e^{-0.35} = x$
   c. $e^{-5.2} = x$ d. $e^{2.18} = x$
   e. $e^x = 2.0138$ f. $e^x = 0.9231$
   g. $e^x = 0.8906$ h. $e^x = 51$
3. Find the compounded amount for the following investments.
   a. $1000 compounded continuously for 3 years at 6 percent.
   b. $30,000 compounded continuously for 5 years at 5 percent.
   c. $10,000 compounded continuously for 40 years at 8 percent.
   d. $200,000 compounded continuously for 9 months at 5.2 percent.

4. An amount is invested at 6 percent compounded continuously. In how many years does the amount triple?

5. At what rate of interest compounded continuously must money be deposited if the amount is to double in 7 years?

6. How much would a person have to invest today at 6 percent compounded continuously if he wants to have $1000 1 year from today? (*Note:* An important consideration of an economist is how much a dollar x years from now is worth today.)

7. Assume that in a specific culture the number of bacteria (A) present after t hours is given by $A = 1000e^{0.15t}$.

    a. How many bacteria are present after $\frac{1}{2}$ hour?

    b. How long will it take to have 40,000 bacteria?

8. The number of bacteria in a culture at the start of an experiment is about 10^5. Four hours later the population has increased to about 10^7.

    a. Approximate the size of the population after 6 hours.

    b. When was the number of bacteria 250,000?

9. In 1960 the population of a certain town was 1000, and in 1970 it was 4000. What population can the city planning commission expect in 1990 if this growth rate continues?

10. The amount (A) of a certain radioactive element remaining after t minutes is given by $A = A_0e^{-0.03t}$. How much of the element would remain after 1 hour if 50 g was present initially?

11. Physicists tell us that the half-life of an isotope of strontium is 25 years. If a research center owns 10 g of this isotope, how much will it have in 10 years?

12. A radioactive substance decays from 10 g to 6 g in 5 days. Find the half-life of the substance.

13. Specimens in geology and archaeology are dated by considering the exponential decay of a particular radioactive element found in the specimen. This technique is very reliable because radioactive disintegration is not affected by conditions such as pressure and temperature, which change the rate of ordinary chemical reactions. For dating to an age of about 60,000 years the researcher considers the concentration of carbon 14, an isotope of carbon. In the cells of all living plants and animals, there is a fixed ratio of carbon 14 to ordinary stable carbon. However, when the plant or animal dies, the carbon 14 decreases according to the law of exponential decay.

    a. The half-life of carbon 14 is about 5600 years. Determine k, the rate of decay, for this element.

    b. What percentage of the carbon 14 is left in a specimen of bone that is 2000 years old?

(*Note:* The radiocarbon dating method was developed in 1955. For specimens older than 60,000 years, too much carbon 14 has disintegrated for an accurate measurement. In such cases the researcher considers the concentration of other radioactive substances. For example, long periods of geologic time are frequently determined by measuring the presence of uranium 238 which has a half-life of about 4.5 billion years. Findings based on these reliable radioactive "clocks" are upsetting many of the traditional theories about the development of man that were based on previous dating techniques.)

14. The temperature of a six-pack of beer (bought from a distributor on a hot summer day) is 90°F. The beer is placed in a refrigerator with a constant temperature of 40°F. If the beer cools to 60°F in 1 hour, when will the beer reach the more thirst-quenching temperature of 45°F?

15. A coroner examined the body of a murder victim at 9 A.M. and determines its temperature to be 88°F. An hour later the body temperature is down to 84°F. If the temperature of the room in which the body was found is 68°F and if the victim's body temperature was 98°F when he died, approximate the time of the murder.

16. The following formula (derived in calculus) is used to compute the values of e^x in Table 2 of the Appendix.

$$e^x = 1 + x + \frac{x^2}{2!} + \frac{x^3}{3!} + \cdots + \frac{x^n}{n!} + \cdots$$

where $n! = 1 \cdot 2 \cdot 3 \cdots n$ (for example, $4! = 1 \cdot 2 \cdot 3 \cdot 4$).

a. Use the first four terms of this formula to approximate $e^{0.2}$. Compare your result with the value in Table 2.

b. Use the first five terms of this formula to approximate e to the nearest hundredth.

9-3 INVERSE FUNCTIONS

Consider Figure 9.2, which illustrates a few of the assignments in the function $f(x) = x + 2$. Now consider Figure 9.3, which illustrates a few of the assignments in the function $g(x) = x - 2$.

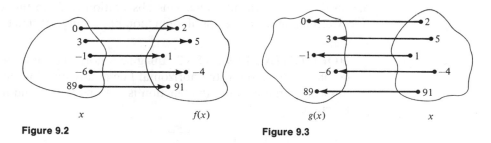

Figure 9.2   Figure 9.3

Note that the functions f and g have reverse assignments. For example, $f(0) = 2$ and $g(2) = 0$. Two functions with exactly reverse assignments are called *inverse functions*. Thus, the inverse of $f(x) = x + 2$ is $g(x) = x - 2$, and vice versa. In the notation of ordered pairs the reverse assignments of inverse functions means when (a, b) is a correspondence in a function (b, a) is a correspondence in the inverse function. Since the components in the ordered pairs are reversed, the domain of a function f is the range of its inverse function and the range of f is the domain of its inverse.

EXAMPLE 1: A function f consists of the correspondences $(7, -2)$, $(9, 1)$, and $(-3, 2)$.
a. Find the correspondences in the inverse function.
b. Compare the domain and range of the two functions.

SOLUTION:

a. Since the inverse function reverses the assignments, the inverse function (say g) consists of the correspondences $(-2, 7)$, $(1, 9)$, and $(2, -3)$.

b. The domain of f is 7, 9, -3, and the range of f is -2, 1, 2. The domain of g is -2, 1, 2, and the range of g is 7, 9, -3. Thus, the domain of f is the range of g and the range of f is the domain of g.

When we reverse the assignments in a function the result is not always a function. For example, consider the function $y = f(x) = x^2$. Since $f(2) = 4$ and $f(-2) = 4$, an inverse of f would have to assign both 2 and -2 to the number 4. The definition of a function excludes this type of assignment. Thus, there is no function that precisely reverses the assignments of the function $y = x^2$. This example suggests the following guideline for judging if the inverse of a function is a function.

A function is *one to one* when each x value in the domain is assigned a different y value so that no two ordered pairs have the same second component. Only the inverse of a one-to-one function is a function.

EXAMPLE 2: Consider the function: $(4, -3)(5, -3)(6, -3)$. Is the inverse of this function a function?

SOLUTION: Since -3 appears as the second component in more than one ordered pair, the function is not one to one. Thus, the inverse of this function is not a function.

In a one-to-one function no two ordered pairs have the same second component. Graphically this means that a one-to-one function cannot contain two points that lie on the same horizontal line. This observation leads to the following simple test for determining from a graph if a function has an inverse function.

HORIZONTAL LINE TEST: Imagine a horizontal line sweeping down the graph of a function. If the horizontal line at any position intersects the graph in more than one point, the function is not one to one and its inverse is not a function.

EXAMPLE 3: Which of the functions graphed in Figure 9.4 has an inverse function?

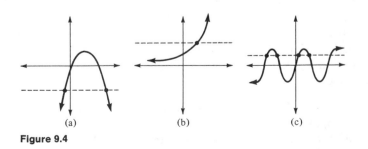

(a)   (b)   (c)

Figure 9.4

SOLUTION:
a. This function does not have an inverse function since a horizontal line may intersect the graph at two points.
b. This function has an inverse function because no horizontal line intersects the graph at more than one point. (*Note:* This is the graph of an exponential function, so these functions have inverse functions. The inverse of the exponential function is discussed in Section 9-4.)
c. This function does not have an inverse function, since a horizontal line may intersect the graph at many points.

Algebraically, if a function has an inverse function and is defined by an equation, we find the rule for the inverse function by interchanging the x and y variables. For example, the inverse of the function defined by

$$y = 2x - 5$$

is defined by

$$x = 2y - 5$$

which, when solved for y, becomes

$$y = \tfrac{1}{2}x + \tfrac{5}{2}$$

EXAMPLE 4: If $f(x) = -3x + 7$, find:
a. $f^{-1}(x)$
b. $f^{-1}(-5)$.
(*Note:* The symbol f^{-1} is used to denote the inverse function of f. The -1 is *not* an exponent.)

SOLUTION:
a. The inverse of the function

$$y = f(x) = -3x + 7$$

is found by interchanging the x and y variables. Thus,

$$x = -3y + 7$$

Solving for y, we have

$$y = f^{-1}(x) = -\tfrac{1}{3}x + \tfrac{7}{3}$$

b. To find $f^{-1}(-5)$, replace x by -5.

$$f^{-1}(-5) = -\tfrac{1}{3}(-5) + \tfrac{7}{3}$$
$$= 4$$

Notice that $f(4) = -3(4) + 7 = -5$.

EXERCISES 9–3

In Exercises 1–6 find the inverse of the function that consists of the given correspondences. Also, find the domain and range of both functions.

1. (1, 7), (2, 8), (3, 9)   **2.** $(4, -1), (8, -2), (12, -3)$

3. $(-2, 10), (0, 0), (5, -1)$ **4.** $(-7, 1), (-4, 11), (1, 13)$

5. (4, 4), (5, 5), (6, 6)   **6.** $(1, 1), (2, \frac{1}{2}), (\frac{1}{2}, 2)$

In Exercises 7–10 determine if the function with the given correspondences has an inverse function.

7. (1, 4) (2, 4) (3, 4)   **8.** $(-4, 4) (-5, 5) (-6, 6)$

9. (0, 0) (5, 0.2) (0.2, 5)   **10.** $(-1, 1) (0, 0) (1, 1)$

In Exercises 11–20 determine which of the functions graphed has an inverse function.

11.

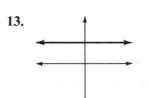

12.

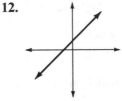

13.

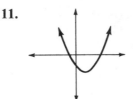

14.

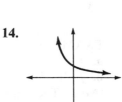

15.

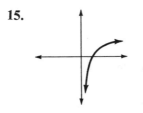

16.

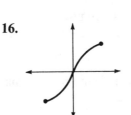

17.

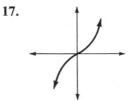

18.

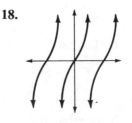

19. 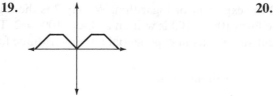   **20.**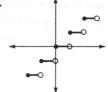

In Exercises 21–30 find the inverse function and then determine the value of the given expression on the right.

21. If $f(x) = x - 10; f^{-1}(1)$.

22. If $g(x) = -x + 5; g^{-1}(0)$.

23. If $g(x) = 7x - 3; g^{-1}(-1)$.

24. If $f(x) = \frac{1}{3}x - \frac{7}{3}; f^{-1}(2)$.

25. If $h(x) = x^3; h^{-1}(-8)$.

26. If $h(x) = \sqrt[3]{x} - 4; h^{-1}(1)$.

27. If $f(x) = \frac{1}{x}; f^{-1}(-4)$.

28. If $f(x) = \frac{1}{x - 2}; f^{-1}(3)$.

29. If $f(x) = 2^x; f^{-1}(8)$.

30. If $f(x) = 9^x; f^{-1}(3)$.

Because inverse functions make reverse assignments, each function undoes the effect of the other. Thus, when two inverse functions are joined by composition, $f^{-1}[f(x)] = x$ for all x in the domain of f and $f[f^{-1}(x)] = x$ for all x in the domain of f^{-1}. In Exercises 31–34 use this criterion to verify that the given functions are inverses of each other.

31. $f(x) = x + 5, f^{-1}(x) = x - 5$

32. $f(x) = 3x - 2, f^{-1}(x) = \frac{1}{3}x + \frac{2}{3}$

33. $f(x) = \sqrt[3]{x}, f^{-1}(x) = x^3$

34. $f(x) = \frac{1}{x + 4}, f^{-1}(x) = \frac{1 - 4x}{x}$

9–4 LOGARITHMIC FUNCTIONS

The logarithmic function is defined by a rule that reverses the assignments of the exponential function. We start with the exponential function $y = b^x$ and obtain its inverse by interchanging the variables x and y. The resulting equation $x = b^y$ is a rule that assigns to each positive number exactly one exponent. For example, suppose that $b = 10$.

If x equals	$x = 10^y$	y equals the exponent
100	$100 = 10^2$	2
1	$1 = 10^0$	0
0.1	$0.1 = 10^{-1}$	-1
10	$10 = 10^1$	1

The logarithm (or log) of a number is the exponent to which a fixed base is raised to obtain the number. In the statement $10^2 = 100$, we call 10 the base, 100 the number, and 2 the exponent or logarithm. We say 2 is the log to the base 10 of 100. In logarithmic form $10^2 = 100$ is written as $\log_{10} 100 = 2$. The following table illustrates equivalent statements in exponential and logarithmic form:

Exponential form	Logarithmic form
$4^2 = 16$	$\log_4 16 = 2$
$10^0 = 1$	$\log_{10} 1 = 0$
$9^{1/2} = 3$	$\log_9 3 = \frac{1}{2}$
$2^{-3} = \frac{1}{8}$	$\log_2 \frac{1}{8} = -3$
$b^y = x$	$\log_b x = y$

The last example in the table shows the form in which the rule of the logarithmic function is usually stated:

$$y = \log_b x$$

The fixed base (b), as in the exponential function, may be any positive real number except 1. Since a positive base raised to any power is positive, we may only substitute positive numbers for x. Thus, the domain of the logarithmic functions is the positive real numbers. We expect this result because the domain and range of inverse functions are interchanged. Thus, the domain of the logarithmic functions is the range of the exponential functions (positive real numbers) and the range of the logarithmic functions is the domain of the exponential functions (real numbers).

To illustrate that a logarithmic function reverses the assignments of an exponential function consider Figure 9.5 in which $y = 2^x$ and $y = \log_2 x$ are sketched on the same axes.

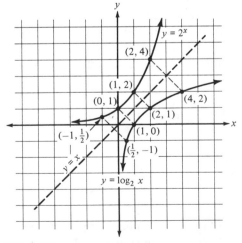

Figure 9.5

$y = 2^x$

x	-3	-2	-1	0	1	2	3
y	$\frac{1}{8}$	$\frac{1}{4}$	$\frac{1}{2}$	1	2	4	8

$y = \log_2 x$ or $x = 2^y$

x	$\frac{1}{8}$	$\frac{1}{4}$	$\frac{1}{2}$	1	2	4	8
y	-3	-2	-1	0	1	2	3

An examination of the tables of correspondences and the two graphs shows that the functions have reverse assignments. Note in Figure 9.5 that the graphs of inverse functions are related in that each one is the reflection of the other across the line $y = x$.

EXERCISES 9–4

In Exercises 1–10 express in logarithmic form.

1. $3^2 = 9$

2. $2^3 = 8$

3. $(\frac{1}{2})^2 = \frac{1}{4}$

4. $(\frac{1}{3})^3 = \frac{1}{27}$

5. $4^{-2} = \frac{1}{16}$

6. $(\frac{1}{2})^{-3} = 8$

7. $25^{1/2} = 5$

8. $8^{-1/3} = \frac{1}{2}$

9. $7^0 = 1$

10. $10^{-4} = 0.0001$

In Exercises 11–20 express in exponential form.

11. $\log_5 5 = 1$

12. $\log_2 32 = 5$

13. $\log_{1/3} \frac{1}{9} = 2$

14. $\log_{1/2} \frac{1}{16} = 4$

15. $\log_2 \frac{1}{4} = -2$

16. $\log_{1/4} 4 = -1$

17. $\log_{49} 7 = \frac{1}{2}$

18. $\log_{27} \frac{1}{3} = -\frac{1}{3}$

19. $\log_{100} 1 = 0$

20. $\log_{10} 0.001 = -3$

In Exercises 21–30 find the value of each expression.

21. $\log_3 9$

22. $\log_2 16$

23. $\log_4 4$

24. $\log_5 1$

25. $\log_3 \frac{1}{3}$

26. $\log_{10} 0.01$

27. $\log_3 \sqrt{3}$

28. $\log_9 3$

29. $\log_{10}(\log_5 5)$

30. $\log_2(\log_4 2)$

In Exercises 31–40 determine the value of the unknown by inspection or by writing in exponential form.

31. $\log_2 8 = y$

32. $\log_4 2 = y$

33. $\log_5 x = -1$

34. $\log_{10} x = -5$

35. $\log_b 125 = 3$

36. $\log_b 10 = \frac{1}{2}$

37. $\log_{10} 10^3 = y$

38. $\log_{10} 10^{2.4} = y$

39. $\log_b b = 1$

40. $\log_b 1 = 0$

In Exercises 41–43 evaluate each function for the given element of the domain.

41. $f(x) = \log_2 x; f(8); f(0.5); f(1)$

42. $g(x) = \log_9 x; g(3); g(27); g(\frac{1}{81})$

43. $h(x) = \log_{10} x; h(10{,}000); h(1); h(0.0000001)$

In Exercises 44–47 graph the function.

44. $y = \log_3 x$ **45.** $y = \log_{10} x$

46. $y = \log_{1/2} x$ **47.** $y = \log_{1/4} x$

In Exercises 48–51 what is the domain of each function?

48. $f(x) = \log_{10}(2x - 1)$ **49.** $g(x) = \log_2(x^2 - 9)$

50. $h(x) = \log_5(-x)$ **51.** $f(x) = \log_{10}|x - 1|$

9–5 PROPERTIES OF LOGARITHMS

Since a logarithm is an exponent, the properties of logarithms may be derived from the following properties of exponents:

1. $b^y \cdot b^z = b^{y+z}$.

2. $b^y/b^z = b^{y-z}$ $(b \neq 0)$.

3. $(b^y)^z = b^{yz}$.

To establish our first property consider the product of two positive numbers x_1 and x_2. Let

$$x_1 = b^y \quad \text{or} \quad y = \log_b x_1$$

and

$$x_2 = b^z \quad \text{or} \quad z = \log_b x_2$$

then

$$x_1 \cdot x_2 = b^y \cdot b^z = b^{y+z}$$

Writing the last equation in logarithmic form, we have

$$\log_b x_1 x_2 = y + z$$

so

$$\log_b x_1 x_2 = \log_b x_1 + \log_b x_2$$

Through similar procedures we may establish two other properties of logarithms and our results are as follows:

PROPERTIES OF LOGARITHMS: If b, x_1, x_2 are positive, with $b \neq 1$ and k any real number, then

$$\log_b x_1 x_2 = \log_b x_1 + \log_b x_2$$

$$\log_b \frac{x_1}{x_2} = \log_b x_1 - \log_b x_2$$

$$\log_b(x_1)^k = k \log_b x_1$$

An understanding of these three properties is essential to the study of the theory and applications of logarithms. Consider how these properties are used in the following examples.

EXAMPLE 1: Express each statement as the sum or difference of simpler logarithmic statements.

STATEMENT	SOLUTION
a. $\log_{10}(4 \cdot 5)$	$= \log_{10} 4 + \log_{10} 5$
b. $\log_2(\frac{3}{7})$	$= \log_2 3 - \log_2 7$
c. $\log_4 \sqrt{5}$	$= \log_4(5)^{1/2} = \frac{1}{2}\log_4 5$
d. $\log_{10}(xy/z)$	$= (\log_{10} x + \log_{10} y) - \log_{10} z$
e. $\log_{10} 2\pi\sqrt{L/g}$	$= \log_{10} 2 + \log_{10} \pi + \log_{10} \sqrt{L/g}$
	$= \log_{10} 2 + \log_{10} \pi + \frac{1}{2}\log_{10}(L/g)$
	$= \log_{10} 2 + \log_{10} \pi + \frac{1}{2}(\log_{10} L - \log_{10} g)$

EXAMPLE 2: Express as a single logarithm with coefficient 1.

STATEMENT	SOLUTION
a. $\log_{10} 9 + \log_{10} 2$	$- \log_{10}(9 \cdot 2) = \log_{10} 18$
b. $\log_5 35 - \log_5 7$	$= \log_5(\frac{35}{7}) = \log_5 5$ (or 1)
c. $2 \log_4 5$	$= \log_4(5)^2 = \log_4 25$
d. $3 \log_b x + 2 \log_b y$	$= \log_b x^3 + \log_b y^2 = \log_b(x^3 \cdot y^2)$
e. $\frac{1}{2}(\log_{10} x - 3 \log_{10} y)$	$= \frac{1}{2}(\log_{10} x - \log_{10} y^3) = \frac{1}{2}(\log_{10}(x/y^3))$
	$= \log_{10}(x/y^3)^{1/2} = \log_{10} \sqrt{x/y^3}$

EXAMPLE 3: Solve for x: $\log_3(x^2 - 4) - \log_3(x + 2) = 2$.

SOLUTION:

$$\log_3(x^2 - 4) - \log_3(x + 2) = 2$$

$$\log_3\left(\frac{x^2 - 4}{x + 2}\right) = 2$$

$$\log_3(x - 2) = 2$$

$$3^2 = x - 2$$

$$11 = x$$

If we substitute 11 in our original equation, both log expressions are defined. Thus, the solution is 11.

EXAMPLE 4: If $\log_{10} 2 = 0.3010$ and $\log_{10} 3 = 0.4771$, simplify the expression.

EXPRESSION SOLUTION

a. $\log_{10} 9 = \log_{10} 3^2 = 2 \log_{10} 3 = 2(0.4771) = 0.9542$

b. $\log_{10} 12 = \log_{10}(2^2 \cdot 3) = \log_{10} 2^2 + \log_{10} 3$

$\qquad = 2 \log_{10} 2 + \log_{10} 3$

$\qquad = 2(0.3010) + 0.4771 = 1.0791$

c. $\log_{10} \sqrt{2} = \log_{10}(2)^{1/2} = \frac{1}{2} \log_{10} 2 = \frac{1}{2}(0.3010) = 0.1505$

EXERCISES 9–5

In Exercises 1–18 express each logarithm as the sum or difference of simpler logarithms.

1. $\log_{10}(7 \cdot 5)$

2. $\log_b xyz$

3. $\log_6(\frac{3}{5})$

4. $\log_b(x/y)$

5. $\log_5(11)^{16}$

6. $\log_b x^{16}$

7. $\log_2 \sqrt{3}$

8. $\log_b \sqrt{x}$

9. $\log_{10} \sqrt[5]{16}$

10. $\log_b \sqrt[5]{x}$

11. $\log_4(4^2 \cdot 3^3)$

12. $\log_b x^2 y^3$

13. $\log_b \sqrt{xy}$

14. $\log_b \sqrt{x/y}$

15. $\log_b \sqrt[3]{x^5}$

16. $\log_b(x)^\pi$

17. $\log_b \sqrt[4]{xy^2/z^3}$

18. $\log_{10} \sqrt{s(s-a)(s-b)(s-c)}$

In Exercises 19–32 express each statement as a single logarithm with coefficient 1.

19. $\log_2 3 + \log_2 4$

20. $\log_b x + \log_b y$

21. $\log_4 20 - \log_4 5$

22. $\log_b y - \log_b x$

23. $2 \log_7 3$

24. $3 \log_b z$

25. $\frac{1}{2} \log_{10} 9 - 3 \log_{10} 2$

26. $3 \log_b x - \frac{1}{2} \log_b z$

27. $\frac{1}{3} \log_b x + \frac{2}{3} \log_b y$

28. $2 \log_b x + \log_b(x + y)$

29. $\log_b(x^2 - 1) - \log_b(x + 1)$

30. $\log_b(y - 2) + \log_b y - 2 \log_b x$

31. $\frac{1}{2}[\log_b x - (5 \log_b y - 3 \log_b z)]$

32. $\frac{1}{2}[(\log_b x - 5 \log_b y) - 3 \log_b z]$

In Exercises 33–40 solve each equation.

33. $\log_{10} x + \log_{10} 2 = 1$

34. $\log_3 x - \log_3 4 = 2$

35. $\log_2(x^2 - 1) = 3$

36. $\log_5(x + 1) - \log_5 x = 1$

37. $\log_6(x + 1) + \log_6 x = 1$

38. $\log_2(x + 1) + \log_2(x + 4) = 2$

39. $\log_b 2x = \log_b 4x - \log_b 2$

40. $\log_{10}(x^2 - x) = \log_{10} x + \log_{10}(x - 1)$

41. If $\log_{10} 2 = 0.3010$ and $\log_{10} 3 = 0.4771$, find:

    a. $\log_{10} 6$

    b. $\log_{10} 4$

    c. $\log_{10} \sqrt{3}$

    d. $\log_{10} 2/3$

    e. $\log_{10} 27$

    f. $\log_{10} 72$

    g. $\log_{10} \sqrt{\frac{1}{3}}$

    h. $\log_{10} \sqrt[5]{\frac{2}{3}}$

    i. $\log_{10}(0.5)$

    j. $\log_{10} 5$

42. Give specific counterexamples to *disprove* each of the following statements.

    a. $\log_b xy = (\log_b x)(\log_b y)$

    b. $(\log_b x)(\log_b y) = \log_b x + \log_b y$

c. $\log_b \dfrac{x}{y} = \dfrac{\log_b x}{\log_b y}$ **d.** $\dfrac{\log_b x}{\log_b y} = \log_b x - \log_b y$

e. $\log_b x^k = (\log_b x)^k$ **f.** $(\log_b x)^k = k \log_b x$

43. Prove the following properties of logarithms.

a. $\log_b \dfrac{x_1}{x_2} = \log_b x_1 - \log_b x_2$ **b.** $\log_b(x_1)^k = k \log_b x_1$

44. What restrictions are placed on x_1, x_2, b, and k in Exercise 43?

45. If $f(x) = 10^x$ and $f^{-1}(x) = \log_{10} x$, verify that $f^{-1}[f(x)] = x$ for all x in the domain of f.

46. Examine the following line of reasoning: $3 > 2$. If we multiply both sides of the inequality by $\log_{10}(\frac{1}{2})$, we have

$$3 \log_{10}(\tfrac{1}{2}) > 2 \log_{10}(\tfrac{1}{2})$$
$$\log_{10}(\tfrac{1}{2})^3 > \log_{10}(\tfrac{1}{2})^2$$
$$\log_{10}(\tfrac{1}{8}) > \log_{10}(\tfrac{1}{4})$$

Thus,

$$\tfrac{1}{8} > \tfrac{1}{4}$$

Our conclusion is incorrect. What went wrong?

9–6 LOGARITHMS TO THE BASE 10

Logarithms may be used to simplify tedious numerical computations. This application is less important now that electronic calculators are available, but it is interesting to study the principle on which these machines (and a slide rule) operate. Also, certain equations that contain exponents require logarithms for their solution.

Since numbers are expressed decimally, we work with logarithms to the base 10. In Section 6-2 we showed how to express any positive number N in scientific notation. That is,

$$N = m \times 10^k \qquad \text{where } 1 \le m < 10 \text{ and } k \text{ is some integer}$$

Finding the logarithm to the base 10 of N, we get

$$\log_{10} N = \log_{10}(m \times 10^k) = \log_{10} m + \log_{10} 10^k$$
$$= \log_{10} m + k \log_{10} 10$$

Since $\log_{10} 10 = 1$, we have

$$\log_{10} N = (\log_{10} m) + k$$

Thus, $\log_{10} N$ is the sum of two numbers: an integer, k, called the *characteristic*; and $\log_{10} m$, called the *mantissa*. Since $1 \le m < 10$, we find the mantissa by consulting Table 3 in the Appendix, which gives approximations of the logarithms (to the base 10) of numbers between 1 and 10. Consider the excerpt from this table shown in Figure 9.6.

m	0	1	2	3	4	5	6	7	8	9
5.5	.7404	.7412	.7419	.7427	.7435	.7443	.7451	.7459	.7466	.7474
5.6	.7482	.7490	.7497	.7505	.7513	.7520	.7528	.7536	.7543	.7551
5.7	.7559	.7566	.7574	.7582	.7589	.7597	.7604	.7612	.7619	.7627
5.8	.7634	.7642	.7649	.7657	.7664	.7672	.7679	.7686	.7694	.7701
5.9	.7709	.7716	.7723	.7731	.7738	.7745	.7752	.7760	.7767	.7774

Figure 9.6

The first column on the left contains the first two significant digits of m, and the top row contains the third significant digit of m. For example, to find $\log_{10} 5.68$ we look at the intersection of the row opposite 5.6 and the column headed by 8. Thus,

$$\log_{10} 5.68 \approx 0.7543$$

The following examples should clarify the procedure for finding the logarithm to the base 10 of any positive number. In most cases the equal sign will be used to relate logarithms with the understanding that these logarithms are only approximations.

EXAMPLE 1: Find $\log_{10} 427$.

SOLUTION:

$$\log_{10} 427 = \log_{10}(4.27 \times 10^2)$$
$$= \log_{10} 4.27 + \log_{10} 10^2$$

Since $\log_{10} 10^2 = 2$ and $\log_{10} 4.27 \approx 0.6304$ (Table 3), we have

$$\log_{10} 427 = 0.6304 + 2 = 2.6304$$

EXAMPLE 2: Find $\log_{10} 80$.

SOLUTION:

$$\log_{10} 80. = \log_{10}(8.00 \times 10^1)$$
$$= \log_{10} 8.00 + \log_{10} 10^1$$

Since $\log_{10} 10^1 = 1$ and $\log_{10} 8.00 \approx 0.9031$ (Table 3), we have

$$\log_{10} 80 = 0.9301 + 1 = 1.9301$$

EXAMPLE 3: Find $\log_{10} 0.485$.

SOLUTION: $\log_{10} 0.485 = \log_{10}(4.85 \times 10^{-1}) = \log_{10} 4.85 + \log_{10} 10^{-1}$

Since $\log_{10} 10^{-1} = -1$ and $\log_{10} 4.85 \approx 0.6857$ (Table 3), we have

$$\log_{10} 0.485 = 0.6857 + (-1) \text{ or } -0.3143$$

EXAMPLE 4: Find $\log_{10} 0.00123$.

SOLUTION:

$$\log_{10} 0.00123 = \log_{10}(1.23 \times 10^{-3})$$
$$= \log_{10} 1.23 + \log_{10} 10^{-3}$$

Since $\log_{10} 10^{-3} = -3$ and $\log_{10} 1.23 \approx 0.0899$ (Table 3), we have

$$\log_{10} 0.00123 = 0.0899 + (-3) \quad \text{or} \quad -2.9101$$

To perform computations it is necessary to be able to reverse the above procedure and find N if we know $\log_{10} N$. The following examples illustrate this procedure.

EXAMPLE 5: If $\log_{10} N = 2.5250$, find N.

SOLUTION: We know $N = m \times 10^k$, where $1 \leq m < 10$ and k is the characteristic.

$$\log_{10} N = 2.5250 = 0.5250 + 2$$

Since $\log_{10} 3.35 \approx 0.5250$ (Table 3), we have $m = 3.35$. Since 2 is the characteristic, we have $k = 2$. Thus,

$$N = 3.35 \times 10^2 = 335$$

EXAMPLE 6: If $\log_{10} N = 0.9990 + (-3)$, find N.

SOLUTION: We know that $N = m \times 10^k$, where $1 \leq m < 10$ and k is the characteristic.

$$\log_{10} N = 0.9990 + (-3)$$

Since $\log_{10} 9.98 \approx 0.9990$ (Table 3), we have $m = 9.98$. Since -3 is the characteristic, we have $k = -3$. Thus,

$$N = 9.98 \times 10^{-3} = 0.00998$$

EXAMPLE 7: If $\log_{10} N = -1.4567$, find N.

SOLUTION: We must be careful because -1.4567 *does not equal* $0.4567 + (-1)$. We must have a positive mantissa to use Table 3, so we change the form of -1.4567 as follows:

$$-1.4567 = (-1.4567 + 2) + (-2)$$
$$= 0.5433 + (-2)$$

We know that $N = m \times 10^k$, where $1 \leq m < 10$ and k is the characteristic since $\log_{10} 3.49 \approx 0.5433$ (Table 3), $m = 3.49$. Since -2 is the characteristic, $k = -2$.

Thus,

$$N = 3.49 \times 10^{-2} = 0.0349$$

(*Note:* In this section we wish to show the principle on which logarithms operate. Accuracy in our results is not our primary concern. If desired, greater accuracy may be obtained in many cases by linear interpolation, an approximation method discussed in Section 3-2 with regard to the trigonometric functions.)

The following examples illustrate the use of logarithms in difficult computations. The properties of logarithms discussed in Section 9-5 are essential to this work.

EXAMPLE 8: Compute $[(343)(0.678)]^{10}$.

SOLUTION:

$$\begin{aligned}
N &= [(343)(0.678)]^{10} \\
\log_{10} N &= \log_{10}[(343)(0.678)]^{10} \\
&= 10(\log_{10} 343 + \log_{10} 0.678) \\
&= 10\{2.5353 + [0.8312 + (-1)]\} \\
&= 10(2.3665) \\
&= 23.665
\end{aligned}$$

We know that $N = m \times 10^k$, where $1 \le m < 10$ and k is the characteristic. Since $\log_{10} 4.62 \approx 0.6650$ (Table 3), $m = 4.62$. Since 23 is the characteristic, $k = 23$. Thus,

$$\begin{aligned}
N &= 4.62 \times 10^{23} \\
&= 462{,}000{,}000{,}000{,}000{,}000{,}000{,}000
\end{aligned}$$

EXAMPLE 9: Find an approximate value for $2^{\sqrt{2}}$. (*Note:* $\sqrt{2} \approx 1.41$.)

SOLUTION:

$$\begin{aligned}
N &= 2^{\sqrt{2}} \\
\log_{10} N &= \log_{10} 2^{\sqrt{2}} = \sqrt{2} \log_{10} 2
\end{aligned}$$

Referring to Table 3, we have $\log_{10} 2 \approx 0.3010$.

$$\begin{aligned}
\log_{10} N &= \sqrt{2}(0.3010) = (1.41)(0.3010) \\
&= 0.4244
\end{aligned}$$

We know that $N = m \times 10^k$, where $1 \le m < 10$ and k is the characteristic. Since $\log_{10} 2.66 \approx 0.4244$ (Table 3), $m = 2.66$. Since 0 is the characteristic, $k = 0$. Thus,

$$N = 2.66 \times 10^0 = 2.66$$

EXAMPLE 10: Compute $\sqrt[5]{7.13/22.1}$.

SOLUTION:

$$N = \sqrt[5]{\frac{7.13}{22.1}}$$

$$\log_{10} N = \log_{10}\left(\frac{7.13}{22.1}\right)^{1/5}$$

$$= \tfrac{1}{5}(\log_{10} 7.13 - \log_{10} 22.1)$$
$$= \tfrac{1}{5}(0.8531 - 1.3444)$$
$$= \tfrac{1}{5}(-0.4913)$$
$$= -0.0983$$

To find N we must write the logarithm in a form in which the mantissa is positive.

$$\log_{10} N = (-0.0983 + 1) + (-1)$$
$$= 0.9017 + (-1)$$

We know that $N = m \times 10^k$, where $1 \leq m < 10$ and k is the characteristic, since $\log_{10} 7.97 \approx 0.9017$ (Table 3), $m = 7.97$. Since -1 is the characteristic, $k = -1$. Thus,

$$N = 7.97 \times 10^{-1} = 0.797$$

The following examples illustrate the use of logarithms in solving equations that contain exponents.

EXAMPLE 11: Solve for x: $3^x = 8$.

SOLUTION:

$$3^x = 8$$
$$\log_{10} 3^x = \log_{10} 8$$
$$x \log_{10} 3 = \log_{10} 8$$

$$x = \frac{\log_{10} 8}{\log_{10} 3} = \frac{0.9031}{0.4771}$$

$$x = 1.893$$

EXAMPLE 12: Solve for x: $x^{1.2} = 23$.

SOLUTION:

$$x^{1.2} = 23$$
$$\log_{10} x^{1.2} = \log_{10} 23$$
$$1.2 \log_{10} x = 1.3617$$

$$\log_{10} x = \frac{1.3617}{1.2} = 1.1348$$

We know that $x = m \times 10^k$, where $1 \leq m < 10$ and k is the characteristic. Since $\log_{10} 1.36 \approx 0.1348$ (Table 3), $m = 1.36$. Since 1 is the characteristic, $k = 1$. Thus,

$$x = 1.36 \times 10^1 = 13.6$$

You are probably aware of the concept of an acid since it is associated with such familiar and diverse topics as indigestion, shampoo, swimming pools, and car batteries. In chemistry a logarithm is used to define pH (hydrogen potential), which is a convenient measure of the acidity of a solution. Briefly, let's see why pH is defined as a logarithm.

When an atom of hydrogen loses an electron, it is called a hydrogen ion. Since this atom is positively charged, the hydrogen ion is symbolized H^+. The electrical imbalance in an ion greatly affects chemical reactions and it is the concentration of hydrogen ions (symbolized $[H^+]$) in a solution that determines its acidity. However, hydrogen ion concentrations are very small numbers. For example, a weak acid solution might have a concentration (measured in moles/liter) of only 1 part H^+ in 10,000, or 1/10,000, or 10^{-4}. When writing such small numbers the exponential form is obviously helpful. In 1909 a further simplification, pH notation, was introduced. With this notation only the exponent is considered as follows:

$$\text{If } [H^+] = 10^{-4} \quad \text{then} \quad pH = 4$$
$$[H^+] = 10^{-8.3} \quad \text{then} \quad pH = 8.3$$
$$[H^+] = 10^{-x} \quad \text{then} \quad pH = x$$

Since pH is defined as an exponent, we determine pH by finding a logarithm. That is, from

$$[H^+] = 10^{-pH}$$

we may write

$$-pH = \log_{10}[H^+]$$
$$pH = -\log_{10}[H^+]$$

Thus, pH is defined as the negative of the logarithm of the hydrogen ion concentration. A solution with $pH = 7$ is called neutral. In acids $pH < 7$; bases or alkalies have $pH > 7$.

EXAMPLE 13: The hydrogen ion concentration in a sample of human blood is 4.4×10^{-8}. Determine the pH of the sample.

SOLUTION:

$$pH = -\log_{10}[H^+]$$
$$= -\log_{10}(4.4 \times 10^{-8}) = -(\log_{10} 4.4 + \log_{10} 10^{-8})$$

Since $\log_{10} 10^{-8} = -8$ and $\log_{10} 4.4 \approx 0.6435$ (Table 3), we have

$$pH = -[0.6435 + (-8)] = -(-7.3565)$$
$$pH \approx 7.4$$

(*Note:* Variations from this pH of more than 0.1 indicate serious problems.)

EXAMPLE 14: The pH of tomato juice is about 4.5. Determine $[H^+]$.

SOLUTION:

$$pH = -\log_{10}[H^+]$$

Replacing pH by 4.5 and solving for $\log_{10}[H^+]$, we have

$$\log_{10}[H^+] = -4.5$$

We must have a positive mantissa to use Table 3, so we change the form of -4.5, as follows:

$$-4.5 = (-4.5 + 5) + (-5)$$
$$= 0.5 + (-5)$$

We know that $[H^+] = m \times 10^k$, where $1 \leq m < 10$ and k is the characteristic. Since $\log_{10} 3.16 \approx 0.5$ (Table 3), $m \approx 3.2$. Since -5 is the characteristic, $k = -5$. Thus,

$$[H^+] = 3.2 \times 10^{-5}$$

EXERCISES 9–6

In Exercises 1–10 find the logarithm to the base 10 of the number.

1. 2.34   2. 23.4
3. 0.234   4. 0.0234
5. 2340   6. 187000
7. 90   8. 0.9
9. 0.111   10. 0.000717

In Exercises 11–20 find N if $\log_{10} N$ equals the number.

11. 0.8189   12. 2.8189
13. $0.8189 + (-1)$ 14. $0.8189 + (-3)$
15. 1.5559   16. 3.9803
17. -1.4123 18. -2.6712
19. -0.5173 20. -4.2990

In Exercises 21–30 compute by means of logarithms.

21. $(413)(2.74)$ 22. $(0.0123)(12.7)$

23. $\dfrac{908}{14.3}$ 24. $\dfrac{19.2}{604}$

25. $(3.2)^5$ 26. $(0.000491)^4$
27. $\sqrt{97}$ 28. $\sqrt[5]{0.4}$

29. $\dfrac{(514)(29.6)}{0.02}$ 30. $\dfrac{17.9}{(49.6)(0.0821)}$

31. $\sqrt{(16.1)(2.5)}$ 32. $\sqrt[3]{(123)(0.22)^2}$

33. $\sqrt[6]{\dfrac{195}{7.6}}$

34. $\sqrt{\dfrac{22.2}{459}}$

35. $\dfrac{(1.8)^2(5.6)^3}{\sqrt{104}}$

36. $\dfrac{\sqrt[3]{17}}{(671)(0.08)^5}$

37. $3^{\sqrt{3}}$ (*Note:* $\sqrt{3} \approx 1.73$.)

38. 2^{π} (*Note:* $\pi \approx 3.14$.)

39. $(12.1)^{-5}$

40. The equal-monthly-payment formula is used by banks when they finance many common loans. The problem is to find the equal monthly installment, E, which will pay off a loan of P dollars with interest in n months. The monthly interest rate is r and interest is charged at each payment period only on the unpaid balance. Each payment first pays off the interest charge with the balance reducing the principal. As the principal becomes smaller, the interest charge decreases while the portion that reduces the principal increases. The formula used by the bank is

$$E = \frac{Pr}{1 - (1 + r)^{-n}}$$

Suppose that you obtain a $1000 home improvement loan for 24 months with 12 percent annual interest charged on the unpaid balance. What is the amount of the equal monthly installment that will pay off this debt? [*Note:* If the annual interest rate is 12 percent, the monthly interest rate (r) is 12 percent/12, or 1 percent.]

In Exercises 41–50 use logarithms to solve the equation for x.

41. $2^x = 10$

42. $5^x = 100$

43. $4^{x+1} = 17$

44. $8^{x-1} = 25$

45. $(1.06)^x = 2$

46. $2^{-x} = 9$

47. $x^{1.7} = 20$

48. $x^{2.5} = 1000$

49. $x^{-0.6} = 12$

50. $x^{-4.1} = 47.6$

51. How long will it take for money invested at 5 percent compounded annually to double?

52. If you were offered $1 million or a penny that doubled in value on each day in the month of November, which option would you accept? What is the value of the penny by the end of the month?

In Exercises 53–56 determine the pH of the solution with the given hydrogen ion concentration.

53. $[H^+] = 10^{-7}$
(water)

54. $[H^+] = 4.2 \times 10^{-3}$
(5 percent vinegar)

55. $[H^+] = 2.5 \times 10^{-4}$
(orange juice)

56. $[H^+] = 3.3 \times 10^{-8}$
(swimming pool water)

In Exercises 57–60 determine the hydrogen ion concentration of the solution with the given pH.

57. pH $= 0$ (pure sulfuric acid)

58. pH $= 8.9$ (seawater)

59. pH $= 2.1$ (stomach juices)

60. pH $= 11.3$ (household ammonia)

The most widely used unit in communications engineering is the decibel. This name is derived from the basic unit of loudness, the bel (in honor of Alexander

Graham Bell), with decibel meaning one-tenth bel. The decibel is defined as a logarithm because the ear perceives changes in sound levels logarithmically. That is, the ear is very perceptible to changes in loudness at low sound levels; while at high sound levels the intensity of the sound must be greatly increased for the ear to perceive change. The formula for decibels (dB) is

$$dB = 10 \log_{10} \frac{P_2}{P_1}$$

where P_1 and P_2 represent the powers required to produce sound levels 1 and 2. Use this formula in Exercises 61–64. (*Note:* There are many other important measures that are defined as the logarithm of a ratio. The Richter scale, which is a measure of the magnitude of an earthquake, and the formula for the magnitude or brightness of a star are common examples.)

61. Determine the difference in loudness (in decibels) between sounds 1 and 2 if the power ratio (P_2/P_1) between the sounds is:

    a. 1   **b.** 10   **c.** 100   **d.** 1000

As the power ratio is progressively multiplied by 10, how does the loudness in decibels increase?

62. What is the decibel gain of an amplifier with a 3-watt input and a 30-watt output?

63. In a normal conversation the power ratio (P_2/P_1) between the highest and lowest sound volumes is about 300 to 1. What is the range of speech in decibels? (*Note:* The decibel scale for loudness ranges from 0 dB for a sound at the threshold of hearing to about 140 dB for an airplane engine. Normal conversation is rated at about 60 dB.)

64. When rating an amplifier from a radio or public address system, P_1 assumes the arbitrary reference level value of 6 milliwatts (or 0.006 watt). What is the gain in decibels of a 60-watt amplifier? That is, find dB when $P_2 = 60$ watts and $P_1 = 0.006$ watt.

9–7 LOGARITHMS TO BASES OTHER THAN 10

We may change a logarithm in one base to a logarithm in another base by the following procedure. Suppose that $y = \log_b x$ and we wish to change to base a. This expression in exponential form is

$$b^y = x$$

Taking the logarithm to the base a of both sides of the equation, we have

$$\log_a b^y = \log_a x$$
$$y \log_a b = \log_a x$$

$$y = \frac{\log_a x}{\log_a b}$$

Since $y = \log_b x$, we have

$$\log_b x = \frac{\log_a x}{\log_a b}$$

EXAMPLE 1: Use logarithms to the base 10 to determine:

a. $\log_2 7$ **b.** $\log_5 0.043$

SOLUTION:

a. $\log_2 7 = \dfrac{\log_{10} 7}{\log_{10} 2}$ **b.** $\log_5 0.043 = \dfrac{\log_{10} 0.043}{\log_{10} 5}$

$\phantom{\textbf{a.} \log_2 7} = \dfrac{0.8451}{0.3010} \approx 2.808$ $= \dfrac{0.6335 + (-2)}{0.6990}$

$$= \dfrac{-1.3665}{0.6990} \approx -1.955$$

In Section 9-2 we showed that the most important exponential function is $y = e^x$, where e is an irrational number equal to about 2.718. Since logarithm and exponential functions are closely related, it follows that $y = \log_e x$ is the most important logarithmic function. Logarithms to the base e are called *natural logarithms* and $\log_e x$ is usually written as ln x. These logarithms may be used to solve most of the exponential equations we encountered in Section 9-2. One method of determining the natural log of a number is to change the expression to base 10 logarithms.

EXAMPLE 2: Use logarithms to the base 10 to determine ln 170.

SOLUTION:

$$\ln 170 = \frac{\log_{10} 170}{\log_{10} e} = \frac{\log_{10} 170}{\log_{10} 2.72} = \frac{2.2304}{0.4346}$$

$\ln 170 \approx 5.132$

A second method, which is illustrated in the following example, is more accurate and uses a table of natural logarithms (Table 4 in the Appendix).

EXAMPLE 3: Find ln 170.

SOLUTION:

$$\ln 170 = \ln(1.70 \times 10^2)$$
$$= \ln 1.7 + \ln 10^2$$
$$= \ln 1.7 + (2)\ln 10$$

Referring to Table 4 we have, ln $1.7 \approx 0.5306$ and ln $10 \approx 2.3026$.

$$\ln 170 = 0.5306 + 2(2.3026)$$
$$\ln 170 \approx 5.136$$

EXAMPLE 4: Find ln 0.45.

SOLUTION:

$$\ln 0.45 = \ln(4.5 \times 10^{-1})$$
$$= \ln 4.5 + \ln 10^{-1}$$
$$= \ln 4.5 + (-1)\ln 10$$

Referring to Table 4, we have $\ln 4.5 \approx 1.5041$ and $\ln 10 \approx 2.3026$. Thus,

$$\ln 0.45 = 1.5041 + (-1)(2.3026)$$
$$= -0.7985$$

EXAMPLE 5: If $\ln x = 0.95$, find x.

SOLUTION: $\ln x = 0.95$. Writing this statement in exponential form, we have

$$e^{0.95} = x$$
$$2.5857 = x \quad \text{(Table 2)}$$

EXERCISES 9–7

In Exercises 1–16 use logarithms to the base 10 to determine each logarithm.

1. $\log_2 9$	**2.** $\log_4 31$
3. $\log_3 5$	**4.** $\log_5 3$
5. $\log_{17} 11$	**6.** $\log_{100} 572$
7. $\log_4 0.012$	**8.** $\log_6 0.735$
9. $\log_{1/2} 19$	**10.** $\log_{1/3} \frac{2}{3}$
11. $\ln 5$	**12.** $\ln 6.2$
13. $\ln 756$	**14.** $\ln 93$
15. $\ln 0.47$	**16.** $\ln 0.003$

In Exercises 17–26 use Table 4 to determine the natural logarithm.

17. $\ln 7$	**18.** $\ln 3.8$
19. $\ln 42$	**20.** $\ln 123$
21. $\ln 6500$	**22.** $\ln 100$
23. $\ln 0.56$	**24.** $\ln 0.009$
25. $\ln 0.00044$	**26.** $\ln 0.292$

In Exercises 27–34 find x if $\ln x$ equals the number.

27. 0.19	**28.** 4.6
29. 2.3	**30.** 0.55
31. -0.24	**32.** -7.5
33. -3.7	**34.** -0.09

9–8 GRAPHS ON LOGARITHMIC PAPER

Technical workers frequently analyze data that range over a large collection of values. In addition, the analysis often requires them to find a formula that fits a set of data. Here we study a scheme that is useful in such situations.

On logarithmic graph paper the axes are laid off at distances proportional to the logarithms of numbers. Consider Figure 9.7, in which we construct an axis by marking on the paper the following numbers:

$$\log_{10} 1 = 0 \qquad\qquad \log_{10} 6 = 0.778$$
$$\log_{10} 2 = 0.301 \qquad \log_{10} 7 = 0.845$$
$$\log_{10} 3 = 0.477 \qquad \log_{10} 8 = 0.903$$
$$\log_{10} 4 = 0.602 \qquad \log_{10} 9 = 0.954$$
$$\log_{10} 5 = 0.699 \qquad \log_{10} 10 = 1 \qquad \text{etc.}$$

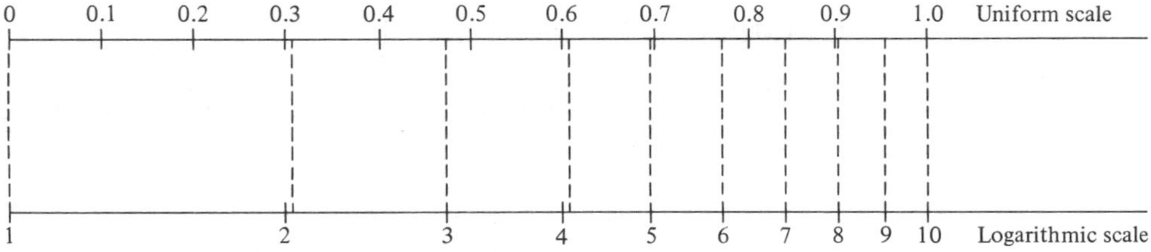

Figure 9.7

On the logarithmic scale we write 1 as shorthand for $\log_{10} 1$. Since $\log_{10} x$ is undefined for $x \leq 0$, only positive numbers may be marked off on logarithmic paper. Using this paper we may obtain accuracy in a graph in which the variables range over a large collection of values.

Logarithmic paper is especially useful when graphing functions of the form $y = ax^m$, which are called *power functions*. By taking the logarithm to the base 10 of a power function, we have

$$\log y = \log ax^m$$
$$\qquad = \log a + m \log x$$

If we let $\log y = Y$, $\log x = X$, and $\log a = b$, the equation becomes

$$Y = b + mX$$

Thus, on logarithmic paper the graph of a power function is a straight line with slope m and Y intercept b. It also follows that if the graph of experimental data is a straight line on logarithmic paper, the variables are related by an equation of the form $y = ax^m$. The slope m is measured with a ruler and the coefficient a is the intercept on the vertical axis at $x = 1$.

EXAMPLE 1: Use logarithmic paper to graph $y = 3x^2$.

SOLUTION: First construct a table of correspondences:

x	1	2	3	4	5
y	3	12	27	48	75

As shown in Figure 9.8, the graph is a straight line. Using a ruler you may verify that $m = \Delta y/\Delta x = 2$. The intercept at $x = 1$ is 3. These readings agree with the constants in the equation $y = 3x^2$. In this example the y axis is scaled in two cycles

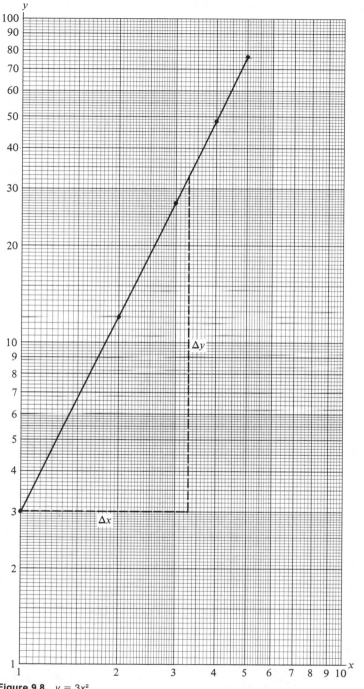

Figure 9.8 $y = 3x^2$

from log 1 = 0 to log 10 = 1 to log 100 = 2. Logarithmic paper may contain any number of cycles. A cycle ranges between any two numbers whose logarithms are consecutive integers.

EXAMPLE 2: Determine the equation relating the variables in the following table:

x	1.0	1.5	2.0	2.5	3.0	4.0	5.0	6.0	7.0	9.0
y	16	21	24	27	29	33	38	41	44	50

SOLUTION: Consider Figure 9.9. The vertical axis requires one cycle which ranges from log 10 = 1 to log 100 = 2. Since the graph is approximately a straight line, the equation relating the variables is a power function. Using a ruler, $m = \Delta y/\Delta x \approx 0.5$. The intercept at $x = 1$ is about 17. Thus, the power function is $y \approx 17x^{0.5}$ or $y \approx 17\sqrt{x}$.

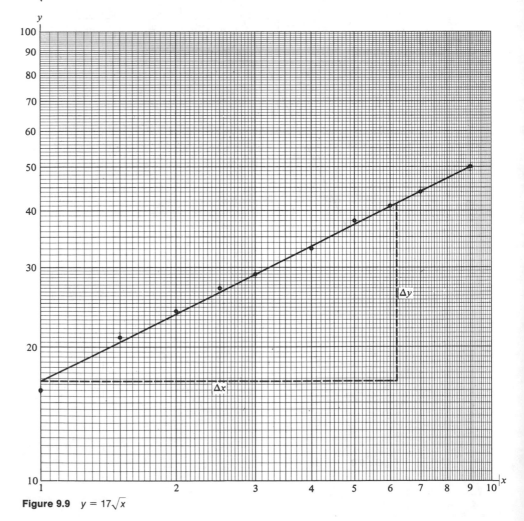

Figure 9.9 $y = 17\sqrt{x}$

On semilogarithmic graph paper one axis is scaled as on logarithmic paper and the other axis is scaled as on Cartesian coordinate paper. This paper is used when only one variable ranges over a large collection of values. Semilogarithmic paper is especially useful when graphing exponential functions since the graph is a straight line.

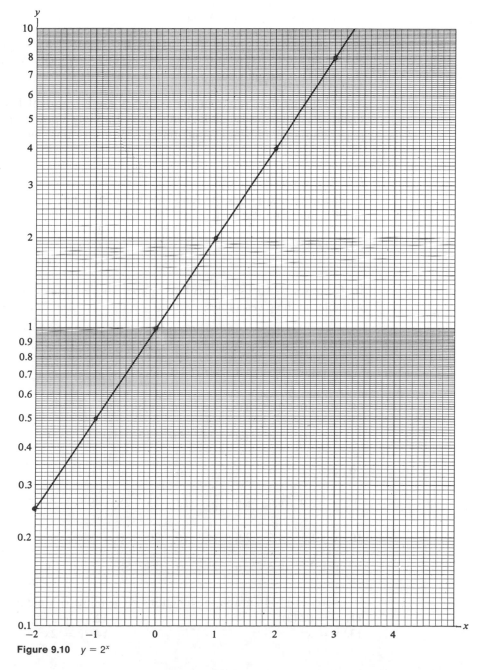

Figure 9.10 $y = 2^x$

EXAMPLE 3: Use semilogarithmic paper to graph $y = 2^x$.

SOLUTION: First, construct a table of correspondences:

x	-2	-1	0	1	2	3
y	0.25	0.5	1	2	4	8

The vertical axis requires two logarithmic cycles. The horizontal axis is equally spaced and is scaled with positive and negative numbers. As shown in Figure 9.10, the graph is a straight line.

Although the graphs in the examples are all straight lines, other curves may result. We obtain a straight line only when graphing a power function on logarithmic paper and an exponential function on semilogarithmic paper.

EXERCISES 9–8

In Exercises 1–6 graph the functions on logarithmic paper. In 1–4 find the slope and the intercept at $x = 1$.

1. $y = x^2$ **2.** $y = 2x^3$
3. $y = 4\sqrt{x}$ **4.** $xy = 1$
5. $y = 2^x$ **6.** $y = 2(3)^x$

In Exercises 7–10 use logarithmic paper to determine the equation relating the variables in the given tables.

7.

x	1	2	3	4	5	6	7
y	7.0	5.5	4.8	4.3	4.0	3.7	3.5

8.

x	1.0	2.0	3.0	4.0	5.0	6.0	7.0
y	5.0	1.4	.66	.39	.26	.19	.14

9.

x	1	3	5	7	9	25	30	70
y	3.0	6.0	8.6	11	13	25	28	49

10.

x	1.5	4.5	7.5	11	16	20	30
y	2.1	8.4	16	25	40	53	88

In Exercises 11–14 graph the functions on semilogarithmic paper.
11. $y = 0.5(3)^x$ **12.** $y = 2^{-x}$
13. $y = x^2$ **14.** $y = 2x^3$

Sample Test: Chapter 9

1. If $y = (\frac{1}{4})^x$, fill in the missing component of each of the following ordered pairs: $(3, \square)$ $(\square, 4)$ $(\square, 1)$ $(\frac{3}{2}, \square)$ $(-2, \square)$.

2. If $500 is invested at 6 percent compounded annually, find a formula for the value (y) of the investment at the end of t years.

3. $3000 is invested at 5 percent compounded continuously. How much is the investment worth in 4 years?

4. A radioactive substance decays from 10 g to 8 g in 11 days. Find the half-life of the substance.

5. If $f = (3, -1)(4, -2)$, find f^{-1}. Determine the domain and range of each function.

6. Determine if the function has an inverse function.
   a. $f = (3, 1)(0, 0)(1, 3)$
   b. The function graphed in the figure

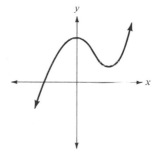

7. If $g(x) = \frac{1}{2}x - \frac{5}{2}$, find:
   a. $g^{-1}(x)$
   b. $g^{-1}(-1)$

8. Sketch on the same axes the graphs of $f(x) = (\frac{1}{2})^x$ and $g(x) = \log_{1/2} x$. What is the domain and range of each function?

9. a. Express in logarithmic form: $4^{-1} = \frac{1}{4}$.
   b. Express in exponential form: $\log_2 8 = 3$.

10. a. Solve for x: $\log_5 5^7 = x$.
    b. If $f(x) = \log_4 x$, find $f(1)$.

11. What is the domain of $f(x) = \ln(3x - 7)$?

12. a. Express as the sum or difference of simpler logarithms:

$$\log_b \sqrt{x^3/y}$$

   b. Express as a single logarithm with coefficient 1:

$$2 \log_b x + \frac{1}{3} \log_b y$$

13. Solve for x:

$$\log_{10}(3x - 2) - \log_{10}(x + 4) = 1$$

14. Use $\log_{10} 2 = 0.3010$ and $\log_{10} 3 = 0.4771$ to find $\log_{10} \sqrt[3]{6}$.

15. True or false; if false, give a specific counterexample.

    a. $\dfrac{\log_b x}{\log_b y} = \log_b x - \log_b y$

    b. $\log_b \dfrac{x}{y} = \log_b x - \log_b y$

    c. $\log_b \dfrac{x}{y} = \dfrac{\log_b x}{\log_b y}$

16. Solve for x:
    a. $\log_{10} 0.123 = x$
    b. $\log_{10} x = -2.4157$

17. Compute by means of logarithms: $\sqrt{78.4}$.

18. Use logarithms to solve the equation $x^{0.1} = 5.67$.

19. Use logarithms to base 10 to evaluate $\log_4 9$.

20. Use the table of natural logarithms to determine $\ln 13.2$.

Chapter 10

Trigonometric functions of real numbers

10−1 RADIANS

In discussing the trigonometric functions in Chapter 3 we measured angles in degrees. However, in many applications of these functions and in calculus a different angle measure, called a *radian*, is more useful. We define the radian measure of an angle by first placing the vertex of the angle at the center of a circle. Let s be the length of the intercepted arc and let r be the radius. Then

$$\theta = \frac{s}{r}$$

is the radian measure of the angle (see Figure 10.1). Equivalently, an angle measuring 1 radian intercepts an arc equal in length to the radius of the circle (see Figure 10.2).

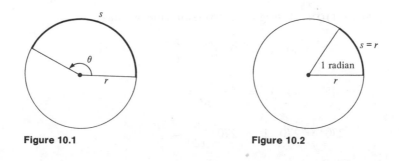

Figure 10.1   **Figure 10.2**

EXAMPLE 1: A central angle in a circle whose radius is 3 in. intercepts an arc of 12 in. Find the radian measure of the angle.

SOLUTION: Substituting $r = 3$ in. and $s = 12$ in. in the formula

$$\theta = \frac{s}{r}$$

we have

$$\theta = \frac{12 \text{ in.}}{3 \text{ in.}} = 4$$

Thus, the radian measure of the angle is 4. Note that r and s are measured in the same unit, which cancels in the ratio. Thus, the radian measure of an angle is a number without dimension. Although the word "radian" is often added, an angle measure with no units means radian measure.

We may find the relation between degrees and radians by considering the measure of an angle that makes one complete rotation. In degrees, the measure of the angle is 360. Since the circumference of a circle is $2\pi r$, the radian measure is

$$\theta = \frac{s}{r} = \frac{2\pi r}{r} = 2\pi$$

Thus

$$360° = 2\pi \text{ radians}$$

From this relation we derive the following conversion rules between degrees and radians:

$$1° = \frac{\pi}{180} \text{ radian} \approx 0.0175 \text{ radian}$$

and

$$1 \text{ radian} = \frac{180°}{\pi} \approx 57.3°$$

EXAMPLE 2: Express 30°, 45°, and 270° in terms of radians.

SOLUTION: Using the conversion rule, we have

$$30° = 30 \cdot 1° = 30 \cdot \frac{\pi}{180} = \frac{\pi}{6}$$

$$45° = 45 \cdot 1° = 45 \cdot \frac{\pi}{180} = \frac{\pi}{4}$$

$$270° = 270 \cdot 1° = 270 \cdot \frac{\pi}{180} = \frac{3\pi}{2}$$

EXAMPLE 3: Express $\pi/3$, π, and $7\pi/5$ radians in terms of degrees.

SOLUTION: Using the conversion rule, we have

$$\frac{\pi}{3} = \frac{\pi}{3} \cdot 1 = \frac{\pi}{3} \cdot \frac{180°}{\pi} = 60°$$

$$\pi = \pi \cdot 1 = \pi \cdot \frac{180°}{\pi} = 180°$$

$$\frac{7\pi}{5} = \frac{7\pi}{5} \cdot 1 = \frac{7\pi}{5} \cdot \frac{180°}{\pi} = 252°$$

To illustrate an application where radians are used over degrees, consider the problem of determining the area of the shaded section in Figure 10.3. If you do not remember from geometry, notice intuitively that the area (A) is proportional to the central angle (θ). That is,

$A = k\theta$ where k is a constant

We may find k since we know that $A = \pi r^2$ when the angle (θ) is a complete rotation of 2π radians. Thus,

$$\pi r^2 = k \cdot 2\pi$$
$$\tfrac{1}{2}r^2 = k$$

The formula (when θ is expressed in radians) is then

$$A = \tfrac{1}{2}r^2\theta$$

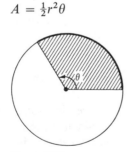

Figure 10.3

EXAMPLE 4: In a circle of radius 12 in., find the area of a sector whose central angle is 120°.

SOLUTION: To use the formula we must first convert the angle to radians.

$$120° = 120 \cdot \frac{\pi}{180} = \frac{2\pi}{3}$$

Now, substituting $r = 12$ and $\theta = 2\pi/3$ in the formula

$$A = \tfrac{1}{2}r^2\theta$$

we have

$$A = \frac{1}{2}(12)^2 \cdot \frac{2\pi}{3}$$

$$= 48\pi \text{ in.}^2$$

One of the most important applications of radians concerns linear and angular velocity. If a point P moves along the circumference of a circle at a constant speed, then the linear velocity, v, is given by the formula

$$v = \frac{s}{t}$$

where s is the distance traveled by the point (or the length of the arc transversed) and t is the time required to travel this distance. During the same time interval the radius of the circle connected to point P swings through θ angular units. Thus, the angular velocity, ω (Greek lowercase omega), is given by

$$\omega = \frac{\theta}{t}$$

We can relate linear and angular velocity through the familiar formula

$$s = \theta r$$

Dividing both sides of this equation by t, we obtain

$$\frac{s}{t} = \frac{\theta}{t} r$$

Substituting v for s/t and ω for θ/t yields

$$v = \omega r$$

Thus, the linear velocity is equal to the product of the angular velocity and the radius.

Since $s = \theta r$ is valid only when θ is measured in radians, the angular velocity ω must be expressed in radians per unit of time when using the formula $v = \omega r$. However, in many common applications the angular velocity is expressed in revolutions per minute. For these problems remember to convert ω to radians (rad) per unit of time through the relationship 1 rpm $= 2\pi$ rad/minute before using the formula.

EXAMPLE 5: A circular saw blade 12 in. in diameter rotates at 1600 rpm. Find:
a. the angular velocity in radians per second.
b. the linear velocity in inches per second at which the teeth would strike a piece of wood.

SOLUTION:
a. Since 1 rpm $= 2\pi$ rad/minute, we have

$$\omega = 1600 \text{ rpm} = 1600 \left(2\pi \, \frac{\text{rad}}{\text{minute}} \right) = 3200\pi \, \frac{\text{rad}}{\text{minute}}$$

To convert to rad/second we use 1 minute = 60 second, as follows:

$$\omega = 3200\pi \frac{\text{rad}}{\text{minute}} \frac{1 \text{ minute}}{60 \text{ seconds}} = \frac{160\pi}{3} \frac{\text{rad}}{\text{second}}$$

b. Since ω is expressed in radians per unit time, we find v by the formula $v = \omega r$, with $r = 6$ in.

$$v = \omega r$$

$$= \frac{160\pi}{3} \frac{\text{rad}}{\text{second}} \cdot 6 \text{ in.}$$

$$= 320\pi \frac{\text{in.}}{\text{second}}$$

Remember that the radian measure of an angle is a number without dimension, so the word "radian" does not appear in the units for linear velocity.

EXERCISES 10–1

In Exercises 1–8 complete the table by replacing each question mark with the appropriate number.

	The radius is	The intercepted arc is	The central angle is
1.	20 ft	100 ft	?
2.	8 in.	?	3
3.	?	56 yd	7
4.	5.2 meters	5.2 meters	?
5.	?	4.8 meters	$\frac{1}{2}$
6.	11 ft	?	3.2
7.	1 unit	5 units	?
8.	1 unit	π units	?

9. The radius of a wheel is 20 in. When the wheel moves 110 in., through how many radians does a point on the wheel turn? How many revolutions are made by the wheel?

10. The radius of a wheel is 16 in. Find the number of radians through which a point on the circumference turns when the wheel moves a distance of 2 ft. How many revolutions are made by the wheel?

11. Find the area of a sector of a circle whose central angle is $\pi/4$ and radius is 10 in.

12. Find the area of a sector of a circle whose central angle is 120° and radius is 5 in.

13. In a circle of radius 2 ft, the arc length of a sector is 8 ft. Find the area of the sector.

14. In a circle of radius 5 ft, the arc length of a sector is 3 ft. Find the area of the sector.

In Exercises 15–35 express each angle in radian measure.

15. 30°   **16.** 45°   **17.** 60°
18. 90°   **19.** 120°   **20.** 135°
21. 150°   **22.** 180°   **23.** 210°
24. 225°   **25.** 240°   **26.** 270°
27. 300°   **28.** 315°   **29.** 330°
30. 360°   **31.** 200°   **32.** 75°
33. 20°   **34.** 162°   **35.** 305°

In Exercises 36–47 express each angle in degree measure.

36. $\dfrac{\pi}{4}$ **37.** $\dfrac{\pi}{3}$ **38.** $\dfrac{\pi}{6}$

39. $\dfrac{2\pi}{9}$ **40.** $\dfrac{11\pi}{36}$ **41.** $\dfrac{2\pi}{3}$

42. $\dfrac{13\pi}{10}$ **43.** $\dfrac{7\pi}{9}$ **44.** $\dfrac{2\pi}{15}$

45. $\dfrac{12\pi}{5}$ **46.** $\dfrac{\pi}{12}$ **47.** $\dfrac{7\pi}{18}$

In Exercises 48–53 find, to the nearest degree, the number of degrees in each angle. (*Note:* Use $\pi \approx 3.14$.)

48. 1   **49.** 2   **50.** 6
51. 4   **52.** 3.5   **53.** 5.8

54. What is the linear velocity in in./second of an object that moves along the circumference of a circle of radius 12 in. with an angular velocity of 7 rad/second?

55. What is the angular velocity in rad/second of an object that moves along the circumference of a circle of radius 9 in. with a linear velocity of 54 in./second?

56. What is the angular velocity in rad/second of the minute hand of a clock?

57. A phonograph record 12 in. in diameter is being played at $33\frac{1}{3}$ rpm. Find:
    a. The angular velocity in rad/minute.
    b. The linear velocity in in./minute of a point on the circumference of the record.

58. A pulley 20 in. in diameter makes 400 rpm. Find:
    a. The angular velocity in rad/second.
    b. The speed in in./second of the belt that drives the pulley.
(*Note:* The speed of a point on the circumference of the pulley is the same as the speed of the belt.)

59. In a dynamo an armature 12 in. in diameter makes 1500 rpm. Find the linear velocity in in./second of the tip of the armature.

60. A car is traveling at 60 mi/hour (88 ft/second) on tires 30 in. in diameter. Find the angular velocity of the tires in rad/second.

61. Arcs of circles are used in many styles of molding that can be found in homes. Consider the figure shown here. If $AB = DE = \frac{1}{2}BC = \frac{1}{2}CD$ and $AB = \frac{1}{2}$ in., what is the length of the curved portion of the molding?

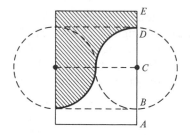

62. In architecture, arcs of circles are basic to the construction of many beautiful designs. For example, consider the quatrefoil in the figure. If the side of the square measures 4 in., determine:

   a. The perimeter.
   b. The area of the quatrefoil (the shaded portion of the figure).

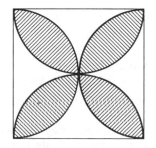

63. The formula $s = \theta r$ may be used to estimate certain distances when θ is a small angle (to about 10°). For example, consider the illustration. An observer measures the angle from earth to opposite ends of the sun's diameter to be 0.53°. Since θ is a small angle, the arc length s is a good approximation for the diameter d. If the distance from the earth to the sun is 93,000,000 mi, determine the diameter of the sun to two significant digits.

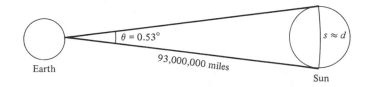

10–2 TRIGONOMETRIC FUNCTIONS OF REAL NUMBERS

Many natural phenomena repeat over definite periods of time. The waves broadcast by a radio station, the rhythmic motion of the heart, alternating electric current, weather-related issues such as air pollution, and the economic pattern of expansion, retrenchment, recession, and recovery—all these occur in cycles. The trigonometric functions, also being repetitive, are very useful in analyzing such periodic phenomena. However, the independent variable in these applications is the time and not angles.

Thus, we need to define the trigonometric functions in terms of real numbers and not degrees. Since radians measure angles in terms of real numbers, they are the starting point for our discussion of the trigonometry of real numbers.

Consider Figure 10.4, in which the central angle θ is measured in radians. It is convenient to label the radius of the circle as 1 unit so that $r = 1$. Then,

$$\theta = \frac{s}{r}$$

$$= \frac{s}{1}$$

$$\theta = s$$

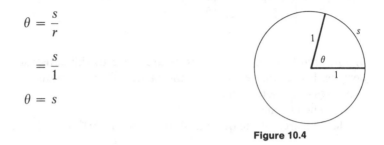

Figure 10.4

In a unit circle the same real number measures both the central angle θ, and the intercepted arc s. Thus, we may base the definitions of the trigonometric functions on either an angle or an arc length. The results are valid for both interpretations. Since angles are meaningless in periodic phenomena, it is useful to emphasize the interpretation of a real number as the measure of an arc length s. We liberally interpret arc length as the distance traveled by a point as it moves around the unit circle repeating its behavior every 2π units (why?).

To illustrate more forcefully the correspondence between real numbers and arc lengths, consider a circle of radius 1 unit with its center at the origin of the Cartesian coordinate system. The equation of this circle is $x^2 + y^2 = 1$. Through the point $(1, 0)$ and parallel to the y axis, we draw a real number line, labeled s. The zero point of s coincides with the point $(1, 0)$ on the circle. Units are marked off in the same scale as the y axis (see Figure 10.5).

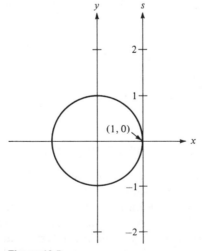

Figure 10.5

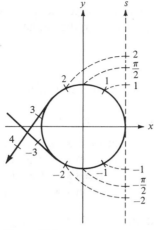

Figure 10.6

If the positive half of s is wrapped around the circle counterclockwise and the negative half of s is wrapped around the circle clockwise, we establish a one-to-one correspondence between real numbers and arc lengths of the circle (see Figure 10.6). Thus, each numbered point on s coincides with exactly one point on the circle and the length of an arc may be read from this curved s axis. We now relate this discussion to trigonometry.

> Consider a point (x, y) on the unit circle $x^2 + y^2 = 1$ at arc length s from $(1, 0)$. We define the cosine of s to be the x coordinate of the point and the sine of s to be the y coordinate.
>
> $$\cos s = x$$
> $$\sin s = y$$

Note in Figure 10.7 that our definitions are consistent with the definitions of sine and cosine in terms of the sides and hypotenuse of a right triangle. The domain of both functions is all the real numbers, since the arc length s is determined by wrapping

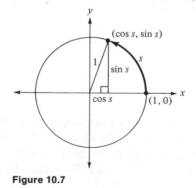

Figure 10.7

a real number line around the unit circle. The x and y coordinates in a unit circle vary between -1 and 1 (inclusive). Thus, for the range we have

$$-1 \leq \cos s \leq 1$$

$$-1 \leq \sin s \leq 1$$

The remaining trigonometric functions are defined as ratios of the sine and (or) cosine.

Name of Function	Abbreviation	Ratio
tangent of s	$\tan s$	$\sin s/\cos s = y/x \quad (x \neq 0)$
cotangent of s	$\cot s$	$\cos s/\sin s = x/y \quad (y \neq 0)$
secant of s	$\sec s$	$1/\cos s = 1/x \quad (x \neq 0)$
cosecant of s	$\csc s$	$1/\sin s = 1/y \quad (y \neq 0)$

Using the definitions we find the values of the trigonometric functions by determining the rectangular coordinates (x, y) of points on the unit circle. For certain real numbers these coordinates are easy to find. For example, let us determine the values of the trigonometric functions of zero. The coordinates for an arc length of zero are $(1, 0)$ (see Figure 10.8). Thus,

$$\sin 0 = y = 0 \quad \leftarrow\text{reciprocals}\rightarrow \quad \csc 0 = \frac{1}{\sin 0} = \frac{1}{0} \quad \text{undefined}$$

$$\cos 0 = x = 1 \quad \leftarrow\text{reciprocals}\rightarrow \quad \sec 0 = \frac{1}{\cos 0} = \frac{1}{1} = 1$$

$$\tan 0 = \frac{\sin 0}{\cos 0} = \frac{0}{1} = 0 \quad \leftarrow\text{reciprocals}\rightarrow \quad \cot 0 = \frac{\cos 0}{\sin 0} = \frac{1}{0} \quad \text{undefined}$$

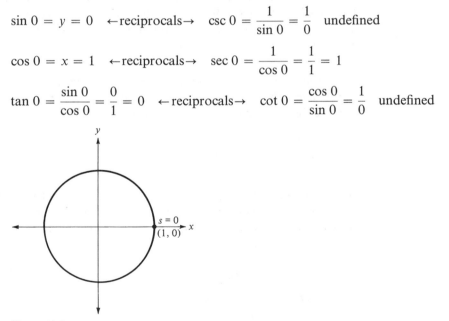

Figure 10.8

We can repeat this procedure for other arc lengths that terminate at one of the axes. Since the circumference of a circle of radius r is $2\pi r$, the circumference of a unit circle is 2π. The x or y axes may then intersect the unit circle at arc lengths of 0, $\pi/2$, π, and $3\pi/2$. Consider Figure 10.9, which illustrates the points of intersection for these arc lengths.

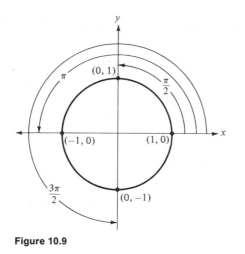

Figure 10.9

Using these coordinates we may determine the trigonometric values of these numbers; the results are summarized in the following table:

s	$\sin s$	$\csc s$	$\cos s$	$\sec s$	$\tan s$	$\cot s$
0	0	undefined	1	1	0	undefined
$\pi/2$	1	1	0	undefined	undefined	0
π	0	undefined	-1	-1	0	undefined
$3\pi/2$	-1	-1	0	undefined	undefined	0

Other numbers terminate at one of the axes, but their trigonometric values are the same as one of the four listed. For example, the trigonometric values of 2π are the same as the trigonometric values of 0 since both numbers are assigned the point $(1, 0)$ on the unit circle. The numbers 4π, 6π, -2π, and -4π are also assigned this point. *A basic fact in our development is that in laying off a length 2π, we pass around the circle and return to our original point.* Thus, the x and y coordinates repeat themselves at intervals of 2π, and for any trigonometric function f, we have

$$f(s + 2\pi k) = f(s) \quad \text{where} \quad k \text{ is an integer}$$

For example,

$$f(4\pi) = f(0 + 2\pi(2)) = f(0)$$
$$f(-2\pi) = f(0 + 2\pi(-1)) = f(0)$$

This observation is very important because it means that if we determine the values of the trigonometric functions in the interval $0 \le s < 2\pi$, we know their values for all real s.

EXAMPLE 1: Find $\sin 7\pi$, $\cos 12\pi$, $\tan(11\pi/2)$, $\cot(17\pi/2)$, $\sec(-5\pi)$, and $\csc(-5\pi/2)$.

SOLUTION: Since the values of the trigonometric functions repeat themselves at multiples of 2π, we have

$$\sin 7\pi = \sin(\pi + 6\pi) = \sin[\pi + 2\pi(3)] = \sin \pi = 0$$

$$\cos 12\pi = \cos(0 + 12\pi) = \cos[0 + 2\pi(6)] = \cos 0 = 1$$

$$\tan \frac{11\pi}{2} = \tan 5\tfrac{1}{2}\pi = \tan\left(\frac{3\pi}{2} + 4\pi\right) = \tan\left[\frac{3\pi}{2} + 2\pi(2)\right]$$

$$= \tan \frac{3\pi}{2} \quad \text{undefined}$$

$$\cot \frac{17\pi}{2} = \cot 8\tfrac{1}{2}\pi = \cot\left(\frac{\pi}{2} + 8\pi\right) = \cot\left[\frac{\pi}{2} + 2\pi(4)\right] = \cot \frac{\pi}{2} = 0$$

$$\sec(-5\pi) = \sec[\pi + (-6\pi)] = \sec[\pi + 2\pi(-3)] = \sec \pi = -1$$

$$\csc\left(\frac{-5\pi}{2}\right) = \csc(-2\tfrac{1}{2}\pi) = \csc\left[\frac{3\pi}{2} + (-4\pi)\right] = \csc\left[\frac{3\pi}{2} + 2\pi(-2)\right]$$

$$= \csc \frac{3\pi}{2} = -1$$

A trigonometric identity is a statement that is true for all real numbers for which the expressions are defined. We simplify our work by developing identities that relate a trigonometric function of a negative number to the same function of a positive number. The symmetry of the unit circle makes these identities easy to derive. Consider Figure 10.10, which illustrates the symmetry for two possible values of s. Note that the numbers s and $-s$ are assigned the same x coordinate, so

$$\cos(-s) = \cos s$$

The y coordinates differ only in their sign. Thus,

$$\sin(-s) = -\sin s$$

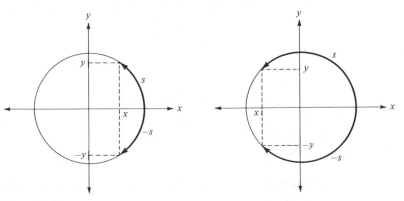

Figure 10.10

The remaining functions are ratios of the sine and cosine:

$$\tan(-s) = \frac{\sin(-s)}{\cos(-s)} = \frac{-\sin s}{\cos s} = -\tan s$$

$$\cot(-s) = -\cot s$$
$$\sec(-s) = \sec s$$
$$\csc(-s) = -\csc s$$

EXAMPLE 2: Find $\cos(-3\pi/2)$ and $\csc(-5\pi/2)$

SOLUTION:

$$\cos\left(\frac{-3\pi}{2}\right) = \cos\frac{3\pi}{2} = 0$$

$$\csc\left(\frac{-5\pi}{2}\right) = -\csc\frac{5\pi}{2} = -\csc 2\tfrac{1}{2}\pi = -\csc\left(\frac{\pi}{2} + 2\pi\right) = -\csc\frac{\pi}{2} = -1$$

The coordinates on the unit circle for $\pi/4$, $\pi/3$, and $\pi/6$ may be found geometrically. First, consider $s = \pi/4$. Since $\pi/4$ is halfway between 0 and $\pi/2$, $\pi/4$ is assigned the midpoint of the arc joining the points $(1, 0)$ and $(0, 1)$ on the unit circle (see Figure 10.11). Thus, the x and y coordinates are equal (that is, $x = y$). Since any point on the circle satisfies the equation $x^2 + y^2 = 1$, we have

$$x^2 + x^2 = 1$$

$$x^2 = \frac{1}{2}$$

$$x = \frac{1}{\sqrt{2}} \quad \text{or} \quad -\frac{1}{\sqrt{2}}$$

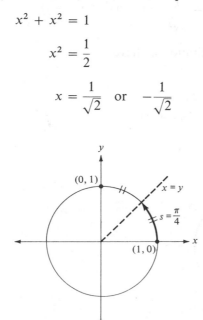

Figure 10.11

and

$$y = \frac{1}{\sqrt{2}} \quad \text{or} \quad -\frac{1}{\sqrt{2}}$$

Since x and y are positive in the first quadrant, we have

$$x = y = \frac{1}{\sqrt{2}}$$

The trigonometric values of $\pi/4$ are then

$$\sin\frac{\pi}{4} = y = \frac{1}{\sqrt{2}} \quad \leftarrow\text{reciprocals}\rightarrow \quad \csc\frac{\pi}{4} = \frac{1}{\sin(\pi/4)} = \sqrt{2}$$

$$\cos\frac{\pi}{4} = x = \frac{1}{\sqrt{2}} \quad \leftarrow\text{reciprocals}\rightarrow \quad \sec\frac{\pi}{4} = \frac{1}{\cos(\pi/4)} = \sqrt{2}$$

$$\tan\frac{\pi}{4} = \frac{\sin(\pi/4)}{\cos(\pi/4)} = \frac{1/\sqrt{2}}{1/\sqrt{2}} = 1 \quad \leftarrow\text{reciprocals}\rightarrow \quad \cot\frac{\pi}{4} = \frac{\cos(\pi/4)}{\sin(\pi/4)} = 1$$

Now consider $s = \pi/3$. Using the symmetry of the circle we can see that if $\pi/3$ is assigned the point (x, y), then $2\pi/3$ is assigned the point $(-x, y)$ (see Figure 10.12). Arcs AB and BC are both of length $\pi/3$. Since equal arcs in a circle subtend equal chords, the distance from $(1, 0)$ to (x, y) is the same as the distance from (x, y) to $(-x, y)$. Therefore, we have

$$\sqrt{(x - 1)^2 + (y - 0)^2} = \sqrt{(x - (-x))^2 + (y - y)^2}$$
$$x^2 + y^2 - 2x + 1 = 4x^2$$

Since $x^2 + y^2 = 1$, by substitution we have

$$(1) - 2x + 1 = 4x^2$$
$$0 = 4x^2 + 2x - 2$$
$$= 2(2x - 1)(x + 1)$$
$$x = \tfrac{1}{2} \quad \text{or} \quad x = -1$$

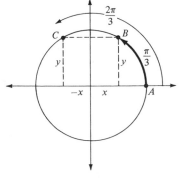

Figure 10.12

Because the point on the circle assigned to $\pi/3$ is in the first quadrant, we have

$$x = \tfrac{1}{2}$$

To find y we replace x by $\tfrac{1}{2}$ in the equation $x^2 + y^2 = 1$.

$$(\tfrac{1}{2})^2 + y^2 = 1$$

$$y^2 = \sqrt{\frac{3}{4}}$$

$$y = \frac{\pm\sqrt{3}}{2}$$

Since the point is in Q_1, we conclude that $y = \sqrt{3}/2$.

Using these coordinates we may determine the trigonometric values of $\pi/3$. In the following table we list these values along with the trigonometric values of $\pi/4$ and $\pi/6$. Exercise 31 asks you to determine the coordinates of the point on the circle from which the trigonometric values of $\pi/6$ were found. (*Note:* You may wish to refer back to the triangle derivations for the exact trigonometric values of 30°, 45°, and 60° in Section 3-5.)

s	$\sin s$	$\csc s$	$\cos s$	$\sec s$	$\tan s$	$\cot s$
$\dfrac{\pi}{3}$	$\dfrac{\sqrt{3}}{2}$	$\dfrac{2}{\sqrt{3}}$	$\dfrac{1}{2}$	2	$\sqrt{3}$	$\dfrac{1}{\sqrt{3}}$
$\dfrac{\pi}{4}$	$\dfrac{1}{\sqrt{2}}$	$\sqrt{2}$	$\dfrac{1}{\sqrt{2}}$	$\sqrt{2}$	1	1
$\dfrac{\pi}{6}$	$\dfrac{1}{2}$	2	$\dfrac{\sqrt{3}}{2}$	$\dfrac{2}{\sqrt{3}}$	$\dfrac{1}{\sqrt{3}}$	$\sqrt{3}$

EXAMPLE 3: Find $\sin(-\pi/3)$, $\cos(-\pi/4)$, $\sin(9\pi/4)$, and $\tan(13\pi/3)$.

SOLUTION: Using the table above and our previous procedure, we have

$$\sin\left(-\frac{\pi}{3}\right) = -\sin\frac{\pi}{3} = -\frac{\sqrt{3}}{2}$$

$$\cos\left(-\frac{\pi}{4}\right) = \cos\frac{\pi}{4} = \frac{1}{\sqrt{2}}$$

$$\sin\frac{9\pi}{4} = \sin 2\tfrac{1}{4}\pi = \sin\left(\frac{\pi}{4} + 2\pi\right) = \sin\frac{\pi}{4} = \frac{1}{\sqrt{2}}$$

$$\tan\frac{13\pi}{3} = \tan 4\tfrac{1}{3}\pi = \tan\left(\frac{\pi}{3} + 4\pi\right) = \tan\frac{\pi}{3} = \sqrt{3}$$

EXERCISES 10–2

In Exercises 1–30 find the function value.

1. $\cos 4\pi$

2. $\sin 100\pi$

3. $\sin 5\pi$

4. $\cos 11\pi$

5. $\tan(-2\pi)$

6. $\cot(-36\pi)$

7. $\cos\left(-\dfrac{\pi}{2}\right)$

8. $\sin\left(-\dfrac{3\pi}{2}\right)$

9. $\sec\dfrac{5\pi}{2}$

10. $\tan\dfrac{19\pi}{2}$

11. $\csc\left(\dfrac{-7\pi}{2}\right)$

12. $\cos\left(\dfrac{-9\pi}{2}\right)$

13. $\sin\left(\dfrac{11\pi}{2}\right)$

14. $\sin\left(\dfrac{-97\pi}{2}\right)$

15. $\cos\left(-\dfrac{\pi}{3}\right)$

16. $\sin\left(-\dfrac{\pi}{4}\right)$

17. $\tan\dfrac{9\pi}{4}$

18. $\cos\dfrac{9\pi}{4}$

19. $\sin\dfrac{13\pi}{6}$

20. $\sec\dfrac{17\pi}{4}$

21. $\cot\dfrac{7\pi}{3}$

22. $\sin\dfrac{19\pi}{3}$

23. $\cos\left(\dfrac{-13\pi}{6}\right)$

24. $\sin\left(\dfrac{-13\pi}{3}\right)$

25. $\csc\left(\dfrac{-31\pi}{3}\right)$

26. $\cos\left(\dfrac{-25\pi}{4}\right)$

27. $\cot\dfrac{55\pi}{3}$

28. $\tan\left(\dfrac{-73\pi}{6}\right)$

29. $\sin\left(\pi-\dfrac{4\pi}{3}\right)$

30. $\cos\left(2\pi-\dfrac{13\pi}{6}\right)$

31. Determine the coordinates of the point on the unit circle assigned to $\pi/6$. (*Hint:* The method is similar to the one given for $\pi/3$. The arc in the circle from $-\pi/6$ to $\pi/6$ is equal to the arc from $\pi/6$ to $\pi/2$.)

32. Consider the unit circle in the following figure in which each of the six trigonometric functions can be represented as a line segment. For example since $OC = 1$ in right triangle OAC we have $\sin\theta = \dfrac{AC}{OC} = \dfrac{AC}{1} = AC$. Notice we obtained the

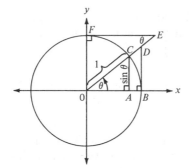

desired line segment by selecting a right triangle where the denominator in the defining ratio is 1. Determine the line segments representing the five remaining trigonometric functions in the figure.

10–3 EVALUATING TRIGONOMETRIC FUNCTIONS

A basic fact in our development has been that in laying off a length 2π, we pass around the unit circle and return to our original point. Thus, if we determine the values of the trigonometric functions in the interval $0 \leq s < 2\pi$, we know their values for all real s. Consider Figure 10.13, which illustrates how the symmetry of the circle may be used to further simplify the evaluation.

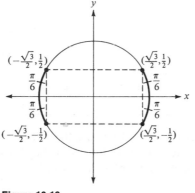

Figure 10.13

Note that the coordinates of the point assigned to $\pi/6$ differ only in sign with the coordinates assigned to $\pi - \pi/6 = 5\pi/6$, $\pi + \pi/6 = 7\pi/6$, and $2\pi - \pi/6 = 11\pi/6$. Thus, for a specific trigonometric function, say sine, we have

$$\sin \frac{\pi}{6} = \left| \sin \frac{5\pi}{6} \right| = \left| \sin \frac{7\pi}{6} \right| = \left| \sin \frac{11\pi}{6} \right| = \frac{1}{2}$$

The number $\pi/6$ is called the *reference number* for $5\pi/6$, $7\pi/6$, and $11\pi/6$. In general, we determine the reference number for s (denoted by s_R) by finding the shortest positive arc length between the point on the circle assigned to s and the x axis. The discussion above indicates that the trigonometric values of s and s_R are related as follows:

> Any trigonometric function of s is equal in absolute value to the same named function of its reference number s_R.

We determine the correct sign by considering the function definitions together with the sign of x and y in the four quadrants. The following table indicates the signs of the functions in the various quadrants:

Function	quadrant$_1$ (Q_1)	quadrant$_2$ (Q_2)	quadrant$_3$ (Q_3)	quadrant$_4$ (Q_4)
$\sin s = y$	$+$	$+$	$-$	$-$
$\csc s = \dfrac{1}{\sin s}$	$+$	$+$	$-$	$-$
$\cos s = x$	$+$	$-$	$-$	$+$
$\sec s = \dfrac{1}{\cos s}$	$+$	$-$	$-$	$+$
$\tan s = \dfrac{\sin s}{\cos s}$	$\dfrac{+}{+} = +$	$\dfrac{+}{-} = -$	$\dfrac{-}{-} = +$	$\dfrac{-}{+} = -$
$\cot s = \dfrac{1}{\tan s}$	$+$	$-$	$+$	$-$

You may wish to remember this table by using the following chart, which we developed in Section 3-1. The underlined first letters in the chart correspond to the first letters in the mnemonic "All students take chemistry."

$$
\left.\begin{array}{r}\underline{\text{sin }} s \\ \csc s \end{array}\right\} + \qquad \underline{\text{All}} \text{ the functions}
$$
$$
\text{others} \} - \qquad \text{are positive}
$$

$$
\left.\begin{array}{r}\underline{\text{tan }} s \\ \cot s \end{array}\right\} + \qquad \left.\begin{array}{r}\underline{\text{cos }} s \\ \sec s \end{array}\right\} +
$$
$$
\text{others} \} - \qquad \text{others} \} -
$$

EXAMPLE 1: Find $\sin(7\pi/4)$.

SOLUTION: First, determine the reference number.

$$s_R = 2\pi - \frac{7\pi}{4} = \frac{\pi}{4} \quad \text{(see Figure 10.14)}$$

Second, determine $\sin(\pi/4)$.

$$\sin \frac{\pi}{4} = \frac{1}{\sqrt{2}}$$

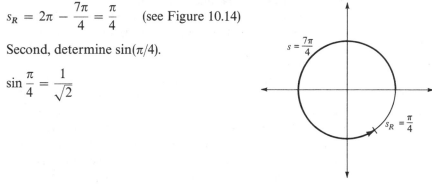

Figure 10.14

Third, determine the correct sign. The point assigned to $7\pi/4$ is in Q_4, where the value of the sine function is negative. Therefore,

$$\sin \frac{7\pi}{4} = -\frac{1}{\sqrt{2}}$$

EXAMPLE 2: Find $\cos(-10\pi/3)$.

SOLUTION: First, simplify the expression to the form $\cos s$, where $0 \le s < 2\pi$.

$$\cos\left(\frac{-10\pi}{3}\right) = \cos \frac{10\pi}{3} = \cos 3\tfrac{1}{3}\pi = \cos\left(\frac{4\pi}{3} + 2\pi\right) = \cos \frac{4\pi}{3}$$

Second, determine s_R.

$$s_R = \frac{4\pi}{3} - \pi = \frac{\pi}{3} \quad \text{(see Figure 10.15)}$$

Third, determine $\cos(\pi/3)$.

$$\cos \frac{\pi}{3} = \frac{1}{2}$$

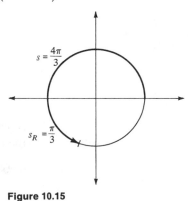

Figure 10.15

Fourth, determine the correct sign. The point assigned to $4\pi/3$ is in Q_3, where the value of the cosine function is negative. Therefore,

$$\cos\left(\frac{-10\pi}{3}\right) = -\frac{1}{2}$$

Until now we have been considering special numbers, in that their trigonometric values may be found *exactly* by geometric means. To approximate the trigonometric values of other numbers we use the same procedure, together with Table 5 in the Appendix. This table lists the trigonometric values for numbers between 0 and 1.57 (or about $\pi/2$). These values are computed through the aid of calculus, and it should be understood that they are *approximations* that are rounded off to four significant digits. In these problems we use 3.14 as an approximation for π.

At this time we also switch our notation. We defined the sine and cosine functions as $x = \cos s$ and $y = \sin s$, where x and y are the coordinates of the point on the unit circle at arc length s from $(1, 0)$. This notation becomes awkward because x and y usually denote the independent and dependent variables, respectively. In keeping with this custom we will from this point on use x, instead of s, to represent the independent variable, so x now represents an arc length. Similarly, we now use y as the dependent variable that represents the function values.

EXAMPLE 3: Approximate sin 2.

SOLUTION: First, determine the reference number.

$$x_R = \pi - 2 \approx 3.14 - 2 = 1.14 \qquad \text{(see Figure 10.16)}$$

Second, determine sin 1.14.

$$\sin 1.14 \approx 0.9086 \qquad \text{(Table 5)}$$

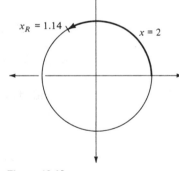

Figure 10.16

Third, determine the correct sign. The point assigned to 2 is in Q_2, where the value of the sine function is positive. Therefore,

$$\sin 2 \approx 0.9086$$

EXAMPLE 4: Approximate $\tan(-7.12)$.

SOLUTION: First, simplify the expression to the form

$$\tan x \quad \text{where} \quad 0 \le x < 6.28 \quad \text{(or } 0 \le x < 2\pi\text{)}$$

$$\tan(-7.12) = -\tan 7.12 = -\tan(0.84 + 6.28) = -\tan 0.84$$

Second, determine tan 0.84.

tan 0.84 ≈ 1.116   (Table 5)

Therefore,

tan(−7.12) = − tan 0.84 ≈ −1.116

EXAMPLE 5: Approximate sec(8π/5).

SOLUTION: First, determine the reference number.

$$x_R = 2\pi - \frac{8\pi}{5} = \frac{2\pi}{5} \approx \frac{2(3.14)}{5} = 1.26 \qquad \text{(see Figure 10.17)}$$

Second, determine sec 1.26.

sec 1.26 ≈ 3.270   (Table 5)

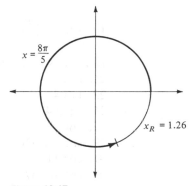

$x = \dfrac{8\pi}{5}$

$x_R = 1.26$

Figure 10.17

Third, determine the correct sign. The point assigned to 8π/5 is in Q_4, where the value of the secant function is positive. Therefore,

$$\sec \frac{8\pi}{5} \approx 3.270$$

EXERCISES 10–3
In Exercises 1–20 find the exact function value.

1. $\cos \dfrac{7\pi}{6}$

2. $\sin \dfrac{5\pi}{3}$

3. $\tan \dfrac{4\pi}{3}$

4. $\sec \dfrac{11\pi}{6}$

5. $\sin \dfrac{7\pi}{4}$

6. $\cos \dfrac{2\pi}{3}$

7. $\csc \dfrac{5\pi}{6}$

8. $\sin \dfrac{3\pi}{4}$

9. $\cos \dfrac{5\pi}{3}$

10. $\cot \dfrac{5\pi}{4}$

11. $\sin \left(\dfrac{-7\pi}{6} \right)$

12. $\cos \left(\dfrac{-4\pi}{3} \right)$

13. $\sec \left(\dfrac{-2\pi}{3} \right)$

14. $\tan \left(\dfrac{-7\pi}{4} \right)$

15. $\cos \dfrac{11\pi}{4}$

16. $\sin \dfrac{8\pi}{3}$

17. $\cot \dfrac{23\pi}{6}$

18. $\csc \dfrac{15\pi}{4}$

19. $\sin \left(\dfrac{-20\pi}{3} \right)$

20. $\cos \left(\dfrac{-23\pi}{6} \right)$

In Exercises 21–40 find the approximate function value.

21. $\cos 2$

22. $\sin 3$

23. $\tan 4$

24. $\sec 5$

25. $\sin 5.41$

26. $\csc 2.23$

27. $\cot 3.71$

28. $\cos 1.84$

29. $\sin(-6.07)$

30. $\tan(-1.69)$

31. $\cos 11.73$

32. $\cot 9.61$

33. $\csc(-7.57)$

34. $\sin(-10.25)$

35. $\sin \dfrac{2\pi}{5}$

36. $\tan \dfrac{9\pi}{5}$

37. $\cos \dfrac{3\pi}{7}$

38. $\sin \dfrac{4\pi}{9}$

39. $\sec \left(\dfrac{-7\pi}{8} \right)$

40. $\cos \left(\dfrac{-7\pi}{10} \right)$

41. The following formulas (derived from calculus) are used to compute the values of $\cos s$ and $\sin s$ in Table 5.

$$\cos s = 1 - \frac{s^2}{2!} + \frac{s^4}{4!} - \frac{s^6}{6!} + \cdots$$

$$\sin s = s - \frac{s^3}{3!} + \frac{s^5}{5!} - \frac{s^7}{7!} + \cdots$$

where $n! = 1 \cdot 2 \cdot 3 \cdots n$ (for example, $4! = 1 \cdot 2 \cdot 3 \cdot 4$). Use the first three terms of these formulas to approximate the following expressions. Compare your results with the values in Table 5.

    a. $\sin 1$

    b. $\cos 1$

    c. $\cos 0$

    d. $\sin 0.5$

10–4 GRAPHS OF SINE AND COSINE FUNCTIONS

A picture or graph of the sine and cosine functions helps us to understand their cyclic behavior. We use the Cartesian coordinate system and associate the arc length values with points on the horizontal or x axis. Remember that we are now using x, instead of s, to represent the independent variable, which is an arc length. The vertical axis, labeled the y axis, is used to represent the function values.

We begin by considering the values of the sine function as we lay off on the x axis a length 2π and pass once around the unit circle. The following table indicates some correspondences of the sine function at this typical interval:

$y = \sin x$ (for $0 \le x \le 2\pi$ at intervals of $\pi/6$)

x	0	$\dfrac{\pi}{6}$	$\dfrac{\pi}{3}$	$\dfrac{\pi}{2}$	$\dfrac{2\pi}{3}$	$\dfrac{5\pi}{6}$	π	$\dfrac{7\pi}{6}$	$\dfrac{4\pi}{3}$	$\dfrac{3\pi}{2}$	$\dfrac{5\pi}{3}$	$\dfrac{11\pi}{6}$	2π
y	0	0.5	0.87	1	0.87	0.5	0	-0.5	-0.87	-1	-0.87	-0.5	0

If we plot these points and join them with a smooth curve, we obtain the graph in Figure 10.18, which describes the essential characteristics of the sine function during one cycle. The plot of $y = \sin x$ starts at the origin, attains a maximum at one-fourth of the cycle length, returns to zero halfway through the cycle, attains a minimum at the three-quarter point, and returns to zero at the end of the cycle. Each time we lay off a length 2π we pass around the circle and repeat this behavior. Thus, the graph of $y = \sin x$ weaves continuously through cycles in both directions, as illustrated in Figure 10.19.

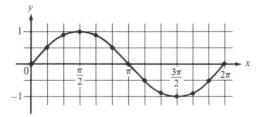

Figure 10.18

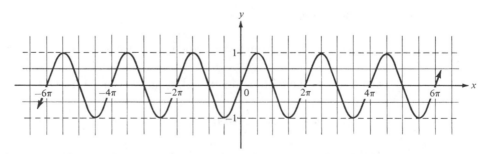

Figure 10.19

When we look at such a graph of the sine function, we see that the function repeats its values at regular intervals of 2π. For this reason the sine function is said to be *periodic*. We define a periodic function as follows:

PERIODIC FUNCTION: A function f is periodic if

$$f(x) = f(x + p)$$

for all x in the domain of f. The smallest positive number p for which this is true is called the *period* of the function.

This definition applies to the sine function, since

$$\sin x = \sin(x + 2\pi)$$

Thus, the sine function is periodic with period 2π.

We can use this information to sketch functions of the form $y = \sin bx$, where b represents a positive real number. These functions are similar to $y = \sin x$ in that they have the same basic shape; they may differ by having a different period. For example, to obtain the graph of $y = \sin 2x$, we multiply x by 2 and take the sine of the resulting number. This means that when $x = \pi$, we evaluate the sine of $2 \cdot \pi$ or $\sin 2\pi$. Thus, when we substitute numbers from 0 to π for x, we evaluate the function from $\sin 0$ to $\sin 2\pi$ and complete one cycle. The period for $y = \sin 2x$ is therefore π. In general, the period of $y = \sin bx$ ($b > 0$) is found by computing $2\pi/b$. Figure 10.20 compares the graphs of $y = \sin x$ and $y = \sin 2x$ for $0 \le x \le 2\pi$.

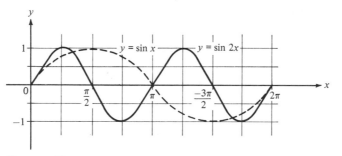

Figure 10.20

EXAMPLE 1: Sketch the graph of $y = \sin \frac{1}{2}x$ for $0 \le x \le 2\pi$.

SOLUTION: Determine the period by substituting $\frac{1}{2}$ for b

$$\text{Period} = \frac{2\pi}{b} = \frac{2\pi}{1/2} = 4\pi$$

If the curve completes one cycle in 4π, the curve completes one-half of a sine wave in 2π (see Figure 10.21).

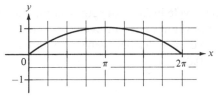

Figure 10.21

A second important characteristic of the graph of $y = \sin x$ is that the greatest y value is 1. When the graph of a periodic function is centered about the x axis, the maximum y value is called the *amplitude* of the function. Thus, the amplitude of $y = \sin x$ is 1.

We can use this information to sketch functions of the form $y = a \sin bx$, where a represents a real number. These functions are similar to $y = \sin bx$ in that they have the same basic shape and period; they may differ by having a different amplitude. For example, to sketch the graph of $y = 3 \sin x$, we obtain values for $\sin x$ and multiply these values by 3. Since the greatest y value that $\sin x$ attains is 1, the greatest y value that $3 \sin x$ attains is 3. Thus, the amplitude of $y = 3 \sin x$ is 3.

In general, since the greatest value that $\sin x$ attains is 1, the greatest value that $a \sin x$ attains is $|a|$. Thus, the amplitude of $y = a \sin bx$ is $|a|$. Figure 10.22 compares the graphs of $y = \sin x$ and $y = 3 \sin x$ for $0 \le x \le 2\pi$.

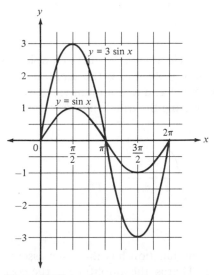

Figure 10.22

EXAMPLE 2: Sketch the graph of $y = 2 \sin 3x$ for $0 \le x \le 2\pi$.

SOLUTION: First determine the amplitude and period.

$$\text{Amplitude} = |a| = |2| = 2$$

$$\text{Period} = \frac{2\pi}{b} = \frac{2\pi}{3}$$

If the curve completes one cycle in $2\pi/3$, the curve completes three cycles in the given interval. Note that b gives the number of cycles in the interval $0 \le x \le 2\pi$. Since the amplitude is 2, the curve oscillates between a maximum value of 2 and a minimum value of -2 (see Figure 10.23).

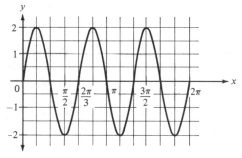

Figure 10.23

EXAMPLE 3: Sketch the graph of $y = -\sin \pi x$ for $0 \le x \le 2\pi$.

SOLUTION: First determine the amplitude and period.

$$\text{Amplitude} = |a| = |-1| = 1$$

$$\text{Period} = \frac{2\pi}{b} = \frac{2\pi}{\pi} = 2$$

If the curve completes one cycle in two, the curve completes slightly more than three cycles in the interval $0 \le x \le 2\pi$. Since the amplitude is 1, the curve oscillates between a maximum value of 1 and a minimum value of -1. Because a is negative, we obtain the graph by inverting the usual sine wave (see Figure 10.24).

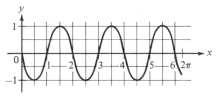

Figure 10.24

The graph of the cosine function has the same essential characteristics as the graph of the sine function. That is, the amplitude of the cosine function is 1 and the period is 2π. This is evidenced by the following table, which indicates some values of the cosine function for the interval from 0 to 2π.

$y = \cos x$ (for $0 \le x \le 2\pi$ at intervals of $\pi/6$)

x	0	$\dfrac{\pi}{6}$	$\dfrac{\pi}{3}$	$\dfrac{\pi}{2}$	$\dfrac{2\pi}{3}$	$\dfrac{5\pi}{6}$	π	$\dfrac{7\pi}{6}$	$\dfrac{4\pi}{3}$	$\dfrac{3\pi}{2}$	$\dfrac{5\pi}{3}$	$\dfrac{11\pi}{6}$	2π
y	1	0.87	0.5	0	-0.5	-0.87	-1	-0.87	-0.5	0	0.5	0.87	1

If we plot these points and join them with a smooth curve, we obtain the graph shown in Figure 10.25. This graph demonstrates that the cosine function completes one cycle in 2π and attains a maximum value of 1. Like that of the sine function, this graph can be reproduced indefinitely in both directions to obtain as much of the graph of the cosine function as desired (see Figure 10.26).

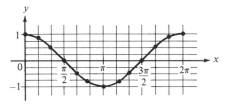

Figure 10.25

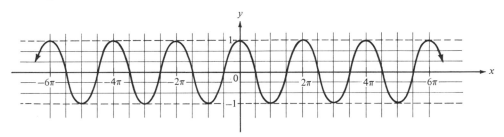

Figure 10.26

The great similarity between the graph of the sine and cosine functions should be apparent. In fact, if we move the vertical axis $\pi/2$ to the left in the graph of the cosine function so that the vertical axis crosses at $-\pi/2$, the resulting graph is the sine function. Thus, the only difference between the two graphs is that one curve leads the other by $\pi/2$. That is, $\cos(x - \pi/2) = \sin x$.

We graph cosine functions of the form $y = a \cos bx$ in a manner similar to $y = a \sin bx$; that is, we find the amplitude by computing $|a|$ and the period by evaluating $2\pi/b$. The difference is that the graph of $y = a \cos bx$ resembles the basic cosine wave, which attains a maximum or minimum height at $x = 0$. A common error in these problems is to draw the graph of a sine wave, which has a height of 0 when $x = 0$.

EXAMPLE 4: Sketch the graph of $y = \frac{1}{2} \cos 2x$ for $0 \le x \le 2\pi$.

SOLUTION: First, determine the amplitude and period.

$$\text{Amplitude} = |a| = \left|\frac{1}{2}\right| = \frac{1}{2}$$

$$\text{Period} = \frac{2\pi}{b} = \frac{2\pi}{2} = \pi$$

If the curve completes one cycle in π, the curve will complete two cycles in the given interval. Since the amplitude is $\frac{1}{2}$, the curve oscillates between a maximum value of $\frac{1}{2}$ and a minimum value of $-\frac{1}{2}$, as shown in Figure 10.27.

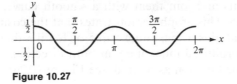

Figure 10.27

EXAMPLE 5: Sketch one cycle of the graph of $y = -4 \cos 10x$.

SOLUTION: First, determine the amplitude and period.

$$\text{Amplitude} = |a| = |-4| = 4$$

$$\text{Period} = \frac{2\pi}{b} = \frac{2\pi}{10} = \frac{\pi}{5}.$$

The curve completes one cycle in $\pi/5$ and attains a maximum value of 4 and a mini-mum value of -4. Since a is a negative number, we obtain the graph shown in Figure 10.28 by inverting the usual cosine wave and starting at the minimum point.

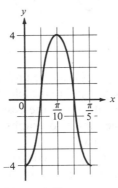

Figure 10.28

Finally, let us determine the effect of the constant c in a function of the form $y = a \sin(bx + c)$. Consider $y = \sin(x - \pi/2)$. As $x - \pi/2$ ranges from 0 to 2π, the curve completes one sine wave.

$$x - \frac{\pi}{2} = 0 \quad \text{when} \quad x = \frac{\pi}{2}$$

$$x - \frac{\pi}{2} = 2\pi \quad \text{when} \quad x = \frac{5\pi}{2}$$

Thus, the function completes one cycle in the interval from $\pi/2$ to $5\pi/2$. The period of the function is 2π and the amplitude is 1. For comparison, one cycle of the graphs of $y = \sin x$ and $y = \sin(x - \pi/2)$ is given in Figure 10.29.

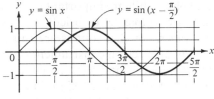

Figure 10.29

The complete graph of $y = \sin(x - \pi/2)$ is sketched by continuing the pattern in Figure 10.29 to the left and to the right. Notice that the graph of $y = \sin(x - \pi/2)$ may be obtained by shifting the graph of $y = \sin x$ to the right $\pi/2$ units. The sine wave then starts a cycle at $\pi/2$ instead of 0 and we call $\pi/2$ the *phase shift*.

In general, the constant c in the function $y = a \sin(bx + c)$ causes a shift of the graph of $y = a \sin bx$. The shift is of distance $|c/b|$ and is to the left if $c > 0$ and to the right if $c < 0$. The phase shift is given by $-c/b$. Similar remarks hold for functions of the form $y = a \cos(bx + c)$.

EXAMPLE 6: Graph one cycle of the function $y = 3 \cos(2x + \pi/2)$. Indicate the amplitude, period, and phase shift.

SOLUTION: Determine the amplitude and period.

$$\text{Amplitude} = |a| = |3| = 3$$

$$\text{Period} = \frac{2\pi}{b} = \frac{2\pi}{2} = \pi$$

The function completes one cosine cycle as $2x + \pi/2$ varies from 0 to 2π.

$$2x + \frac{\pi}{2} = 0 \quad \text{when} \quad x = -\frac{\pi}{4}$$

$$2x + \frac{\pi}{2} = 2\pi \quad \text{when} \quad x = \frac{3\pi}{4}$$

Thus, the function completes one cycle in the interval from $-\pi/4$ to $3\pi/4$ (see Figure 10.30). This interval checks with the computed period since $3\pi/4 - (-\pi/4) = \pi$. A cycle starts at $-\pi/4$, so $-\pi/4$ is the phase shift. We may verify the phase shift since

$$\frac{-c}{b} = \frac{-\pi/2}{2} = -\frac{\pi}{4}$$

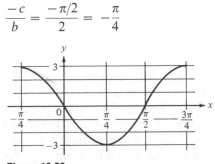

Figure 10.30

Amplitude, period, and phase shift are important considerations when analyzing any periodic phenomena. For example, let us briefly discuss two familiar concepts: musical sounds and radio waves. Musical sounds are caused by regular vibrations that have a definite period. On an electronic instrument called an *oscilloscope*, which changes sounds to electrical impulses and then to light waves, the sound from a tuning fork has the shape illustrated in Figure 10.31a. The period of the wave depends upon the pitch of the sound. With higher notes the pitch or frequency of the sound increases, producing a wave that has a smaller period (Figure 10.31b). Man can detect frequencies between about 50 and 15,000 vibrations (cycles) per second. However, some animals, such as the bat, can hear frequencies as high as 120,000 hertz (cycles per second). The amplitude of the wave depends upon the intensity of the sound. Since man hears best at a frequency of about 3500 hertz, the loudness of a sound depends upon both intensity and frequency. Although most musical sounds are very complex (such as the sound produced by the piano in Figure 10.31c), the French mathematician Joseph Fourier, in about 1800, showed that *any periodic function is the sum of simple sine functions.* Thus, all these sounds can be graphed and analyzed by some combination of sine waves.

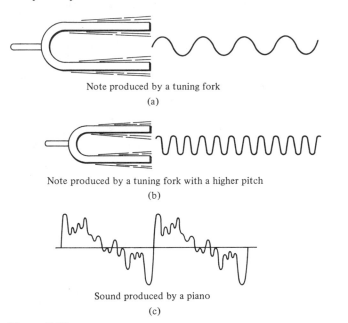

Note produced by a tuning fork

(a)

Note produced by a tuning fork with a higher pitch

(b)

Sound produced by a piano

(c)

Figure 10.31

Radio waves, used to transmit information at the speed of light (186,000 mi/second), are produced by oscillations in an electric current. Each current cycle produces a single radio wave. Special electronic equipment is needed to produce the alternating current since ordinary stations broadcast between 500,000 and 1,500,000 radio waves per second. Each station is licensed to broadcast a fixed number of radio waves per second, called its *frequency.* You receive their program by adjusting or tuning your set to accept this frequency.

Information is imposed on a radio wave as follows: A carrier wave (Figure 10.32a) is produced at a transmitting station which makes the licensed number of cycles per second. This carrier wave is then modulated by the program current from the broadcasting site. In amplitude-modulated (AM) broadcasting, the amplitude of the carrier is made to vary according to the message; the wavelength remains constant (see Figure 10.32b). With frequency modulation (FM), the amplitude of the wave remains constant and the wavelength varies (see Figure 10.32c).

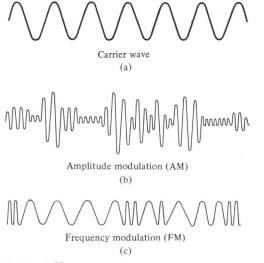

Carrier wave
(a)

Amplitude modulation (AM)
(b)

Frequency modulation (FM)
(c)

Figure 10.32

FM broadcasting is superior to AM in that it produces better fidelity of sound and is relatively free from static and interference. However, AM stations are more numerous since they are less expensive and have a greater broadcasting range. In television, an AM signal is used to transmit the picture but an FM signal carries the sound. In radio, television, and other similar forms of communication, it is important to remember that the sound or picture is not transmitted directly. Rather, the information is changed to electrical impulses, broadcast as radio waves, and then converted and reproduced in the receiving set.

EXERCISES 10–4

In Exercises 1–10 state the amplitude and the period, and sketch the curve over the interval $-p \le x \le p$, where p is the period of the function.

1. $y = 2 \sin x$ **2.** $y = 3 \cos 2x$

3. $y = -3 \cos x$ **4.** $y = -4 \sin \frac{1}{3}x$

5. $y = 2 \sin 3x$ **6.** $y = \frac{1}{2} \cos 4x$

7. $y = -\cos 18x$ **8.** $y = -6 \sin \frac{x}{4}$

9. $y = 10 \cos \pi x$ **10.** $y = 110 \sin 120\pi x$

In Exercises 11–20 state the amplitude and the period, and sketch the curve over the interval $0 \leq x \leq 2\pi$.

11. $y = 3 \cos 4x$

12. $y = 2 \cos \dfrac{x}{4}$

13. $y = -\sin \dfrac{x}{2}$

14. $y = -3 \sin 2x$

15. $y = \frac{1}{2} \sin 3x$

16. $y = 1.5 \cos \frac{1}{3}x$

17. $y = \sin \dfrac{\pi}{2} x$

18. $y = 2 \cos \pi x$

19. $y = -2 \cos \dfrac{\pi}{3} x$

20. $y = -0.9 \sin \dfrac{\pi}{4} x$

In Exercises 21–30 state the amplitude, the period, and the phase shift, and sketch one cycle of the function.

21. $y = \sin\left(x + \dfrac{\pi}{2}\right)$

22. $y = 2 \sin(x - \pi)$

23. $y = \cos\left(x - \dfrac{\pi}{4}\right)$

24. $y = 3 \cos\left(x + \dfrac{\pi}{3}\right)$

25. $y = \frac{1}{2} \cos\left(2x + \dfrac{\pi}{4}\right)$

26. $y = -\sin\left(\dfrac{x}{2} - \pi\right)$

27. $y = -\cos\left(\dfrac{x}{4} + \dfrac{\pi}{2}\right)$

28. $y = \sin(x - 1)$

29. $y = \sin(\pi x - \pi)$

30. $y = 1.2 \cos\left(2\pi x - \dfrac{\pi}{2}\right)$

In Exercises 31–40 find an equation for the curves with the given single cycle. The equations will be of the form $y = a \sin bx$ or $y = a \cos bx$.

31.

32.

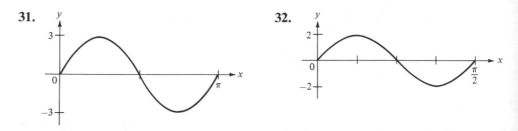

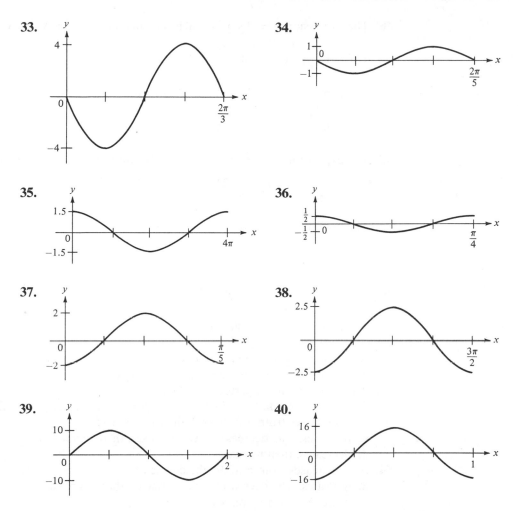

Exercises 41–59 are multiple choice questions. Answer each question by indicating the letter that precedes the expression that best completes the statement.

41. The amplitude of the function $y = 4 \sin 3x$ is
   a. 1,   **b.** 12,   **c.** 3,   **d.** 4

42. The maximum value of $y = 3 \cos \frac{1}{2}x$ is
   a. 1,   **b.** $\frac{1}{2}$,   **c.** 3,   **d.** $\frac{3}{2}$

43. The minimum value of $y = 2 \cos 3x$ is
   a. 0,   **b.** -3,   **c.** 3,   **d.** -2

44. The maximum value of $y = 2 + \sin x$ is
   a. 1,   **b.** 2,   **c.** 3,   **d.** 4

45. The expression $y = 3 \sin \frac{1}{2}x$ reaches its maximum value when x equals:

   a. 0,   **b.** $\dfrac{\pi}{2}$,   **c.** π,   **d.** $\dfrac{3\pi}{2}$

46. The expression $y = 2 \sin 3x$ reaches its maximum value when x equals:

    a. 0,   **b.** $\dfrac{\pi}{6}$,   **c.** $\dfrac{\pi}{4}$,   **d.** $\dfrac{2\pi}{3}$

47. The expression $y = 3 \cos \frac{1}{2}x$ reaches its minimum value when x equals:

    a. 0,   **b.** $\dfrac{\pi}{2}$,   **c.** π,   **d.** 2π

48. The expression $y = 2 \cos 3x$ reaches its minimum value when x equals:

    a. 0,   **b.** $\dfrac{\pi}{3}$,   **c.** $\dfrac{\pi}{2}$,   **d.** $\dfrac{2\pi}{3}$

49. The period of the curve $y = 2 \sin x$ is:

    a. 0,   **b.** $\dfrac{\pi}{2}$,   **c.** π,   **d.** 2π

50. The period of the curve $y = 3 \cos 2x$ is:

    a. $\dfrac{\pi}{2}$,   **b.** π,   **c.** 2π,   **d.** 4π

51. A function having the period π is:
    a. $y = 2 \cos x$,   **b.** $y = \cos \frac{1}{2}x$,   **c.** $y = \frac{1}{2} \cos x$,   **d.** $y = \cos 2x$

52. A function having a period of $\pi/2$ is:
    a. $y = \frac{1}{4} \sin x$,   **b.** $y = 4 \sin 2x$,   **c.** $y = 2 \sin 4x$,   **d.** $y = 4 \sin \frac{1}{4}x$

53. As x increases from $\pi/2$ to $3\pi/2$, then $y = \sin x$:
    a. increases,   **b.** decreases,   **c.** increases, then decreases,
    d. decreases, then increases

54. As x increases from π to 2π, the cosine of x:
    a. increases,   **b.** decreases,   **c.** increases, then decreases,
    d. decreases, then increases

55. As x increases in the interval from $\pi/2$ to $3\pi/2$, the value of $y = \cos x$ will:
    a. increase,   **b.** decrease,   **c.** increase, then decrease,
    d. decrease, then increase

56. As x increases from $-\pi/2$ to 0, $\sin x$:
    a. increases from 0 to 1,   **b.** decreases from 0 to -1,   **c.** increases from -1 to 0,
    d. decreases from 1 to 0

57. $y = \sin x$ and $y = \cos x$ both increase in:
    a. Quadrant 1,   **b.** Quadrant 2,   **c.** Quadrant 3,   **d.** Quadrant 4

58. Between $x = 0$ and $x = 2\pi$ the graphs of $y = \sin x$ and $y = \cos x$ have in common:
    a. no points,   **b.** one point,   **c.** two points,   **d.** four points

59. When graphs of $y = \sin x$ and $y = \cos x$ are drawn on the same axes, how many times do they intersect in the interval $\pi/2 \le x \le \pi$?
    a. 0,   **b.** 1,   **c.** 2,   **d.** 3

60. For any periodic function the amplitude is defined as $(M - m)/2$, where M is the maximum function value and m is the minimum function value. Use this definition to do the following problems.

    a. Show the amplitude of $y = \sin x$ is 1.

    b. Show the amplitude of $y = a \sin x$ is $|a|$.

    c. Show the amplitude of $y = 3 + \sin x$ is 1.

    d. What are the amplitude and period of the function shown in the figure?

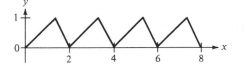

10-5 GRAPHS OF THE OTHER TRIGONOMETRIC FUNCTIONS

Although the tangent function is periodic, its behavior differs dramatically from that of the sine and cosine. To see this difference, compare the graph of $y = \tan x$ to the smooth weaving curves of these functions. We begin by constructing the following table, which lists some correspondences of the tangent function in the interval $0 \leq x \leq 2\pi$:

$y = \tan x = \sin x/\cos x$ (for $0 \leq x \leq 2\pi$ at intervals of $\pi/6$)

x	0	$\dfrac{\pi}{6}$	$\dfrac{\pi}{3}$	$\dfrac{\pi}{2}$	$\dfrac{2\pi}{3}$	$\dfrac{5\pi}{6}$	π	$\dfrac{7\pi}{6}$	$\dfrac{4\pi}{3}$	$\dfrac{3\pi}{2}$	$\dfrac{5\pi}{3}$	$\dfrac{11\pi}{6}$	2π
y	0	0.6	1.7	und.	-1.7	-0.6	0	0.6	1.7	und.	-1.7	-0.6	0

Unlike the sine and cosine, the tangent function is not defined for all real numbers. That is, $y = \tan x = \sin x/\cos x$ is undefined when $\cos x = 0$. Thus, we must exclude from the domain of this function $\pi/2$, $3\pi/2$, and any x for which $x = (\pi/2) + k\pi$ (k any integer). As x approaches $\pi/2$, $\sin x$ approaches 1 and $\cos x$ approaches 0. This means that $\tan x$ becomes very large as x gets close to $\pi/2$. Consider Table 5 in the Appendix, which verifies this observation. On the basis of the table of correspondences and the preceding discussion, the graph of $y = \tan x$ is presented in Figure 10.33.

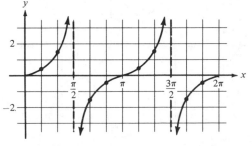

Figure 10.33

Using this graph we can see that the tangent function differs from the sine and cosine in the following important respects:

1. As was mentioned, the domain of the sine and cosine is all the real numbers. The tangent function excludes $x = (\pi/2) + k\pi$ (k any integer).
2. The values of the sine and cosine vary between -1 and 1. The range of the tangent function is all the real numbers.
3. The sine and cosine are periodic with period 2π. Careful consideration of Figure 10.33 shows that the tangent function repeats its values after π units. Thus, the period of the tangent function is π.

From this discussion, the value of graphs in analyzing the behavior of various functions should be clear.

The other three trigonometric functions may be graphed as the reciprocals of the sine, cosine, or tangent. That is,

$$\csc x = \frac{1}{\sin x} \qquad \sec x = \frac{1}{\cos x} \qquad \tan x = \frac{1}{\cot x}$$

In Figures 10.34–10.36 we first graph the sine, cosine, and tangent as light curves. By obtaining the reciprocals of various y values, we then graph the cosecant, secant, and cotangent. Note the following about a function and its reciprocal function:

1. As one function increases, the other decreases, and vice versa.
2. The two functions always have the same sign.
3. When one function is zero, the other is undefined.
4. When the value of the function is 1 or -1, the reciprocal function has the same value.

The dashed lines in these figures are called asymptotes (see Section 8-11).

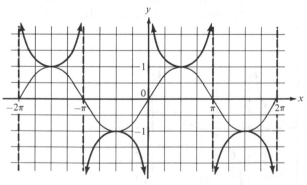

Figure 10.34 $y = \csc x$

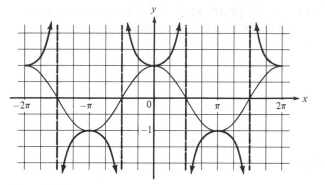

Figure 10.35 $y = \sec x$

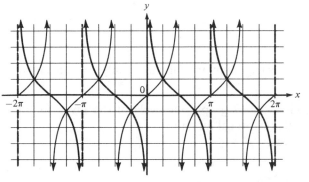

Figure 10.36 $y = \cot x$

EXERCISES 10–5

In Exercises 1–3 complete the table for the function and then sketch the curve from these points for the interval $0 \le x \le 2\pi$.

x	0	$\dfrac{\pi}{6}$	$\dfrac{\pi}{3}$	$\dfrac{\pi}{2}$	$\dfrac{2\pi}{3}$	$\dfrac{5\pi}{6}$	π	$\dfrac{7\pi}{6}$	$\dfrac{4\pi}{3}$	$\dfrac{3\pi}{2}$	$\dfrac{5\pi}{3}$	$\dfrac{11\pi}{6}$	2π
y													

1. $y = \cot x$
2. $y = \sec x$
3. $y = \csc x$

In Exercises 4–6 use Figures 10.34–10.36 to determine the domain, range, and period of the function.

4. $y = \cot x$
5. $y = \sec x$
6. $y = \csc x$

In Exercises 7–12 complete the table by determining if the function is increasing or decreasing in the interval. Use graphs.

	$0 < x < \dfrac{\pi}{2}$	$\dfrac{\pi}{2} < x < \pi$	$\pi < x < \dfrac{3\pi}{2}$	$\dfrac{3\pi}{2} < x < 2\pi$
7. $\sin x$				
8. $\cos x$				
9. $\tan x$				
10. $\cot x$				
11. $\sec x$				
12. $\csc x$				

13. Of the six trigonometric functions, only $y = \sin x$ and $y = \cos x$ have amplitudes. Why?

In Exercises 14–17 graph one cycle of the given functions by first sketching the appropriate reciprocal function.

14. $y = \sec 2x$

15. $y = 3 \csc \dfrac{x}{2}$

16. $y = 2 \csc \pi x$

17. $y = -\sec \left(x + \dfrac{\pi}{4} \right)$

10–6 TRIGONOMETRIC EQUATIONS

In solving a trigonometric equation we are looking for *all* the values of x that satisfy the given equation. Before we attempt to find a procedure for writing all the solutions to a particular equation, let us first establish a method for finding solutions between 0 and 2π. (*Note:* A review of Section 10-3 may be helpful.)

EXAMPLE 1: Find the exact values of x ($0 \le x < 2\pi$) for which the equation $\sin x = -\frac{1}{2}$ is a true statement.

SOLUTION: First, determine the quadrant that contains the point assigned to x.

$\sin x = -\frac{1}{2}$ which is a negative number

The point assigned to x could be in either Q_3 or Q_4 since the sine function is negative in both quadrants.

Second, determine the reference number (x_R).

$$\sin \frac{\pi}{6} = \tfrac{1}{2} \qquad \text{(from Section 10-2)}$$

Therefore, the reference number is $\pi/6$. (*Note:* When the reference number is $\pi/3$, $\pi/4$, or $\pi/6$, we may read exact values from the table in Section 10-2. Otherwise, we need Table 5 in the Appendix.)

Third, determine the appropriate values of x (Figure 10.37).

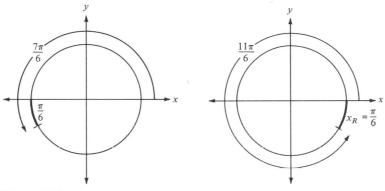

Figure 10.37

$\pi + \pi/6 = 7\pi/6$ is the number in Q_3 with a reference number of $\pi/6$

$2\pi - \pi/6 = 11\pi/6$ is the number in Q_4 with a reference number of $\pi/6$

Therefore, $7\pi/6$ and $11\pi/6$ make the equation a true statement.

EXAMPLE 2: Solve the equation $2 \sin x \cos x - \sin x = 0$ for $0 \leq x < 2\pi$.

SOLUTION:

$$2 \sin x \cos x - \sin x = 0$$

$$\sin x(2 \cos x - 1) = 0$$

Note that we have found two factors whose product is zero. Hence, the original equation will be satisfied whenever either factor is zero, and we treat each factor separately from this point on.

$$\sin x = 0$$

The sine function is 0 when $x = 0$ and $x = \pi$.

$$2 \cos x - 1 = 0$$
$$\cos x = \tfrac{1}{2}$$

1. Since $\cos x$ is positive, the point assigned to x could be in Q_1 or Q_4.

2. Since $\cos \pi/3 = \tfrac{1}{2}$, the reference number is $\pi/3$.

3. $\pi/3$ is the number in Q_1 with a reference number of $\pi/3$ (Figure 10.38).

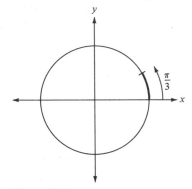

Figure 10.38

$2\pi - \pi/3 = 5\pi/3$ is the number in Q_4 with a reference number of $\pi/3$. (Figure 10.39).

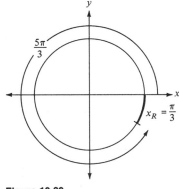

Figure 10.39

Thus, 0, $\pi/3$, π, and $5\pi/3$ are the solutions of the equation $(0 \leq x < 2\pi)$.

Once we have the solutions of a trigonometric equation that are between 0 and 2π, we can determine all the solutions, since the trigonometric functions are periodic. In laying off a length 2π, we pass around the unit circle and return to our original point. Thus, we generate all the solutions to an equation by adding multiples of 2π to the solutions that are in the interval $0 \leq x < 2\pi$.

EXAMPLE 3: Approximate all the solutions to the equation $4 \cos x + 1 = 0$.

SOLUTION: First, solve the equation for $\cos x$.

$4 \cos x + 1 = 0$

$\cos x = -\frac{1}{4} = -0.2500$

Second, determine the quadrant that contains the point assigned to x.

$\cos x = -0.2500$ which is a negative number

The point assigned to x could be in either Q_2 or Q_3 since the cosine function is negative in both quadrants.

Third, determine the reference number.

$\cos 1.32 \approx 0.2500$ (from Table 5)

Therefore, the reference number is 1.32.

Fourth, determine the appropriate values of x (Figure 10.40).

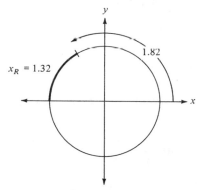 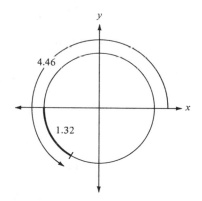

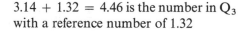

Figure 10.40

3.14 − 1.32 = 1.82 is the number in Q_2 with a reference number of 1.32

3.14 + 1.32 = 4.46 is the number in Q_3 with a reference number of 1.32

Thus,

$1.82 + k2\pi$ $4.46 + k2\pi$

where k is an integer, generates all the solutions to the equation.

EXAMPLE 4: Find the values of x $(0 \leq x < 2\pi)$ for which the equation $\sin 3x = 1/\sqrt{2}$ is a true statement.

SOLUTION: Solve the equation for $3x$.
(1) Since $\sin 3x$ is positive the point assigned to $3x$ could be in Q_1 or Q_2.
(2) Since $\sin \pi/4 = 1/\sqrt{2}$ the reference number is $\pi/4$ (Figure 10.41).

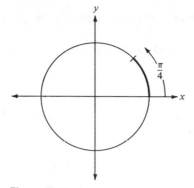
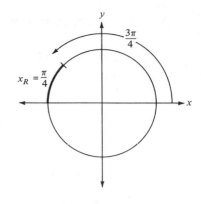

Figure 10.41

(3) $\pi/4$ is the number in Q_1 with a reference number of $\pi/4$

$\pi - \pi/4 = 3\pi/4$ is the number in Q_2 with a reference number of $\pi/4$

Thus,

$$3x = \frac{\pi}{4} + k2\pi \quad \text{or} \quad 3x = \frac{3\pi}{4} + k2\pi$$

from which we have

$$x = \frac{\pi}{12} + k\frac{2\pi}{3} \quad \text{or} \quad x = \frac{\pi}{4} + k\frac{2\pi}{3}$$

(4) To obtain solutions in the interval $0 \le x < 2\pi$, use 0, 1, and 2 as replacements for k.

$$\frac{\pi}{12} + (0)\frac{2\pi}{3} = \frac{\pi}{12} \qquad \frac{\pi}{4} + (0)\frac{2\pi}{3} = \frac{\pi}{4}$$

$$\frac{\pi}{12} + (1)\frac{2\pi}{3} = \frac{3\pi}{4} \qquad \frac{\pi}{4} + (1)\frac{2\pi}{3} = \frac{11\pi}{12}$$

$$\frac{\pi}{12} + (2)\frac{2\pi}{3} = \frac{17\pi}{12} \qquad \frac{\pi}{4} + (2)\frac{2\pi}{3} = \frac{19\pi}{12}$$

Thus, $\pi/12$, $\pi/4$, $3\pi/4$, $11\pi/12$, $17\pi/12$, and $19\pi/12$ are the solutions of the equation $(0 \le x < 2\pi)$.

EXERCISES 10–6
In Exercises 1–16 solve for x in the interval $0 \le x < 2\pi$.

1. $\sin x = \dfrac{\sqrt{3}}{2}$

2. $\sin x = -\dfrac{\sqrt{3}}{2}$

3. $\cos x = -\frac{1}{2}$

4. $\tan x = 1$

5. $\tan x = -1$

6. $\sec x = 2$

7. $\sin x = 0.1219$

8. $\sin x = -0.1219$

9. $\tan x = -3.145$

10. $\tan x = 3.145$

11. $\sqrt{2}\cos x = 1$

12. $\csc x - \sqrt{2} = 0$

13. $\sin x + 2 = 0$

14. $\cos x - 3 = 0$

15. $3\tan x - 1 = 0$

16. $2\tan x + 7 = 0$

In Exercises 17–20 find all solutions to the equation.

17. $4\sin x - 1 = 1$

18. $10\cos x - 2 = 0$

19. $4\csc x + 9 = 0$

20. $5\cot x + 3 = 0$

In Exercises 21–36 solve for x in the interval $0 \le x < 2\pi$.

21. $2\sin^2 x - 1 = 0$

22. $3\tan^2 x - 1 = 0$

23. $\cos^2 x = \cos x$

24. $\sin^3 x = \sin x$

25. $2\cos x \sin x - \cos x = 0$

26. $\tan x \cos x = \tan x$

27. $\tan^2 x + 4\tan x - 21 = 0$

28. $2\sin^2 x - 5\sin x = 3$

29. $\cos 3x = 0$

30. $\sin 4x = 1$

31. $\sin 2x = \frac{1}{2}$

32. $\sin \frac{1}{2}x = \frac{1}{2}$

33. $\tan \frac{1}{3}x = 1$

34. $\csc \frac{1}{4}x = 2$

35. $\sin^2 2x = \sin 2x$

36. $\cot^2 2x + 2\cot 2x + 1 = 0$

In Exercises 37–40 find all solutions to the given equations. Use the quadratic formula.

37. $\cos^2 x + \cos x - 1 = 0$

38. $3\sin^2 x - \sin x - 1 = 0$

39. $\sin^2 x + \sin x + 1 = 0$

40. $2\tan^2 x + \tan x - 5 = 0$

10–7 INVERSE TRIGONOMETRIC FUNCTIONS

The concept of an inverse function was introduced in Section 9-3. Recall three facts:

1. Two functions with exactly reverse assignments are inverse functions.

2. The domain of a function f is the range of its inverse function and the range of f is the domain of its inverse.

3. A function is one to one when each x value in the domain is assigned a different y value so that no two ordered pairs have the same second component. Only the inverse of a one-to-one function is a function.

Now consider the graph of $y = \sin x$ in Figure 10.42. Because the sine function is periodic, many x values are assigned the same y value. For example,

$$\sin 0 = \sin \pi = \sin 2\pi = \sin(-\pi) = 0$$

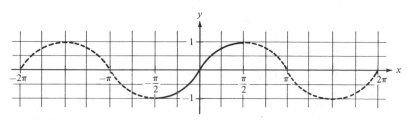

Figure 10.42

Thus, the inverse of the sine function is not a function. The so-called *inverse sine function* is defined by restricting the domain of $y = \sin x$ to the interval $-\pi/2 \le x \le \pi/2$. Note in Figure 10.42 that each x value is assigned a different y value in this limited version of the sine function.

We find the rule for the inverse sine function by interchanging the x and y variables. Thus, the inverse of the function defined by

$$y = \sin x \qquad -\frac{\pi}{2} \le x \le \frac{\pi}{2}$$

is defined by

$$x = \sin y \qquad -\frac{\pi}{2} \le y \le \frac{\pi}{2}$$

which is written in inverse notation as

$$y = \arcsin x \quad \text{or} \quad \sin^{-1} x$$

For example, since $\sin(\pi/2) = 1$, we write

$$\frac{\pi}{2} = \arcsin 1 \quad \text{or} \quad \frac{\pi}{2} = \sin^{-1} 1$$

Both expressions are read "$\pi/2$ is the arc (or number) whose sine is 1." Recall that the -1 in the symbol $\sin^{-1}$ is not an exponent.

Because a function and its inverse interchange their domain and range for $y = \arcsin x$, we have

$$-1 \le x \le 1 \quad \text{and} \quad -\frac{\pi}{2} \le y \le \frac{\pi}{2}$$

Consider Figure 10.43, which gives the graph of $y = \arcsin x$. The dashed lines indicate how the graph would continue if we did not restrict the domain of the sine function.

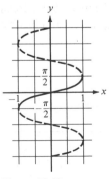

Figure 10.43

Through similar considerations we may define inverses for the other five trigonometric functions. We restrict the domain of the trigonometric functions and hence the range of the inverse as follows:

$$y = \arcsin x \qquad -\frac{\pi}{2} \le y \le \frac{\pi}{2}$$

$$y = \arccos x \qquad 0 \le y \le \pi$$

$$y = \arctan x \qquad -\frac{\pi}{2} < y < \frac{\pi}{2}$$

$$y = \operatorname{arccot} x \qquad 0 < y < \pi$$

$$y = \operatorname{arcsec} x \qquad 0 \le y \le \pi \quad y \ne \frac{\pi}{2}$$

$$y = \operatorname{arccsc} x \qquad -\frac{\pi}{2} \le y \le \frac{\pi}{2} \quad y \ne 0$$

A consideration of the graphs of $y = \cos x$, $y = \tan x$, $y = \cot x$, $y = \sec x$, and $y = \csc x$ would show why the respective intervals were chosen.

EXAMPLE 1: Find $\arcsin \frac{1}{2}$.

SOLUTION: Let $y = \arcsin \frac{1}{2}$; then $\frac{1}{2} = \sin y$. We solve this equation and find the number y in the interval $-\pi/2 \le y \le \pi/2$ whose sine is $\frac{1}{2}$. Since $\sin \pi/6 = \frac{1}{2}$, $\arcsin \frac{1}{2} = \pi/6$.

EXAMPLE 2: Find $\cos^{-1}(-\frac{1}{2})$.

SOLUTION: Let $y = \cos^{-1}(-\frac{1}{2})$; then $-\frac{1}{2} = \cos y$. We solve this equation and find the number y in the interval $0 \le y \le \pi$ whose cosine is $-\frac{1}{2}$. Since $\cos(2\pi/3) = -\frac{1}{2}$, $\cos^{-1}(-\frac{1}{2}) = 2\pi/3$.

EXAMPLE 3: Find $\arctan(-1)$.

SOLUTION: Let $y = \arctan(-1)$; then $-1 = \tan y$. We solve this equation and find the number y in the interval $-\pi/2 < y < \pi/2$ whose tangent is -1. Since $\tan(-\pi/4) = -1$, $\arctan(-1) = -\pi/4$.

EXAMPLE 4: Find $\operatorname{arccot} 0.5312$.

SOLUTION: Let $y = \operatorname{arccot} 0.5312$; then $0.5312 = \cot y$. We solve this equation and find the number y in the interval $0 < y < \pi$ whose cotangent is 0.5312. Table 5 in the Appendix indicates that $\cot 1.08 \approx 0.5312$. Thus, $\operatorname{arccot} 0.5312 \approx 1.08$.

EXAMPLE 5: Find sin(arccos 0).

SOLUTION: First determine arccos 0. Since $\cos(\pi/2) = 0$, we have

$$\frac{\pi}{2} = \text{arccos } 0$$

Now replace arccos 0 by $\pi/2$ in the original expression.

$$\sin(\text{arccos } 0) = \sin\frac{\pi}{2} = 1$$

EXAMPLE 6: Find $\tan(\sin^{-1}\frac{4}{5})$.

SOLUTION: It is useful to interpret $\sin^{-1}\frac{4}{5}$ as the measure of an angle in a right triangle. Let $\theta = \sin^{-1}\frac{4}{5}$ and sketch the triangle in Figure 10.44. The length of the side opposite θ is 4 and the hypotenuse is 5. The remaining side of the triangle is found by the Pythagorean relationship

$$x = \sqrt{5^2 - 4^2}$$
$$= \sqrt{9}$$
$$= 3$$

Thus,

$$\tan\theta = \tan(\sin^{-1}\tfrac{4}{5}) = \tfrac{4}{3}$$

Figure 10.44 $\theta = \sin^{-1}\frac{4}{5}$

EXERCISES 10–7

In Exercises 1–30 evaluate the expression.

1. $\arccos\frac{1}{2}$

2. $\arccos(-\frac{1}{2})$

3. $\sin^{-1}\left(-\dfrac{\sqrt{3}}{2}\right)$

4. $\arcsin\dfrac{\sqrt{3}}{2}$

5. arctan 1

6. $\text{arccot}(-1)$

7. $\arcsin(-1)$

8. $\text{arcsec}(-1)$

9. $\arcsin\dfrac{1}{\sqrt{2}}$

10. $\tan^{-1}\sqrt{3}$

11. $\csc^{-1}(-2)$

12. $\arccos\left(-\dfrac{1}{\sqrt{2}}\right)$

13. arcsin 0.3124

14. $\sin^{-1}(-0.3124)$

15. $\cos^{-1}(-0.5509)$

16. arccos 0.5509

17. arctan 1.758

18. $\tan^{-1}(-1.758)$

19. $\sec^{-1}(-\frac{5}{4})$

20. $\text{arccot}\frac{2}{3}$

21. cos(arcsin 0)   **22.** sin(arccos 1)

23. $\sin\left(\cos^{-1}\dfrac{\sqrt{3}}{2}\right)$ **24.** $\cos[\sin^{-1}(-\tfrac{1}{2})]$

25. cos(arccos 0)   **26.** $\cos^{-1}(\cos 0)$
27. arcsin(sin 1)   **28.** sin(arcsin 1)
29. tan[arcsin(−0.5518)]   **30.** csc(arccos 0.0129)

In Exercises 31–36 evaluate the expression. Use right triangles.
31. $\sin(\tan^{-1}\tfrac{3}{4})$ **32.** $\cos[\arcsin(-\tfrac{1}{3})]$
33. $\tan(\arcsin\tfrac{12}{13})$ **34.** $\cot(\cos^{-1}\tfrac{8}{17})$

35. $\cos\left[\operatorname{arccsc}\left(-\dfrac{\sqrt{5}}{2}\right)\right]$ **36.** $\sin\left(\sec^{-1}\dfrac{4}{\sqrt{5}}\right)$

In Exercises 37–41 find the domain and range of the function.
37. $y = \arccos x$
38. $y = \arctan x$
39. $y = \operatorname{arccot} x$
40. $y = \operatorname{arcsec} x$
41. $y = \operatorname{arccsc} x$
42. Sketch the graph of $y = \arccos x$.
43. Sketch the graph of $y = \arctan x$.
44. For what values of x is each statement true?
    a. $\sin(\sin^{-1}x) = x$ **b.** $\sin^{-1}(\sin x) = x$
    c. $\cos(\cos^{-1}x) = x$ **d.** $\cos^{-1}(\cos x) = x$

10-8 TRIGONOMETRIC IDENTITIES

Trigonometry is characterized by many formulas that may be used to simplify trigonometric expressions. For example, the formula

$$\cos(-x) = \cos x$$

made it easier to evaluate the cosine of a negative number. In Section 10-5 we used the formula

$$\csc x = \frac{1}{\sin x}$$

to study the behavior of the cosecant function. These formulas are examples of trigonometric identities. That is, they are true for all values of x for which the expressions are defined.

    We now develop other useful trigonometric identities. Most of these formulas need not be memorized. It is more important that you be able to manipulate the formulas to establish a desired result. In the following examples we first state, and then verify, the particular identity.

EXAMPLE 1: Verify the identity: $\sin^2 x + \cos^2 x = 1$.

SOLUTION: To avoid confusion we return to the original notation of using s to represent the real numbers associated with arc lengths on the unit circle. Then, by definition,

$$x = \cos s \quad \text{and} \quad y = \sin s$$

Any point (x, y) on the unit circle must satisfy the equation $x^2 + y^2 = 1$. Thus, by substitution we have

$$(\cos s)^2 + (\sin s)^2 = 1$$

For convenience, this statement is usually written with x as the independent variable. Thus,

$$\sin^2 x + \cos^2 x = 1$$

EXAMPLE 2: Verify the identity:

$$\cos(x_1 + x_2) = \cos x_1 \cos x_2 - \sin x_1 \sin x_2$$

SOLUTION: As in Example 1, we first prove the identity by using s as the independent variable. The derivation is quite long, but once established, this formula yields many important results. Consider the coordinates of the points on the unit circle in Figure 10.45. The arc lengths from $(1, 0)$ to $[\cos(s_1 + s_2), \sin(s_1 + s_2)]$, and from $[\cos(-s_1), \sin(-s_1)]$ to $(\cos s_2, \sin s_2)$ are both of length $s_1 + s_2$. Equal arcs in a circle subtend equal chords. Thus, by the distance formula we have

$$\sqrt{[\cos(s_1 + s_2) - 1]^2 + [\sin(s_1 + s_2) - 0]^2}$$
$$= \sqrt{(\cos s_2 - \cos s_1)^2 + [\sin s_2 - (-\sin s_1)]^2}$$

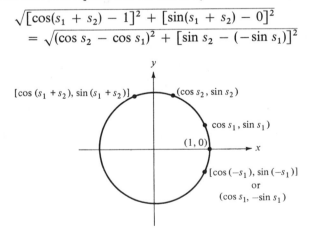

Figure 10.45

After squaring both sides to remove radicals and performing the indicated operations, we obtain

$$\cos^2(s_1 + s_2) - 2\cos(s_1 + s_2) + 1 + \sin^2(s_1 + s_2)$$
$$= \cos^2 s_2 - 2\cos s_1 \cos s_2 + \cos^2 s_1 + \sin^2 s_2 + 2\sin s_1 \sin s_2 + \sin^2 s_1$$

Because $\sin^2 s + \cos^2 s = 1$, we regroup the terms as follows:

$$[\sin^2(s_1 + s_2) + \cos^2(s_1 + s_2)] - 2\cos(s_1 + s_2) + 1$$
$$= (\sin^2 s_2 + \cos^2 s_2) + (\sin^2 s_1 + \cos^2 s_1) - 2\cos s_1 \cos s_2 + 2\sin s_1 \sin s_2$$

Upon substituting 1 for the appropriate expressions, we have

$$1 - 2\cos(s_1 + s_2) + 1 = 1 + 1 - 2\cos s_1 \cos s_2 + 2\sin s_1 \sin s_2$$
$$-2\cos(s_1 + s_2) = -2\cos s_1 \cos s_2 + 2\sin s_1 \sin s_2$$
$$\cos(s_1 + s_2) = \cos s_1 \cos s_2 - \sin s_1 \sin s_2$$

For convenience, we let x represent the independent variable. Thus,

$$\cos(x_1 + x_2) = \cos x_1 \cos x_2 - \sin x_1 \sin x_2$$

Note that

$$\cos(x_1 + x_2) \neq \cos x_1 + \cos x_2$$

EXAMPLE 3: Verify the identity:

$$\cos(x_1 - x_2) = \cos x_1 \cos x_2 + \sin x_1 \sin x_2$$

SOLUTION: Rewrite $\cos(x_1 - x_2)$ as $\cos[x_1 + (-x_2)]$ and use the formula from Example 2.

$$\cos[x_1 + (-x_2)] = \cos x_1 \cos(-x_2) - \sin x_1 \sin(-x_2)$$

Since $\cos(-x_2) = \cos x_2$ and $\sin(-x_2) = -\sin x_2$, we have

$$\cos[x_1 + (-x_2)] = \cos x_1 \cos x_2 - \sin x_1(-\sin x_2)$$

or

$$\cos(x_1 - x_2) = \cos x_1 \cos x_2 + \sin x_1 \sin x_2$$

EXAMPLE 4: Verify the identity: $\cos[(\pi/2) - x] = \sin x$.

SOLUTION: Use the formula for the cosine of the difference of two numbers (Example 3).

$$\cos\left(\frac{\pi}{2} - x\right) = \cos\frac{\pi}{2}\cos x + \sin\frac{\pi}{2}\sin x$$
$$= 0 \cdot \cos x + 1 \cdot \sin x$$
$$= \sin x$$

EXAMPLE 5: Verify the identity: $\cos 2x = \cos^2 x - \sin^2 x$.

SOLUTION: Write $2x$ as $x + x$ and use the formula for the cosine of the sum of two numbers.

$$\cos 2x = \cos(x + x) = \cos x \cos x - \sin x \sin x = \cos^2 x - \sin^2 x$$

There are two alternative forms for $\cos 2x$. From $\sin^2 x + \cos^2 x = 1$, it follows that $\sin^2 x = 1 - \cos^2 x$. Thus,

$$\cos 2x = \cos^2 x - \sin^2 x$$
$$= \cos^2 x - (1 - \cos^2 x)$$
$$= 2 \cos^2 x - 1$$

The remaining alternative form is

$$\cos 2x = 1 - 2 \sin^2 x$$

EXAMPLE 6: Verify the identity:

$$\cos \frac{x}{2} = \pm \sqrt{\frac{1 + \cos x}{2}}$$

SOLUTION: In Example 5 we obtained the identity

$$\cos 2x = 2 \cos^2 x - 1$$

If we replace x by $x/2$ in this identity, we have

$$\cos 2\left(\frac{x}{2}\right) = 2 \cos^2 \frac{x}{2} - 1$$

$$\cos x = 2 \cos^2 \frac{x}{2} - 1$$

$$\frac{\cos x + 1}{2} = \cos^2 \frac{x}{2}$$

$$\pm \sqrt{\frac{\cos x + 1}{2}} = \cos \frac{x}{2}$$

The correct sign is determined by the quadrant in which $x/2$ lies.

From the examples you should notice how new formulas are derived from previous results. In order for you to grasp this concept, only identities concerned with the cosine function were developed. We now show how the formula for the sine of the sum of two numbers is usually derived. This formula can then be used to derive a list of identities concerning the sine function. The exercises ask you to verify many of these formulas. For reference purposes, a summary of important identities is given in Appendix A-4.

In Example 4 we proved

$$\cos\left(\frac{\pi}{2} - x\right) = \sin x$$

Replacing x by $x_1 + x_2$ gives

$$\sin(x_1 + x_2) = \cos\left[\frac{\pi}{2} - (x_1 + x_2)\right] = \cos\left[\left(\frac{\pi}{2} - x_1\right) - x_2\right]$$

If we now use the formula for the cosine of the difference of two numbers, we obtain

$$\sin(x_1 + x_2) = \cos\left[\left(\frac{\pi}{2} - x_1\right) - x_2\right]$$

$$= \cos\left(\frac{\pi}{2} - x_1\right)\cos x_2 + \sin\left(\frac{\pi}{2} - x_1\right)\sin x_2$$

By the identity in Example 4 we know $\cos(\pi/2 - x_1) = \sin x_1$. In Exercise 6 of this section you are asked to verify $\sin(\pi/2 - x_1) = \cos x_1$. The formula for the sine of the sum of two numbers is then

$$\sin(x_1 + x_2) = \sin x_1 \cos x_2 + \cos x_1 \sin x_2$$

Finally let's consider some numerical examples using the various identities.

EXAMPLE 7: If $\sin x = \frac{3}{5}$ and $\pi/2 < x < \pi$, find:
a. $\cos x$ **b.** $\tan x$ **c.** $\sec x$
d. $\cos 2x$ **e.** $\sin 2x$ **f.** $\cos x/2$

SOLUTION:
a. Since $\pi/2 < x < \pi$ where $\cos x$ is negative, the identity $\sin^2 x + \cos^2 x = 1$, when solved for $\cos x$, becomes

$$\cos x = -\sqrt{1 - \sin^2 x} = -\sqrt{1 - \left(\frac{3}{5}\right)^2} = -\sqrt{\frac{16}{25}} = -\frac{4}{5}$$

b. $\tan x = \dfrac{\sin x}{\cos x} = \dfrac{3/5}{-4/5} = \dfrac{3}{-4}$

c. $\sec x = \dfrac{1}{\cos x} = \dfrac{1}{-4/5} = \dfrac{5}{-4}$

d. $\cos 2x = 1 - 2\sin^2 x = 1 - 2\left(\frac{3}{5}\right)^2 = \dfrac{7}{25}$

e. From Appendix A-4 we have

$$\sin 2x = 2\sin x \cos x = 2\left(\frac{3}{5}\right)\left(-\frac{4}{5}\right) = -\frac{24}{25}$$

f. If $\pi/2 < x < \pi$, then $x/2$ is between $\pi/4$ and $\pi/2$ and $\cos(x/2)$ is positive. Thus,

$$\cos\frac{x}{2} = \sqrt{\frac{\cos x + 1}{2}} = \sqrt{\frac{(-4/5) + 1}{2}} = \sqrt{\frac{1}{10}} \quad \text{or} \quad \frac{\sqrt{10}}{10}$$

EXAMPLE 8: If $\sin x_1 = -\frac{1}{2}$ where $\pi < x_1 < 3\pi/2$ and $\cos x_2 = \sqrt{3}/2$ where $0 < x_2 < \pi/2$, find:
a. $\cos x_1$ **b.** $\sin x_2$ **c.** $\cos(x_1 + x_2)$ **d.** $\sin(x_1 - x_2)$

SOLUTION:

a. Since $\pi < x_1 < 3\pi/2$ where $\cos x_1$ is negative, we have

$$\cos x_1 = -\sqrt{1 - \sin^2 x_1} = -\sqrt{1 - \left(\frac{1}{2}\right)^2} = -\sqrt{\frac{3}{4}} = -\frac{\sqrt{3}}{2}$$

b. Since $0 < x_2 < \pi/2$ where $\sin x_2$ is positive, we have

$$\sin x_2 = \sqrt{1 - \cos^2 x_2} = \sqrt{1 - \left(\frac{\sqrt{3}}{2}\right)^2} = \sqrt{\frac{1}{4}} = \frac{1}{2}$$

c. $\cos(x_1 + x_2) = \cos x_1 \cos x_2 - \sin x_1 \sin x_2$

$$= \left(\frac{-\sqrt{3}}{2}\right)\left(\frac{\sqrt{3}}{2}\right) - \left(-\frac{1}{2}\right)\left(\frac{1}{2}\right)$$

$$= -\frac{3}{4} - \left(-\frac{1}{4}\right)$$

$$= -\frac{1}{2}$$

d. From Appendix A-4 we have

$$\sin(x_1 - x_2) = \sin x_1 \cos x_2 - \cos x_1 \sin x_2$$

$$= \left(-\frac{1}{2}\right)\left(\frac{\sqrt{3}}{2}\right) - \left(\frac{-\sqrt{3}}{2}\right)\left(\frac{1}{2}\right)$$

$$= -\frac{\sqrt{3}}{4} - \left(-\frac{\sqrt{3}}{4}\right)$$

$$= 0$$

EXERCISES 10–8

In Exercises 1–6 verify the identity. Use the formulas for the cosine of either the sum or difference of two numbers.

1. $\cos\left(\dfrac{\pi}{2} + x\right) = -\sin x$

2. $\cos(\pi + x) = -\cos x$

3. $\cos(\pi - x) = -\cos x$

4. $\cos\left(\dfrac{3\pi}{2} - x\right) = -\sin x$

5. $\cos\left(\dfrac{3\pi}{2} + x\right) = \sin x$

6. Show $\sin(\pi/2 - x) = \cos x$. (*Hint:* Write $\cos x$ as $\cos[\pi/2 - (\pi/2 - x)]$ and use the formula for the cosine of the difference of two numbers.)

In Exercises 7 and 8 verify the identity. Use $\sin(x_1 + x_2) = \sin x_1 \cos x_2 + \cos x_1 \sin x_2$.

7. $\sin(x_1 - x_2) = \sin x_1 \cos x_2 - \cos x_1 \sin x_2$

8. $\sin 2x = 2 \sin x \cos x$

In Exercises 9–14 verify the identity. Use the formulas for the sine of either the sum or difference of two numbers.

9. $\sin\left(\dfrac{\pi}{2} - x\right) = \cos x$ **10.** $\sin\left(\dfrac{\pi}{2} + x\right) = \cos x$

11. $\sin(\pi + x) = -\sin x$ **12.** $\sin(\pi - x) = \sin x$

13. $\sin\left(\dfrac{3\pi}{2} - x\right) = -\cos x$ **14.** $\sin\left(\dfrac{3\pi}{2} + x\right) = -\cos x$

15. Verify the identity: $\cos 2x = 1 - 2 \sin^2 x$.

16. Use Exercise 15 and Example 6 to verify the identity:

$$\sin \frac{x}{2} = \pm \sqrt{\frac{1 - \cos x}{2}}$$

In Exercises 17 and 18 use $\tan x = (\sin x/\cos x)$ to verify the identity.

17. $\tan(x_1 + x_2) = \dfrac{\tan x_1 + \tan x_2}{1 - \tan x_1 \tan x_2}$

18. $\tan\left(\dfrac{\pi}{2} + x\right) = -\cot x$

In Exercises 19 and 20 verify the identity. Use the formula for the tangent of the sum of two numbers.

19. $\tan(x_1 - x_2) = \dfrac{\tan x_1 - \tan x_2}{1 + \tan x_1 \tan x_2}$

20. $\tan 2x = \dfrac{2 \tan x}{1 - \tan^2 x}$

21. If $\sin x = \frac{12}{13}$ and $0 < x < \pi/2$, find:

  a. $\cos x$ **b.** $\cot x$

  c. $\csc x$ **d.** $\cos 2x$

  e. $\sin 2x$ **f.** $\cos \dfrac{x}{2}$

  g. $\sin \dfrac{x}{2}$

22. If $\cos x = -0.40$ and $\pi < x < 3\pi/2$, estimate:

  a. $\sin x$ **b.** $\cos 2x$

  c. $\sin 2x$ **d.** $\cos \dfrac{x}{2}$

  e. $\sin \dfrac{x}{2}$

23. If $\sin x_1 = \frac{1}{2}$ where $0 < x_1 < \pi/2$, and $\cos x_2 = -\sqrt{3}/2$ where $\pi/2 < x_2 < \pi$, find:

    a. $\cos x_1$ **b.** $\sin x_2$

    c. $\cos(x_1 + x_2)$ **d.** $\sin(x_1 - x_2)$

24. If $\cos x_1 = \frac{3}{5}$ where $3\pi/2 < x_1 < 2\pi$, and $\sin x_2 = -\frac{5}{13}$, where $3\pi/2 < x_2 < 2\pi$, find:

    a. $\sin x_1$ **b.** $\cos x_2$

    c. $\cos(x_1 - x_2)$ **d.** $\sin(x_1 + x_2)$

10–9 MORE ON TRIGONOMETRIC IDENTITIES

Since all the trigonometric functions are defined in terms of the sine and/or cosine these functions are interrelated. This enables us to change the form of many trigonometric expressions to an expression that is either simpler or more useful for a particular problem. For convenience, we list below the identities used in this section. Except for identity 6, these statements are definitions.

Identity 1. $\sin x \csc x = 1$ or $\sin x = \dfrac{1}{\csc x}$ or $\csc x = \dfrac{1}{\sin x}$

Identity 2. $\cos x \sec x = 1$ or $\cos x = \dfrac{1}{\sec x}$ or $\sec x = \dfrac{1}{\cos x}$

Identity 3. $\tan x \cot x = 1$ or $\tan x = \dfrac{1}{\cot x}$ or $\cot x = \dfrac{1}{\tan x}$

Identity 4. $\tan x = \dfrac{\sin x}{\cos x}$

Identity 5. $\cot x = \dfrac{\cos x}{\sin x}$

Identity 6. $\sin^2 x + \cos^2 x = 1$

Remember, two statements are identical when they have the same value for all x at which the expressions are defined.

    EXAMPLE 1: Show that the expression $\sin x \cot x \sec x$ is identical to 1.

SOLUTION:

$$\sin x \cot x \sec x = \sin x \left(\frac{\cos x}{\sin x}\right) \sec x \qquad \text{identity 5}$$

$$= \sin x \frac{\cos x}{\sin x} \frac{1}{\cos x} \qquad \text{identity 2}$$

$$= 1 \qquad \text{simplify}$$

Thus, $\sin x \cot x \sec x = 1$ for all values of x at which the expression is defined.

EXAMPLE 2: Prove the identity: $\tan^2 x + 1 = \sec^2 x$.

SOLUTION:

$$\tan^2 x + 1 = \frac{\sin^2 x}{\cos^2 x} + 1 \qquad \text{identity 4}$$

$$= \frac{\sin^2 x + \cos^2 x}{\cos^2 x}$$

$$= \frac{1}{\cos^2 x} \qquad \text{identity 6}$$

$$= \sec^2 x \qquad \text{identity 2}$$

Thus $\tan^2 x + 1$ is identical to $\sec^2 x$. This identity, together with $\cot^2 x + 1 = \csc^2 x$, is often used in proving other identities.

In general there is no standard procedure for working with identities. In fact, a given identity can usually be proved in several ways. However, these suggestions should be helpful.

1. Change the more complicated expression in the identity to the same form as the less complicated expression.
2. Until you become more familiar with identities, change all functions to sines and cosines. This procedure might necessitate more algebra in some instances, but it will provide a direct approach to the problem. Gradually, try to make use of the other trigonometric functions.

EXAMPLE 3: Show that the expression $\cos^2 x(1 + \tan^2 x)$ may be replaced by 1.

SOLUTION:

$$\cos^2 x(1 + \tan^2 x) = \cos^2 x \left(1 + \frac{\sin^2 x}{\cos^2 x}\right) \qquad \text{identity 4}$$

$$\left.\begin{array}{l} = \cos^2 x \cdot 1 + \cos^2 x \cdot \dfrac{\sin^2 x}{\cos^2 x} \\[2mm] = \cos^2 x + \sin^2 x \end{array}\right\} \quad \text{simplify}$$

$$= 1 \qquad \text{identity 6}$$

Alternative Solution

$$\cos^2 x(1 + \tan^2 x) = \cos^2 x(\sec^2 x) \qquad \text{Example 2}$$

$$= \cos^2 x \left(\frac{1}{\cos^2 x}\right) \qquad \text{identity 2}$$

$$= 1$$

Thus, $\cos^2 x(1 + \tan^2 x)$ may be replaced by 1 since $\cos^2 x(1 + \tan^2 x) = 1$ is an identity.

EXAMPLE 4: Show that the expression $(\tan^2 x - 1)/(\tan^2 x + 1)$ is identical to the expression $\sin^2 x - \cos^2 x$.

SOLUTION:

$$\frac{\tan^2 x - 1}{\tan^2 x + 1} = \frac{(\sin^2 x/\cos^2 x) - 1}{(\sin^2 x/\cos^2 x) + 1} \qquad \text{identity 4}$$

$$= \frac{\cos^2 x[(\sin^2 x/\cos^2 x) - 1]}{\cos^2 x[(\sin^2 x/\cos^2 x) + 1]}$$

$$= \frac{\sin^2 x - \cos^2 x}{\sin^2 x + \cos^2 x} \qquad\qquad \text{simplify the complex fraction}$$

$$= \frac{\sin^2 x - \cos^2 x}{1} \qquad \text{identity 6}$$

$$= \sin^2 x - \cos^2 x$$

Thus,

$$\frac{\tan^2 x - 1}{\tan^2 x + 1} = \sin^2 x - \cos^2 x$$

is an identity.

EXAMPLE 5: Prove the identity: $(\cos x \csc x)/\cot^2 x = \tan x$.

SOLUTION:

$$\frac{\cos x \csc x}{\cot^2 x} = \frac{\cos x(1/\sin x)}{\cos^2 x/\sin^2 x} \qquad \text{identities 1 and 5}$$

$$= \frac{\sin^2 x(\cos x/\sin x)}{\sin^2 x(\cos^2 x/\sin^2 x)}$$

$$= \frac{\sin x \cos x}{\cos^2 x} \qquad\qquad \text{simplify the complex fraction}$$

$$= \frac{\sin x}{\cos x}$$

$$= \tan x \qquad \text{identity 4}$$

Thus,

$$\frac{\cos x \csc x}{\cot^2 x} = \tan x$$

Remember that we prove an identity by changing the more complicated expression to the same form as the less complicated expression. If both expressions are complicated, you might try to change them both to the same expression. Do *not* attempt to prove an identity by treating it as an equation and using the associated techniques, for this involves assuming what you want to prove.

The following examples illustrate how the solution to a problem can be dependent upon a knowledge of identities.

EXAMPLE 6: For what values of x ($0 \le x < 2\pi$) is the equation $\tan x + \cot x = 2 \csc x$ a true statement?

SOLUTION: $\tan x + \cot x = 2 \csc x$. Using identities 4, 5, and 1 we rewrite an equivalent equation that contains only $\sin x$ and $\cos x$.

$$\frac{\sin x}{\cos x} + \frac{\cos x}{\sin x} = \frac{2}{\sin x}$$

Simplifying the fractional equation, we have

$$\sin x \cos x \left(\frac{\sin x}{\cos x} + \frac{\cos x}{\sin x} \right) = \sin x \cos x \left(\frac{2}{\sin x} \right)$$

$$\sin^2 x + \cos^2 x = 2 \cos x$$

$$1 = 2 \cos x \qquad \text{identity 6}$$

$$\tfrac{1}{2} = \cos x$$

As illustrated in Example 2 of Section 10-6, $\cos x = \tfrac{1}{2}$ when $x = \pi/3$ and $x = 5\pi/3$.

EXERCISES 10–9

In Exercises 1–20 transform each first expression and show that it is identical to the second expression.

1. $\sin x \sec x$; $\tan x$
2. $\tan x \csc x$; $\sec x$
3. $\sin^2 x \csc x$; $\sin x$
4. $\tan x \cot^2 x$; $\cot x$
5. $\cos x \tan x$; $1/\csc x$
6. $\sec x \cot x$; $1/\sin x$
7. $\cos x \tan x \csc x$; 1
8. $\cot x \sec x \sin x$; 1
9. $\sin^2 x \cot^2 x$; $\cos^2 x$
10. $\tan^2 x \csc^2 x$; $\sec^2 x$

11. $\sin^2 x \sec^2 x$; $\tan^2 x$
12. $\dfrac{\cot x}{\csc x}$; $\cos x$

13. $\dfrac{\cos^2 x}{\cot^2 x}$; $\sin^2 x$
14. $\dfrac{\sin x \sec x}{\tan x}$; 1

15. $\dfrac{\cot x \tan x}{\sec x}$; $\cos x$
16. $1 + \cot^2 x$; $\csc^2 x$

17. $(1 - \cos x)(1 + \cos x)$; $1/\csc^2 x$
18. $\sin x(\csc x - \sin x)$; $\cos^2 x$

19. $\dfrac{\cos^2 x}{1 + \sin x}$; $1 - \sin x$
20. $\sin^4 x - \cos^4 x$; $\sin^2 x - \cos^2 x$

In Exercises 21–40 prove that the equation is an identity.

21. $\dfrac{1}{\sin^2 x} - 1 = \cot^2 x$

22. $\csc x \sin x - \dfrac{1}{\sec^2 x} = \sin^2 x$

23. $\dfrac{1 + \tan^2 x}{\csc^2 x} = \tan^2 x$

24. $\dfrac{1 + \tan^2 x}{\tan^2 x} = \csc^2 x$

25. $\sin x \tan x + \cos x = \sec x$

26. $\tan x \csc^2 x - \tan x = \cot x$

27. $\dfrac{\tan^2 x - \sin^2 x}{\tan^2 x} = \sin^2 x$

28. $\dfrac{\sec^2 x + \csc^2 x}{\sec^2 x} = \csc^2 x$

29. $\dfrac{\csc x}{\tan x + \cot x} = \cos x$

30. $\dfrac{\sec x + \csc x}{1 + \tan x} = \csc x$

31. $(1 - \sin x)(\sec x + \tan x) = \cos x$

32. $\sec x \csc x - 2 \cos x \csc x + \cot x = \tan x$

33. $\csc x \sin 2x = 2 \cos x$

34. $\sin 2x / 2 \sin x = \cos x$

35. $\cos^2 x = \dfrac{\cos 2x + 1}{2}$

36. $\sin^2 x = \dfrac{1 - \cos 2x}{2}$

37. $\cos 2x + 2 \sin^2 x = 1$

38. $\sin 2x = \dfrac{2 \cot x}{\csc^2 x}$

39. $\dfrac{1 + \cos 2x}{\sin 2x} = \cot x$

40. $\cos 2x / \sin x + \sin 2x / \cos x = \csc x$

In Exercises 41–54 solve each equation for $0 \le x < 2\pi$ through the aid of the fundamental identities.

41. $\sin x = \cos x$

42. $\sqrt{3} \sin x - \cos x = 0$

43. $\sin x = \tan x$

44. $2 \sin x - \tan x = 0$

45. $\sin^2 x + \cos^2 x = \tan x$

46. $\tan^2 x - \sec^2 x = \cos x$
    (*Note:* $\tan^2 x + 1 = \sec^2 x$.)

47. $\sin x + \cos x \tan x = 1$

48. $\tan x + 4 \cot x = 5$

49. $2 \sin x - \csc x = 1$

50. $1 - \cos^2 x = \sin x$

51. $2 \cos^2 x = 3 \sin x + 3$

52. $2 \sin^2 x = \cos x + 1$

53. $\sec^2 x + 5 \tan x + 4 = 0$

54. $\tan^2 x - 3 \sec x = 9$

Sample Test: Chapter 10

Find:

1. the arc length

2. the area

of a sector of a circle of radius 10 in. that is subtended by a central angle of 60°.

In Exercises 3–6 find the given function value.

3. $\sin 99\pi$

4. $\cos(-1.11)$

5. $\tan \dfrac{56\pi}{3}$

6. $\sec 3$

In Exercises 7 and 8 state the amplitude, the period, and the phase shift, and sketch one cycle of the function $y = -\cos(2x + \pi/2)$.

9. Find the equation for the curve in the figure. The equation will be of the form $y = a \sin bx$ or $y = a \cos bx$.

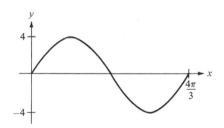

10. In which quadrant do $y = \sin x$ and $y = \cos x$ both decrease?

11. What is the domain of $y = \csc x$?

In Exercises 12 and 13 solve for x in the interval $0 \le x < 2\pi$.

12. $\sin x = -\frac{1}{2}$

13. $4 \sin^2 x - \sin x - 2 = 0$

In Exercises 14 and 15 evaluate the given expression.

14. $\arcsin(-0.4439)$

15. $\sin[\cos^{-1}(-1)]$

16. Use the formula for the sine of the difference of two numbers and simplify the expression $\sin(3\pi - x)$.

17. Start with the formulas for $\sin(x/2)$ and $\cos(x/2)$ and derive a formula for $\tan(x/2)$.

18. If $\sin x = 5/13$ and $\pi/2 < x < \pi$, find $\sin 2x$.

In Exercises 19 and 20 verify the given identity.

19. $\dfrac{\sin x + \tan x}{1 + \sec x} = \sin x$

20. $(\sin x - \cos x)^2 = 1 - \sin 2x$

Chapter II

Conic sections

11–1 INTRODUCTION

Analytical geometry bridges the gap between algebra and geometry. By representing the ordered pairs that satisfy some algebraic equation as points in the Cartesian coordinate system, we generate a geometric picture or graph. The fundamental relationship between an equation and its graph is as follows:

> Every ordered pair that satisfies the equation corresponds to a point in its graph, and every point in the graph corresponds to an ordered pair that satisfies the equation.

We have considered some techniques for sketching the graph of a given equation. However, the reverse question is of major importance. If we are given in geometric terms some object or motion from the physical world, can we describe it with some equation? If so, the powerful methods of algebra can be used to analyze them. Thus, we now consider the problem of determining the equation corresponding to a given geometric condition.

Suppose a point moves in the plane in the following way: Its distance from the point $(2, -1)$ is always 3. It is not hard to tell that such a path forms a circle of radius 3. Let us consider the more difficult problem of determining the equation of this circle.

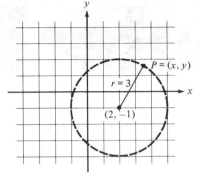

Figure 11.1

We start by making the sketch in Figure 11.1. $P = (x, y)$ represents any point on the given circle. By the distance formula, the distance from $(2, -1)$ to P is

$$d = \sqrt{(x - 2)^2 + [y - (-1)]^2}$$

This distance is the radius of the circle, which is given as 3. Thus

$$3 = \sqrt{(x - 2)^2 + [y - (-1)]^2}$$

Squaring both sides of this equation yields

$$9 = (x - 2)^2 + (y + 1)^2$$

We now simplify as follows:

$$9 = x^2 - 4x + 4 + y^2 + 2y + 1$$
$$0 = x^2 + y^2 - 4x + 2y - 4$$

This equation defines the circle formed by the given geometric condition.

EXAMPLE 1: Find the equation whose graph is formed by a point that moves in the following way: Its distance from the point $(1, 0)$ is always equal to its distance from the line $y = 3$.

SOLUTION: Let $P = (x, y)$ represent any point satisfying the given geometric condition. Then draw the sketch in Figure 11.2. P_1 lies on the line $y = 3$, so we represent

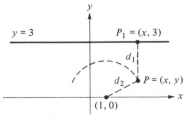

Figure 11.2

the point as $(x, 3)$. Since P_1 and P have the same x coordinate, the distance d_1 is given by

$$d_1 = |y - 3|$$

The distance from P to $(1, 0)$ is given by

$$d_2 = \sqrt{(x - 1)^2 + (y - 0)^2}$$

The given geometric condition states that

$$d_1 = d_2$$

Thus,

$$|y - 3| = \sqrt{(x - 1)^2 + (y - 0)^2}$$

Squaring both sides of the equation yields

$$(y - 3)^2 = (x - 1)^2 + (y - 0)^2$$

We simplify this equation as follows:

$$y^2 - 6y + 9 = x^2 - 2x + 1 + y^2$$
$$0 = x^2 - 2x + 6y - 8$$

The graph of this equation satisfies the given geometric condition.

EXAMPLE 2: Find the equation whose graph is formed by a point that moves in the following way: The sum of its distances from $(2, 0)$ and $(-2, 0)$ is 10.

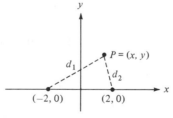

Figure 11.3

SOLUTION: Let $P = (x, y)$ represent any point satisfying the given geometric condition. Then draw the sketch in Figure 11.3. By the distance formula we have

$$d_1 = \sqrt{[x - (-2)]^2 + (y - 0)^2}$$
$$d_2 = \sqrt{(x - 2)^2 + (y - 0)^2}$$

The given geometric condition states that

$$d_1 + d_2 = 10$$

Thus,

$$\sqrt{(x + 2)^2 + y^2} + \sqrt{(x - 2)^2 + y^2} = 10$$

or

$$\sqrt{(x + 2)^2 + y^2} = 10 - \sqrt{(x - 2)^2 + y^2}$$

Square both sides of the equation and simplify.

$$(x + 2)^2 + y^2 = 100 - 20\sqrt{(x - 2)^2 + y^2} + (x - 2)^2 + y^2$$
$$8x = 100 - 20\sqrt{(x - 2)^2 + y^2}$$
$$2x - 25 = -5\sqrt{(x - 2)^2 + y^2}$$

Square both sides of the equation and simplify.

$$4x^2 - 100x + 625 = 25[(x - 2)^2 + y^2]$$
$$4x^2 - 100x + 625 = 25x^2 - 100x + 100 + 25y^2$$
$$0 = 21x^2 + 25y^2 - 525$$

The graph of this equation satisfies the given geometric condition.

EXERCISES 11–1

In Exercises 1–10 find the equation whose graph is formed by a point that moves in the given manner.

1. It is equidistant from $(0, 2)$ and $(-1, 5)$.
2. It is equidistant from $(1, -3)$ and $(-4, 2)$.
3. Its distance from the point $(-2, -3)$ is always 4.
4. Its distance from the point $(4, 0)$ is always 2.
5. Its distance from the point $(0, -2)$ is equal to its distance from the line $x = 4$.
6. Its distance from the point $(3, -1)$ is equal to its distance from the x axis.
7. The sum of its distances from $(1, 0)$ and $(-1, 0)$ is 4.
8. The sum of its distances from $(0, 2)$ and $(0, -2)$ is 8.
9. The difference of its distances from $(0, 3)$ and $(0, -3)$ is 3.
10. The difference of its distances from $(4, 0)$ and $(-4, 0)$ is 6.

11–2 THE CIRCLE

A circle is defined as the collection of points in the plane at a given distance from a fixed point. From this definition we derive the general equation for a circle as follows (see also Figure 11.4):

Let (h, k) be the fixed point, the center of the circle.
Let r be the given distance, the radius of the circle.
Let (x, y) be a point on the circle.

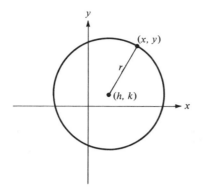

Figure 11.4

Then, by the distance formula, we have

$$\sqrt{(x - h)^2 + (y - k)^2} = r$$

or, upon squaring both sides of the equation,

$$(x - h)^2 + (y - k)^2 = r^2$$

This equation is the standard form of a circle of radius r with center (h, k).

EXAMPLE 1: Find the equation of the circle with center at $(2, -1)$ and radius 4.

SOLUTION: Substituting $h = 2$, $k = -1$, and $r = 4$ in the equation

$$(x - h)^2 + (y - k)^2 = r^2$$

we have

$$(x - 2)^2 + [y - (-1)]^2 = (4)^2$$
$$(x - 2)^2 + (y + 1)^2 = 16$$

EXAMPLE 2: Find the equation of the circle with center at $(4, 1)$ which passes through the origin.

SOLUTION: Substituting $h = 4$ and $k = 1$ into the general equation for a circle, we have

$$(x - 4)^2 + (y - 1)^2 = r^2$$

Since the circle passes through the origin, we find r^2 by substituting the coordinates of the point $(0, 0)$ in the equation.

$$(0 - 4)^2 + (0 - 1)^2 = r^2$$
$$17 = r^2$$

The equation of the circle is then

$$(x - 4)^2 + (y - 1)^2 = 17 \qquad \text{See Figure 11.5}$$

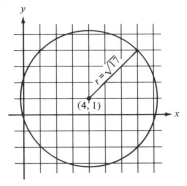

Figure 11.5

EXAMPLE 3: Find the center and radius of the circle

$$x^2 + y^2 + 4x - 6y = 12$$

SOLUTION: We must transform the given equation to the standard form.

$$(x - h)^2 + (y - k)^2 = r^2$$

We start by completing the square in the x terms and the y terms. To do this, first group the equation as

$$(x^2 + 4x \quad) + (y^2 - 6y \quad) = 12$$

Now find one-half of the coefficient of x (2). Square it (4), and add the result to both sides of the equation. Similarly, find one-half of the coefficient of y (-3). Square it (9), and add the result to both sides of the equation. Thus, we have

$$(x^2 + 4x + 4) + (y^2 - 6y + 9) = 12 + 4 + 9$$
$$(x + 2)^2 \quad + \quad (y - 3)^2 \quad = 25$$

By comparing this equation to the standard form, we determine that the center of the circle is at $(-2, 3)$ and the radius is $\sqrt{25}$, or 5 (see Figure 11.6).

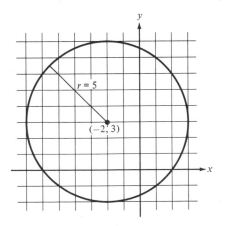

Figure 11.6

EXERCISES 11-2

In Exercises 1-10 find the equation for each of the circles from the given information.
   1. Center at $(-3, 4)$, radius 2.
   2. Center at $(-2, 1)$, radius 3.
   3. Center at $(0, 0)$, radius 1.
   4. Center at $(0, 2)$, radius 5.
   5. Center at $(-4, -1)$, tangent to the line $x = -1$. (*Note:* A line is tangent to a circle if the two intersect at exactly one point. The tangent is perpendicular to the radius at the point of intersection.)
   6. Center at $(1, 4)$, tangent to the line $y = -3$.
   7. Center at $(3, -3)$, passes through the origin.
   8. Center at $(2, -5)$, passes through $(1, 0)$.
   9. Center at $(-1, -1)$, passes through $(6, 2)$.
   10. Center at $(-4, -3)$, passes through $(-1, -5)$.

In Exercises 11-14 find the center and radius of each circle.
   11. $(x - 2)^2 + (y - 5)^2 = 16$ **12.** $(x + 1)^2 + (y + 4)^2 = 49$
   13. $(x - 3)^2 + y^2 = 20$ **14.** $x^2 + (y + 2)^2 = 5$

In Exercises 15-22 find the center and radius of each circle. Sketch the circle in each case.
   15. $x^2 + y^2 = 9$ **16.** $x^2 + y^2 - 4 = 0$
   17. $x^2 + y^2 - 10x - 6y = 15$ **18.** $x^2 + y^2 - 4x + 6y = 23$
   19. $x^2 + y^2 + 8x - 2y - 1 = 0$ **20.** $x^2 + y^2 + 4x - 9 = 0$
   21. $x^2 + y^2 + 7x + 3y + 4 = 0$ **22.** $x^2 + y^2 - 5x - y - 3 = 0$

11-3 THE ELLIPSE

An *ellipse* is defined as a *collection of points in the plane the sum of whose distances from two fixed points is a constant*. From this definition we derive a general equation for an ellipse as follows. For simplicity let the two fixed points, called *foci*, be positioned at $(-c, 0)$ and $(c, 0)$, and let $2a$ represent the constant sum (see Figure 11.7).

$$d_1 + d_2 = 2a$$

Then, by the distance formula, we have

$$\sqrt{(x + c)^2 + (y - 0)^2} + \sqrt{(x - c)^2 + (y - 0)^2} = 2a$$

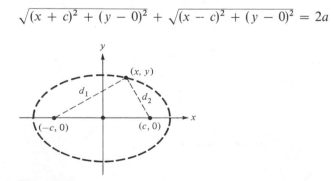

Figure 11.7

or

$$\sqrt{(x + c)^2 + y^2} = 2a - \sqrt{(x - c)^2 + y^2}$$

Now square both sides of the equation and simplify.

$$(x + c)^2 + y^2 = 4a^2 - 4a\sqrt{(x - c)^2 + y^2} + (x - c)^2 + y^2$$
$$xc = a^2 - a\sqrt{(x - c)^2 + y^2}$$
$$a\sqrt{(x - c)^2 + y^2} = a^2 - cx$$

Again square both sides of the equation and simplify.

$$a^2[(x - c)^2 + y^2] = a^4 - 2a^2cx + c^2x^2$$
$$a^2x^2 - 2a^2cx + a^2c^2 + a^2y^2 = a^4 - 2a^2cx + c^2x^2$$
$$a^2x^2 - c^2x^2 + a^2y^2 = a^4 - a^2c^2$$
$$(a^2 - c^2)x^2 + a^2y^2 = a^2(a^2 - c^2)$$
$$\frac{x^2}{a^2} + \frac{y^2}{a^2 - c^2} = 1$$

Since $d_1 + d_2$ is greater than the distance between the foci, it follows that $a > c$ and $a^2 - c^2 > 0$. If we now define

$$b^2 = a^2 - c^2 \quad \text{or} \quad a^2 = b^2 + c^2$$

we obtain the standard form of an ellipse with center at the origin and foci on the x axis.

$$\frac{x^2}{a^2} + \frac{y^2}{b^2} = 1$$

Consider Figure 11.8. By setting $y = 0$, we find that the x intercepts are $(-a, 0)$ and $(a, 0)$. By setting $x = 0$, we find that the y intercepts are $(0, -b)$ and $(0, b)$. The larger segment from $(-a, 0)$ to $(a, 0)$ is called the *major axis*; the *minor axis* is the

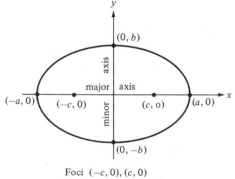

Foci $(-c, 0), (c, 0)$
Vertices $(-a, 0), (a, 0)$

Figure 11.8

segment from $(0, -b)$ to $(0, b)$. The end points of the major axis are called the *vertices of the ellipse.*

Foci: $(-c, 0), (c, 0)$

Vertices: $(-a, 0), (a, 0)$

If the foci are placed on the y axis at $(0, -c)$ and $(0, c)$, the equation of the ellipse is

$$\frac{x^2}{b^2} + \frac{y^2}{a^2} = 1$$

The larger denominator is denoted by a^2. The major axis is then along the y axis, as shown in Figure 11.9. Note that the foci always lie on the major axis.

Foci: $(0, c), (0, -c)$

Vertices: $(0, a), (0, -a)$

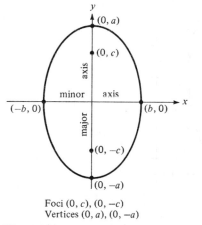

Foci $(0, c), (0, -c)$
Vertices $(0, a), (0, -a)$

Figure 11.9

EXAMPLE 1: Find the coordinates of the foci and the end points of the major and minor axes in the ellipse $4x^2 + y^2 = 36$.

SOLUTION: First, divide both sides of the equation by 36, to obtain the standard form.

$$\frac{x^2}{9} + \frac{y^2}{36} = 1$$

Then

$$a^2 = 36 \qquad a = 6$$
$$b^2 = 9 \qquad b = 3$$

To find c we replace a^2 by 36 and b^2 by 9 in the formula

$$a^2 = b^2 + c^2$$
$$36 = 9 + c^2$$
$$27 = c^2 \quad \text{or} \quad c = \sqrt{27}$$

The major axis is along the y axis since the larger denominator appears in the y term. Thus,

end points of major axis: $(0, -6), (0, 6)$

end points of minor axis: $(-3, 0), (3, 0)$

coordinates of foci: $(0, -\sqrt{27}), (0, \sqrt{27})$

See Figure 11.10.

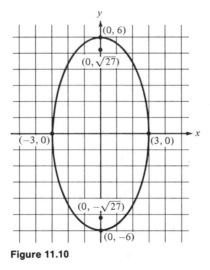

Figure 11.10

EXAMPLE 2: Find the equation of the ellipse with center at the origin, one focus at (3, 0), and one vertex at (4, 0).

SOLUTION: Since c is the distance from the center to a focus, c is 3. Since a is the distance from the center to a vertex, a is 4. To find b^2 replace c by 3 and a by 4 in the formula

$$a^2 = b^2 + c^2$$
$$(4)^2 = b^2 + (3)^2$$
$$7 = b^2$$

Since the major axis is along the x axis, the general equation is

$$\frac{x^2}{a^2} + \frac{y^2}{b^2} = 1$$

Thus,

$$\frac{x^2}{16} + \frac{y^2}{7} = 1$$

is the equation of the ellipse.   **Answer**

If the ellipse is centered at the point (h, k) the standard form is

$$\frac{(x - h)^2}{a^2} + \frac{(y - k)^2}{b^2} = 1$$

if the major axis is parallel to the x axis, and

$$\frac{(x - h)^2}{b^2} + \frac{(y - k)^2}{a^2} = 1$$

if the major axis is parallel to the y axis. As in the case of the circle, the standard form is obtained by completing the square.

EXAMPLE 3: Find the coordinates of the foci and the end points of the major and minor axes in the ellipse $9x^2 + 4y^2 + 18x - 8y - 23 = 0$. Also, sketch the curve.

SOLUTION: First, put the equation in standard form by completing the square.

$$(9x^2 + 18x \qquad) + (4y^2 - 8y \qquad) = 23$$
$$9(x^2 + 2x \qquad) + 4(y^2 - 2y \qquad) = 23$$
$$9(x^2 + 2x + 1) + 4(y^2 - 2y + 1) = 23 + 9(1) + 4(1) = 36$$

Be careful to add 9(1) and 4(1) to the right side of the equation. Now, divide both sides of the equation by 36.

$$\frac{(x + 1)^2}{4} + \frac{(y - 1)^2}{9} = 1$$

From this equation in standard form we conclude that the center is at $(-1, 1)$.

$$a^2 = 9 \qquad a = 3$$
$$b^2 = 4 \qquad b = 2$$

To find c we replace a^2 by 9 and b^2 by 4 in the formula

$$a^2 = b^2 + c^2$$
$$9 = 4 + c^2$$
$$5 = c^2 \quad \text{or} \quad c = \sqrt{5}$$

The major axis is parallel to the y axis since the larger denominator appears in the y term; a is the distance from the center to the vertices. Thus,

end points of major axis: $(-1, -2), (-1, 4)$

The distance from the center to the end points of the minor axis is b. Thus,

  end points of minor axis: $(-3, 1), (1, 1)$

The distance from the center to the foci is c. Thus,

  coordinates of the foci: $(-1, 1 - \sqrt{5}), (-1, 1 + \sqrt{5})$

The ellipse is graphed in Figure 11.11.

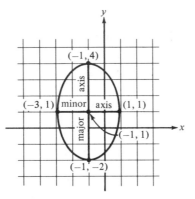

Figure 11.11

Finally, let us consider a few of the ways the ellipse appears in the physical world.

1. In our solar system many bodies revolve in elliptical orbits around a larger body which is located at one focus. For instance, all planets and some comets (such as Halley's comet, which should appear next in 1985 after a 75-year period) travel in elliptical paths with the sun at one focus (see Figure 11.12). The moon and artificial satellites follow a similar path around the earth. The ellipse was first analyzed in detail by the Greek mathematician Apollonius in about 230 B.C. for its aesthetic, not practical, value. However, if this study had not been available, it is likely that Johannes Kepler (A.D. 1600) would have been unable to explain planetary motion scientifically, which would have caused a long delay in the development of astronomy and the subsequent sciences. This episode is just one of many cases in which a purely mathematical investigation later proved to be of great practical value.

2. The ellipse has a reflection property which causes any ray or wave that originates at one focus to strike the ellipse and pass through the other focus. Acoustically, this property means that in a room with an elliptical ceiling, a slight noise made at one focus can be heard at the other focus (see Figure 11.13). However, if you are standing between the foci, you hear nothing. Such rooms are called whispering galleries, and a famous one is located in the Capitol in Washington, D.C.

3. Arches whose main purpose is beauty and not strength are often elliptical in shape. These ornamental arches are usually large structures made of masonry.

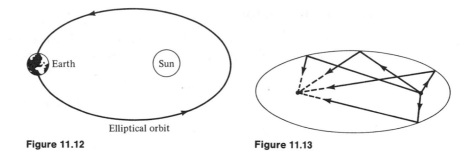

Figure 11.12

Figure 11.13

EXERCISES 11–3

In Exercises 1–14 find the coordinates of the foci and the end points of the major and minor axes. Sketch each ellipse.

1. $\dfrac{x^2}{25} + \dfrac{y^2}{16} = 1$

2. $\dfrac{x^2}{36} + \dfrac{y^2}{100} = 1$

3. $\dfrac{x^2}{1} + \dfrac{y^2}{4} = 1$

4. $\dfrac{x^2}{49} + \dfrac{y^2}{9} = 1$

5. $x^2 + 9y^2 = 36$

6. $9x^2 + 4y^2 = 144$

7. $4x^2 + y^2 = 1$

8. $x^2 + 9y^2 = 25$

9. $\dfrac{(x-1)^2}{9} + \dfrac{(y+3)^2}{25} = 1$

10. $\dfrac{(x+2)^2}{100} + \dfrac{(y-2)^2}{64} = 1$

11. $4(x-3)^2 + 25(y+1)^2 = 100$

12. $25(x+4)^2 + 16y^2 = 400$

13. $4x^2 + 9y^2 + 8x - 54y + 49 = 0$

14. $9x^2 + y^2 - 36x + 2y + 1 = 0$

In Exercises 15–24 find the equation of the ellipse satisfying the conditions.

15. Center at origin, vertex (0, 5), focus (0, 3).

16. Center at origin, vertex (−10, 0), focus (−8, 0).

17. Center at origin, focus (4, 0), length of major axis 12.

18. Center at origin, focus (0, −2), length of minor axis 2.

19. Center at origin, vertex (3, 0), length of minor axis 4.

20. Center at origin, vertex (0, 3), passes through (2, 1).

21. Center at (2, 2), focus (2, −1), vertex (2, −3).

22. Center at (−1, 3), vertex (3, 3), length of minor axis 6.

23. Foci at (5, 3) and (−1, 3), length of major axis 10.

24. Vertices at (4, 1) and (−8, 1), focus at (−4, 1).

25. The earth travels in an elliptical orbit around the sun, which is located at one of the foci. The longest distance between the earth and the sun is 94,500,000 mi and the shortest distance is 91,500,000 mi. Find:

    a. The length of the major axis of the ellipse.

    b. The distance between the foci of the orbit.

26. An elliptical arch has a height of 20 ft and a span of 50 ft. How high is the arch 15 ft each side of center?

11–4 THE HYPERBOLA

A *hyperbola* is defined as the *collection of points in the plane the difference of whose distances from two fixed points is a constant.* From this definition we derive a general equation for a hyperbola, as follows. For simplicity, let the two fixed points, called *foci*, be positioned at $(-c, 0)$ and $(c, 0)$, and let $2a$ represent the constant difference (see Figure 11.14). Since $d_1 < d_2$ or $d_1 > d_2$, by definition we want

$$|d_1 - d_2| = 2a \quad \text{or} \quad d_1 - d_2 = \pm 2a$$

Then, by the distance formula, we have

$$\sqrt{(x + c)^2 + (y - 0)^2} - \sqrt{(x - c)^2 + (y - 0)^2} = \pm 2a$$

or

$$\sqrt{(x + c)^2 + y^2} = \pm 2a + \sqrt{(x - c)^2 + y^2}$$

Now square both sides of the equation and simplify.

$$(x + c)^2 + y^2 = 4a^2 \pm 4a\sqrt{(x - c)^2 + y^2} + (x - c)^2 + y^2$$
$$4cx = 4a^2 \pm 4a\sqrt{(x - c)^2 + y^2}$$
$$cx - a^2 = \pm a\sqrt{(x - c)^2 + y^2}$$

Again square both sides and simplify.

$$c^2x^2 - 2cxa^2 + a^4 = a^2[(x - c)^2 + y^2]$$
$$c^2x^2 - 2cxa^2 + a^4 = a^2x^2 - 2cxa^2 + a^2c^2 + a^2y^2$$
$$c^2x^2 - a^2x^2 - a^2y^2 = a^2c^2 - a^4$$
$$(c^2 - a^2)x^2 - a^2y^2 = a^2(c^2 - a^2)$$
$$\frac{x^2}{a^2} - \frac{y^2}{c^2 - a^2} = 1$$

·If we now define

$$b^2 = c^2 - a^2 \quad \text{or} \quad c^2 = a^2 + b^2$$

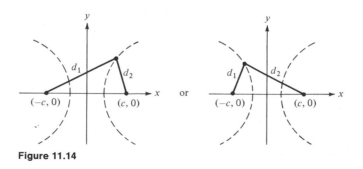

Figure 11.14

we obtain the standard form of a hyperbola, with center at the origin and foci on the x axis.

$$\frac{x^2}{a^2} - \frac{y^2}{b^2} = 1$$

Consider Figure 11.15. By setting $y = 0$ we find that the x intercepts are $(-a, 0)$ and $(a, 0)$. The line segment joining these two points is called the *transverse axis*. The endpoints of the transverse axis are called the *vertices of the hyperbola*. By setting $x = 0$, we find that there are no y intercepts. The line segment from $(0, b)$ to $(0, -b)$ is called the *conjugate axis*. To determine the significance of b, we rewrite $(x^2/a^2) - (y^2/b^2) = 1$ as

$$y = \frac{\pm bx}{a} \sqrt{1 - \frac{a^2}{x^2}}$$

As $|x|$ gets very large, $1 - a^2/x^2$ approaches 1. Thus, the graph of the hyperbola gets close to the lines

$$y = \frac{\pm bx}{a}$$

These lines are called the *asymptotes of the hyperbola*. They are a great aid in sketching the curve. As shown in Figure 11.15, the asymptotes are the diagonals of a rectangle of dimensions $2a$ by $2b$.

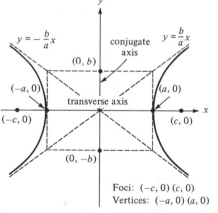

Figure 11.15

EXAMPLE 1: For the hyperbola $4x^2 - y^2 = 36$, find the coordinates of the vertices and the foci. Also, determine the asymptotes and sketch the curve.

SOLUTION: First divide both sides of the equation by 36 to obtain the standard form.

$$\frac{x^2}{9} - \frac{y^2}{36} = 1$$

Then

$$a^2 = 9 \qquad a = 3$$
$$b^2 = 36 \qquad b = 6$$

Note that a^2 is not necessarily the larger denominator. Since a is the distance from the center to the vertices, we have

coordinates of the vertices: $(-3, 0)(3, 0)$

To find c we replace a^2 by 9 and b^2 by 36 in the formula

$$c^2 = a^2 + b^2$$
$$= 9 + 36$$
$$c^2 = 45 \quad \text{or} \quad c = \sqrt{45}$$

The distance from the center to the foci is c. Thus,

coordinates of the foci: $(-\sqrt{45}, 0)(\sqrt{45}, 0)$

For the asymptotes we have

$$y = \frac{\pm bx}{a} = \frac{\pm 6x}{3} = \pm 2x$$

The hyperbola is sketched in Figure 11.16.   **Answer**

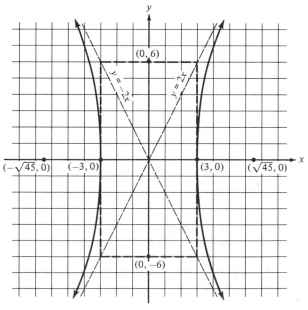

Figure 11.16

If the foci are positioned on the y axis at $(0, -c)$ and $(0, c)$ the standard form of the hyperbola is

$$\frac{y^2}{a^2} - \frac{x^2}{b^2} = 1$$

In this case the asymptotes are given by $y = \pm ax/b$, as shown in Figure 11.17.

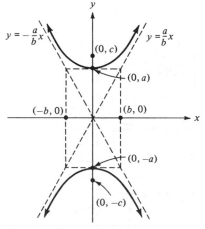

Figure 11.17

EXAMPLE 2: For the hyperbola $25y^2 - 4x^2 = 100$, find the coordinates of the vertices and the foci. Also determine the asymptotes and sketch the curve.

SOLUTION: First, divide both sides of the equation by 100 to obtain the standard form.

$$\frac{y^2}{4} - \frac{x^2}{25} = 1$$

Then

$$a^2 = 4 \qquad a = 2$$
$$b^2 = 25 \qquad b = 5$$

Since a is the distance from the center to the vertices, we have

coordinates of the vertices: $(0, 2)(0, -2)$

To find c we replace a^2 by 4 and b^2 by 25 in the formula

$$c^2 = a^2 + b^2$$
$$= 4 + 25$$
$$c^2 = 29 \quad \text{or} \quad c = \sqrt{29}$$

The distance from the center to the foci is c. Thus,

coordinates of foci: $(0, \sqrt{29})(0, -\sqrt{29})$

For the asymptotes we have

$$y = \frac{\pm ax}{b} = \frac{\pm 2x}{5}$$

The hyperbola is sketched in Figure 11.18.

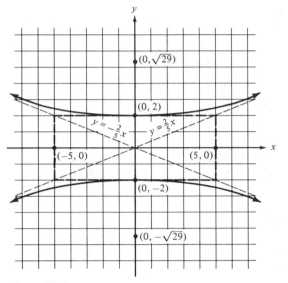

Figure 11.18

EXAMPLE 3: Find the equation of the hyperbola with center at origin one focus at (0, 5) and one vertex at (0, 3).

SOLUTION: Since c is the distance from the center to a focus, c is 5. Since a is the distance from the center to a vertex, a is 3. To find b^2, replace c by 5 and a by 3 in the formula

$$c^2 = a^2 + b^2$$
$$(5)^2 = (3)^2 + b^2$$
$$16 = b^2$$

Since the foci are along the y axis, the general equation is

$$\frac{y^2}{a^2} - \frac{x^2}{b^2} = 1$$

Thus, the equation of the hyperbola is

$$\frac{y^2}{9} - \frac{x^2}{16} = 1 \qquad \textbf{Answer}$$

If the hyperbola is centered at the point (h, k), the standard form is

$$\frac{(x - h)^2}{a^2} - \frac{(y - k)^2}{b^2} = 1$$

when the transverse axis is parallel to the x axis. In this case the asymptotes are given by $y - k = (\pm b/a)(x - h)$. If the transverse axis is parallel to the y axis, the standard form is

$$\frac{(y - k)^2}{a^2} - \frac{(x - h)^2}{b^2} = 1$$

and the asymptotes are given by $y - k = (\pm a/b)(x - h)$.

EXAMPLE 4: For the hyperbola $16y^2 - x^2 - 32y + 4x - 4 = 0$, find the coordinates of the vertices and the foci. Also, determine the asymptotes and sketch the curve.

SOLUTION: First put the equation in standard form by completing the square.

$$(16y^2 - 32y \quad) - (x^2 - 4x \quad) = 4$$
$$16(y^2 - 2y \quad) - (x^2 - 4x \quad) = 4$$
$$16(y^2 - 2y + 1) - (x^2 - 4x + 4) = 4 + 16(1) + (-1)4$$
$$16(y - 1)^2 - (x - 2)^2 = 16$$

Dividing both sides of the equation by 16, we have

$$\frac{(y - 1)^2}{1} - \frac{(x - 2)^2}{16} = 1$$

From this equation in standard form we conclude that the center is at $(2, 1)$.

$$a^2 = 1 \qquad a = 1$$
$$b^2 = 16 \qquad b = 4$$

The transverse axis is parallel to the y axis since the y term is positive; a is the distance from the center to the vertices. Thus,

coordinates of vertices: $(2, 0)(2, 2)$

To find c we replace a^2 by 1 and b^2 by 16 in the formula.

$$c^2 = a^2 + b^2$$
$$= 1 + 16$$
$$c^2 = 17 \quad \text{or} \quad c = \sqrt{17}$$

The distance from the center to the foci is c. Thus,

coordinates of foci: $(2, 1 - \sqrt{17})(2, 1 + \sqrt{17})$

The asymptotes are the diagonals of the rectangle of dimensions $2a$ by $2b$ that is centered at $(2, 1)$. The equations of these asymptotes are

$$y - k = \pm\frac{a}{b}(x - h)$$

$$y - 1 = \pm\frac{1}{4}(x - 2)$$

The hyperbola is graphed in Figure 11.19.

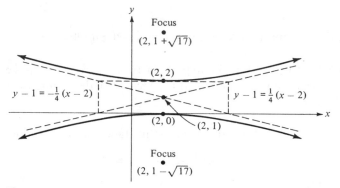

Figure 11.19

EXERCISES 11–4

In Exercises 1–14 find the coordinates of the vertices and the foci, determine the asymptotes, and sketch each curve.

1. $\dfrac{x^2}{16} - \dfrac{y^2}{9} = 1$

2. $\dfrac{x^2}{36} - \dfrac{y^2}{100} = 1$

3. $\dfrac{y^2}{25} - \dfrac{x^2}{16} = 1$

4. $\dfrac{y^2}{1} - \dfrac{x^2}{4} = 1$

5. $x^2 - 9y^2 = 36$

6. $9x^2 - 4y^2 = 144$

7. $4y^2 - x^2 = 1$

8. $y^2 - 9x^2 = 25$

9. $\dfrac{(x + 2)^2}{9} - \dfrac{(y - 3)^2}{25} = 1$

10. $\dfrac{(y - 1)^2}{64} - \dfrac{(x - 2)^2}{36} = 1$

11. $4(y + 1)^2 - 25(x + 3)^2 = 100$

12. $25(x + 4)^2 - 16y^2 = 400$

13. $4x^2 - 9y^2 + 8x - 54y - 113 = 0$

14. $9y^2 - x^2 - 36y + 2x - 1 = 0$

In Exercises 15–24 find the equation of the hyperbola satisfying the conditions.

15. Center at origin, focus $(0, 5)$, vertex $(0, 4)$.

16. Center at origin, focus $(-10, 0)$, vertex $(-8, 0)$.

17. Center at origin, focus $(4, 0)$, length of transverse axis 6.

18. Center at origin, focus $(0, -2)$, length of conjugate axis 2.

19. Center at origin, vertex (3, 0), length of conjugate axis 10.
20. Center at origin, vertex (0, 2), passes through (3, 4).
21. Center at $(-3, -4)$, focus $(2, -4)$, vertex $(0, -4)$.
22. Center at (5, 0), vertex (5, 6), length of conjugate axis 8.
23. Foci at (3, 4) and $(3, -2)$, length of transverse axis 4.
24. Vertices at (4, 1) and $(-10, 1)$, focus (7, 1).

11–5 THE PARABOLA

A *parabola* is defined as the collection of points in the plane equidistant from a given point (called the *focus*) and a given line (called the *directrix*). From this definition we derive a general equation for a parabola as follows. For simplicity, let the directrix be the line $x = -p$ and position the focus at $(p, 0)$. $P = (x, y)$ represents any point on the parabola (see Figure 11.20). Since P_1 and P have the same y coordinate, the distance d_1 is given by

$$d_1 = |x - (-p)| = |x + p|$$

The distance from P to $(p, 0)$ is given by

$$d_2 = \sqrt{(x - p)^2 + (y - 0)^2}$$

The geometric condition for a parabola states that

$$d_1 = d_2$$

Thus,

$$|x + p| = \sqrt{(x - p)^2 + y^2}$$

Squaring both sides of the equation yields

$$(x + p)^2 = (x - p)^2 + y^2$$

Simplifying this equation we have

$$x^2 + 2px + p^2 = x^2 - 2px + p^2 + y^2$$
$$4px = y^2$$

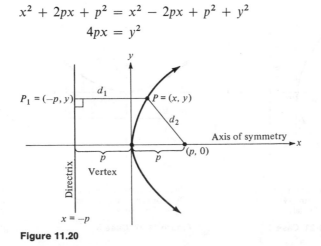

Figure 11.20

This equation is the standard form of a parabola with directrix $x = -p$ and focus at $(p, 0)$. The line through the focus that is perpendicular to the directrix is called the *axis of symmetry*. In this case the axis of symmetry is the x axis. The point on the axis of symmetry that is midway between the focus and the directrix is called the *vertex*. The vertex is the turning point of the parabola. In this case the vertex is the origin. Note that p is the distance from the vertex to the focus and from the vertex to the directrix.

EXAMPLE 1: Find the focus and the directrix of the parabola $y^2 = 12x$.

SOLUTION: Matching the equation $y^2 = 12x$ to the form

$$y^2 = 4px$$

we write

$$y^2 = 12x = 4(3)x$$

Thus,

$$p = 3$$

The focus is on the axis of symmetry (the x axis); p units to the right of the vertex (the origin). Thus,

focus: $(3, 0)$

The directrix is the line p units to the left of the vertex. Thus,

directrix: $x = -3$

There are three other possibilities for a parabola with a directrix that is parallel to the x or y axis and whose vertex is the origin. These cases are illustrated in Figures 11.21–11.23. In all cases $p > 0$.

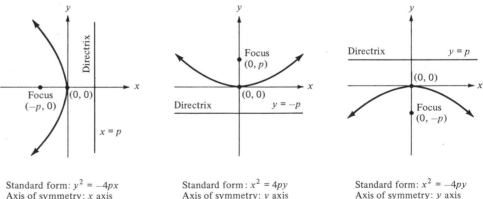

Standard form: $y^2 = -4px$
Axis of symmetry: x axis
opens to the left

Figure 11.21 Case 2

Standard form: $x^2 = 4py$
Axis of symmetry: y axis
opens upward

Figure 11.22 Case 3

Standard form: $x^2 = -4py$
Axis of symmetry: y axis
opens downward

Figure 11.23 Case 4

To summarize, the graphs of the equations $y^2 = \pm 4px$ and $x^2 = \pm 4py$ are parabolas. If the y term is squared, the axis of symmetry is the x axis. The parabola opens to the right when the coefficient of x is positive and to the left if this coefficient is negative. When the x term is squared, the axis of symmetry is the y axis. The parabola opens upward if the coefficient of y is positive and downward when this coefficient is negative. In all cases p gives the distance from the vertex to the focus and from the vertex to the directrix.

EXAMPLE 2: Find the focus and directrix of the parabola $x^2 = -6y$.

SOLUTION: Matching the equation $x^2 = -6y$ to the form in case 4,

$$x^2 = -4py$$

we write

$$x^2 = -6y = -4(\tfrac{3}{2})y$$

Thus,

$$p = \tfrac{3}{2}$$

The focus is on the axis of symmetry (the y axis) p units down from the vertex (the origin). Thus,

focus: $(0, -\tfrac{3}{2})$

The directrix is the line p units up from the vertex. Thus,

directrix: $y = \tfrac{3}{2}$

EXAMPLE 3: A parabola with its vertex at the origin has its focus at $(-5, 0)$. Find the directrix and the equation of the parabola, and sketch the curve.

SOLUTION: This is an example of case 2. The term p, which represents the distance from the vertex $(0, 0)$ to the focus $(-5, 0)$, is 5. Thus,

directrix: $x = 5$

To find the equation of the parabola, replace p by 5 in the form

$$y^2 = -4px$$
$$= -4(5)x$$
$$y^2 = -20x$$

The parabola opens to the left and is graphed in Figure 11.24.

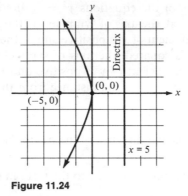

Figure 11.24

If the vertex of the parabola is at the point (h, k) and the directrix is parallel to the x or y axis, there are four possible forms for the parabola. These cases are illustrated in Figures 11.25–11.28. In all cases $p > 0$.

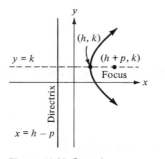

Figure 11.25 Case 1

$$(y - k)^2 = 4p(x - h)$$

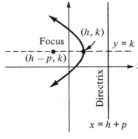

Figure 11.26 Case 2

$$(y - k)^2 = -4p(x - h)$$

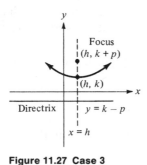

Figure 11.27 Case 3

$$(x - h)^2 = 4p(y - k)$$

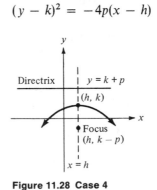

Figure 11.28 Case 4

$$(x - h)^2 = -4p(y - k)$$

EXAMPLE 4: For the parabola $y = x^2 - 2x$, find the vertex, the focus, and the directrix. Also, sketch the curve.

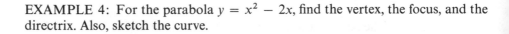

SOLUTION: First put the equation in standard form by completing the square.

$$y = (x^2 - 2x \qquad)$$
$$y + 1 = (x^2 - 2x + 1)$$
$$y + 1 = (x - 1)^2$$

Matching this equation to the form in case 3,

$$4p(y - k) = (x - h)^2$$

we conclude that

$$h = 1 \qquad k = -1 \qquad 4p = 1 \quad \text{or} \quad p = \tfrac{1}{4}$$

Thus,

vertex: $(1, -1)$

The focus is on the axis of symmetry ($x = 1$), p units up from the vertex $(1, -1)$. Thus,

focus: $(1, -\tfrac{3}{4})$

The directrix is the line p units down from the vertex. Thus

directrix: $y = -\tfrac{5}{4}$

The parabola opens upward and is graphed in Figure 11.29.

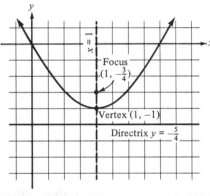

Figure 11.29

In conclusion let us consider a few of the applications of the parabola.

1. The parabola has an important reflection property. Any ray or wave that originates at the focus and strikes the parabola is reflected parallel to the axis of symmetry (see Figure 11.30). For this reason instruments such as a flashlight or searchlight use a parabolic reflector with the bulb located at the focus. The reflector redirects light that would have been wasted parallel to the axis so that a straight beam of light is formed. In an automobile headlight a bulb located at the focus produces

Figure 11.30   Figure 11.31

the high beam. For the low beam the bulb is placed slightly ahead and above the focus. In reverse, the reflection property of a parabola causes any ray or wave that comes into a parabolic reflector parallel to the axis of symmetry to be directed to the focus point. Radar, radio antennas, and reflecting telescopes work on this principle (see Figure 11.31).

2. Arches whose main purpose is strength are usually parabolic in shape and constructed of steel.

3. In a suspension bridge the cables are designed for engineering purposes to hang in a shape that is almost parabolic.

4. All comets in our solar system travel in an orbit that is either an ellipse, a parabola, or a hyperbola. In each case the sun is located at one focus. Only comets with an elliptical orbit reappear.

5. Projectiles, such as a baseball or the water that streams from a fountain, follow a parabolic path.

EXERCISES 11–5

In Exercises 1–14 find the vertex, the focus, the equation of the directrix, and sketch each curve.

1. $y^2 = 4x$ **2.** $y^2 = -8x$
3. $x^2 = -2y$ **4.** $x^2 = 5y$
5. $y^2 + 12x = 0$ **6.** $y^2 - 10x = 0$
7. $4x^2 - 3y = 0$ **8.** $2x^2 + 3y = 0$
9. $(y + 2)^2 = 4(x + 1)$ **10.** $(x - 1)^2 = -6y$
11. $y = x^2 - 4x$ **12.** $x = y^2 + 2y$
13. $x^2 - 2x - 4y - 7 = 0$ **14.** $y^2 + 4y + 3x - 8 = 0$

In Exercises 15–24 find the equation of the parabola satisfying the conditions.

15. Vertex at origin, focus $(0, 3)$.
16. Vertex at origin, focus $(-2, 0)$.
17. Vertex at origin, directrix $x = -\frac{1}{2}$.
18. Vertex at origin, directrix $y = \frac{3}{2}$.
19. Focus $(0, 4)$, directrix $y = -4$.
20. Focus $(-\frac{5}{2}, 0)$, directrix $x = \frac{5}{2}$.
21. Focus $(2, 1)$, directrix $x = -4$.
22. Focus $(-1, -3)$, directrix $y = 5$.
23. Vertex $(3, 0)$, focus $(3, 3)$.
24. Vertex $(1, 4)$, directrix $x = \frac{7}{2}$.

25. Consider the parabolic reflector shown. Choose axes in a convenient position and determine:
  a. The equation of the parabola.
  b. The location of the focus point.

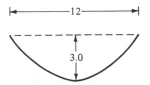

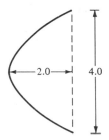

26. Answer the questions in Exercise 25 for the parabolic reflector shown.

27. A parabolic arch has a height of 25 ft and a span of 40 ft. How high is the arch 8 ft each side of the center?

11-6 CONIC SECTIONS

Circles, ellipses, hyperbolas, and parabolas are referred to as *conic sections*. This name was first used by Greek mathematicians, who discovered that these curves result from the intersections of a cone with an appropriate plane. Consider Figure 11.32. The intersection is a circle if the plane is parallel to the base of the cone; a parabola if the plane is parallel to the side of the cone; a hyperbola if the plane is parallel to the axis of the cone; and an ellipse if the plane is parallel to neither the base, the side, nor the axis of the cone.

Each of the conics can be represented by an equation of the form

$$Ax^2 + Cy^2 + Dx + Ey + F = 0$$

In this general equation, the letter B is reserved for an xy term that we do not discuss. The distinguishing characteristics for each conic are summarized below. In these possibilities we do not consider special cases in which the graph degenerates to a line, a point, or no graph at all.

1. A circle: $A = C \neq 0$.
2. An ellipse: $A \neq C$, A and C have the same sign.
3. A hyperbola: A and C have different signs.
4. A parabola: either A or C equals 0, but not both.

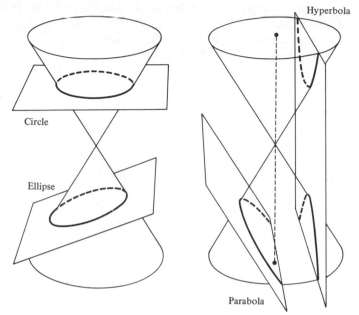

Figure 11.32

EXAMPLE 1: The equation $4x^2 - 9y^2 + 8x - 54y - 113 = 0$ defines a hyperbola since A and C have different signs.

EXAMPLE 2: To determine the conic defined by $2x^2 = y - 3y^2$, we first write the equation as

$$2x^2 + 3y^2 - y = 0$$

In this case A and C have the same sign with $A \neq C$. Thus, the equation defines an ellipse.

EXAMPLE 3: The equation $x^2 + y^2 + 1 = 0$ does not define a circle even though $A = C = 1$. There is no ordered pair of real numbers that satisfies the equation since x^2 and y^2 are never negative, so their sum can never be -1. This equation is an example of one of the special cases that is not considered in the scheme above.

EXERCISES 11–6

In Exercises 1–10 classify each equation as defining either a circle, ellipse, hyperbola, or parabola.

1. $y^2 - 3x + 2y + 1 = 0$

2. $9y^2 - x^2 - 36x + 2y - 1 = 0$

3. $4x^2 + 9y^2 + 8x - 54y - 49 = 0$

4. $x^2 + y^2 + 7x + 3y - 4 = 0$

5. $x^2 = y^2 - 4$

6. $3y^2 = 4x - x^2$

7. $3y = 4x - x^2$

8. $5x^2 = 1 - 5y^2$

9. $2y^2 = 9 - 2x^2$

10. $5x - 2y = y^2$

Sample Test: Chapter 11

1. Find the equation whose graph is formed by a point that moves in the plane in the following way: its distance from $(5, 9)$ is always three times its distance from $(1, 3)$. Which conic section does this equation define?

2. Classify the graph defined by each of the following equations.
   a. $9x^2 = 4y^2 - 54x - 16y + 61$
   b. $x^2 + 4x + 8y = 4$
   c. $7x - 2y = 3$
   d. $16x^2 + 25y^2 - 32x - 284 = 0$
   e. $x^2 + 6x = 3 - y^2$

3. Determine the center and radius of the circle $x^2 + y^2 + 4x - 6y = 12$.

4. Determine the equation of the circle with center at $(4, -3)$ which passes through $(0, -1)$.

5. Find the coordinates of the foci for $25x^2 + 16y^2 = 400$.

6. What is the equation of the ellipse with vertices at $(3, 2)$ and $(-7, 2)$ and a minor axis of length 6?

7. Find the equations of the asymptotes of the hyperbola $y^2/64 - x^2/100 = 1$.

8. Determine the standard equation of the hyperbola whose center is at the origin with one focus at $(-8, 0)$ and a transverse axis of length 14.

9. What are the coordinates of the focus point of the parabola $x^2 = -7y$?

10. Find the equation of the parabola with focus at $(5, 1)$ and directrix $x = -1$.

Appendixes

Appendixes

Appendix A-1

Approximate numbers

Measurements of one kind or another are essential to both scientific and nontechnical work. Weight, distance, time, volume, and temperature are but a few of the quantities for which measurements are required. We measure through the medium of numbers and arbitrary units, such as the foot (British system) or the meter (metric system). These numbers are approximations and can only be as accurate as the measuring instruments allow. For example, if we measure the height of a man to be 68 in., we are saying that his height to the nearest inch is 68 in. This means that his exact height (h) is somewhere in the interval 67.5 in. $\leq h < 68.5$ in. Using a different measuring device we might record the height to be 68.2 in., so that 68.15 in. $\leq h < 68.25$ in.

The exact height is contained in these intervals because the measurements are estimated by rounding off to some decimal place. We round off by considering the digit in the next place to the right of the desired decimal place. If this digit is less than 5, the digit in the desired decimal place remains the same; if the digit is 5 or greater, the digit in the desired decimal place is increased by one. The digits to the right of the desired decimal place are then dropped and replaced by zeros if the zeros are needed to maintain the position of the decimal point. Rounding off is best understood by considering the following table, which indicates how we round off a measurement of 265.307 ft to various places.

Precision desired	Measurement of 265.307 ft is given as
Nearest 100 ft	300 ft
Nearest 10 ft	270 ft
Nearest foot	265 ft
Nearest tenth of a foot	265.3 ft
Nearest hundredth of a foot	265.31 ft

The *precision* of a number refers to the place at which we are rounding off. A number rounded off to the nearest hundredth is more precise than a number rounded off to the nearest tenth, and so on.

Another important consideration is the number of *significant digits* in our approximation. All digits, except the zeros that are required to indicate the position of the decimal point, are significant. The following table illustrates this concept:

Approximate number	Number of significant digits
15.13	Four
0.44	Two
0.003	One; the number is *three* thousandths and only the three is significant. The zeros are needed to indicate the position of the decimal point.
0.0196	Three
307	Three; the zero is *not* used to indicate the position of the decimal point. Zeros between nonzero digits are always significant.
50.007	Five
8.0	Two; the zero is significant because it shows that the number is rounded off to the nearest tenth. The zero is not required to write the number eight and should not be written unless it is significant.
0.2600	Four
230	Two; we do not have sufficient information to determine if the number is rounded off in the tens place or to the nearest unit. Unless stated otherwise, we assume the number to be rounded off in the tens place, so the zero is needed to indicate the position of the decimal point.
2000	One
$200\bar{0}$	Four; the bar is used to avoid ambiguity by indicating that the number is rounded off at the digit below the bar.
$2\bar{0}00$	Two

The *accuracy* of a number is determined by the number of significant digits in the number. Thus, 21 (two significant digits) is a more accurate number than 0.07 (one significant digit). It is important to distinguish between the accuracy and the precision of a number. For example, consider the following pairs of numbers:

21 and 0.07 $\begin{cases} \text{21 is more accurate since it contains two significant digits.} \\ \text{0.07 is more precise since it is written to the nearest hundredth.} \end{cases}$

123 and 12.3 $\begin{cases} \text{same accuracy since both contain three significant digits.} \\ \text{12.3 is more precise since it is written to the nearest tenth.} \end{cases}$

171 and 59 $\begin{cases} \text{171 is more accurate since it contains three significant digits.} \\ \text{same precision since both are written to the nearest integer.} \end{cases}$

517 and 1.703 $\begin{cases} \text{1.703 is more accurate since it contains four significant digits.} \\ \text{1.703 is more precise since it is written to the nearest} \\ \text{thousandth.} \end{cases}$

EXERCISES

In Exercises 1–20 determine the number of significant digits in the approximate number.

1. 123.14
2. 0.59
3. 0.02
4. 0.000491
5. 904
6. 30.03
7. 22.0
8. 0.78000
9. 670
10. 80,000
11. 5.40
12. 540
13. 0.054
14. 54.0
15. 504
16. 540.0
17. 540$\bar{0}$
18. 54$\bar{0}$0
19. 0.010
20. 0.100

In Exercises 21–32 round off the numbers (a) to four significant digits; (b) to two significant digits.

21. 12.3456
22. 46.054
23. 59372
24. 72488
25. 29697
26. 59999
27. 84.096
28. 10.998
29. 0.068547
30. 0.0034581
31. 0.0020006
32. 0.39999

In Exercises 33–44 determine the approximate number in each pair that is (a) more accurate; (b) more precise.

33. 423, 0.004
34. 18, 0.700
35. 402, 40.2
36. 312, 98
37. 6.430, 0.304
38. 0.9, 12.0
39. 2000, 0.0002
40. 1230, 3200
41. 9080, 48.2
42. 4$\bar{0}$00, 4$\bar{0}$
43. 3$\bar{0}$0, 300
44. 888$\bar{0}$, 8.008

Appendix A-2

Operations with approximate numbers

When computing with approximate numbers, we should express our results to a precision or accuracy that is appropriate to the data. We establish guidelines for the result by considering the error involved in computing with such numbers. For example, suppose that $x \approx 123$ and $y \approx 4.27$. Adding these numbers, we have $123 + 4.27 = 127.27$. However, because of rounding off, we know that the exact values of x and y are in the following intervals:

$$122.5 \leq x < 123.5$$
$$4.265 \leq y < 4.275$$

Adding the minimum values to find the minimum possible sum, and the maximum values to find the maximum possible sum, we have

$$122.5 + 4.265 \leq x + y < 123.5 + 4.275$$
$$126.765 \leq x + y < 127.775$$

To the nearest integer the minimum sum is 127 and the maximum sum is 128. Thus, the precision in the result may not be given to more than the nearest integer, and there is even a chance of error if we are that precise with our answer. This example motivates us to the following guideline:

When *adding* or *subtracting* approximate numbers, perform the operation and then round off the result to the *precision* of the least precise number.

Thus if $x \approx 123$ and $y \approx 4.27$, then $x + y \approx 123 + 4.27 = 127.27 \approx 127$. We round off to the nearest integer because 123, the less precise number, is rounded off to the nearest integer.

Before adding or subtracting, it is permissible to round off the more precise numbers since the extra digits are not considered when we round off our result. If this is done, the numbers should be rounded off to one place beyond that of the least precise number.

EXAMPLE 1: Add the approximate numbers: 12.79, 4.3131, 46.2, 9.618.

SOLUTION: 46.2 is the least precise number, so before adding we may round off the other numbers to the nearest hundredth:

$$
\begin{array}{r}
12.79 \\
4.31 \\
46.2 \\
9.62 \\
\hline
72.92
\end{array}
$$

Rounding off to the nearest tenth, the result is 72.9.

EXAMPLE 2: Perform the indicated operation for the approximate numbers: $0.396 - 17.6 + 150$.

SOLUTION: 150 is the least precise number. We assume that 150 is rounded off in the tens place, so before performing the operations, we round off the other numbers to the nearest integer. Thus, we have $0 - 18 + 150 = 132$. Rounding off in the tens place, the result is 130.

To determine the possible error when multiplying or dividing approximate numbers, consider the product of $x \approx 12$ and $y \approx 4.27$:

$$x \cdot y \approx 12(4.27) = 51.24$$

Because of rounding off, we know that the exact values of x and y are in the following intervals:

$$11.5 \le x < 12.5$$
$$4.265 \le y < 4.275$$

Multiplying the minimum values to find the minimum possible product and the maximum values to find the maximum possible product, we have

$$(11.5)(4.265) \le xy < (12.5)(4.275)$$
$$49.0475 \le xy < 53.4375$$

To two significant digits the minimum product is 49 and the maximum product is 53. Thus, the accuracy of the result may not be given to more than two significant digits, and even that accuracy may be wrong. This example motivates us to the following guideline:

> When *multiplying* or *dividing* approximate numbers, perform the operation and then round off the result to the *accuracy* of the least accurate number.

Thus, if $x \approx 12$ and $y \approx 4.27$, then $x \cdot y \approx 12(4.27) = 51.24 \approx 51$. We round off to two significant digits because 12, the less accurate number, has two significant digits.

Before multiplying or dividing, it is permissible to round off the more accurate numbers to one more significant digit than the least accurate number. If you are using a calculator, you may find it easier to key in the number without rounding off. Either way, you will usually obtain the same final result.

Since finding a power or root of a number involves multiplication, we are concerned with accuracy in such problems. The result should contain the same number of significant digits as the given approximate number.

EXAMPLE 3: Multiply the approximate numbers: (11.5)(0.042649).

SOLUTION: 11.5 is the less accurate number, so before multiplying we may round off the other number to four significant digits: $(11.5)(0.4265) = 0.490475$. Rounding off to three significant digits, the result is 0.490.

EXAMPLE 4: Perform the indicated operations for the approximate numbers:

$$\frac{(215.1)(7.63)}{0.0002}$$

SOLUTION: 0.0002 is the least accurate number. We round off the other numbers to two significant digits and then perform the operation:

$$\frac{220(7.6)}{0.0002} = 8,360,000$$

Rounding off to one significant digit, the result is 8,000,000.

EXAMPLE 5: Find the approximate value of $\sqrt{79.1}$.

SOLUTION: Using a calculator the initial result is: $\sqrt{79.1} \approx 8.893818078$. Since 79.1 contains three significant digits, the final result is 8.89.

Exact numbers

Although most of the numbers in scientific work are approximate, there are some numbers that are exact. Exact numbers are the result of a counting process or a definition, and are commonly seen in formulas. The following are examples of situations in which the numbers are exact:

1. There are 25 students in the class (counting process).
2. Of 100 light bulbs tested, 4 were defective (counting process).
3. There are 24 hours in 1 day (definition).
4. A bank pays 6 percent interest (definition).
5. $P = 4s$ (definition: formula for the perimeter of a square).
6. $A = \frac{1}{2}ab$ (definition: formula for the area of a triangle).

Since these numbers are exact, no error is involved when using them in computations. Only approximate numbers must be considered in determining the accuracy or precision of a result.

EXERCISES

Perform the indicated operations for the approximate numbers.

1. $16.27 + 2.1515 - 4.3$
2. $15 - 3.043 + 103.1$
3. $0.002 - 11.0 + 3.59$
4. $0.200 + 4.6 + 120$
5. $0.3047 + 0.8 + 0.092 + 0.69$
6. $214.32 + 1000 + 0.75 + 3.2$
7. $(2.1)(0.57892)$
8. $(0.08)(489.0)$
9. $(16.4)(2001)$
10. $(7.00)(0.094265)$
11. $408 \div 0.002$
12. $602 \div 0.200$

13. $\dfrac{(799.4)(8.6)}{0.04}$
14. $\dfrac{(0.460)(126.1)}{2200}$

15. $9.714 + (2.3)(0.812)$
16. $(4.50)(0.1234) - 5.8$

17. $\dfrac{4.07}{0.06} + (0.8715)(20)$
18. $5(0.0861) - \dfrac{0.90}{123}$

19. $\sqrt{95.0}$
20. $\sqrt{0.040}$

Find the value of the variable on the left side of each of the following formulas with the aid of a calculator. (*Note:* $\pi \approx 3.14159$.)

21. $P = 4s$ $\qquad$ $s = 12.75$ ft
22. $A = \frac{1}{2}ab$ $\qquad$ $a = 123.5$ meters $\qquad$ $b = 43.9$ meters
23. $d = rt$ $\qquad$ $r = 24.5$ mi/hour $\qquad$ $t = 11.2$ hours
24. $P = 2(L + W)$ $\qquad$ $L = 22.13$ cm $\qquad$ $W = 38.61$ cm

25. $A = \dfrac{a + b + c}{3}$ $\qquad$ $a = 111.72$ ft $\qquad$ $b = 81.67$ ft

$\qquad\qquad\qquad\qquad c = 238.42$ ft

26. $Z = \dfrac{x - m}{S}$ $\qquad$ $m = 12.17$ $\qquad$ $S = 1.76$

$\qquad\qquad\qquad\qquad x = 15.21$

27. $A = S^2$ $\qquad$ $S = 4.325$ cm

28. $A = \pi r^2$ $\qquad$ $r = 16.625$ ft

29. $V = \frac{1}{3}\pi r^2 h$ $\qquad$ $r = 4.5$ meters $\qquad$ $h = 9.25$ meters

30. $L = g\left(\dfrac{t}{2\pi}\right)^2$ $\qquad$ $g = 32.0$ ft/second^2 $\qquad$ $t = 7.25$ seconds

Appendix A-3

Geometric formulas

1. Rectangle
    Area: $A = LW$
    Perimeter: $P = 2L + 2W$

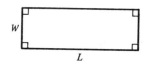

2. Square
    Area: $A = s^2$
    Perimeter: $P = 4s$

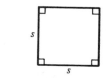

3. Parallelogram
    Area: $A = bh$

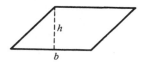

4. Trapezoid
    Area: $A = \frac{1}{2}h(a + b)$

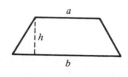

5. Triangle

Area: $A = \frac{1}{2}bh$

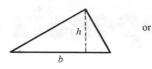 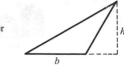

or

Area: $A = \sqrt{s(s-a)(s-b)(s-c)}$,
where $s = \frac{1}{2}(a + b + c)$
Angle sum: $A + B + C = 180°$

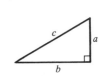

6. Right Triangle

Pythagorean theorem: $a^2 + b^2 = c^2$

7. Circle

Area: $A = \pi r^2$
Circumference: $c = \pi d = 2\pi r$

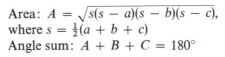

Appendix A-4

Trigonometric identities

1. $\sin x \csc x = 1$ or $\sin x = \dfrac{1}{\csc x}$ or $\csc x = \dfrac{1}{\sin x}$

2. $\cos x \sec x = 1$ or $\cos x = \dfrac{1}{\sec x}$ or $\sec x = \dfrac{1}{\cos x}$

3. $\tan x \cot x = 1$ or $\tan x = \dfrac{1}{\cot x}$ or $\cot x = \dfrac{1}{\tan x}$

4. $\tan x = \dfrac{\sin x}{\cos x}$

5. $\cot x = \dfrac{\cos x}{\sin x}$

6. $\sin^2 x + \cos^2 x = 1$
7. $\tan^2 x + 1 = \sec^2 x$
8. $\cot^2 x + 1 = \csc^2 x$
9. $\sin(-x) = -\sin x$
10. $\cos(-x) = \cos x$
11. $\tan(-x) = -\tan x$
12. $\cos(x_1 + x_2) = \cos x_1 \cos x_2 - \sin x_1 \sin x_2$
13. $\cos(x_1 - x_2) = \cos x_1 \cos x_2 + \sin x_1 \sin x_2$

14. $\sin(x_1 + x_2) = \sin x_1 \cos x_2 + \cos x_1 \sin x_2$

15. $\sin(x_1 - x_2) = \sin x_1 \cos x_2 - \cos x_1 \sin x_2$

16. $\tan(x_1 + x_2) = \dfrac{\tan x_1 + \tan x_2}{1 - \tan x_1 \tan x_2}$

17. $\sin 2x = 2 \sin x \cos x$

18. **a.** $\cos 2x = \cos^2 x - \sin^2 x$

    b. $\cos 2x = 2 \cos^2 x - 1$

    c. $\cos 2x = 1 - 2 \sin^2 x$

19. $\sin \dfrac{x}{2} = \pm \sqrt{\dfrac{1 - \cos x}{2}}$

20. $\cos \dfrac{x}{2} = \pm \sqrt{\dfrac{1 + \cos x}{2}}$

Table 1 Trigonometric Functions of Angles

Angle	sin	cos	tan	cot	sec	csc	
0°00′	.0000	1.0000	.0000	—	1.000	—	90°00′
10	.0029	1.0000	.0029	343.8	1.000	343.8	50
20	.0058	1.0000	.0058	171.9	1.000	171.9	40
30	.0087	1.0000	.0087	114.6	1.000	114.6	30
40	.0116	.9999	.0116	85.94	1.000	85.95	20
50	.0145	.9999	.0145	68.75	1.000	68.76	10
1°00′	.0175	.9998	.0175	57.29	1.000	57.30	89°00′
10	.0204	.9998	.0204	49.10	1.000	49.11	50
20	.0233	.9997	.0233	42.96	1.000	42.98	40
30	.0262	.9997	.0262	38.19	1.000	38.20	30
40	.0291	.9996	.0291	34.37	1.000	34.38	20
50	.0320	.9995	.0320	31.24	1.001	31.26	10
2°00′	.0349	.9994	.0349	28.64	1.001	28.65	88°00′
10	.0378	.9993	.0378	26.43	1.001	26.45	50
20	.0407	.9992	.0407	24.54	1.001	24.56	40
30	.0436	.9990	.0437	22.90	1.001	22.93	30
40	.0465	.9989	.0466	21.47	1.001	21.49	20
50	.0494	.9988	.0495	20.21	1.001	20.23	10
3°00′	.0523	.9986	.0524	19.08	1.001	19.11	87°00′
10	.0552	.9985	.0553	18.07	1.002	18.10	50
20	.0581	.9983	.0582	17.17	1.002	17.20	40
30	.0610	.9981	.0612	16.35	1.002	16.38	30
40	.0640	.9980	.0641	15.60	1.002	15.64	20
50	.0669	.9978	.0670	14.92	1.002	14.96	10
4°00′	.0698	.9976	.0699	14.30	1.002	14.34	86°00′
10	.0727	.9974	.0729	13.73	1.003	13.76	50
20	.0756	.9971	.0758	13.20	1.003	13.23	40
30	.0785	.9969	.0787	12.71	1.003	12.75	30
40	.0814	.9967	.0816	12.25	1.003	12.29	20
50	.0843	.9964	.0846	11.83	1.004	11.87	10
5°00′	.0872	.9962	.0875	11.43	1.004	11.47	85°00′
10	.0901	.9959	.0904	11.06	1.004	11.10	50
20	.0929	.9957	.0934	10.71	1.004	10.76	40
30	.0958	.9954	.0963	10.39	1.005	10.43	30
40	.0987	.9951	.0992	10.08	1.005	10.13	20
50	.1016	.9948	.1022	9.788	1.005	9.839	10
6°00′	.1045	.9945	.1051	9.514	1.006	9.567	84°00′
10	.1074	.9942	.1080	9.255	1.006	9.309	50
20	.1103	.9939	.1110	9.010	1.006	9.065	40
30	.1132	.9936	.1139	8.777	1.006	8.834	30
40	.1161	.9932	.1169	8.556	1.007	8.614	20
50	.1190	.9929	.1198	8.345	1.007	8.405	10
7°00′	.1219	.9925	.1228	8.144	1.008	8.206	83°00′
10	.1248	.9922	.1257	7.953	1.008	8.016	50
20	.1276	.9918	.1287	7.770	1.008	7.834	40
30	.1305	.9914	.1317	7.596	1.009	7.661	30
40	.1334	.9911	.1346	7.429	1.009	7.496	20
50	.1363	.9907	.1376	7.269	1.009	7.337	10
8°00′	.1392	.9903	.1405	7.115	1.010	7.185	82°00′
10	.1421	.9899	.1435	6.968	1.010	7.040	50
20	.1449	.9894	.1465	6.827	1.011	6.900	40
30	.1478	.9890	.1495	6.691	1.011	6.765	30
40	.1507	.9886	.1524	6.561	1.012	6.636	20
50	.1536	.9881	.1554	6.435	1.012	6.512	10
9°00′	.1564	.9877	.1584	6.314	1.012	6.392	81°00′
	cos	sin	cot	tan	csc	sec	Angle

Table 1 (*Continued*)

Angle	sin	cos	tan	cot	sec	csc	
9°00′	.1564	.9877	.1584	6.314	1.012	6.392	81°00′
10	.1593	.9872	.1614	6.197	1.013	6.277	50
20	.1622	.9868	.1644	6.084	1.013	6.166	40
30	.1650	.9863	.1673	5.976	1.014	6.059	30
40	.1679	.9858	.1703	5.871	1.014	5.955	20
50	.1708	.9853	.1733	5.769	1.015	5.855	10
10°00′	.1736	.9848	.1763	5.671	1.015	5.759	80°00′
10	.1765	.9843	.1793	5.576	1.016	5.665	50
20	.1794	.9838	.1823	5.485	1.016	5.575	40
30	.1822	.9833	.1853	5.396	1.107	5.487	30
40	.1851	.9827	.1883	5.309	1.018	5.403	20
50	.1880	.9822	.1914	5.226	1.018	5.320	10
11°00′	.1908	.9816	.1944	5.145	1.019	5.241	79°00′
10	.1937	.9811	.1974	5.066	1.019	5.164	50
20	.1965	.9805	.2004	4.989	1.020	5.089	40
30	.1994	.9799	.2035	4.915	1.020	5.016	30
40	.2022	.9793	.2065	4.843	1.021	4.945	20
50	.2051	.9787	.2095	4.773	1.022	4.876	10
12°00′	.2079	.9781	.2126	4.705	1.022	4.810	78°00′
10	.2108	.9775	.2156	4.638	1.023	4.745	50
20	.2136	.9769	.2186	4.574	1.024	4.682	40
30	.2164	.9763	.2217	4.511	1.024	4.620	30
40	.2193	.9757	.2247	4.449	1.025	4.560	20
50	.2221	.9750	.2278	4.390	1.026	4.502	10
13°00′	.2250	.9744	.2309	4.331	1.026	4.445	77°00′
10	.2278	.9737	.2339	4.275	1.027	4.390	50
20	.2306	.9730	.2370	4.219	1.028	4.336	40
30	.2334	.9724	.2401	4.165	1.028	4.284	30
40	.2363	.9717	.2432	4.113	1.029	4.232	20
50	.2391	.9710	.2462	4.061	1.030	4.182	10
14°00′	.2419	.9703	.2493	4.011	1.031	4.134	76°00′
10	.2447	.9696	.2524	3.962	1.031	4.086	50
20	.2476	.9689	.2555	3.914	1.032	4.039	40
30	.2504	.9681	.2586	3.867	1.033	3.994	30
40	.2532	.9674	.2617	3.821	1.034	3.950	20
50	.2560	.9667	.2648	3.776	1.034	3.906	10
15°00′	.2588	.9659	.2679	3.732	1.035	3.864	75°00′
10	.2616	.9652	.2711	3.689	1.036	3.822	50
20	.2644	.9644	.2742	3.647	1.037	3.782	40
30	.2672	.9636	.2773	3.606	1.038	3.742	30
40	.2700	.9628	.2805	3.566	1.039	3.703	20
50	.2728	.9621	.2836	3.526	1.039	3.665	10
16°00′	.2756	.9613	.2867	3.487	1.040	3.628	74°00′
10	.2784	.9605	.2899	3.450	1.041	3.592	50
20	.2812	.9596	.2931	3.412	1.042	3.556	40
30	.2840	.9588	.2962	3.376	1.043	3.521	30
40	.2868	.9580	.2994	3.340	1.044	3.487	20
50	.2896	.9572	.3026	3.305	1.045	3.453	10
17°00′	.2924	.9563	.3057	3.271	1.046	3.420	73°00′
10	.2952	.9555	.3089	3.237	1.047	3.388	50
20	.2979	.9546	.3121	3.204	1.048	3.356	40
30	.3007	.9537	.3153	3.172	1.049	3.326	30
40	.3035	.9528	.3185	3.140	1.049	3.295	20
50	.3062	.9520	.3217	3.108	1.050	3.265	10
18°00′	.3090	.9511	.3249	3.078	1.051	3.236	72°00′
	cos	sin	cot	tan	csc	sec	Angle

Table 1 (*Continued*)

Angle	sin	cos	tan	cot	sec	csc	
18°00′	.3090	.9511	.3249	3.078	1.051	3.236	72°00′
10	.3118	.9502	.3281	3.047	1.052	3.207	50
20	.3145	.9492	.3314	3.018	1.053	3.179	40
30	.3173	.9483	.3346	2.989	1.054	3.152	30
40	.3201	.9474	.3378	2.960	1.056	3.124	20
50	.3228	.9465	.3411	2.932	1.057	3.098	10
19°00′	.3256	.9455	.3443	2.904	1.058	3.072	71°00′
10	.3283	.9446	.3476	2.877	1.059	3.046	50
20	.3311	.9436	.3508	2.850	1.060	3.021	40
30	.3338	.9426	.3541	2.824	1.061	2.996	30
40	.3365	.9417	.3574	2.798	1.062	2.971	20
50	.3393	.9407	.3607	2.773	1.063	2.947	10
20°00′	.3420	.9397	.3640	2.747	1.064	2.924	70°00′
10	.3448	.9387	.3673	2.723	1.065	2.901	50
20	.3475	.9377	.3706	2.699	1.066	2.878	40
30	.3502	.9367	.3739	2.675	1.068	2.855	30
40	.3529	.9356	.3772	2.651	1.069	2.833	20
50	.3557	.9346	.3805	2.628	1.070	2.812	10
21°00′	.3584	.9336	.3839	2.605	1.071	2.790	69°00′
10	.3611	.9325	.3872	2.583	1.072	2.769	50
20	.3638	.9315	.3906	2.560	1.074	2.749	40
30	.3665	.9304	.3939	2.539	1.075	2.729	30
40	.3692	.9293	.3973	2.517	1.076	2.709	20
50	.3719	.9283	.4006	2.496	1.077	2.689	10
22°00′	.3746	.9272	.4040	2.475	1.079	2.669	68°00′
10	.3773	.9261	.4074	2.455	1.080	2.650	50
20	.3800	.9250	.4108	2.434	1.081	2.632	40
30	.3827	.9239	.4142	2.414	1.082	2.613	30
40	.3854	.9228	.4176	2.394	1.084	2.595	20
50	.3881	.9216	.4210	2.375	1.085	2.577	10
23°00′	.3907	.9205	.4245	2.356	1.086	2.559	67°00′
10	.3934	.9194	.4279	2.337	1.088	2.542	50
20	.3961	.9182	.4314	2.318	1.089	2.525	40
30	.3987	.9171	.4348	2.300	1.090	2.508	30
40	.4014	.9159	.4383	2.282	1.092	2.491	20
50	.4041	.9147	.4417	2.264	1.093	2.475	10
24°00′	.4067	.9135	.4452	2.246	1.095	2.459	66°00′
10	.4094	.9124	.4487	2.229	1.096	2.443	50
20	.4120	.9112	.4522	2.211	1.097	2.427	40
30	.4147	.9100	.4557	2.194	1.099	2.411	30
40	.4173	.9088	.4592	2.177	1.100	2.396	20
50	.4200	.9075	.4628	2.161	1.102	2.381	10
25°00′	.4226	.9063	.4663	2.145	1.103	2.366	65°00′
10	.4253	.9051	.4699	2.128	1.105	2.352	50
20	.4279	.9038	.4734	2.112	1.106	2.337	40
30	.4305	.9026	.4770	2.097	1.108	2.323	30
40	.4331	.9013	.4806	2.081	1.109	2.309	20
50	.4358	.9001	.4841	2.066	1.111	2.295	10
26°00′	.4384	.8988	.4877	2.050	1.113	2.281	64°00′
10	.4410	.8975	.4913	2.035	1.114	2.268	50
20	.4436	.8962	.4950	2.020	1.116	2.254	40
30	.4462	.8949	.4986	2.006	1.117	2.241	30
40	.4488	.8936	.5022	1.991	1.119	2.228	20
50	.4514	.8923	.5059	1.977	1.121	2.215	10
27°00′	.4540	.8910	.5095	1.963	1.122	2.203	63°00′
	cos	sin	cot	tan	csc	sec	Angle

505

Table 1 (*Continued*)

Angle	sin	cos	tan	cot	sec	csc	
27°00′	.4540	.8910	.5095	1.963	1.122	2.203	63°00′
10	.4566	.8897	.5132	1.949	1.124	2.190	50
20	.4592	.8884	.5169	1.935	1.126	2.178	40
30	.4617	.8870	.5206	1.921	1.127	2.166	30
40	.4643	.8857	.5243	1.907	1.129	2.154	20
50	.4669	.8843	.5280	1.894	1.131	2.142	10
28°00′	.4695	.8829	.5317	1.881	1.133	2.130	62°00′
10	.4720	.8816	.5354	1.868	1.134	2.118	50
20	.4746	.8802	.5392	1.855	1.136	2.107	40
30	.4772	.8788	.5430	1.842	1.138	2.096	30
40	.4797	.8774	.5467	1.829	1.140	2.085	20
50	.4823	.8760	.5505	1.816	1.142	2.074	10
29°00′	.4848	.8746	.5543	1.804	1.143	2.063	61°00′
10	.4874	.8732	.5581	1.792	1.145	2.052	50
20	.4899	.8718	.5619	1.780	1.147	2.041	40
30	.4924	.8704	.5658	1.767	1.149	2.031	30
40	.4950	.8689	.5696	1.756	1.151	2.020	20
50	.4975	.8675	.5735	1.744	1.153	2.010	10
30°00′	.5000	.8660	.5774	1.732	1.155	2.000	60°00′
10	.5025	.8646	.5812	1.720	1.157	1.990	50
20	.5050	.8631	.5851	1.709	1.159	1.980	40
30	.5075	.8616	.5890	1.698	1.161	1.970	30
40	.5100	.8601	.5930	1.686	1.163	1.961	20
50	.5125	.8587	.5969	1.675	1.165	1.951	10
31°00′	.5150	.8572	.6009	1.664	1.167	1.942	59°00′
10	.5175	.8557	.6048	1.653	1.169	1.932	50
20	.5200	.8542	.6088	1.643	1.171	1.923	40
30	.5225	.8526	.6128	1.632	1.173	1.914	30
40	.5250	.8511	.6168	1.621	1.175	1.905	20
50	.5275	.8496	.6208	1.611	1.177	1.896	10
32°00′	.5299	.8480	.6249	1.600	1.179	1.887	58°00′
10	.5324	.8465	.6289	1.590	1.181	1.878	50
20	.5348	.8450	.6330	1.580	1.184	1.870	40
30	.5373	.8434	.6371	1.570	1.186	1.861	30
40	.5398	.8418	.6412	1.560	1.188	1.853	20
50	.5422	.8403	.6453	1.550	1.190	1.844	10
33°00′	.5446	.8387	.6494	1.540	1.192	1.836	57°00′
10	.5471	.8371	.6536	1.530	1.195	1.828	50
20	.5495	.8355	.6577	1.520	1.197	1.820	40
30	.5519	.8339	.6619	1.511	1.199	1.812	30
40	.5544	.8323	.6661	1.501	1.202	1.804	20
50	.5568	.8307	.6703	1.492	1.204	1.796	10
34°00′	.5592	.8290	.6745	1.483	1.206	1.788	56°00′
10	.5616	.8274	.6787	1.473	1.209	1.781	50
20	.5640	.8258	.6830	1.464	1.211	1.773	40
30	.5664	.8241	.6873	1.455	1.213	1.766	30
40	.5688	.8225	.6916	1.446	1.216	1.758	20
50	.5712	.8208	.6959	1.437	1.218	1.751	10
35°00′	.5736	.8192	.7002	1.428	1.221	1.743	55°00′
10	.5760	.8175	.7046	1.419	1.223	1.736	50
20	.5783	.8158	.7089	1.411	1.226	1.729	40
30	.5807	.8141	.7133	1.402	1.228	1.722	30
40	.5831	.8124	.7177	1.393	1.231	1.715	20
50	.5854	.8107	.7221	1.385	1.233	1.708	10
36°00′	.5878	.8090	.7265	1.376	1.236	1.701	54°00′
	cos	sin	cot	tan	csc	sec	Angle

Table 1 (*Continued*)

Angle	sin	cos	tan	cot	sec	csc	
36°00′	.5878	.8090	.7265	1.376	1.236	1.701	54°00′
10	.5901	.8073	.7310	1.368	1.239	1.695	50
20	.5925	.8056	.7355	1.360	1.241	1.688	40
30	.5948	.8039	.7400	1.351	1.244	1.681	30
40	.5972	.8021	.7445	1.343	1.247	1.675	20
50	.5995	.8004	.7490	1.335	1.249	1.668	10
37°00′	.6018	.7986	.7536	1.327	1.252	1.662	53°00′
10	.6041	.7969	.7581	1.319	1.255	1.655	50
20	.6065	.7951	.7627	1.311	1.258	1.649	40
30	.6088	.7934	.7673	1.303	1.260	1.643	30
40	.6111	.7916	.7720	1.295	1.263	1.636	20
50	.6134	.7898	.7766	1.288	1.266	1.630	10
38°00′	.6157	.7880	.7813	1.280	1.269	1.624	52°00′
10	.6180	.7862	.7860	1.272	1.272	1.618	50
20	.6202	.7844	.7907	1.265	1.275	1.612	40
30	.6225	.7826	.7954	1.257	1.278	1.606	30
40	.6248	.7808	.8002	1.250	1.281	1.601	20
50	.6271	.7790	.8050	1.242	1.284	1.595	10
39°00′	.6293	.7771	.8098	1.235	1.287	1.589	51°00′
10	.6316	.7753	.8146	1.228	1.290	1.583	50
20	.6338	.7735	.8195	1.220	1.293	1.578	40
30	.6361	.7716	.8243	1.213	1.296	1.572	30
40	.6383	.7698	.8292	1.206	1.299	1.567	20
50	.6406	.7679	.8342	1.199	1.302	1.561	10
40°00′	.6428	.7660	.8391	1.192	1.305	1.556	50°00′
10	.6450	.7642	.8441	1.185	1.309	1.550	50
20	.6472	.7623	.8491	1.178	1.312	1.545	40
30	.6494	.7604	.8541	1.171	1.315	1.540	30
40	.6517	.7585	.8591	1.164	1.318	1.535	20
50	.6539	.7566	.8642	1.157	1.322	1.529	10
41°00′	.6561	.7547	.8693	1.150	1.325	1.524	49°00′
10	.6583	.7528	.8744	1.144	1.328	1.519	50
20	.6604	.7509	.8796	1.137	1.332	1.514	40
30	.6626	.7490	.8847	1.130	1.335	1.509	30
40	.6648	.7470	.8899	1.124	1.339	1.504	20
50	.6670	.7451	.8952	1.117	1.342	1.499	10
42°00′	.6691	.7431	.9004	1.111	1.346	1.494	48°00′
10	.6713	.7412	.9057	1.104	1.349	1.490	50
20	.6734	.7392	.9110	1.098	1.353	1.485	40
30	.6756	.7373	.9163	1.091	1.356	1.480	30
40	.6777	.7353	.9217	1.085	1.360	1.476	20
50	.6799	.7333	.9271	1.079	1.364	1.471	10
43°00′	.6820	.7314	.9325	1.072	1.367	1.466	47°00′
10	.6841	.7294	.9380	1.066	1.371	1.462	50
20	.6862	.7274	.9435	1.060	1.375	1.457	40
30	.6884	.7254	.9490	1.054	1.379	1.453	30
40	.6905	.7234	.9545	1.048	1.382	1.448	20
50	.6926	.7214	.9601	1.042	1.386	1.444	10
44°00′	.6947	.7193	.9657	1.036	1.390	1.440	46°00′
10	.6967	.7173	.9713	1.030	1.394	1.435	50
20	.6988	.7153	.9770	1.024	1.398	1.431	40
30	.7009	.7133	.9827	1.018	1.402	1.427	30
40	.7030	.7112	.9884	1.012	1.406	1.423	20
50	.7050	.7092	.9942	1.006	1.410	1.418	10
45°00′	.7071	.7071	1.000	1.000	1.414	1.414	45°00′
	cos	sin	cot	tan	csc	sec	Angle

Table 2 Exponential Functions

x	e^x	e^{-x}	x	e^x	e^{-x}
0.00	1.0000	1.0000	1.5	4.4817	0.2231
0.01	1.0101	0.9901	1.6	4.9530	0.2019
0.02	1.0202	0.9802	1.7	5.4739	0.1827
0.03	1.0305	0.9705	1.8	6.0496	0.1653
0.04	1.0408	0.9608	1.9	6.6859	0.1496
0.05	1.0513	0.9512	2.0	7.3891	0.1353
0.06	1.0618	0.9418	2.1	8.1662	0.1225
0.07	1.0725	0.9324	2.2	9.0250	0.1108
0.08	1.0833	0.9231	2.3	9.9742	0.1003
0.09	1.0942	0.9139	2.4	11.023	0.0907
0.10	1.1052	0.9048	2.5	12.182	0.0821
0.11	1.1163	0.8958	2.6	13.464	0.0743
0.12	1.1275	0.8869	2.7	14.880	0.0672
0.13	1.1388	0.8781	2.8	16.445	0.0608
0.14	1.1503	0.8694	2.9	18.174	0.0550
0.15	1.1618	0.8607	3.0	20.086	0.0498
0.16	1.1735	0.8521	3.1	22.198	0.0450
0.17	1.1853	0.8437	3.2	24.533	0.0408
0.18	1.1972	0.8353	3.3	27.113	0.0369
0.19	1.2092	0.8270	3.4	29.964	0.0334
0.20	1.2214	0.8187	3.5	33.115	0.0302
0.21	1.2337	0.8106	3.6	36.598	0.0273
0.22	1.2461	0.8025	3.7	40.447	0.0247
0.23	1.2586	0.7945	3.8	44.701	0.0224
0.24	1.2712	0.7866	3.9	49.402	0.0202
0.25	1.2840	0.7788	4.0	54.598	0.0183
0.30	1.3499	0.7408	4.1	60.340	0.0166
0.35	1.4191	0.7047	4.2	66.686	0.0150
0.40	1.4918	0.6703	4.3	73.700	0.0136
0.45	1.5683	0.6376	4.4	81.451	0.0123
0.50	1.6487	0.6065	4.5	90.017	0.0111
0.55	1.7333	0.5769	4.6	99.484	0.0101
0.60	1.8221	0.5488	4.7	109.95	0.0091
0.65	1.9155	0.5220	4.8	121.51	0.0082
0.70	2.0138	0.4966	4.9	134.29	0.0074
0.75	2.1170	0.4724	5.0	148.41	0.0067
0.80	2.2255	0.4493	5.5	244.69	0.0041
0.85	2.3396	0.4274	6.0	403.43	0.0025
0.90	2.4596	0.4066	6.5	665.14	0.0015
0.95	2.5857	0.3867	7.0	1096.6	0.0009
1.0	2.7183	0.3679	7.5	1808.0	0.0006
1.1	3.0042	0.3329	8.0	2981.0	0.0003
1.2	3.3201	0.3012	8.5	4914.8	0.0002
1.3	3.6693	0.2725	9.0	8103.1	0.0001
1.4	4.0552	0.2466	10.0	22026	0.00005

Table 3 Common Logarithms

m	0	1	2	3	4	5	6	7	8	9
1.0	.0000	.0043	.0086	.0128	.0170	.0212	.0253	.0294	.0334	.0374
1.1	.0414	.0453	.0492	.0531	.0569	.0607	.0645	.0682	.0719	.0755
1.2	.0792	.0828	.0864	.0899	.0934	.0969	.1004	.1038	.1072	.1106
1.3	.1139	.1173	.1206	.1239	.1271	.1303	.1335	.1367	.1399	.1430
1.4	.1461	.1492	.1523	.1553	.1584	.1614	.1644	.1673	.1703	.1732
1.5	.1761	.1790	.1818	.1847	.1875	.1903	.1931	.1959	.1987	.2014
1.6	.2041	.2068	.2095	.2122	.2148	.2175	.2201	.2227	.2253	.2279
1.7	.2304	.2330	.2355	.2380	.2405	.2430	.2455	.2480	.2504	.2529
1.8	.2553	.2577	.2601	.2625	.2648	.2672	.2695	.2718	.2742	.2765
1.9	.2788	.2810	.2833	.2856	.2878	.2900	.2923	.2945	.2967	.2989
2.0	.3010	.3032	.3054	.3075	.3096	.3118	.3139	.3160	.3181	.3201
2.1	.3222	.3243	.3263	.3284	.3304	.3324	.3345	.3365	.3385	.3404
2.2	.3424	.3444	.3464	.3483	.3502	.3522	.3541	.3560	.3579	.3598
2.3	.3617	.3636	.3655	.3674	.3692	.3711	.3729	.3747	.3766	.3784
2.4	.3802	.3820	.3838	.3856	.3874	.3892	.3909	.3927	.3945	.3962
2.5	.3979	.3997	.4014	.4031	.4048	.4065	.4082	.4099	.4116	.4133
2.6	.4150	.4166	.4183	.4200	.4216	.4232	.4249	.4265	.4281	.4298
2.7	.4314	.4330	.4346	.4362	.4378	.4393	.4409	.4425	.4440	.4456
2.8	.4472	.4487	.4502	.4518	.4533	.4548	.4564	.4579	.4594	.4609
2.9	.4624	.4639	.4654	.4669	.4683	.4698	.4713	.4728	.4742	.4757
3.0	.4771	.4786	.4800	.4814	.4829	.4843	.4857	.4871	.4886	.4900
3.1	.4914	.4928	.4942	.4955	.4969	.4983	.4997	.5011	.5024	.5038
3.2	.5051	.5065	.5079	.5092	.5105	.5119	.5132	.5145	.5159	.5172
3.3	.5185	.5198	.5211	.5224	.5237	.5250	.5263	.5276	.5289	.5302
3.4	.5315	.5328	.5340	.5353	.5366	.5378	.5391	.5403	.5416	.5428
3.5	.5441	.5453	.5465	.5478	.5490	.5502	.5514	.5527	.5539	.5551
3.6	.5563	.5575	.5587	.5599	.5611	.5623	.5635	.5647	.5658	.5670
3.7	.5682	.5694	.5705	.5717	.5729	.5740	.5752	.5763	.5775	.5786
3.8	.5798	.5809	.5821	.5832	.5843	.5855	.5866	.5877	.5888	.5899
3.9	.5911	.5922	.5933	.5944	.5955	.5966	.5977	.5988	.5999	.6010
4.0	.6021	.6031	.6042	.6053	.6064	.6075	.6085	.6096	.6107	.6117
4.1	.6128	.6138	.6149	.6160	.6170	.6180	.6191	.6201	.6212	.6222
4.2	.6232	.6243	.6253	.6263	.6274	.6284	.6294	.6304	.6314	.6325
4.3	.6335	.6345	.6355	.6365	.6375	.6385	.6395	.6405	.6415	.6425
4.4	.6435	.6444	.6454	.6464	.6474	.6484	.6493	.6503	.6513	.6522
4.5	.6532	.6542	.6551	.6561	.6571	.6580	.6590	.6599	.6609	.6618
4.6	.6628	.6637	.6646	.6656	.6665	.6675	.6684	.6693	.6702	.6712
4.7	.6721	.6730	.6739	.6749	.6758	.6767	.6776	.6785	.6794	.6803
4.8	.6812	.6821	.6830	.6839	.6848	.6857	.6866	.6875	.6884	.6893
4.9	.6902	.6911	.6920	.6928	.6937	.6946	.6955	.6964	.6972	.6981
5.0	.6990	.6998	.7007	.7016	.7024	.7033	.7042	.7050	.7059	.7067
5.1	.7076	.7084	.7093	.7101	.7110	.7118	.7126	.7135	.7143	.7152
5.2	.7160	.7168	.7177	.7185	.7193	.7202	.7210	.7218	.7226	.7235
5.3	.7243	.7251	.7259	.7267	.7275	.7284	.7292	.7300	.7308	.7316
5.4	.7324	.7332	.7340	.7348	.7356	.7364	.7372	.7380	.7388	.7396

Table 3 (*Continued*)

m	0	1	2	3	4	5	6	7	8	9
5.5	.7404	.7412	.7419	.7427	.7435	.7443	.7451	.7459	.7466	.7474
5.6	.7482	.7490	.7497	.7505	.7513	.7520	.7528	.7536	.7543	.7551
5.7	.7559	.7566	.7574	.7582	.7589	.7597	.7604	.7612	.7619	.7627
5.8	.7634	.7642	.7649	.7657	.7664	.7672	.7679	.7686	.7694	.7701
5.9	.7709	.7716	.7723	.7731	.7738	.7745	.7752	.7760	.7767	.7774
6.0	.7782	.7789	.7796	.7803	.7810	.7818	.7825	.7832	.7839	.7846
6.1	.7853	.7860	.7868	.7875	.7882	.7889	.7896	.7903	.7910	.7917
6.2	.7924	.7931	.7938	.7945	.7952	.7959	.7966	.7973	.7980	.7987
6.3	.7993	.8000	.8007	.8014	.8021	.8028	.8035	.8041	.8048	.8055
6.4	.8062	.8069	.8075	.8082	.8089	.8096	.8102	.8109	.8116	.8122
6.5	.8129	.8136	.8142	.8149	.8156	.8162	.8169	.8176	.8182	.8189
6.6	.8195	.8202	.8209	.8215	.8222	.8228	.8235	.8241	.8248	.8254
6.7	.8261	.8267	.8274	.8280	.8287	.8293	.8299	.8306	.8312	.8319
6.8	.8325	.8331	.8338	.8344	.8351	.8357	.8363	.8370	.8376	.8382
6.9	.8388	.8395	.8401	.8407	.8414	.8420	.8426	.8432	.8439	.8445
7.0	.8451	.8457	.8463	.8470	.8476	.8482	.8488	.8494	.8500	.8506
7.1	.8513	.8519	.8525	.8531	.8537	.8543	.8549	.8555	.8561	.8567
7.2	.8573	.8579	.8585	.8591	.8597	.8603	.8609	.8615	.8621	.8627
7.3	.8633	.8639	.8645	.8651	.8657	.8663	.8669	.8675	.8681	.8686
7.4	.8692	.8698	.8704	.8710	.8716	.8722	.8727	.8733	.8739	.8745
7.5	.8751	.8756	.8762	.8768	.8774	.8779	.8785	.8791	.8797	.8802
7.6	.8808	.8814	.8820	.8825	.8831	.8837	.8842	.8848	.8854	.8859
7.7	.8865	.8871	.8876	.8882	.8887	-8893	.8899	.8904	.8910	.8915
7.8	.8921	.8927	.8932	.8938	.8943	.8949	.8954	.8960	.8965	.8971
7.9	.8976	.8982	.8987	.8993	.8998	.9004	.9009	.9015	.9020	.9025
8.0	.9031	.9036	.9042	.9047	.9053	.9058	.9063	.9069	.9074	.9079
8.1	.9085	.9090	9096	.9101	.9106	.9112	.9117	.9122	.9128	.9133
8.2	.9138	.9143	.9149	.9154	.9159	.9165	.9170	.9175	.9180	.9186
8.3	.9191	.9196	.9201	.9206	.9212	.9217	.9222	.9227	.9232	.9238
8.4	.9243	.9249	.9253	.9258	.9263	.9269	.9274	.9279	.9284	.9289
8.5	.9294	.9299	.9304	.9309	.9315	.9320	.9325	.9330	.9335	.9340
8.6	.9345	.9350	.9355	.9360	.9365	.9370	.9375	.9380	.9385	.9390
8.7	.9395	.9400	.9405	.9410	.9415	.9420	.9425	.9430	.9435	.9440
8.8	.9445	.9450	.9455	.9460	.9465	.9469	.9474	.9479	.9484	.9489
8.9	.9494	.9499	.9504	.9509	.9513	.9518	.9523	.9528	.9533	.9538
9.0	.9542	.9547	.9552	.9557	.9562	.9566	.9571	.9576	.9581	.9586
9.1	.9590	.9595	.9600	.9605	.9609	.9614	.9619	.9624	.9628	.9633
9.2	.9638	.9643	.9647	.9652	.9657	.9661	.9666	.9671	.9675	.9680
9.3	.9685	.9689	.9694	.9699	.9703	.9708	.9713	.9717	.9722	.9727
9.4	.9731	.9736	.9741	.9745	.9750	.9754	.9759	.9763	.9768	.9773
9.5	.9777	.9782	.9786	.9791	.9795	.9800	.9805	.9809	.9814	.9818
9.6	.9823	.9827	.9832	.9836	.9841	.9845	.9850	.9854	.9859	.9863
9.7	.9868	.9872	.9877	.9881	.9886	.9890	.9894	.9899	.9903	.9908
9.8	.9912	.9917	.9921	.9926	.9930	.9934	.9939	.9943	.9948	.9952
9.9	.9956	.9961	.9965	.9969	.9974	.9978	.9983	.9987	.9991	.9996

Table 4 Natural Logarithms (base e)

	0	1	2	3	4	5	6	7	8	9
1.0	0.0000	0.0100	0.0198	0.0296	0.0392	0.0488	0.0583	0.0677	0.0770	0.0862
1.1	0.0953	0.1044	0.1133	0.1222	0.1310	0.1398	0.1484	0.1570	0.1655	0.1740
1.2	0.1823	0.1906	0.1989	0.2070	0.2151	0.2231	0.2311	0.2390	0.2469	0.2546
1.3	0.2624	0.2700	0.2776	0.2852	0.2927	0.3001	0.3075	0.3148	0.3221	0.3293
1.4	0.3365	0.3436	0.3507	0.3577	0.3646	0.3716	0.3784	0.3853	0.3920	0.3988
1.5	0.4055	0.4121	0.4187	0.4253	0.4318	0.4383	0.4447	0.4511	0.4574	0.4637
1.6	0.4700	0.4762	0.4824	0.4886	0.4947	0.5008	0.5068	0.5128	0.5188	0.5247
1.7	0.5306	0.5365	0.5423	0.5481	0.5539	0.5596	0.5653	0.5710	0.5766	0.5822
1.8	0.5878	0.5933	0.5988	0.6043	0.6098	0.6152	0.6206	0.6259	0.6313	0.6366
1.9	0.6419	0.6471	0.6523	0.6575	0.6627	0.6678	0.6729	0.6780	0.6831	0.6881
2.0	0.6932	0.6981	0.7031	0.7080	0.7129	0.7178	0.7227	0.7275	0.7324	0.7372
2.1	0.7419	0.7467	0.7514	0.7561	0.7608	0.7655	0.7701	0.7747	0.7793	0.7839
2.2	0.7885	0.7930	0.7975	0.8020	0.8065	0.8109	0.8154	0.8198	0.8242	0.8286
2.3	0.8329	0.8373	0.8416	0.8459	0.8502	0.8544	0.8587	0.8629	0.8671	0.8713
2.4	0.8755	0.8796	0.8838	0.8879	0.8920	0.8961	0.9002	0.9042	0.9083	0.9123
2.5	0.9163	0.9203	0.9243	0.9282	0.9322	0.9361	0.9400	0.9439	0.9478	0.9517
2.6	0.9555	0.9594	0.9632	0.9670	0.9708	0.9746	0.9783	0.9821	0.9858	0.9895
2.7	0.9933	0.9969	1.0006	1.0043	1.0080	1.0116	1.0152	1.0188	1.0225	1.0260
2.8	1.0296	1.0332	1.0367	1.0403	1.0438	1.0473	1.0508	1.0543	1.0578	1.0613
2.9	1.0647	1.0682	1.0716	1.0750	1.0784	1.0818	1.0852	1.0886	1.0919	1.0953
3.0	1.0986	1.1019	1.1053	1.1086	1.1119	1.1151	1.1184	1.1217	1.1249	1.1282
3.1	1.1314	1.1346	1.1378	1.1410	1.1442	1.1474	1.1506	1.1537	1.1569	1.1600
3.2	1.1632	1.1663	1.1694	1.1725	1.1756	1.1787	1.1817	1.1848	1.1878	1.1909
3.3	1.1939	1.1969	1.2000	1.2030	1.2060	1.2090	1.2119	1.2149	1.2179	1.2208
3.4	1.2238	1.2267	1.2296	1.2326	1.2355	1.2384	1.2413	1.2442	1.2470	1.2499
3.5	1.2528	1.2556	1.2585	1.2613	1.2641	1.2669	1.2698	1.2726	1.2754	1.2782
3.6	1.2809	1.2837	1.2865	1.2892	1.2920	1.2947	1.2975	1.3002	1.3029	1.3056
3.7	1.3083	1.3110	1.3137	1.3164	1.3191	1.3218	1.3244	1.3271	1.3297	1.3324
3.8	1.3350	1.3376	1.3403	1.3429	1.3455	1.3481	1.3507	1.3533	1.3558	1.3584
3.9	1.3610	1.3635	1.3661	1.3686	1.3712	1.3737	1.3762	1.3788	1.3813	1.3838
4.0	1.3863	1.3888	1.3913	1.3938	1.3962	1.3987	1.4012	1.4036	1.4061	1.4085
4.1	1.4110	1.4134	1.4159	1.4183	1.4207	1.4231	1.4255	1.4279	1.4303	1.4327
4.2	1.4351	1.4375	1.4398	1.4422	1.4446	1.4469	1.4493	1.4516	1.4540	1.4563
4.3	1.4586	1.4609	1.4633	1.4656	1.4679	1.4702	1.4725	1.4748	1.4771	1.4793
4.4	1.4816	1.4839	1.4861	1.4884	1.4907	1.4929	1.4951	1.4974	1.4996	1.5019
4.5	1.5041	1.5063	1.5085	1.5107	1.5129	1.5151	1.5173	1.5195	1.5217	1.5239
4.6	1.5261	1.5282	1.5304	1.5326	1.5347	1.5369	1.5390	1.5412	1.5433	1.5454
4.7	1.5476	1.5497	1.5518	1.5539	1.5560	1.5581	1.5602	1.5623	1.5644	1.5665
4.8	1.5686	1.5707	1.5728	1.5748	1.5769	1.5790	1.5810	1.5831	1.5851	1.5872
4.9	1.5892	1.5913	1.5933	1.5953	1.5974	1.5994	1.6014	1.6034	1.6054	1.6074
5.0	1.6094	1.6114	1.6134	1.6154	1.6174	1.6194	1.6214	1.6233	1.6253	1.6273
5.1	1.6292	1.6312	1.6332	1.6351	1.6371	1.6390	1.6409	1.6429	1.6448	1.6467
5.2	1.6487	1.6506	1.6525	1.6544	1.6563	1.6582	1.6601	1.6620	1.6639	1.6658
5.3	1.6677	1.6696	1.6715	1.6734	1.6752	1.6771	1.6790	1.6808	1.6827	1.6845
5.4	1.6864	1.6882	1.6901	1.6919	1.6938	1.6956	1.6974	1.6993	1.7011	1.7029

Table 4 (*Continued*)

	0	1	2	3	4	5	6	7	8	9
5.5	1.7047	1.7066	1.7084	1.7102	1.7120	1.7138	1.7156	1.7174	1.7192	1.7210
5.6	1.7228	1.7246	1.7263	1.7281	1.7299	1.7317	1.7334	1.7352	1.7370	1.7387
5.7	1.7405	1.7422	1.7440	1.7457	1.7475	1.7492	1.7509	1.7527	1.7544	1.7561
5.8	1.7579	1.7596	1.7613	1.7630	1.7647	1.7664	1.7681	1.7699	1.7716	1.7733
5.9	1.7750	1.7766	1.7783	1.7800	1.7817	1.7834	1.7851	1.7868	1.7884	1.7901
6.0	1.7918	1.7934	1.7951	1.7967	1.7984	1.8001	1.8017	1.8034	1.8050	1.8066
6.1	1.8083	1.8099	1.8116	1.8132	1.8148	1.8165	1.8181	1.8197	1.8213	1.8229
6.2	1.8245	1.8262	1.8278	1.8294	1.8310	1.8326	1.8342	1.8358	1.8374	1.8390
6.3	1.8405	1.8421	1.8437	1.8453	1.8469	1.8485	1.8500	1.8516	1.8532	1.8547
6.4	1.8563	1.8579	1.8594	1.8610	1.8625	1.8641	1.8656	1.8672	1.8687	1.8703
6.5	1.8718	1.8733	1.8749	1.8764	1.8779	1.8795	1.8810	1.8825	1.8840	1.8856
6.6	1.8871	1.8886	1.8901	1.8916	1.8931	1.8946	1.8961	1.8976	1.8991	1.9006
6.7	1.9021	1.9036	1.9051	1.9066	1.9081	1.9095	1.9110	1.9125	1.9140	1.9155
6.8	1.9169	1.9184	1.9199	1.9213	1.9228	1.9242	1.9257	1.9272	1.9286	1.9301
6.9	1.9315	1.9330	1.9344	1.9359	1.9373	1.9387	1.9402	1.9416	1.9430	1.9445
7.0	1.9459	1.9473	1.9488	1.9502	1.9516	1.9530	1.9544	1.9559	1.9573	1.9587
7.1	1.9601	1.9615	1.9629	1.9643	1.9657	1.9671	1.9685	1.9699	1.9713	1.9727
7.2	1.9741	1.9755	1.9769	1.9782	1.9796	1.9810	1.9824	1.9838	1.9851	1.9865
7.3	1.9879	1.9892	1.9906	1.9920	1.9933	1.9947	1.9961	1.9974	1.9988	2.0001
7.4	2.0015	2.0028	2.0042	2.0055	2.0069	2.0082	2.0096	2.0109	2.0122	2.0136
7.5	2.0149	2.0162	2.0176	2.0189	2.0202	2.0215	2.0229	2.0242	2.0255	2.0268
7.6	2.0281	2.0295	2.0308	2.0321	2.0334	2.0347	2.0360	2.0373	2.0386	2.0399
7.7	2.0412	2.0425	2.0438	2.0451	2.0464	2.0477	2.0490	2.0503	2.0516	2.0528
7.8	2.0541	2.0554	2.0567	2.0580	2.0592	2.0605	2.0618	2.0631	2.0643	2.0656
7.9	2.0669	2.0681	2.0694	2.0707	2.0719	2.0732	2.0744	2.0757	2.0769	2.0782
8.0	2.0794	2.0807	2.0819	2.0832	2.0844	2.0857	2.0869	2.0882	2.0894	2.0906
8.1	2.0919	2.0931	2.0943	2.0956	2.0968	2.0980	2.0992	2.1005	2.1017	2.1029
8.2	2.1041	2.1054	2.1066	2.1078	2.1090	2.1102	2.1114	2.1126	2.1138	2.1150
8.3	2.1163	2.1175	2.1187	2.1199	2.1211	2.1223	2.1235	2.1247	2.1259	2.1270
8.4	2.1282	2.1294	2.1306	2.1318	2.1330	2.1342	2.1353	2.1365	2.1377	2.1389
8.5	2.1401	2.1412	2.1424	2.1436	2.1448	2.1459	2.1471	2.1483	2.1494	2.1506
8.6	2.1518	2.1529	2.1541	2.1552	2.1564	2.1576	2.1587	2.1599	2.1610	2.1622
8.7	2.1633	2.1645	2.1656	2.1668	2.1679	2.1691	2.1702	2.1713	2.1725	2.1736
8.8	2.1748	2.1759	2.1770	2.1782	2.1793	2.1804	2.1815	2.1827	2.1838	2.1849
8.9	2.1861	2.1872	2.1883	2.1894	2.1905	2.1917	2.1928	2.1939	2.1950	2.1961
9.0	2.1972	2.1983	2.1994	2.2006	2.2017	2.2028	2.2039	2.2050	2.2061	2.2072
9.1	2.2083	2.2094	2.2105	2.2116	2.2127	2.2138	2.2148	2.2159	2.2170	2.2181
9.2	2.2192	2.2203	2.2214	2.2225	2.2235	2.2246	2.2257	2.2268	2.2279	2.2289
9.3	2.2300	2.2311	2.2322	2.2332	2.2343	2.2354	2.2364	2.2375	2.2386	2.2396
9.4	2.2407	2.2418	2.2428	2.2439	2.2450	2.2460	2.2471	2.2481	2.2492	2.2502
9.5	2.2513	2.2523	2.2534	2.2544	2.2555	2.2565	2.2576	2.2586	2.2597	2.2607
9.6	2.2618	2.2628	2.2638	2.2649	2.2659	2.2670	2.2680	2.2690	2.2701	2.2711
9.7	2.2721	2.2732	2.2742	2.2752	2.2762	2.2773	2.2783	2.2793	2.2803	2.2814
9.8	2.2824	2.2834	2.2844	2.2854	2.2865	2.2875	2.2885	2.2895	2.2905	2.2915
9.9	2.2925	2.2935	2.2946	2.2956	2.2966	2.2976	2.2986	2.2996	2.3006	2.3016

$\ln 10 \approx 2.3026$

Table 5   Trigonometric Functions of Real Numbers

Real Number x or θ radians	θ degrees	sin x or sin θ	csc x or csc θ	tan x or tan θ	cot x or cot θ	sec x or sec θ	cos x or cos θ
0.00	0°00′	0.0000	No value	0.0000	No value	1.000	1.000
.01	0°34′	.0100	100.0	.0100	100.0	1.000	1.000
.02	1°09′	.0200	50.00	.0200	49.99	1.000	0.9998
.03	1°43′	.0300	33.34	.0300	33.32	1.000	0.9996
.04	2°18′	.0400	25.01	.0400	24.99	1.001	0.9992
0.05	2°52′	0.0500	20.01	0.0500	19.98	1.001	0.9988
.06	3°26′	.0600	16.68	.0601	16.65	1.002	.9982
.07	4°01′	.0699	14.30	.0701	14.26	1.002	.9976
.08	4°35′	.0799	12.51	.0802	12.47	1.003	.9968
.09	5°09′	.0899	11.13	.0902	11.08	1.004	.9960
0.10	5°44′	0.0998	10.02	0.1003	9.967	1.005	0.9950
.11	6°18′	.1098	9.109	.1104	9.054	1.006	.9940
.12	6°53′	.1197	8.353	.1206	8.293	1.007	.9928
.13	7°27′	.1296	7.714	.1307	7.649	1.009	.9916
.14	8°01′	.1395	7.166	.1409	7.096	1.010	.9902
0.15	8°36′	0.1494	6.692	0.1511	6.617	1.011	0.9888
.16	9°10′	.1593	6.277	.1614	6.197	1.013	.9872
.17	9°44′	.1692	5.911	.1717	5.826	1.015	.9856
.18	10°19′	.1790	5.586	.1820	5.495	1.016	.9838
.19	10°53′	.1889	5.295	.1923	5.200	1.018	.9820
0.20	11°28′	0.1987	5.033	0.2027	4.933	1.020	0.9801
.21	12°02′	.2085	4.797	.2131	4.692	1.022	.9780
.22	12°36′	.2182	4.582	.2236	4.472	1.025	.9759
.23	13°11′	.2280	4.386	.2341	4.271	1.027	.9737
.24	13°45′	.2377	4.207	.2447	4.086	1.030	.9713
0.25	14°19′	0.2474	4.042	0.2553	3.916	1.032	0.9689
.26	14°54′	.2571	3.890	.2660	3.759	1.035	.9664
.27	15°28′	.2667	3.749	.2768	3.613	1.038	.9638
.28	16°03′	.2764	3.619	.2876	3.478	1.041	.9611
.29	16°37′	.2860	3.497	.2984	3.351	1.044	.9582
0.30	17°11′	0.2955	3.384	0.3093	3.233	1.047	0.9553
.31	17°46′	.3051	3.278	.3203	3.122	1.050	.9523
.32	18°20′	.3146	3.179	.3314	3.018	1.053	.9492
.33	18°54′	.3240	3.086	.3425	2.920	1.057	.9460
.34	19°29′	.3335	2.999	.3537	2.827	1.061	.9428
0.35	20°03′	0.3429	2.916	0.3650	2.740	1.065	0.9394
.36	20°38′	.3523	2.839	.3764	2.657	1.068	.9359
.37	21°12′	.3616	2.765	.3879	2.578	1.073	.9323
.38	21°46′	.3709	2.696	.3994	2.504	1.077	.9287
.39	22°21′	.3802	2.630	.4111	2.433	1.081	.9249
0.40	22°55′	0.3894	2.568	0.4228	2.365	1.086	0.9211
.41	23°29′	.3986	2.509	.4346	2.301	1.090	.9171
.42	24°04′	.4078	2.452	.4466	2.239	1.095	.9131
.43	24°38′	.4169	2.399	.4586	2.180	1.100	.9090
.44	25°13′	.4259	2.348	.4708	2.124	1.105	.9048
0.45	25°47′	0.4350	2.299	0.4831	2.070	1.111	0.9004
.46	26°21′	.4439	2.253	.4954	2.018	1.116	.8961
.47	26°56′	.4529	2.208	.5080	1.969	1.122	.8916
.48	27°30′	.4618	2.166	.5206	1.921	1.127	.8870
.49	28°04′	.4706	2.125	.5334	1.875	1.133	.8823
0.50	28°39′	0.4794	2.086	0.5463	1.830	1.139	0.8776

Table 5 (*Continued*)

Real Number x or θ radians	θ degrees	sin x or sin θ	csc x or csc θ	tan x or tan θ	cot x or cot θ	sec x or sec θ	cos x or cos θ
0.50	28°39′	0.4794	2.086	0.5463	1.830	1.139	0.8776
.51	29°13′	.4882	2.048	.5594	1.788	1.146	.8727
.52	29°48′	.4969	2.013	.5726	1.747	1.152	.8678
.53	30°22′	.5055	1.978	.5859	1.707	1.159	.8628
.54	30°56′	.5141	1.945	.5994	1.668	1.166	.8577
0.55	31°31′	0.5227	1.913	0.6131	1.631	1.173	0.8525
.56	32°05′	.5312	1.883	.6269	1.595	1.180	.8473
.57	32°40′	.5396	1.853	.6410	1.560	1.188	.8419
.58	33°14′	.5480	1.825	.6552	1.526	1.196	.8365
.59	33°48′	.5564	1.797	.6696	1.494	1.203	.8309
0.60	34°23′	0.5646	1.771	0.6841	1.462	1.212	0.8253
.61	34°57′	.5729	1.746	.6989	1.431	1.220	.8196
.62	35°31′	.5810	1.721	.7139	1.401	1.229	.8139
.63	36°06′	.5891	1.697	.7291	1.372	1.238	.8080
.64	36°40′	.5972	1.674	.7445	1.343	1.247	.8021
0.65	37°15′	0.6052	1.652	0.7602	1.315	1.256	0.7961
.66	37°49′	.6131	1.631	.7761	1.288	1.266	.7900
.67	38°23′	.6210	1.610	.7923	1.262	1.276	.7838
.68	38°58′	.6288	1.590	.8087	1.237	1.286	.7776
.69	39°32′	.6365	1.571	.8253	1.212	1.297	.7712
0.70	40°06′	0.6442	1.552	0.8423	1.187	1.307	0.7648
.71	40°41′	.6518	1.534	.8595	1.163	1.319	.7584
.72	41°15′	.6594	1.517	.8771	1.140	1.330	.7518
.73	41°50′	.6669	1.500	.8949	1.117	1.342	.7452
.74	42°24′	.6743	1.483	.9131	1.095	1.354	.7385
0.75	42°58′	0.6816	1.467	0.9316	1.073	1.367	0.7317
.76	43°33′	.6889	1.452	.9505	1.052	1.380	.7248
.77	44°07′	.6961	1.436	.9697	1.031	1.393	.7179
.78	44°41′	.7033	1.422	.9893	1.011	1.407	.7109
.79	45°16′	.7104	1.408	1.009	.9908	1.421	.7038
0.80	45°50′	0.7174	1.394	1.030	0.9712	1.435	0.6967
.81	46°25′	.7243	1.381	1.050	.9520	1.450	.6895
.82	46°59′	.7311	1.368	1.072	.9331	1.466	.6822
.83	47°33′	.7379	1.355	1.093	.9146	1.482	.6749
.84	48°08′	.7446	1.343	1.116	.8964	1.498	.6675
0.85	48°42′	0.7513	1.331	1.138	0.8785	1.515	0.6600
.86	49°16′	.7578	1.320	1.162	.8609	1.533	.6524
.87	49°51′	.7643	1.308	1.185	.8437	1.551	.6448
.88	50°25′	.7707	1.297	1.210	.8267	1.569	.6372
.89	51°00′	.7771	1.287	1.235	.8100	1.589	.6294
0.90	51°34′	0.7833	1.277	1.260	0.7936	1.609	0.6216
.91	52°08′	.7895	1.267	1.286	.7774	1.629	.6137
.92	52°43′	.7956	1.257	1.313	.7615	1.651	.6058
.93	53°17′	.8016	1.247	1.341	.7458	1.673	.5978
.94	53°51′	.8076	1.238	1.369	.7303	1.696	.5898
0.95	54°26′	0.8134	1.229	1.398	0.7151	1.719	0.5817
.96	55°00′	.8192	1.221	1.428	.7001	1.744	.5735
.97	55°35′	.8249	1.212	1.459	.6853	1.769	.5653
.98	56°09′	.8305	1.204	1.491	.6707	1.795	.5570
.99	56°43′	.8360	1.196	1.524	.6563	1.823	.5487
1.00	57°18′	0.8415	1.188	1.557	0.6421	1.851	0.5403
1.01	57°52′	.8468	1.181	1.592	.6281	1.880	.5319
1.02	58°27′	.8521	1.174	1.628	.6142	1.911	.5234
1.03	59°01′	.8573	1.166	1.665	.6005	1.942	.5148
1.04	59°35′	.8624	1.160	1.704	.5870	1.975	.5062
1.05	60°10′	0.8674	1.153	1.743	0.5736	2.010	0.4976

Table 5 (*Continued*)

Real Number x or θ radians	θ degrees	sin x or sin θ	csc x or csc θ	tan x or tan θ	cot x or cot θ	sec x or sec θ	cos x or cos θ
1.05	60°10′	0.8674	1.153	1.743	0.5736	2.010	0.4976
1.06	60°44′	.8724	1.146	1.784	.5604	2.046	.4889
1.07	61°18′	.8772	1.140	1.827	.5473	2.083	.4801
1.08	61°53′	.8820	1.134	1.871	.5344	2.122	.4713
1.09	62°27′	.8866	1.128	1.917	.5216	2.162	.4625
1.10	63°02′	0.8912	1.122	1.965	0.5090	2.205	0.4536
1.11	63°36′	.8957	1.116	2.014	.4964	2.249	.4447
1.12	64°10′	.9001	1.111	2.066	.4840	2.295	.4357
1.13	64°45′	.9044	1.106	2.120	.4718	2.344	.4267
1.14	65°19′	.9086	1.101	2.176	.4596	2.395	.4176
1.15	65°53′	0.9128	1.096	2.234	0.4475	2.448	0.4085
1.16	66°28′	.9168	1.091	2.296	.4356	2.504	.3993
1.17	67°02′	.9208	1.086	2.360	.4237	2.563	.3902
1.18	67°37′	.9246	1.082	2.247	.4120	2.625	.3809
1.19	68°11′	.9284	1.077	2.498	.4003	2.691	.3717
1.20	68°45′	0.9320	1.073	2.572	0.3888	2.760	0.3624
1.21	69°20′	.9356	1.069	2.650	.3773	2.833	.3530
1.22	69°54′	.9391	1.065	2.733	.3659	2.910	.3436
1.23	70°28′	.9425	1.061	2.820	.3546	2.992	.3342
1.24	71°03′	.9458	1.057	2.912	.3434	3.079	.3248
1.25	71°37′	0.9490	1.054	3.010	0.3323	3.171	0.3153
1.26	72°12′	.9521	1.050	3.113	.3212	3.270	.3058
1.27	72°46′	.9551	1.047	3.224	.3102	3.375	.2963
1.28	72°20′	.9580	1.044	3.341	.2993	3.488	.2867
1.29	73°55′	.9608	1.041	3.467	.2884	3.609	.2771
1.30	74°29′	0.9636	1.038	3.602	0.2776	3.738	0.2675
1.31	75°03′	.9662	1.035	3.747	.2669	3.878	.2579
1.32	75°38′	.9687	1.032	3.903	.2562	4.029	.2482
1.33	76°12′	.9711	1.030	4.072	.2456	4.193	.2385
1.34	76°47′	.9735	1.027	4.256	.2350	4.372	.2288
1.35	77°21′	0.9757	1.025	4.455	0.2245	4.566	0.2190
1.36	77°55′	.9779	1.023	4.673	.2140	4.779	.2092
1.37	78°30′	.9799	1.021	4.913	.2035	5.014	.1994
1.38	79°04′	.9819	1.018	5.177	.1931	5.273	.1896
1.39	79°38′	.9837	1.017	5.471	.1828	5.561	.1798
1.40	80°13′	0.9854	1.015	5.798	0.1725	5.883	0.1700
1.41	80°47′	.9871	1.013	6.165	.1622	6.246	.1601
1.42	81°22′	.9887	1.011	6.581	.1519	6.657	.1502
1.43	81°56′	.9901	1.010	7.055	.1417	7.126	.1403
1.44	82°30′	.9915	1.009	7.602	.1315	7.667	.1304
1.45	83°05′	0.9927	1.007	8.238	0.1214	8.299	0.1205
1.46	83°39′	.9939	1.006	8.989	.1113	9.044	.1106
1.47	84°13′	.9949	1.005	9.887	.1011	9.938	.1006
1.48	84°48′	.9959	1.004	10.98	.0910	11.03	.0907
1.49	85°22′	.9967	1.003	12.35	.0810	12.39	.0807
1.50	85°57′	0.9975	1.003	14.10	0.0709	14.14	0.0707
1.51	86°31′	.9982	1.002	16.43	.0609	16.46	.0608
1.52	87°05′	.9987	1.001	19.67	.0508	19.69	.0508
1.53	87°40′	.9992	1.001	24.50	.0408	24.52	.0408
1.54	88°14′	.9995	1.000	32.46	.0308	32.48	.0308
1.55	88°49′	0.9998	1.000	48.08	0.0208	48.09	0.0208
1.56	89°23′	.9999	1.000	92.62	.0108	92.63	.0108
1.57	89°57′	1.000	1.000	1256	.0008	1256	.0008

Table 6 Squares, Square Roots, and Prime Factors

No.	Sq.	Sq. Rt.	Factors	No.	Sq.	Sq. Rt.	Factors
1	1	1.000		51	2,601	7.141	$3 \cdot 17$
2	4	1.414	2	52	2,704	7.211	$2^2 \cdot 13$
3	9	1.732	3	53	2,809	7.280	53
4	16	2.000	2^2	54	2,916	7.348	$2 \cdot 3^3$
5	25	2.236	5	55	3,025	7.416	$5 \cdot 11$
6	36	2.449	$2 \cdot 3$	56	3,136	7.483	$2^3 \cdot 7$
7	49	2.646	7	57	3,249	7.550	$3 \cdot 19$
8	64	2.828	2^3	58	3,364	7.616	$2 \cdot 29$
9	81	3.000	3^2	59	3,481	7.681	59
10	100	3.162	$2 \cdot 5$	60	3,600	7.746	$2^2 \cdot 3 \cdot 5$
11	121	3.317	11	61	3,721	7.810	61
12	144	3.464	$2^2 \cdot 3$	62	3,844	7.874	$2 \cdot 31$
13	169	3.606	13	63	3,969	7.937	$3^2 \cdot 7$
14	196	3.742	$2 \cdot 7$	64	4,096	8.000	2^6
15	225	3.873	$3 \cdot 5$	65	4,225	8.062	$5 \cdot 13$
16	256	4.000	2^4	66	4,356	8.124	$2 \cdot 3 \cdot 11$
17	289	4.123	17	67	4,489	8.185	67
18	324	4.243	$2 \cdot 3^2$	68	4,624	8.246	$2^2 \cdot 17$
19	361	4.359	19	69	4,761	8.307	$3 \cdot 23$
20	400	4.472	$2^2 \cdot 5$	70	4,900	8.367	$2 \cdot 5 \cdot 7$
21	441	4.583	$3 \cdot 7$	71	5,041	8.426	71
22	484	4.690	$2 \cdot 11$	72	5,184	8.485	$2^3 \cdot 3^2$
23	529	4.796	23	73	5,329	8.544	73
24	576	4.899	$2^3 \cdot 3$	74	5,476	8.602	$2 \cdot 37$
25	625	5.000	5^2	75	5,625	8.660	$3 \cdot 5^2$
26	676	5.099	$2 \cdot 13$	76	5,776	8.718	$2^2 \cdot 19$
27	729	5.196	3^3	77	5,929	8.775	$7 \cdot 11$
28	784	5.292	$2^2 \cdot 7$	78	6,084	8.832	$2 \cdot 3 \cdot 13$
29	841	5.385	29	79	6,241	8.888	79
30	900	5.477	$2 \cdot 3 \cdot 5$	80	6,400	8.944	$2^4 \cdot 5$
31	961	5.568	31	81	6,561	9.000	3^4
32	1,024	5.657	2^5	82	6,724	9.055	$2 \cdot 41$
33	1,089	5.745	$3 \cdot 11$	83	6,889	9.110	83
34	1,156	5.831	$2 \cdot 17$	84	7,056	9.165	$2^2 \cdot 3 \cdot 7$
35	1,225	5.916	$5 \cdot 7$	85	7,225	9.220	$5 \cdot 17$
36	1,296	6.000	$2^2 \cdot 3^2$	86	7,396	9.274	$2 \cdot 43$
37	1,369	6.083	37	87	7,569	9.327	$3 \cdot 29$
38	1,444	6.164	$2 \cdot 19$	88	7,744	9.381	$2^3 \cdot 11$
39	1,521	6.245	$3 \cdot 13$	89	7,921	9.434	89
40	1,600	6.325	$2^3 \cdot 5$	90	8,100	9.487	$2 \cdot 3^2 \cdot 5$
41	1,681	6.403	41	91	8,281	9.539	$7 \cdot 13$
42	1,764	6.481	$2 \cdot 3 \cdot 7$	92	8,464	9.592	$2^2 \cdot 23$
43	1,849	6.557	43	93	8,649	9.644	$3 \cdot 31$
44	1,936	6.633	$2^2 \cdot 11$	94	8,836	9.695	$2 \cdot 47$
45	2,025	6.708	$3^2 \cdot 5$	95	9,025	9.747	$5 \cdot 19$
46	2,116	6.782	$2 \cdot 23$	96	9,216	9.798	$2^5 \cdot 3$
47	2,209	6.856	47	97	9,409	9.849	97
48	2,304	6.928	$2^4 \cdot 3$	98	9,604	9.899	$2 \cdot 7^2$
49	2,401	7.000	7^2	99	9,801	9.950	$3^2 \cdot 11$
50	2,500	7.071	$2 \cdot 5^2$	100	10,000	10.000	$2^2 \cdot 5^2$

Table 7 Pascal's Triangle

n	r=1	2	3	4	5	6	7	8	9	10	11	12	13	14	15	16	17	18	19	20	21
1	1	1																			
2	1	2	1																		
3	1	3	3	1																	
4	1	4	6	4	1																
5	1	5	10	10	5	1															
6	1	6	15	20	15	6	1														
7	1	7	21	35	35	21	7	1													
8	1	8	28	56	70	56	28	8	1												
9	1	9	36	84	126	126	84	36	9	1											
10	1	10	45	120	210	252	210	120	45	10	1										
11	1	11	55	165	330	462	462	330	165	55	11	1									
12	1	12	66	220	495	792	924	792	495	220	66	12	1								
13	1	13	78	286	715	1287	1716	1716	1287	715	286	78	13	1							
14	1	14	91	364	1001	2002	3003	3432	3003	2002	1001	364	91	14	1						
15	1	15	105	455	1365	3003	5005	6435	6435	5005	3003	1365	455	105	15	1					
16	1	16	120	560	1820	4368	8008	11440	12870	11440	8008	4368	1820	560	120	16	1				
17	1	17	136	680	2380	6188	12376	19448	24310	24310	19448	12376	6188	2380	680	136	17	1			
18	1	18	153	816	3060	8568	18564	31824	43758	48620	43758	31824	18564	8568	3060	816	153	18	1		
19	1	19	171	969	3876	11628	27132	50388	75582	92378	92378	75582	50388	27132	11628	3876	969	171	19	1	
20	1	20	190	1140	4845	15504	38760	77520	125970	167960	184756	167960	125970	77520	38760	15504	4845	1140	190	20	1

Answers to Odd-Numbered Problems

Answers to Odd-Numbered Problems

Exercises 1–1

Number	Real number	Rational number	Irrational number	Integer	Positive integer	None of these
1. -19	✓	✓		✓		
3. 1	✓	✓		✓	✓	
5. π	✓		✓			
7. $-25/3$	✓	✓				
9. $\sqrt{7}$	✓		✓			
11. $\sqrt{-4}$						✓
13. $6/9$	✓	✓				
15. $1/\sqrt{2}$	✓		✓			
17. $0.\overline{3}$	✓	✓				
19. 7%	✓	✓				

21. $0.8\overline{0}$ **23.** $0.\overline{3}$ **25.** $0.\overline{45}$ **27.** $6.1\overline{6}$ **29.** $1.\overline{428571}$
31. False, $\frac{1}{2}$ **33.** False, $\frac{1}{2}$ **35.** False, $\frac{1}{2}$ **37.** True **39.** False, 0

Exercises 1–2

1. Comm. (Add.) **3.** Assoc. (Mult.) **5.** Prop. of 1 **7.** Comm. (Mult.)
9. Dist. **11.** Assoc. (Mult.) **13.** Comm. (Add.) **15.** Comm. (Add.)
17. Exist. of Reciprocals **19.** Comm. (Mult.) **21.** -7 **23.** $\frac{2}{101}$
25. 0 **27.** $\frac{1}{25}$ **29.** $\frac{17}{5}$ **31.** No reciprocal
33. $a(b + c)[\text{total area}] = ab[\text{area 1}] + ac[\text{area 2}]$ **35.** Comm. (Mult.)
37. Comm. (Mult.) **39.** Comm. (Mult.) **41.** Comm. (Mult.)

Exercises 1–3

1. 4 **3.** 3.148 **5.** 0 **7.** -7 **9.** -5 **11.** -2 **13.** $\frac{14}{5}$
15. -0.54 **17.** -4.1 **19.** -66 **21.** 2 **23.** $-\frac{1}{32}$ **25.** $\frac{20}{9}$
27. 0 **29.** -0.04 **31.** 10.8 **33.** -2 **35.** -14 **37.** 6
39. -40 **41.** -3 **43.** -32 **45.** $\frac{13}{16}$ **47.** 10 **49.** 1
51. -52 **53.** -30 **55.** -12 **57.** -15 **59.** 14 **61.** $\frac{1}{2}$
63. 1 **65.** $-\frac{1}{2}$ **67.** Undefined **69.** 0 **71.** 20 **73.** 18
75. 80 **77.** 25 **79.** 9 **81.** Even-positive, Odd-negative
83. $y = 1$, x any real number or $x = 0$, y any real number **85.** Lose \$300

Exercises 1–4

1. $19a$ **3.** $14a - 8b$ **5.** $3xy + 5cd$ **7.** $-3x^3 + 8y^2$ **9.** $2x - 5y$
11. $-16a + 11b$ **13.** $10x - 7y$ **15.** $2x^3 + 13x^2y^2 + 9xy^3 - 3x^3y$
17. $7y^3 - 3y^2 + 4y + 6$ **19.** $b^3 - 3b^2 + b + 5$ **21.** $12a + 14$
23. $2a + 5b$ **25.** 6

Exercises 1–5

1. 3 **3.** 6 **5.** -1 **7.** 3 **9.** 2 **11.** -6
13. Any real number **15.** No solution **17.** 0 **19.** No solution
21. 0 **23.** 5 **25.** -5 **27.** 3 **29.** $\frac{36}{7}$ **31.** 2 **33.** -5
35. 45 **37.** 1100, 3300, 1110 **39.** 1200 in.^{2} **41.** \$350 **43.** $11.\overline{1}\%$
45. 57 mi/hour, 65 mi/hour **47.** 5:20 P.M. **49.** 1.25 qt **51.** 24 lb

53. $\dfrac{2(x + 3) + 6}{2} - x = 6$; $\dfrac{3x + (x + 1) + 7}{4} - 2 = x$

Exercises 1–6

1. 144 **3.** 144 **5.** 32 **7.** 11 **9.** 6 **11.** $m = F/a$

13. $a = \dfrac{2d}{t^2}$ **15.** $m = \dfrac{P}{gh}$ **17.** $b = 3A - a - c$ **19.** $L = \dfrac{2s}{n} - a$

21. $t = \dfrac{(a/p) - 1}{r}$ **23.** $R^2 = \dfrac{A}{\pi} + r^2$ **25.** $r = \dfrac{a - s}{-s}$

27 $r = \dfrac{2E - IR}{2I}$ **29.** $A = \dfrac{D}{n - 1}$

Exercises 1–7

1. 3:2   **3.** 4:5   **5.** 5:2   **7.** 5:4   **9.** 1:2   **11.** 12.5 lb
13. 9 hours   **15.** 63 g   **17.** $7500   **19.** $2160   **21.** 44 ft   **23.** 95
25. 324 gal   **27.** 92 mi   **29.** 25 qt
31. **(a)** 15 ft   **(b)** 12/5 ft   **(c)** 25/4 mi

Exercises 1–8

1. $<$ **3.** $<$ **5.** $<$ **7.** $<$ **9.** $<$ **11.** $=$ **13.** $<$
15. $>$ **17.** $>$ **19.** $>$ **21.** $x > 8$ **23.** $x > 2$ **25.** $x \le 1$
27. $x < -\frac{9}{2}$ **29.** $x \le -1$ **31.** $x < -6$ **33.** $x < \frac{12}{7}$ **35.** $x > -2$
37. $x \le -3$ **39.** $x > -8$ **41.** Any real number   **43.** $x > 0$
45. No solution   **47.** $y \le 0$ **49.** No solution   **51.** True   **53.** False
55. True   **57.** True   **59.** False   **61.** True   **63.** Figure K-1
65. Figure K-2   **67.** $x > 2$ and $x < 4$; $2 \le x \le 4$ **69.** $x < 2$ or $x > 4$
71. $\$0 \le a \le \$1500$ **73.** $0 < x < 40$
75. $0 \le 150x + 250y \le 10{,}000$; $x, y \ge 0$
77. $0 \le 90x + 50y + 300z \le 2000$; $x, y, z \ge 0$
79. $0 \le Bx + Cy \le A$; $x, y \ge 0$

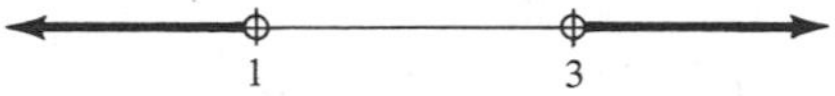

Figure K-1    Figure K-2

SAMPLE TEST: CHAPTER 1

1. False; π **2.** $1.\overline{18}$ **3.** Commutative property of addition   **4.** 7

5. -55 **6.** $\dfrac{5}{-7}$ **7.** $3x - 3$ **8.** 2   **9.** 0   **10.** 20   **11.** $306

12. $\dfrac{40}{7}$ qt   **13.** $L = \dfrac{v}{n}$ **14.** $y = x - z - 10P$ **15.** 82   **16.** 51.3 mi

17. **(a)** $<$ **(b)** $<$ **18.** $x < -2$ **19.** $0 \le x$ **20.** $x \le 0$ or $x > 6$

Exercises 2–1 (There are many possibilities for 1–10)

1. $(-2, -2), (-1, -1), (0, 0), (1, 1), (2, 2)$ **3.** $(-2, 6), (-1, 5), (0, 4), (1, 3), (2, 2)$
5. $(-2, -1), (-1, -2), (0, -1), (1, 2), (2, 7)$
7. $(1, 60), (2, 120), (3, 180), (4, 240), (1/2, 30)$

9. $(32, 0), (41, 5), (50, 10), \left(0, \dfrac{-160}{9}\right), (-4, -20)$ **11.** $(-2, -5), (1, 4)$

13. $(1, 1), (-1, 1)$ **15.** Function **17.** Function **19.** Not a function
21. $r > 0$ **23.** All real numbers **25.** All real numbers except -4
27. $-2, -1, 1, 2$ **29.** $x \geq 0$ **31.** $y = 2x$ **33.** $y = x - 3$
35. $a = s^2$ **37.** $y = x$ **39.** $y = 10x + 2$ **41.** $y = 2x - 1$

43. $A = s^2; s > 0$ **45.** $L = \dfrac{A}{5}; A \geq 25$ **47.** $s = \dfrac{P}{4}; P > 0$

49. $d = 40t; t \geq 0$ **51.** $e = 8n; n \geq 0$ **53.** $e = 0.06p; p \geq 0$
55. $c = 5x + 400; x \geq 0$, where x is an integer
57. $A = 10,000 - 50n; 0 \leq n \leq 200$
59. $t = 310 + 0.19(i - 2000); 2000 \leq i \leq 4000$

61. $c = \begin{cases} 1.75 & \text{if } 0 \leq n \leq 12 \\ 1.75 + 0.0382(n - 12) & \text{if } 12 < n \leq 48; \end{cases} \quad 0 \leq n \leq 48$

Exercises 2–2

1. $\frac{70}{3}$ **3.** 12 **5.** 3 **7.** $\frac{8}{9}$ **9.** 60 **11.** 49 **13.** 4 in.
15. 2,000,000 tons **17.** 3 in. **19.** 600 rpm **21.** $500\pi/3$ cubic units
23. 181 lb **25.** Exposure time is multiplied by 4 **27.** $\frac{5}{3}$ atm
29. Force is multiplied by 24

Exercises 2–3

1. $-2, -3, 2$ **3.** $11, 5, -3$ **5.** $19, 4, 7$
7. $\frac{2}{5}, -2$, undefined **9.** (a) 3; (b) 1; (c) -1
11. (a) 2; (b) -10; (c) 10; (d) -4; (e) 3; (f) $\frac{1}{2}$; (g) 4; (h) 4; (i) 2;
(j) $x^2 + 1$
13. (a) 3; (b) -1; (c) No **15.** (a) 10 cents; (b) 30 cents; (c) 20 cents
17. (a) 1; (b) 1; (c) -1 **19.** (a) 3; (b) 0; (c) 3

Exercises 2–4

1. Figure K-3 **3.** $(-1, -1)$ **5.** $(2, 4), (\frac{1}{2}, 1), (-1, -2)$, other possibilities
7. (a) 0 (b) 0 **9.** Figure K-4 **11.** Figure K-5 **13.** Figure K-6
15. Figure K-7 **17.** Figure K-8 **19.** Figure K-9 **21.** Figure K-10
23. Figure K-11 **25.** Figure K-12 **27.** Figure K-13 **29.** Figure K-14
31. Figure K-15 **33.** Figure K-16 **35.** Figure K-17 **37.** Figure K-18
39. Figure K-19 **41.** Figure K-20 **43.** Figure K-21
45. (a) function; (b) not a function; (c) not a function; (d) function;
(e) not a function; (f) function; (g) function; (h) not a function;
(i) function; (j) function
47. (a) 40 percent; (b) 100 percent; (c) 23

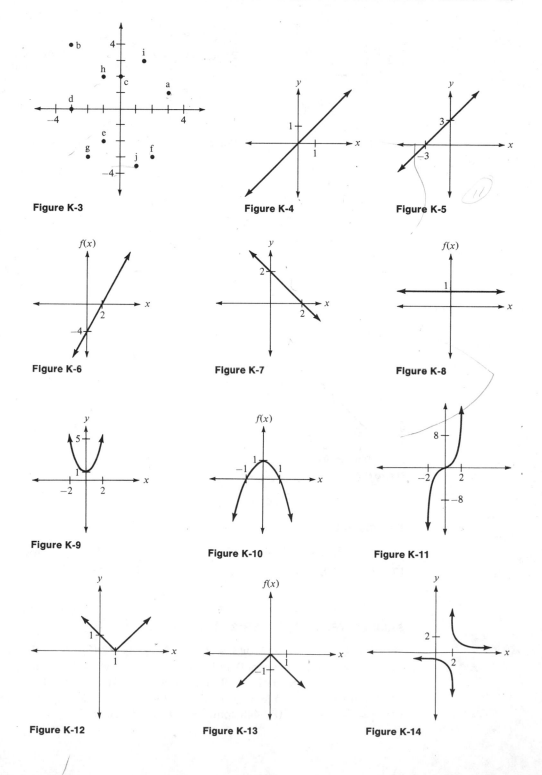

Figure K-3

Figure K-4

Figure K-5

Figure K-6

Figure K-7

Figure K-8

Figure K-9

Figure K-10

Figure K-11

Figure K-12

Figure K-13

Figure K-14

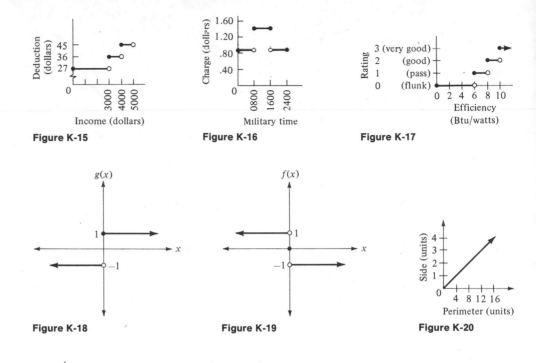

Figure K-15   Figure K-16   Figure K-17

Figure K-18   Figure K-19   Figure K-20

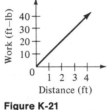

Figure K-21

Exercises 2–5

1. 4   **3.** 2   **5.** 4   **7.** 5   **9.** $\sqrt{2}$ **11.** 13   **13.** 5   **15.** 13

17. 10   **19.** $\sqrt{40}$ **21.** $\sqrt{2}$ **23.** $\sqrt{74}$ **25.** $\sqrt{32}$

SAMPLE TEST: CHAPTER 2

1. $(0, 4)(\frac{5}{2}, -1)(-1, 6)(2, 0)(3, -2)$ **2.** $(1, 2)(2, 2)(3, 2)$ **3.** Yes
4. All real numbers except 0 and -1 **5.** $y = x^2 + 1$ **6.** $C = 2\pi r$
7. $A = d^2/2$ **8.** Yes   **9.** Undefined   **10.** 22   **11.** $-\frac{1}{2}$ **12.** -3
13. Figure K-22   **14.** $x \geq 0$ **15.** $(3, -3)$ **16.** Figure K-23
17. $y = \frac{5}{12}x; \frac{25}{3}$ **18.** 448 rpm   **19.** $\frac{9}{4}$ **20.** $\sqrt{29}$

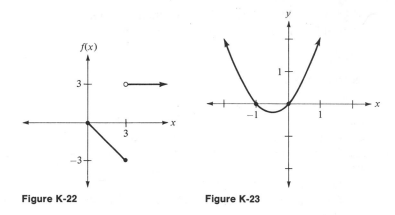

Figure K-22   **Figure K-23**

Exercises 3–1

	sin θ	csc θ	cos θ	sec θ	tan θ	cot θ
1.	$-4/5$	$5/-4$	$3/5$	$5/3$	$-4/3$	$3/-4$
3.	$5/13$	$13/5$	$-12/13$	$13/-12$	$5/-12$	$-12/5$
5.	$-4/5$	$5/-4$	$3/5$	$5/3$	$-4/3$	$3/-4$
7.	$1/\sqrt{2}$	$\sqrt{2}$	$1/\sqrt{2}$	$\sqrt{2}$	1	1
9.	$2/\sqrt{5}$	$\sqrt{5}/2$	$-1/\sqrt{5}$	$-\sqrt{5}$	-2	$1/-2$

11. Q_4
13. Q_3
15. Q_2

	sin θ	csc θ	cos θ	sec θ	tan θ	cot θ
17.	$-3/5$	$5/-3$	$-4/5$	$5/-4$	$3/4$	$4/3$
19.	$-\sqrt{3}/2$	$2/-\sqrt{3}$	$1/2$	2	$-\sqrt{3}$	$1/-\sqrt{3}$
21.	$-3/4$	$-4/3$	$\sqrt{7}/-4$	$-4/\sqrt{7}$	$3/\sqrt{7}$	$\sqrt{7}/3$
23.	$1/3$	3	$\sqrt{8}/3$	$3/\sqrt{8}$	$1/\sqrt{8}$	$\sqrt{8}$
25.	$\sqrt{8}/3$	$3/\sqrt{8}$	$-1/3$	-3	$-\sqrt{8}$	$-1/\sqrt{8}$

27. 4/3   **29.** 5/7   **31.** Undefined   **33.** 3   **35.** $\sin \theta = 1/\csc \theta$
37. $\tan \theta = 1/\cot \theta$ **39.** $\cot \theta = 1/\tan \theta$

Exercises 3–2

1. 0.6799   **3.** 5.396   **5.** 2.381   **7.** 0.5000   **9.** 0   **11.** 1
13. 2   **15.** 0.1398   **17.** 0.2141   **19.** 1.000   **21.** 0.9999   **23.** 4.072
25. 0.9437   **27.** 30°00′   **29.** 20°20′   **31.** 1°10′   **33.** 81°00′
35. No solution   **37.** No solution   **39.** 33°43′   **41.** 52°35′
43. 3°35′   **45.** 85°47′   **47.** 21°48′   **49.** 45°28′
51. (a) 0.7071;   **(b)** 0.5;
     (c) The function does not change uniformly. The angles must be close for the
          error to be small.
53. $\csc \theta = r/y$ and $y \le r$ **57. (a)** 21°00′;   **(b)** 48°40′

Exercises 3–3

	a	b	c	A	B
1.		87 ft	1̄00 ft		60°
3.	13 ft	7.5 ft			30°
5.	8.1 ft	24 ft		19°	
7.	2̄0 ft	14 ft			35°
9.		85.1 ft	121 ft		44°30′
11.	379 ft	497 ft		37°20′	
13.		975 ft	975 ft	1°50′	
15.		14 ft		24°	66°
17.			1.4 ft	45°	45°
19.	23.1 ft			62°30′	27°30′

21. $\sin A = \frac{15}{17}$; $\csc A = \frac{17}{15}$; $\cos A = \frac{8}{17}$; $\sec A = \frac{17}{8}$; $\tan A = \frac{15}{8}$; $\cot A = \frac{8}{15}$
23. 36   **25.** 4   **27.** 25 ft   **29.** 52 yd   **31.** 89 ft   **33.** 4̄00 ft
35. 10°   **37.** 77 ft   **39.** 50°   **41.** 22 ft   **43.** 120 in.   **45.** 59 in.
47. 9°40′   **49.** 55°, 55°, 70°   **51.** 88 ft   **53.** 16 ohms, 15°
55. 48 mi   **57.** 430,000 mi

Exercises 3–4

1. **(a)** 13 lb; **(b)** 23°   **3.** **(a)** 18 lb; **(b)** 34°
5. **(a)** 16 mi/hour; **(b)** 18°   **7.** **(a)** 510 mi; **(b)** 81°
9. **(a)** 13°; **(b)** 310 mi/hour   **11.** v, $5\overline{0}$ lb; h, 87 lb   **13.** v, 8.2 lb; h, 16 lb
15. **(a)** 340 mi/hour; **(b)** $9\overline{0}$ mi/hour   **17.** 34 lb   **19.** 26 lb

Exercises 3–5

1. 70°   **3.** 85°   **5.** 56°50′   **7.** 49°30′   **9.** 35°   **11.** 30°
13. 22°30′   **15.** cos 73°   **17.** cot 89°   **19.** tan 11°51′   **21.** $\frac{3}{2}$
23. 3   **25.** $\frac{15}{2}$ **27.** −1   **29.** 1   **31.** 1   **33.** 1   **35.** 0
37. $1/\sqrt{2}$ **39.** Undefined   **41.** 1   **43.** 4   **45.** 1

Exercises 3–6

1. −1/2   **3.** $2/\sqrt{3}$ **5.** −1   **7.** 1/2   **9.** $-1/\sqrt{2}$ **11.** $\sqrt{3}/2$
13. $2/\sqrt{3}$ **15.** $-\sqrt{3}$ **17.** $-1/\sqrt{2}$ **19.** $-1/\sqrt{2}$ **21.** $-\sqrt{3}/2$
23. $\sqrt{3}$ **25.** 1   **27.** $2/\sqrt{3}$ **29.** 1   **31.** −0.5299   **33.** 3.487
35. 1.923   **37.** −0.8557   **39.** −0.5354   **41.** 0.0494   **43.** −1.360
45. −0.9983   **47.** −0.6111   **49.** 0.3346   **51.** 0.1564   **53.** 1.150
55. 0.4848   **57.** −0.6787   **59.** 1.881

Exercises 3–7

1. 60°, 300°   **3.** 135°, 315°   **5.** 45°, 135°   **7.** 150°, 210°
9. 30°, 210°   **11.** 225°, 315°   **13.** 60°, 240°   **15.** 240°, 300°
17. 210°, 330°   **19.** 45°, 225°   **21.** 7°00′, 173°00′   **23.** 120°50′, 239°10′
25. 22°10′, 202°10′   **27.** 161°00′, 199°00′   **29.** 72°20′, 252°20′
31. 101°20′, 281°20′   **33.** 131°50′, 228°10′   **35.** 74°00′, 254°00′
37. 206°20′, 333°40′   **39.** No solution
41. 405°, 765°, 1125°, −315°, −675°
43. 382°10′, 742°10′, 1102°10′, −337°50′, −697°50′
45. 480°, 840°, 1200°, −240°, −600°   **47.** −390°, −750°, −1120°, 330°, 690°
49. −460°50′, −820°50′, −1180°50′, 259°10′, 619°10′
51. 45° + $k360$°; 135° + $k360$°   **53.** 199°30′ + $k360$°; 340°30′ + $k360$°
55. 113°30′ + $k360$°; 246°30′ + $k360$°   **57.** No solution
59. 180° + $k360$°
Note: In Exercises 41–50 other solutions are possible; in Exercises 51–60 k may be any integer.

Exercises 3–8

	a	b	c	A	B	C
1.		35 ft	48 ft			105°
3.	11 ft		43 ft		78°	
5.		140 ft	1̄00 ft	20°		
7.		34.6 ft	23.5 ft			41°20′
9.		243 ft	152 ft	33°00′		
11.			110 ft		26°	109°
13.			57 ft	42°		108°
			12 ft	138°		12°
15.	26 ft			8°	22°	
17.	165 ft			164°00′		8°50′
	17.4 ft			1°40′		171°10′

19.	No triangle satisfies the data.

21. 433 ft   **23.** 95 mi   **25.** 190 yd   **27.** 401 ft   **29.** 23 lb

Exercises 3–9

	a	b	c	A	B	C
1.			14 ft	49°	71°	
3.		27.4 ft		162°30′		7°10′
5.			78 ft	26°	34°	
7.			23.1 ft	55°50′	28°30′	
9.				104°	29°	47°
11.				128°	20°	32°
13.				46°30′	57°50′	75°40′
15.				37°40′	93°30′	48°50′

17. (a) $4\overline{0}$ lb; **(b)** $12°$ **19. (a)** 321 lb; **(b)** $5°50'$ **21.** $30°, 56°, 86°$
23. 64 ft   **25.** 99 mi
27. Since $\cos 90° = 0$, $c^2 = a^2 + b^2 - 2ab \cos 90°$ becomes $c^2 = a^2 + b^2$.

SAMPLE TEST: CHAPTER 3

1. $-\frac{3}{5}$ **2.** $135°$ **3.** 0.8408 **4.** 3 **5.** $-\sqrt{2}$ **6.** $\frac{1}{2}$

7. -0.0581 **8.** $210°, 330°$ **9.** $\left.\begin{array}{c} 71°30' + k360° \\ 251°30' + k360° \end{array}\right\}$ where k is any integer

10. $460°, 820°, 1180°, -260°, -620°$ **11.** $10°$ **12.** d **13.** $\frac{8}{17}$
14. $38°40'$ **15.** 21 lb   **16.** 5.3 **17.** False   **18.** $109°$
19. $\sqrt{37} \approx 6.1$ **20. (a)** 19; **(b)** 344

Exercises 4–1

1. $2^5 = 32$ **3.** $3^3 = 27$ **5.** $3^6 = 729$ **7.** $10^2 = 100$
9. $2^6 \cdot 3^2 = 576$ **11.** $2^{11} = 2048$ **13.** 2^{x+y+5} **15.** 4^{ab} **17.** 2^{n-1}
19. $3^3/(5 \cdot 7^2) = \frac{27}{245}$ **21. (a)** c^{12}; **(b)** $\cos^{12}\theta$ **23. (a)** t^5; **(b)** $\tan^5\theta$
25. (a) c^{a+b}; **(b)** $\cos^{a+b}\theta$ **27. (a)** s^{7a}; **(b)** $\sin^{7a}\theta$
29. (a) c^5; **(b)** $\cos^5\theta$ **31. (a)** $1/t^3$; **(b)** $1/\tan^3\theta$
33. (a) s^{a-1}; **(b)** $\sin^{a-1}\theta$ **35. (a)** $1/c$; **(b)** $1/\cos\theta$
37. (a) t^8; **(b)** $\tan^8\theta$ **39. (a)** $9s^2$; **(b)** $9\sin^2\theta$ **41. (a)** t^4; **(b)** $\tan^4\theta$
43. (a) $3^4 t^{12} s^8$; **(b)** $3^4 \tan^{12}\theta \sin^8\theta$ **45. (a)** $50t^5$; **(b)** $50\tan^5\theta$
47. (a) $108c^3 s^2 t^5$; **(b)** $108\cos^3\theta \sin^2\theta \tan^5\theta$
49. (a) $6^3 c^{18} t^8$; **(b)** $6^3 \cos^{18}\theta \tan^8\theta$ **51. (a)** $-27t^4$; **(b)** $-27\tan^4\theta$
53. (a) $-s/81t$; **(b)** $-\sin\theta/81\tan\theta$
55. (a) $-c^4/4s^5 t^8$; **(b)** $-\cos^4\theta/4\sin^5\theta \tan^8\theta$ **57.** $V = \frac{1}{6}\pi d^3$

Exercises 4–2

1. (a) $5s + 5t$; **(b)** $5\sin\theta + 5\tan\theta$
3. (a) $-2c + 2s$; **(b)** $-2\cos\theta + 2\sin\theta$
5. (a) $2s^4 - 14s^2 + 2s$; **(b)** $2\sin^4\theta - 14\sin^2\theta + 2\sin\theta$
7. (a) $-5t^4 + 5t^3 + 5t^2$; **(b)** $-5\tan^4\theta + 5\tan^3\theta + 5\tan^2\theta$
9. (a) $2c^3 t - 3c^2 t^2 + 4c^2 t$; **(b)** $2\cos^3\theta \tan\theta - 3\cos^2\theta \tan^2\theta + 4\cos^2\theta \tan\theta$
11. (a) $-8s^2 tc + 2st^2 c - 14stc^2$;
   (b) $-8\sin^2\theta \tan\theta \cos\theta + 2\sin\theta \tan^2\theta \cos\theta - 14\sin\theta \tan\theta \cos^2\theta$
13. (a) $4s^2 t - 5st^2$; **(b)** $4\sin^2\theta \tan\theta - 5\sin\theta \tan^2\theta$
15. (a) $t^4 s^2 + t^2 s^4$; **(b)** $\tan^4\theta \sin^2\theta + \tan^2\theta \sin^4\theta$
17. (a) $2c - 7$; **(b)** $2\cos\theta - 7$
19. (a) $4s + 3t$; **(b)** $4\sin\theta + 3\tan\theta$
21. (a) $2s^2 - 3s + 4$; **(b)** $2\sin^2\theta - 3\sin\theta + 4$
23. (a) $t^2 + 7t + 12$; **(b)** $\tan^2\theta + 7\tan\theta + 12$
25. (a) $c^2 - 16$; **(b)** $\cos^2\theta - 16$

27. (a) $6t^2 - 11t + 4$; **(b)** $6 \tan^2 \theta - 11 \tan \theta + 4$
29. (a) $12t^2 + 28tc - 5c^2$; **(b)** $12 \tan^2 \theta + 28 \tan \theta \cos \theta - 5 \cos^2 \theta$
31. (a) $c^2 - 4c + 4$; **(b)** $\cos^2 \theta - 4 \cos \theta + 4$
33. (a) $16c^2 - 8cs + s^2$; **(b)** $16 \cos^2 \theta - 8 \cos \theta \sin \theta + \sin^2 \theta$
35. (a) $s^3 + s^2 - 21s + 4$; **(b)** $\sin^3 \theta + \sin^2 \theta - 21 \sin \theta + 4$
37. (a) $3c^4 + 2c^3 - 3c^2 - 5c - 2$; **(b)** $3 \cos^4 \theta + 2 \cos^3 \theta - 3 \cos^2 \theta - 5 \cos \theta - 2$
39. (a) $t^3 - c^3$; **(b)** $\tan^3 \theta - \cos^3 \theta$
41. (a) $8t^3 + 2t^2c - 17tc^2 + 4c^3$;
    (b) $8 \tan^3 \theta + 2 \tan^2 \theta \cos \theta - 17 \tan \theta \cos^2 \theta + 4 \cos^3 \theta$
43. (a) $-6t^3 + 5t^2 + 3t - 2$; **(b)** $-6 \tan^3 \theta + 5 \tan^2 \theta + 3 \tan \theta - 2$
45. (a) $c^2 + 7c + 10$; **(b)** $\cos^2 \theta + 7 \cos \theta + 10$
47. (a) $s^2 + s - 42$; **(b)** $\sin^2 \theta + \sin \theta - 42$
49. (a) $t^2 - 36$; **(b)** $\tan^2 \theta - 36$
51. (a) $4s^2 - 25$; **(b)** $4 \sin^2 \theta - 25$
53. (a) $3t^2 - 5t - 28$; **(b)** $3 \tan^2 \theta - 5 \tan \theta - 28$
55. (a) $5t^2 - 16t + 12$; **(b)** $5 \tan^2 \theta - 16 \tan \theta + 12$
57. (a) $-3s^2 + 17s + 28$; **(b)** $-3 \sin^2 \theta + 17 \sin \theta + 28$
59. (a) $s^2 - 6s + 9$; **(b)** $\sin^2 \theta - 6 \sin \theta + 9$
61. (a) $16t^2 - 24t + 9$; **(b)** $16 \tan^2 \theta - 24 \tan \theta + 9$ **63.** 25 in.

Exercises 4–3

1. $(x + y)^6 = x^6 + 6x^5y + 15x^4y^2 + 20x^3y^3 + 15x^2y^4 + 6xy^5 + y^6$
3. $(x - y)^5 = x^5 - 5x^4y + 10x^3y^2 - 10x^2y^3 + 5xy^4 - y^5$
5. $(x + h)^4 = x^4 + 4x^3h + 6x^2h^2 + 4xh^3 + h^4$
7. $(x - 1)^7 = x^7 - 7x^6 + 21x^5 - 35x^4 + 35x^3 - 21x^2 + 7x - 1$
9. $(2x + y)^3 = 8x^3 + 12x^2y + 6xy^2 + y^3$
11. $(3c - 4d)^4 = (3c)^4 + 4(3c)^3(-4d) + 6(3c)^2(-4d)^2 + 4(3c)(-4d)^3 + (-4d)^4$
      $= 81c^4 - 432c^3d + 864c^2d^2 - 768cd^3 + 256d^4$
13. $(x + y)^{15} \doteq x^{15} + 15x^{14}y + 105x^{13}y^2 + 455x^{12}y^3 + \cdots$
15. $(x - 3y)^{12} = x^{12} + 12x^{11}(-3y) + 66x^{10}(-3y)^2 + 220x^9(-3y)^3 + \cdots$
      $= x^{12} - 36x^{11}y + 594x^{10}y^2 - 5940x^9y^3 + \cdots$
17. $18x^{17}y$ **19.** $78(3x)^2(-y)^{11} = 702x^2y^{11}$ **21.** 1, 21, 210, 1330, 5985

Exercises 4–4

1. (a) $7(s + c)$; **(b)** $7(\sin \theta + \cos \theta)$ **3. (a)** $s(c + t)$; **(b)** $\sin \theta(\cos \theta + \tan \theta)$
5. (a) $7s(3s - 2)$; **(b)** $7 \sin \theta(3 \sin \theta - 2)$
7. (a) $3st(3st + 1)$; **(b)** $3 \sin \theta \tan \theta(3 \sin \theta \tan \theta + 1)$
9. (a) $3ct(5t - 7c)$; **(b)** $3 \cos \theta \tan \theta(5 \tan \theta - 7 \cos \theta)$
11. (a) $s(2st^2 + 4c - 5sc^2)$; **(b)** $\sin \theta(2 \sin \theta \tan^2 \theta + 4 \cos \theta - 5 \sin \theta \cos^2 \theta)$
13. (a) $(s - 5)(c + t)$; **(b)** $(\sin \theta - 5)(\cos \theta + \tan \theta)$
15. (a) $(c - s)(t + s)$; **(b)** $(\cos \theta - \sin \theta)(\tan \theta + \sin \theta)$
17. (a) $(s + 8)(s - 8)$; **(b)** $(\sin \theta + 8)(\sin \theta - 8)$

19. (a) $(6c + 1)(6c - 1)$; **(b)** $(6 \cos \theta + 1)(6 \cos \theta - 1)$
21. (a) $(2t + 5s)(2t - 5s)$; **(b)** $(2 \tan \theta + 5 \sin \theta)(2 \tan \theta - 5 \sin \theta)$
23. (a) $(6s^3 + t^2)(6s^3 - t^2)$; **(b)** $(6 \sin^3 \theta + \tan^2 \theta)(6 \sin^3 \theta - \tan^2 \theta)$
25. (a) $(3s + tc)(3s - tc)$; **(b)** $(3 \sin \theta + \tan \theta \cos \theta)(3 \sin \theta - \tan \theta \cos \theta)$
27. (a) $(s + 4)(s + 1)$; **(b)** $(\sin \theta + 4)(\sin \theta + 1)$
29. (a) $(t - 2)(t - 1)$; **(b)** $(\tan \theta - 2)(\tan \theta - 1)$
31. (a) $(s + 4)(s - 3)$; **(b)** $(\sin \theta + 4)(\sin \theta - 3)$
33. (a) $(t - 5)(t - 4)$; **(b)** $(\tan \theta - 5)(\tan \theta - 4)$
35. (a) $(3 - s)(2 + s)$; **(b)** $(3 - \sin \theta)(2 + \sin \theta)$
37. (a) $(3s + 1)(s - 3)$ **(b)** $(3 \sin \theta + 1)(\sin \theta - 3)$
39. (a) $(7c + 1)(c - 2)$ **(b)** $(7 \cos \theta + 1)(\cos \theta - 2)$
41. (a) $(3s + 2)(2s + 1)$; **(b)** $(3 \sin \theta + 2)(2 \sin \theta + 1)$
43. (a) $(9t + 2)(t - 3)$; **(b)** $(9 \tan \theta + 2)(\tan \theta - 3)$
45. (a) $(5s - 12)(4s + 1)$; **(b)** $(5 \sin \theta - 12)(4 \sin \theta + 1)$
47. (a) $(s + 4c)(s + c)$; **(b)** $(\sin \theta + 4 \cos \theta)(\sin \theta + \cos \theta)$
49. (a) $(4c - t)(c - 2t)$; **(b)** $(4 \cos \theta - \tan \theta)(\cos \theta - 2 \tan \theta)$
51. (a) $(4t - 3)^2$; **(b)** $(4 \tan \theta - 3)^2$ **53. (a)** $(2s - 1)^2$; **(b)** $(2 \sin \theta - 1)^2$
55. (a) $(3s - 2c)^2$; **(b)** $(3 \sin \theta - 2 \cos \theta)^2$
57. (a) $7(s + 3)(s - 3)$; **(b)** $7(\sin \theta + 3)(\sin \theta - 3)$
59. (a) $s(1 + t)(1 - t)$; **(b)** $\sin \theta(1 + \tan \theta)(1 - \tan \theta)$
61. (a) $(t^2 + 1)(t + 1)(t - 1)$; **(b)** $(\tan^2 \theta + 1)(\tan \theta + 1)(\tan \theta - 1)$
63. (a) $3(s - 4)(s + 2)$; **(b)** $3(\sin \theta - 4)(\sin \theta + 2)$
65. (a) $s(s - 4)(s - 2)$ **(b)** $\sin \theta(\sin \theta - 4)(\sin \theta - 2)$
67. (a) $3c(c - 3)(c - 1)$; **(b)** $3 \cos \theta(\cos \theta - 3)(\cos \theta - 1)$
69. (a) $(t^2 + 5)(t + 1)(t - 1)$; **(b)** $(\tan^2 \theta + 5)(\tan \theta + 1)(\tan \theta - 1)$
71. 1, factorable **73.** 13, not factorable **75.** 81, factorable
77. -44, not factorable **79.** -4, not factorable

Exercises 4–5

1. (a) $4, -1$; **(b)** -1 **3.** $0, -5$ **5. (a)** $\frac{5}{2}, -\frac{3}{5}$; **(b)** $-\frac{3}{5}$
7. (a) $0, 5$; **(b)** 0 **9. (a)** $0, \frac{5}{4}$; **(b)** 0 **11.** $2, -2$
13. (a) $2, 1$; **(b)** 1 **15.** $4, -2$ **17. (a)** $10, -2$; **(b)** No solution
19. $4, -1$ **21.** 1 **23. (a)** $\frac{1}{3}, 5$; **(b)** $\frac{1}{3}$ **25.** $\frac{1}{5}, 3$ **27.** $\frac{2}{3}, 1$
29. $-\frac{3}{2}$ **31. (a)** $-\frac{1}{2}, 2$; **(b)** $-\frac{1}{2}$ **33. (a)** $0, 2, -2$; **(b)** 0
35. $0, -7, -2$ **37.** $0, 1, -1$ **39.** $71°30', 98°10', 251°30', 278°10'$
41. $30°, 150°, 221°50', 318°10'$ **43.** $0°, 90°, 180°, 270°$
45. $63°30' + k360°; 101°20' + k360°; 243°30' + k360°; 281°20' + k360°$
47. $0° + k360°; 90° + k360°; 270° + k360°$
49. $56°20' + k360°; 170°30' + k360°; 236°20' + k360°; 350°30' + k360°$
51. 6 ft, 4 ft **53.** $\frac{1}{2}$ ft **55.** 2 seconds
57. (a) 6 seconds; **(b)** 5 seconds, 1 second;
  (c) Attains the given height going up and coming down.
59. \$5, \$3, low unit price produces high sales volume, and vice versa.
Note: In Exercises 45–50 k may be any integer.

SAMPLE TEST: CHAPTER 4

1. 3^4 **2.** 2^{2x} **3.** $4 \sin^6 \theta$ **4.** $\dfrac{72y^2z^3}{x^4}$ **5.** $3c^2t^2 - 2ct^3 + 4ct^2$

6. $\cos^2 \theta - 4 \sin \theta$ **7.** $t^2 - 6t + 9$ **8.** $6 \sin^2 \theta - \sin \theta - 2$
9. $16 + 12x - 2x^2 - 2x^3$
10. $x^6 + 6x^5h + 15x^4h^2 + 20x^3h^3 + 15x^2h^4 + 6xh^5 + h^6$ **11.** $-220x^2y^9$
12. $(x - 6)(x - 2)$ **13.** $(\cos \theta + 4)^2$ **14.** $2(2c + 3t)(c - 2t)$
15. $t^3(1 + x)(1 - x)$ **16.** -47, no   **17.** $0, -2$ **18.** $1, -4$
19. $0, \frac{2}{3}, -4$ **20.** $76°, 135°$

Exercises 5–1

1. $\frac{3}{4}$ **3.** $\frac{4}{11}$ **5. (a)** $\dfrac{1}{2s}$; **(b)** $\dfrac{1}{2 \sin \theta}$ **7. (a)** $\dfrac{s + 2}{s + 3}$; **(b)** $\dfrac{\sin \theta + 2}{\sin \theta + 3}$

9. (a) $\dfrac{1}{s + 1}$; **(b)** $\dfrac{1}{\sin \theta + 1}$ **11. (a)** $\dfrac{1}{3t + 1}$; **(b)** $\dfrac{1}{3 \tan \theta + 1}$

13. -1 **15. (a)** $\dfrac{-1}{s + 1}$; **(b)** $\dfrac{-1}{\sin \theta + 1}$ **17. (a)** $\dfrac{8 - c}{2}$; **(b)** $\dfrac{8 - \cos \theta}{2}$

19. (a) $\dfrac{s + 1}{s - 4}$; **(b)** $\dfrac{\sin \theta + 1}{\sin \theta - 4}$ **21. (a)** $\dfrac{c - 4}{c + 4}$; **(b)** $\dfrac{\cos \theta - 4}{\cos \theta + 4}$

23. (a) $\dfrac{7 - t}{7 + t}$; **(b)** $\dfrac{7 - \tan \theta}{7 + \tan \theta}$ **25. (a)** $\dfrac{5(c + 1)}{2(c - 6)}$; **(b)** $\dfrac{5(\cos \theta + 1)}{2(\cos \theta - 6)}$

27. (a) $\dfrac{3t(2t - 5)}{2(t - 4)}$; **(b)** $\dfrac{3 \tan \theta(2 \tan \theta - 5)}{2(\tan \theta - 4)}$ **29.** -1 **31.** 2

33. $0, -\frac{1}{3}$ **35.** 4 **37.** All real numbers except -3
39. All real numbers except 1
41. If $a = b$, then $a - b = 0$. Thus, we cannot divide both sides of the equation by $a - b$.

Exercises 5–2

1. $\frac{15}{56}$ **3. (a)** $\dfrac{2t}{3}$; **(b)** $\dfrac{2 \tan \theta}{3}$ **5. (a)** $\dfrac{c}{4s^2}$; **(b)** $\dfrac{\cos \theta}{4 \sin^2 \theta}$

7. (a) $18sct^2$;   **(b)** $18 \sin \theta \cos \theta \tan^2 \theta$

9. (a) $\dfrac{2}{c(c - 3)}$; **(b)** $\dfrac{2}{\cos \theta(\cos \theta - 3)}$ **11. (a)** $\dfrac{st}{6(s - 2)}$; **(b)** $\dfrac{\sin \theta \tan \theta}{6(\sin \theta - 2)}$

13. (a) $\dfrac{(c - 1)(c - 2)}{(c + 2)(c + 1)}$; **(b)** $\dfrac{(\cos \theta - 1)(\cos \theta - 2)}{(\cos \theta + 2)(\cos \theta + 1)}$

15. (a) $\dfrac{s-4}{s-2}$; **(b)** $\dfrac{\sin\theta - 4}{\sin\theta - 2}$ **17. (a)** $\dfrac{(t-1)^2}{(t+1)^2}$; **(b)** $\dfrac{(\tan\theta - 1)^2}{(\tan\theta + 1)^2}$

19. (a) $-4(s+2)$;   **(b)** $-4(\sin\theta + 2)$ **21.** $\frac{9}{10}$ **23. (a)** $\dfrac{1}{s}$; **(b)** $\dfrac{1}{\sin\theta}$

25. (a) $\dfrac{1}{s}$; **(b)** $\dfrac{1}{\sin\theta}$ **27.** $\frac{10}{3}$ **29. (a)** $\dfrac{c-1}{c-3}$; **(b)** $\dfrac{\cos\theta - 1}{\cos\theta - 3}$

31. (a) $\dfrac{4(s-1)}{3s(s+1)}$; **(b)** $\dfrac{4(\sin\theta - 1)}{3\sin\theta(\sin\theta + 1)}$

33. (a) $\dfrac{(t+1)(t-1)^2}{7t}$; **(b)** $\dfrac{(\tan\theta + 1)(\tan\theta - 1)^2}{7\tan\theta}$ **35.** $\frac{1}{3}$

37. (a) $\dfrac{-(c-1)}{c(s+1)}$; **(b)** $\dfrac{-(\cos\theta - 1)}{\cos\theta(\sin\theta + 1)}$ **39.** $-\frac{3}{8}$

Exercises 5–3

1. 80   **3.** 196   **5.** 114   **7.** 210   **9.** 2160
11. 31, 37, 41, 43, 47, 51, 53, 59, 61, 67   **13. (a)** $4t^3$; **(b)** $4\tan^3\theta$
15. (a) $18ct$; **(b)** $18\cos\theta\tan\theta$ **17. (a)** $90s^4tc^3$; **(b)** $90\sin^4\theta\tan\theta\cos^3\theta$
19. (a) $c^3(c+1)^3$; **(b)** $\cos^3\theta(\cos\theta + 1)^3$
21. (a) $4(s+2)^2$; **(b)** $4(\sin\theta + 2)^2$
23. (a) $(t+c)(t-c)$; **(b)** $(\tan\theta + \cos\theta)(\tan\theta - \cos\theta)$
25. (a) $-(t-2)$; **(b)** $-(\tan\theta - 2)$
27. (a) $(2s-1)(s+1)$; **(b)** $(2\sin\theta - 1)(\sin\theta + 1)$
29. (a) $6(2t+3)(2t-3)^2$; **(b)** $6(2\tan\theta + 3)(2\tan\theta - 3)^2$

Exercises 5–4

1. $\frac{7}{11}$ **3. (a)** $\dfrac{-6}{s}$; **(b)** $\dfrac{-6}{\sin\theta}$ **5. (a)** $\dfrac{3s}{tc}$; **(b)** $\dfrac{3\sin\theta}{\tan\theta\cos\theta}$ **7.** $\frac{1}{2}$

9. (a) $\dfrac{c-1}{c+2}$; **(b)** $\dfrac{\cos\theta - 1}{\cos\theta + 2}$ **11. (a)** $\dfrac{-1}{t-1}$; **(b)** $\dfrac{-1}{\tan\theta - 1}$

13. (a) $\dfrac{3s-2}{s-6}$; **(b)** $\dfrac{3\sin\theta - 2}{\sin\theta - 6}$ **15. (a)** $\dfrac{13t}{8}$; **(b)** $\dfrac{13\tan\theta}{8}$

17. (a) $\dfrac{9t}{5}$; **(b)** $\dfrac{9\tan\theta}{5}$ **19. (a)** $\dfrac{1-t^2}{t}$; **(b)** $\dfrac{1-\tan^2\theta}{\tan\theta}$

21. (a) $\dfrac{7}{12s}$; **(b)** $\dfrac{7}{12\sin\theta}$ **23. (a)** $\dfrac{25t-12}{10t^2}$; **(b)** $\dfrac{25\tan\theta - 12}{10\tan^2\theta}$

25. (a) $\dfrac{5c^2 - 3c + 2}{c^3}$; (b) $\dfrac{5\cos^2\theta - 3\cos\theta + 2}{\cos^3\theta}$

27. (a) $\dfrac{13c^2 + 4}{10c}$; (b) $\dfrac{13\cos^2\theta + 4}{10\cos\theta}$ **29.** (a) $\dfrac{2s + 47}{10s}$; (b) $\dfrac{2\sin\theta + 47}{10\sin\theta}$

31. (a) $\dfrac{52c^2 + c - 12}{30c}$; (b) $\dfrac{52\cos^2\theta + \cos\theta - 12}{30\cos\theta}$

33. (a) $\dfrac{s - 2}{6(3s + 2)}$; (b) $\dfrac{\sin\theta - 2}{6(3\sin\theta + 2)}$ **35.** (a) $\dfrac{7t}{4(t - 3)}$; (b) $\dfrac{7\tan\theta}{4(\tan\theta - 3)}$

37. (a) $\dfrac{3c^2 - 3c + 9}{c(c - 3)}$; (b) $\dfrac{3\cos^2\theta - 3\cos\theta + 9}{\cos\theta(\cos\theta - 3)}$

39. (a) $\dfrac{8c}{(c + s)(c - s)}$; (b) $\dfrac{8\cos\theta}{(\cos\theta + \sin\theta)(\cos\theta - \sin\theta)}$

41. (a) $\dfrac{6t + 11}{(2t - 3)(2t + 3)}$; (b) $\dfrac{6\tan\theta + 11}{(2\tan\theta - 3)(2\tan\theta + 3)}$

43. (a) $\dfrac{-24s^2 + 29s - 5}{4s - 1}$; (b) $\dfrac{-24\sin^2\theta + 29\sin\theta - 5}{4\sin\theta - 1}$

45. (a) $\dfrac{2c^2 - 2c + 1}{c(c - 1)}$; (b) $\dfrac{2\cos^2\theta - 2\cos\theta + 1}{\cos\theta(\cos\theta - 1)}$

47. (a) $\dfrac{6s^2 + 12s + 1}{(s - 5)(s + 2)(s + 1)}$; (b) $\dfrac{6\sin^2\theta + 12\sin\theta + 1}{(\sin\theta - 5)(\sin\theta + 2)(\sin\theta + 1)}$

49. (a) $\dfrac{7t - 4}{t^2 - 1}$; (b) $\dfrac{7\tan\theta - 4}{\tan^2\theta - 1}$ **51.** (a) $\dfrac{t + 1}{t - 1}$; (b) $\dfrac{\tan\theta + 1}{\tan\theta - 1}$

Exercises 5–5

1. $\frac{65}{72}$ **3.** $-\frac{33}{34}$ **5.** (a) $\dfrac{t + 1}{t - 1}$; (b) $\dfrac{\tan\theta + 1}{\tan\theta - 1}$

7. (a) $\dfrac{s^2 - 5}{5s}$; (b) $\dfrac{\sin^2\theta - 5}{5\sin\theta}$ **9.** (a) $\dfrac{s + c}{c}$; (b) $\dfrac{\sin\theta + \cos\theta}{\cos\theta}$

11. (a) $\dfrac{s^2 - c^2}{s^2 + c^2}$; (b) $\dfrac{\sin^2\theta - \cos^2\theta}{\sin^2\theta + \cos^2\theta}$

13. (a) $\dfrac{s^2 + s + 2}{s^2 - s - 2}$; (b) $\dfrac{\sin^2\theta + \sin\theta + 2}{\sin^2\theta - \sin\theta - 2}$ **15.** (a) $\dfrac{4c - 3}{3c - 1}$; (b) $\dfrac{4\cos\theta - 3}{3\cos\theta - 1}$

17. (a) $\dfrac{t^2 + 3t - 4}{4}$; (b) $\dfrac{\tan^2\theta + 3\tan\theta - 4}{4}$ **19.** (a) s; (b) $\sin\theta$

21. 24 mi/hour

Exercises 5–6

1. 12   **3.** $\frac{26}{5}$ **5.** 12   **7.** $-\frac{1}{10}$ **9.** 1   **11.** -1 **13.** -3

15. $-\frac{5}{2}$ **17.** 2   **19.** 1   **21.** 3   **23.** No solution   **25.** No solution

27. $\frac{3}{4}$ **29.** 1   **31.** $L_1 = \dfrac{L_2 W_2}{W_1}$ **33.** $v = v_0 + at$ **35.** $r = \dfrac{S - a}{S - L}$

37. $Z_2 = \dfrac{Z Z_1}{Z_1 - Z}$ **39.** $a = \dfrac{fb}{b - f}$ **41.** \$60,000   **43.** $\frac{18}{5}$ hr

45. 12 mi/hour   **47.** 16   **49.** He cannot qualify

SAMPLE TEST: CHAPTER 5

1. -1 **2.** $\dfrac{2x + y}{x + y}$ **3.** $\dfrac{\cos\theta - 1}{\cos\theta + 1}$ **4.** $0, \frac{1}{3}$ **5.** $\dfrac{2c}{3t^3}$

6. $\dfrac{\sin\theta - 3}{\sin\theta + 6}$ **7.** $\dfrac{c^2 + 1}{4c}$ **8.** $72 s^2 t^2 c^3$ **9.** $(t - 2)^2(t + 2)$

10. $\dfrac{5 - c}{c - 1}$ **11.** $\dfrac{2y + 3x}{xy}$ **12.** $\dfrac{6\sin^2\theta - 5}{15\sin\theta}$ **13.** $\dfrac{7t^2 + 8t}{(t + 2)(t + 3)(t - 1)}$

14. $\dfrac{s - 1}{s + 1}$ **15.** $\dfrac{\cos^3\theta}{1 + 2\cos\theta}$ **16.** $1 - 2x^2$ **17.** -1 **18.** $6, -1$

19. 5   **20.** $F_1 = \dfrac{F_2 d_2}{d_1}$

Exercises 6–1

1. $\frac{1}{9}$ **3.** 1   **5.** 8   **7.** $\frac{1}{4}$ **9.** $\frac{3}{4}$ **11.** $-\frac{1}{8}$ **13.** 64   **15.** $\frac{1}{16}$

17. 412   **19.** 0.00204   **21.** 5   **23.** 8   **25.** $\frac{1}{16}$ **27.** $\frac{1}{2}$ **29.** $\frac{3}{4}$

31. $1/x^3$ **33.** $4/x^2$ **35.** 1   **37.** $1/x^4 y$ **39.** $1/y^2$ **41.** x^2

43. $5x$ **45.** $x^3 y/45$ **47.** $x^2 y^6$ **49.** $\dfrac{(1/x) + (1/y)}{1/(x + y)}$ **51.** x^2

53. $-6/x^5$ **55.** $-16/x^5$ **57.** $x^4/27$ **59.** xz^2/y^2 **61.** 10

63. $\dfrac{2y^2 + 3y - 4}{y^2}$ **65.** $\dfrac{8x + 5}{(3x + 1)^2}$ **67.** $y + x$ **69.** $\dfrac{xy}{y + x}$

Exercises 6–2

1. 7.4   **3.** -2 **5.** 1.23×10^7 **7.** 45   **9.** 0.000000908

11. 4.2×10^1 **13.** 3.4251×10^4 **15.** 5.9×10^{12} mi   **17.** 1.26×10^6

19. 5.3×10^{-23} g   **21.** 92,000   **23.** 42.1   **25.** 580,000,000 mi

27. 6,280,000,000,000,000,000   **29.** 0.00000000000000000000000003 g

31. 5.58×10^7 mi   **33.** 3.6×10^{-25}

Exercises 6–3

1. 5   **3.** $\frac{1}{2}$ **5.** 1   **7.** 2   **9.** -3 **11.** 2   **13.** $\frac{1}{2}$ **15.** $\frac{5}{3}$
17. $-\frac{1}{2}$ **19.** 5   **21.** -2 **23.** 27   **25.** 32   **27.** -32 **29.** 9
31. $\frac{1}{16}$ **33.** $\frac{27}{8}$ **35.** $\frac{28}{27}$ **37.** $\frac{9}{2}$ **39.** 16   **41.** 4   **43.** -27

45. 4   **47.** $\frac{1}{64}$ **49.** $\frac{1}{5}$ **51.** $x^{1/3}; \sqrt[3]{x}$ **53.** $\dfrac{1}{x^{7/6}}; \dfrac{1}{\sqrt[6]{x^7}}$ **55.** $2xy^2$

57. $\dfrac{a}{2b^3}$ **59.** $\dfrac{x^{1/2}y^2}{z^{2/3}}; \dfrac{y^2\sqrt{x}}{\sqrt[3]{z^2}}$ **61.** $\dfrac{y^{5/2}}{x}; \dfrac{\sqrt{y^5}}{x}$ **63.** $\dfrac{125x^3}{343y^{24}}$

65. $\frac{4}{9}x^{5/2}y^{7/12}; \frac{4}{9}\sqrt{x^5}\sqrt[12]{y^7}$ **67.** $2^{7/12}; \sqrt[12]{2^7}$ **69.** $16^{2/3}; \sqrt[3]{16^2}$
71. Since the base (-3) is negative and the exponent $(\frac{1}{2})$ has an even denominator, the number is not a real number. In such a case, the laws of exponents do not necessarily hold.

Exercises 6–4

1. $3\sqrt{5}$ **3.** $5\sqrt{6}$ **5.** $6\sqrt{3}$ **7.** $-6\sqrt{2}$ **9.** $-7\sqrt{2}$ **11.** $3\sqrt[3]{2}$
13. $2\sqrt[4]{3}$ **15.** $10\sqrt{2}$ **17.** $-\sqrt{6}$ **19.** $y\sqrt{x}$ **21.** $2x^3y^2\sqrt{3x}$
23. $5xy^7\sqrt{10xy}$ **25.** $xy\sqrt[4]{9x^2y^3}$ **27.** $\sqrt{5}/2$ **29.** $\sqrt{2}/2$ **31.** $2\sqrt{14}/7$
33. $\sqrt{10}$ **35.** $-8\sqrt{15}/3$ **37.** $\sqrt{22}/4$ **39.** $\sqrt[3]{4}/2$ **41.** $\sqrt[4]{10}/2$
43. $\sqrt{xy}/y$ **45.** $3\sqrt{x}/x$ **47.** $\sqrt[3]{xy^2}/y$ **49.** $\sqrt{30}/6$

Exercises 6–5

1. $11\sqrt{2}$ **3.** $-3\sqrt{3}$ **5.** $5\sqrt{3}$ **7.** $-7\sqrt{2}$ **9.** $-14\sqrt{11}$
11. $3\sqrt{2} - 2\sqrt{3}$ **13.** $13\sqrt{3}$ **15.** $7\sqrt{2}$ **17.** $27\sqrt{14}/14$ **19.** $2\sqrt{6}/3$
21. $37\sqrt{6}/6$ **23.** $9\sqrt[3]{3}$ **25.** $-29\sqrt[3]{3}/3$ **27.** $8\sqrt{2x}$ **29.** $(2y - 6x)\sqrt{5x}$

31. $5x\sqrt{2xy}/2$ **33.** $\left(\dfrac{2}{y} - \dfrac{2}{x^2} + \dfrac{5x}{4}\right)\sqrt{2xy}$ **35.** $\left(\dfrac{1}{y} - 1\right)\sqrt[4]{xy^3}$

37. $2 + \sqrt{7}$

Exercises 6–6

1. $\sqrt{15}$ **3.** $2\sqrt{3}$ **5.** 9   **7.** $12\sqrt{15}$ **9.** -12 **11.** 2   **13.** $\frac{3}{2}$
15. 4   **17.** -4 **19.** $3x\sqrt{10}$ **21.** $-24x^2y\sqrt{6y}$ **23.** $2xy\sqrt[3]{3xy}$
25. 14   **27.** $40\sqrt{2} - 30$ **29.** -1 **31.** 7   **33.** -68
35. $3 + 2\sqrt{2}$ **37.** $16 + 8\sqrt{3}$ **39.** $373 + 70\sqrt{22}$
41. $2 - \sqrt{3} + 2\sqrt{2} - \sqrt{6}$ **43.** $30 + 8\sqrt{6} + 20\sqrt{15} + 16\sqrt{10}$
45. $22 + 8\sqrt{14}$ **47.** $x - 2\sqrt{xy} + y$ **49.** $x + y$ **51.** 0   **53.** 0
55. 6   **57.** 0   **59.** $11 + 4\sqrt{6} - \sqrt{2} - \sqrt{3}$

Exercises 6–7

1. 2   **3.** $2\sqrt{2}$ **5.** $\frac{8}{3}$ **7.** $\sqrt{30}/3$ **9.** $\sqrt{6x}/x$ **11.** -2

13. $6\sqrt{10}/25$ **15.** $-3x\sqrt{2}/2$ **17.** $-\sqrt[3]{4}/4$ **19.** $\sqrt[4]{56x}/2$ **21.** $2\sqrt{3}/3$

23. $7\sqrt{2}/12$ **25.** $3 + \sqrt{2}/2$ **27.** $-\sqrt{5x}/3x$ **29.** $\sqrt[3]{4}/2$

31. $3\sqrt{2} - 3$ **33.** $\dfrac{\sqrt{3} + \sqrt{5}}{-2}$ **35.** $2\sqrt{3} + 3$ **37.** $3 - 2\sqrt{2}$

39. $\dfrac{41 + 9\sqrt{15}}{-233}$ **41.** $\dfrac{x - \sqrt{xy}}{x - y}$

Exercises 6–8

1. -1 **3.** No solution   **5.** $\frac{25}{3}$ **7.** 6   **9.** -9 **11.** 4

13. -2 **15.** 3   **17.** 0   **19.** $-3, 1$ **21.** No solution   **23.** 0, 4

25. $\frac{1}{2}$ **27.** $x \geq 0$

Exercises 6–9

1. $5i$ **3.** $8i$ **5.** $i\sqrt{22}$ **7.** $i\sqrt{3}$ **9.** $2 - 8i\sqrt{2}$ **11.** $\frac{3}{2}i$

13. $\dfrac{3i\sqrt{2}}{2}$ **15.** $\dfrac{3x\sqrt{7}}{2}i$ **17.** $2i$ **19.** $2i\sqrt{2}$ **21.** $9i$ **23.** 0

25. $6i\sqrt{2}$ **27.** $5 - 6i$ **29.** $6 + 4i$ **31.** -6 **33.** -4 **35.** $\frac{2}{5}$

37. $10 - 10i$ **39.** 13   **41.** $-3 - 4i$ **43.** $\frac{80}{9} - 2i$ **45.** $\frac{1}{5} - \frac{2}{5}i$

47. $-\frac{4}{17} - \frac{1}{17}i$ **49.** $\frac{7}{10} + \frac{1}{10}i$ **51.** 0   **53.** 0   **55.** 0

57. $48 + 42i$

59. (a) $-i$; (b) 1; (c) i; (d) -1; (e) $-i$; (f) i; (g) i; (h) 1

Exercises 6–10

1. $3(\cos 0° + i \sin 0°)$ **3.** $2(\cos 270° + i \sin 270°)$

5. $\sqrt{2}(\cos 315° + i \sin 315°)$ **7.** $5(\cos 36°50' + i \sin 36°50')$

9. $\sqrt{13}(\cos 213°40' + i \sin 213°40')$ **11.** $2(\cos 120° + i \sin 120°)$

13. $2(\cos 315° + i \sin 315°)$ **15.** $0 + 3i$ **17.** $-4 + 0i$

19. $-\dfrac{\sqrt{3}}{2} + \dfrac{3}{2}i$ **21.** $-\sqrt{2} - \sqrt{2}i$ **23.** $0.6157 + 0.7880i$ **25.** $16 + 0i$

27. $16 + 16\sqrt{3}i$ **29.** $1 + 0i$

31. $1 + 0i$ **33.** $2(\cos 67°30' + i \sin 67°30')$
$-\dfrac{1}{2} + \dfrac{\sqrt{3}}{2}i$ $2(\cos 157°30' + i \sin 157°30')$
$2(\cos 247°30' + i \sin 247°30')$
$2(\cos 337°30' + i \sin 337°30')$
$-\dfrac{1}{2} - \dfrac{\sqrt{3}}{2}i$

35. $\sqrt[5]{2}(\cos 12° + i \sin 12°)$

$\sqrt[5]{2}(\cos 84° + i \sin 84°)$

$\sqrt[5]{2}(\cos 156° + i \sin 156°)$

$\sqrt[5]{2}(\cos 228° + i \sin 228°)$

$\sqrt[5]{2}(\cos 300° + i \sin 300°)$

37. $1 + \sqrt{3}i$

$-2 + 0i$

$1 - \sqrt{3}i$

39. $1 + 0i$

$0 + i$

$-1 + 0i$

$0 - i$

SAMPLE TEST: CHAPTER 6

1. 4 **2.** $\frac{1}{16}$ **3.** $-\frac{19}{8}$ **4.** $-\frac{7}{10}$ **5.** $4\sqrt{3}$ **6.** $7\sqrt{3}/6$

7. $\dfrac{\sqrt{6} + 2}{2}$ **8.** $\frac{3}{10} - \frac{1}{10}i$ **9.** -45 **10.** $-4 + 2i$ **11.** $x^{1/6}$

12. $1/a$ **13.** 10 **14.** $11 + 6\sqrt{2}$ **15.** 197,000,000 mi²

16. 4.0×10^{-5} cm **17.** 4 **18.** 8 **19.** $\sqrt{2}(\cos 225° + i \sin 225°)$

20. $2 - 2\sqrt{3}i$

Exercises 7–1

1. $(0, -3)(1, -1)(-1, -1)(2, 5)(-2, 5)$ other possibilities **3.** a, d
5. Function **7.** Not a function **9.** Not a function **11.** $r > 0$
13. All real numbers **15.** All real numbers **17.** $x > 0$
19. $-3 \le x \le 3$ **21.** All real numbers **23.** $-2, -1, 0$
25. Figure K-24 **27.** Figure K-25 **29.** Figure K-26
31. Figure K-27 **33.** b, d **35.** $y = x - 1$ **37.** $y = 4x - 2$
39. $A = \pi(d/2)^2; d > 0$ **41.** $A = (P/4)^2; P > 0$
43. $t = 690 + 0.21(i - 4000); 4000 \le i \le 6000$

45. $c = \begin{cases} 2 & \text{if } 0 \le n \le 5 \\ 2 + 0.31(n - 5) & \text{if } 5 < n \le 100; \end{cases} \quad 0 \le n \le 100$

47. $y = 1000 - x; A = (1000 - x)x; 0 < x < 1000$
49. $V = x(8 - 2x)(15 - 2x); 0 < x < 4$

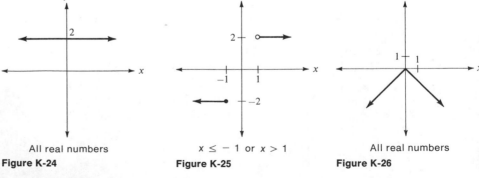

All real numbers
Figure K-24

$x \le -1$ or $x > 1$
Figure K-25

All real numbers
Figure K-26

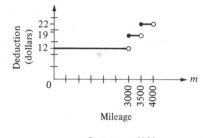

$$0 \le m < 4000$$

Figure K-27

Exercises 7–2

1. 2, 1, 0   **3.** 2, 8, -10 **5.** Undefined, -6, $\dfrac{1}{x_0 + h} - 5$

7. -2, undefined, $\dfrac{a + 2}{a - 1}$ **9. (a)** 4; **(b)** -2; **(c)** -5

11. (a) 3; **(b)** 13; **(c)** 39; **(d)** -2; **(e)** 3; **(f)** undefined; **(g)** 49; **(h)** 7;
  (i) 9; **(j)** $(2x - 1)^2$
13. $f(a + b) = c(a + b) = ca + cb = f(a) + f(b)$
15. (a) $f(x) = x^2$; **(b)** $f(x) = x^2 + x$; **(c)** $f(x) = x$; **(d)** $f(x) = x + 1$
17. (a) 20 cents; **(b)** 10 cents; **(c)** 30 cents   **19. (a)** 0; **(b)** 2; **(c)** 2
21. (a) $\frac{1}{2}$; **(b)** 0; **(c)** π **23. (a)** -1; **(b)** 1; **(c)** 1; **(d)** -1
25. (a) 0; **(b)** 1; **(c)** -2; **(d)** -4; **(e)** Figure K-28

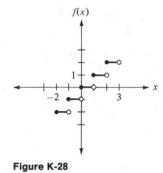

Figure K-28

Exercises 7–3

(*Note:* Unless stated otherwise, the domain of the function is all real numbers.)
  1. $(f + g)(x) = 3x - 1$
   $(f - g)(x) = x + 1$
   $(f \cdot g)(x) = 2x^2 - 2x$

   $\left(\dfrac{f}{g}\right)(x) = \dfrac{2x}{x - 1}$; all real numbers except 1

   $(f \circ g)(x) = 2x - 2$
   $(g \circ f)(x) = 2x - 1$

3. $(f + g)(x) = 3x - 2$
$(f - g)(x) = 5x - 8$
$(f \cdot g)(x) = -4x^2 + 17x - 15$

$\left(\dfrac{f}{g}\right)(x) = \dfrac{4x - 5}{-x + 3}$; all real numbers except 3

$(f \circ g)(x) = -4x + 7$
$(g \circ f)(x) = -4x + 8$

5. $(f + g)(x) = x^2 + 1$
$(f - g)(x) = x^2 - 1$
$(f \cdot g)(x) = x^2$

$\left(\dfrac{f}{g}\right)(x) = x^2$

$(f \circ g)(x) = 1$
$(g \circ f)(x) = 1$

7. $(f + g)(x) = x^3 + \sqrt[3]{x}$
$(f - g)(x) = x^3 - \sqrt[3]{x}$
$(f \cdot g)(x) = \sqrt[3]{x^{10}}$

$\left(\dfrac{f}{g}\right)(x) = \sqrt[3]{x^8}$; all real numbers except 0

$(f \circ g)(x) = x$
$(g \circ f)(x) = x$

9. $f + g = (2, 3)(3, 5)$
$f - g = (2, 5)(3, 5)$
$f \cdot g = (2, -4)(3, 0)$

$\dfrac{f}{g} = (2, -4)$

$f \circ g = (3, 2)(4, 3)(5, 4)$
$g \circ f = (0, -1)(1, 0)(2, 1)(3, 2)$

11. (a) -2; **(b)** $4\sqrt{x} - 4x$; **(c)** 2; **(d)** $4x - 4\sqrt{x}$
13. (a) -3; **(b)** 20; **(c)** undefined; **(d)** 4 **15.** $g(x) = 4x - 1; f(x) = x^3$
17. $g(x) = 3 - x; f(x) = 2x^4$ **19.** $g(x) = 2x + 1; f(x) = \sqrt[3]{x}$
21. $D_f: x \geq 7; D_g: x \leq 2$; there is no value for x at which both functions are defined.

SAMPLE TEST: CHAPTER 7

1. $(1, 7)(2, 7)(3, 7)$ **2.** Yes **3.** $-2 < x < 2$ **4.** $(1, 5)(2, 4)$
5. 3 **6.** Figure K-29 **7.** $x \leq -2$ or $x > 2$ **8.** Yes
9. D: all real numbers; $R: y \leq 3$ **10.** $y = 6$

11. $e = \begin{cases} 500 & \text{if } 0 \leq a \leq 2000 \\ 500 + 0.09(a - 2000) & \text{if } a > 2000 \end{cases}$

12. $A = x\sqrt{100 - x^2}; 0 < x < 10$ **13.** -21 **14.** -7 **15.** $\frac{4}{3}$
16. Even **17.** $6x_1 + 3h$ **18.** $(f \cdot g)(x) = x$; all real numbers except 0
19. 1 **20.** $f(x) = 3x^2, g(x) = (1 - x)$

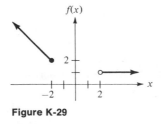

Figure K-29

Exercises 8–1

1. 1 **3.** $\frac{4}{7}$ **5.** $-\frac{11}{2}$ **7.** 0 **9.** Undefined **11.** -1 **13.** $\frac{1}{2}$
15. (a) -1; **(b)** $\frac{1}{2}$; **(c)** 0; **(d)** undefined; **(e)** 3 **17.** Figure K-30

19. $\frac{1}{6}$; for each 6-lb increase of the earth weight of an object, the moon weight increases 1 lb

21. -32; for each second (up to 10 seconds) the velocity decreases by 32 ft/second

23. 0.12; for each brochure the cost increases $0.12

25. 7; for each mile the cost increases $7

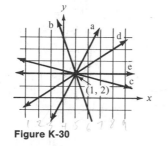

Figure K-30

Exercises 8–2

1. $y = 2x + 3$ **3.** $y = -3x - 1$ **5.** $y = -5$ **7.** 1; (0, 7)

9. 5; (0, 0)   **11.** 0; (0, −2)   **13.** $-\frac{2}{3}$; (0, $-\frac{2}{3}$)   **15.** 6; (0, 7)

17. $y = 5x - 17$ **19.** $y = \frac{1}{2}x + 1$ **21.** $y = -\frac{1}{2}x$ **23.** $y = -x + 5$

25. $y = -x - 6$ **27.** $f(x) = \frac{2}{3}x - 2$ **29.** $f(x) = -\frac{4}{3}x - \frac{13}{3}$

31. $f(x) = 1$ **33.** -1 **35.** $-\frac{11}{3}$ **37.** $y = \frac{1}{25}x + 7$; $13

39. (a) $y = 9x + 58$; (b) $94; (c) $58; (d) $9

41. (a) $y = 0.6x + 40$; (b) $55; (c) $40; (d) $0.60

43. (a) $y = 0.1x + 550$; (b) $1450; (c) $550; (d) 10 percent

45. (a) $v = -32t - 220$; (b) -348 ft/second; (c) -220 ft/second;
   (d) + ball rising, − ball falling

47. m

Exercises 8–3

1. $y = x + 1$; $y = -x + 3$ **3.** $y = -2x + 6$; $y = \frac{1}{2}x - \frac{3}{2}$

5. $y = -\frac{1}{7}x$; $y = 7x$ **7.** $y = \frac{5}{3}x - \frac{14}{3}$; $y = -\frac{3}{5}x - \frac{12}{5}$

9. $y = -\frac{1}{4}x + \frac{13}{4}$; $y = 4x - 18$ **11.** $y = \frac{3}{2}x - \frac{13}{12}$; $y = -\frac{2}{3}x$

13. $y = -\frac{3}{2}x + 10$ **15.** $y = -x - 9$ **17.** $y = \frac{5}{7}x + \frac{6}{7}$

19. $y = \frac{7}{3}x - \frac{16}{3}$

Exercises 8–4

1. $(-2, -4)$ **3.** $(-\frac{3}{2}, -\frac{1}{2})$ **5.** $(0, 4)$ **7.** $(-6, -6)$ **9.** Dependent

11. $(4, 21)$ **13.** $(-1, 1)$ **15.** No solution, inconsistent   **17.** $(5, 1)$

19. $(\frac{67}{41}, -\frac{155}{41})$ **21.** 46, 24   **23.** 140°, 40°   **25.** 200 g

27. $v_0 = 100$ ft/second; $a = -32$ ft/second^2 **29.** $F_1 = 12$ lb; $F_2 = 6$ lb

31. 3 mi/hour, 9 mi/hour   **33.** $8000 at 6 percent, $2000 at 11 percent

35. More than 1000   **37.** 260 gadgets, 220 widgets   **39.** 80 percent

Exercises 8–5

1. -10 **3.** 2 **5.** 26 **7.** -3 **9.** 97 **11.** 0 **13.** -52
15. 17 **17.** 0 **19.** 254 **21.** (4, 21) **23.** $(-1, 1)$
25. No solution, inconsistent **27.** (5, 1) **29.** $(\frac{67}{41}, -\frac{155}{41})$
31. $x = 1, y = 2, z = 0$ **33.** $x = -3, y = 1, z = 4$
35. $x = \frac{1}{2}, y = -\frac{3}{2}, z = 1$ **37.** $x = 2, y = 2, z = -2$
39. No unique solution **41.** $I_1 = -\frac{8}{73}$ amp; $I_2 = \frac{55}{73}$ amp; $I_3 = -\frac{47}{73}$ amp
43. $I_1 = \frac{73}{101}$ amp; $I_2 = -\frac{35}{101}$ amp; $I_3 = -\frac{38}{101}$ amp

Exercises 8–6

1. Figure K-31 **3.** Figure K-32 **5.** Figure K-33 **7.** Figure K-34
9. Figure K-35 **11.** Figure K-36 **13.** Figure K-37 **15.** Figure K-38
17. Figure K-39 **19.** Figure K-40 **21.** All real numbers
23. 144 ft, 6 seconds **25.** 360 watts **27.** 10, 10 **29.** 10, 10
31. 200 yd, 100 yd **33.** 3 in.

35. $P = 2L + 2w$; therefore, $L = \dfrac{P - 2w}{2}$; area $= \left(\dfrac{P - 2w}{2}\right)w = \dfrac{P}{2}w - w^2$;

$w = \dfrac{-b}{2a} = \dfrac{-P/2}{2(-1)} = \dfrac{P}{4}$; therefore, rectangle is a square

37. \$15 **39.** 25 **41.** $(2, -2)(-1, 1)$ **43.** $(5, 12)(-1, 0)$
45. $(1, 1)(-1, 1)$

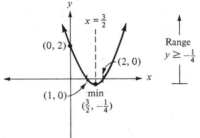

Figure K-31

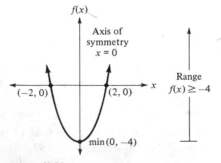

Figure K-32

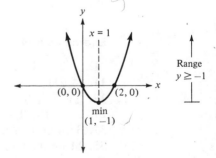

Figure K-33

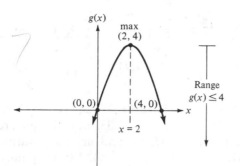

Figure K-34

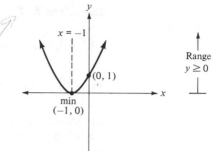

Figure K-35

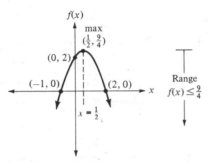

Figure K-36

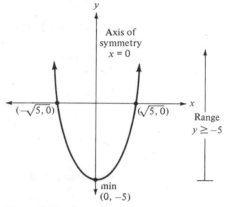

Figure K-37

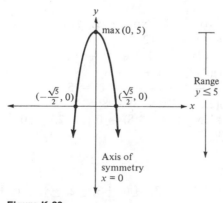

Figure K-38

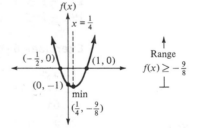

Figure K-39

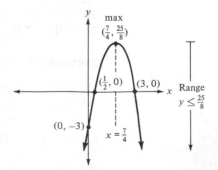

Figure K-40

Exercises 8–7

1. $-4, -1$ **3.** $\dfrac{-1 \pm \sqrt{13}}{2}$ **5.** $1 \pm i$ **7.** $3, 3$ **9.** $2, -2$

11. $0, \frac{2}{3}$ **13.** $\dfrac{2 \pm \sqrt{10}}{3}$ **15.** $\dfrac{-1 \pm i\sqrt{14}}{5}$ **17.** $\dfrac{1 \pm i\sqrt{2}}{3}$ **19.** $-1, \frac{3}{4}$

21. $(5, 0)(-1, 0)$ **23.** None   **25.** $(2, 0)$ **27.** $(2 + \sqrt{7}, 0)(2 - \sqrt{7}, 0)$
29. $(\frac{2}{3}, 0)(-\frac{3}{2}, 0)$ **31.** -15, imaginary   **33.** 0; real, equal, rational
35. 136; real, unequal, irrational   **37.** 84; real, unequal, irrational
39. -3; imaginary   **41.** -3; no points   **43.** 13; two points, irrational
45. 4; two points, rational   **47.** 0; one point, rational

49. 41; two points, irrational   **51.** 7 **53.** 3.8 in., 7.8 in.   **55.** $\dfrac{\sqrt{10}}{4}$ second

57. 0.5 second   **59.** 11.1 in.

61. $x_1 + x_2 = \dfrac{-b + \sqrt{b^2 - 4ac}}{2a} + \dfrac{-b - \sqrt{b^2 - 4ac}}{2a} = -\dfrac{2b}{2a} = -\dfrac{b}{a}$

$\quad x_1 \cdot x_2 = \dfrac{-b + \sqrt{b^2 - 4ac}}{2a} \cdot \dfrac{-b - \sqrt{b^2 - 4ac}}{2a} = \dfrac{b^2 - (b^2 - 4ac)}{4a^2} = \dfrac{4ac}{4a^2} = \dfrac{c}{a}$

Exercises 8–8

1. $x < -1$ or $x > 5$ **3.** $-1 < x < 5$ **5.** $x \le -1$ or $x \ge 2$
7. $x \le -2$ or $x \ge 2$ **9.** $0 < x < 3$ **11.** $-2 \le x \le \frac{3}{2}$
13. $-1 < x < \frac{5}{3}$ **15.** All real numbers except -1 **17.** No solution
19. $-1 \le x \le 1$

Exercises 8–9

1. Yes   **3.** Yes   **5.** No   **7.** No   **9.** Yes
11. $x^2 + 7x - 2 = (x + 5)(x + 2) - 12$
13. $6x^3 - 3x^2 + 14x - 7 = (2x - 1)(3x^2 + 7) + 0$
15. $3x^4 + x - 2 = (x^2 - 1)(3x^2 + 3) + x + 1$
17. $x^3 - 5x^2 + 2x - 3 = (x - 1)(x^2 - 4x - 2) - 5$
19. $2x^3 + x - 5 = (x + 1)(2x^2 - 2x + 3) - 8$
21. $x^4 - 16 = (x - 2)(x^3 + 2x^2 + 4x + 8) + 0$ **23.** $0, 0$ **25.** $-9, 87$
27. $0, 0$ **29.** $-\frac{176}{27}, -\frac{10}{3}$

Exercises 8–10

1. $\pm 20, \pm 10, \pm 5, \pm 4, \pm 2, \pm 1$ **3.** $\pm 6, \pm 3, \pm 2, \pm \frac{3}{2}, \pm 1, \pm \frac{3}{4}, \pm \frac{1}{2}, \pm \frac{1}{4}$
5. $3+, 0-$ **7.** $1+, 3-$ **9.** $4, -1, -2$ **11.** $-\frac{1}{3}, \pm i\sqrt{5}$
13. $\frac{1}{2}, \frac{1}{2}, \frac{1}{2}$ **15.** $-1, 2, \dfrac{1 \pm i\sqrt{79}}{8}$ **17.** $-2, \dfrac{-1}{2}, \dfrac{-1 \pm i\sqrt{35}}{2}$

19. Let $Q(x)$ and r represent the quotient and remainder when $P(x)$ is divided by $x - b$. Then, $P(x) = (x - b)Q(x) + r$. By the remainder theorem, $P(b) = r$. If b is a zero of $P(x)$, then $P(b) = 0$. Thus, $P(x) = (x - b)Q(x) + 0$ and $x - b$ is a factor of $P(x)$.

Exercises 8–11

1. $x = \frac{3}{2}$ **3.** $x = 6, x = -1$ **5.** Figure K-41   **7.** Figure K-42
9. Figure K-43   **11.** Figure K-44   **13.** Figure K-45   **15.** Figure K-46
17. Figure K-47   **19.** Figure K-48   **21.** Figure K-49

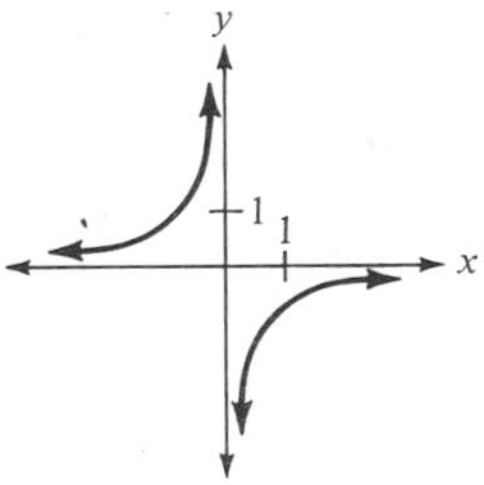

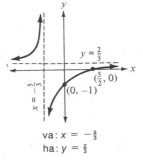

va: $x = 0$
ha: $y = 0$

Figure K-41

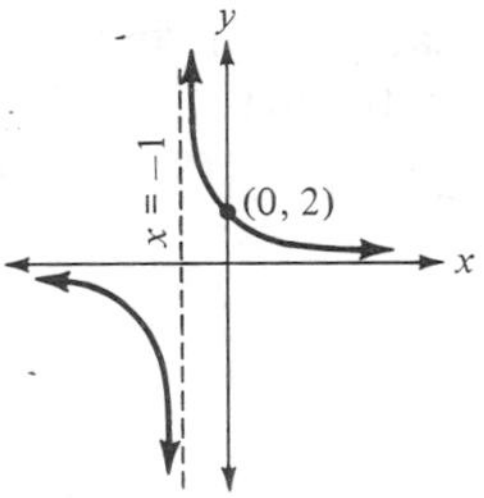

va: $x = -1$
ha: $y = 0$

Figure K-42

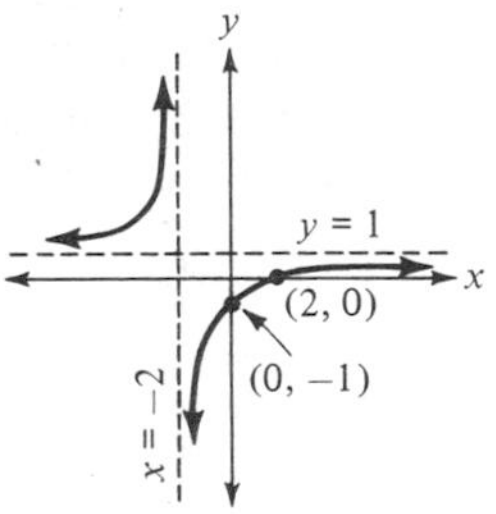

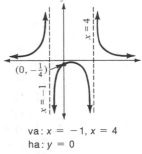

va: $x = -2$
ha: $y = 1$

Figure K-43

va: $x = -\frac{5}{3}$
ha: $y = \frac{2}{3}$

Figure K-44

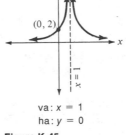

va: $x = 1$
ha: $y = 0$

Figure K-45

va: $x = -1, x = 4$
ha: $y = 0$

Figure K-46

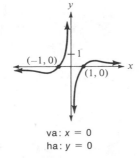

va: $x = 0$
ha: $y = 0$

Figure K-47

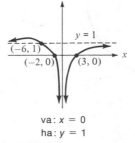

va: $x = 0$
ha: $y = 1$

Figure K-48

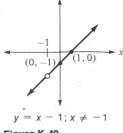

$y = x - 1; x \neq -1$

Figure K-49

SAMPLE TEST: CHAPTER 8

1. (a) No **(b)** No **(c)** Yes **(d)** No **(e)** Yes   **2.** $f(x) = -3x + 3$
3. $m = \frac{3}{2}; (0, -3)$ **4.** $y = -2x - 5$ **5–6.** $(-1, -5)$

7. $\left(\dfrac{1 - \sqrt{21}}{2}, 0\right)\left(\dfrac{1 + \sqrt{21}}{2}, 0\right)$ Figure K-50

8. -47; the graph does not cross the x axis (no x intercepts)   **9.** $y \le 4$
10. $500   **11.** $x \ge 2$ or $x \le -3$
12. $2x^4 - x + 7 = (x^2 + 2)(2x^2 - 4) + (-x + 15)$ **13.** 41
14. $\pm 2, \pm 1, \pm \frac{2}{3}, \pm \frac{1}{3}$ **15.** $4+, 0-$ **16.** $\pm\sqrt{3}$
17. Yes; there can be no real zeros greater than 2 only if the numbers in the bottom row of the synthetic division all have the same sign.
18. $x = 0, x = -2$ **19.** $y = \frac{2}{3}; (\frac{13}{2}, \frac{2}{3})$ **20.** Figure K-51

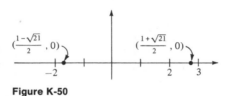

Figure K-50

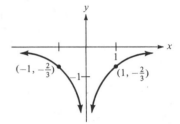

Figure K-51

Exercises 9–1

1. Figure K-52   **3.** Figure K-53   **5.** Figure K-54   **7.** Figure K-55
9. Figure K-56   **11.** Figure K-57   **13.** Figure K-58   **15.** Figure K-59
17. Figure K-60   **19.** Figure K-61
21. (a) 4; **(b)** 3; **(c)** $\frac{1}{3}$; **(d)** 5; **(e)** 16; **(f)** 100; **(g)** $\frac{1}{8}$; **(h)** 9; **(i)** $\frac{1}{8}$;
    (j) 27
23. If $f(x) = b^x$, then $f(x_1 + x_2) = b^{x_1 + x_2} = b^{x_1} \cdot b^{x_2} = f(x_1)f(x_2)$
25. (a) $y = 100(1.05)^t$; **(b)** $115.76   **27. (a)** $f(t) = 10,000(0.8)^t$; **(b)** $5120
29. (a) $y = 16(\frac{1}{2})^{2t}$; **(b)** 4 g

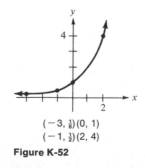

$(-3, \frac{1}{8})(0, 1)$
$(-1, \frac{1}{2})(2, 4)$

Figure K-52

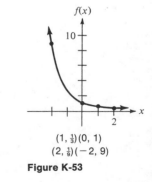

$(1, \frac{1}{3})(0, 1)$
$(2, \frac{1}{9})(-2, 9)$

Figure K-53

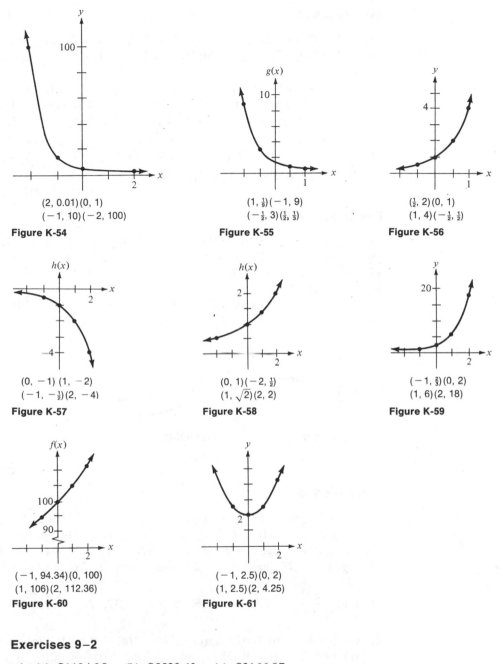

(2, 0.01)(0, 1)
(−1, 10)(−2, 100)
Figure K-54

(1, $\frac{1}{9}$)(−1, 9)
(−$\frac{1}{2}$, 3)($\frac{1}{2}$, $\frac{1}{3}$)
Figure K-55

($\frac{1}{2}$, 2)(0, 1)
(1, 4)(−$\frac{1}{2}$, $\frac{1}{2}$)
Figure K-56

(0, −1) (1, −2)
(−1, −$\frac{1}{2}$)(2, −4)
Figure K-57

(0, 1)(−2, $\frac{1}{2}$)
(1, $\sqrt{2}$)(2, 2)
Figure K-58

(−1, $\frac{2}{3}$)(0, 2)
(1, 6)(2, 18)
Figure K-59

(−1, 94.34)(0, 100)
(1, 106)(2, 112.36)
Figure K-60

(−1, 2.5)(0, 2)
(1, 2.5)(2, 4.25)
Figure K-61

Exercises 9–2

1. (a) \$1194.05; **(b)** \$5522.43; **(c)** \$3166.07
3. (a) \$1197.20; **(b)** \$38,520; **(c)** \$245,330; **(d)** \$208,160   **5.** 10 percent
7. (a) 1083; **(b)** 24.7 hr   **9.** 66,686   **11.** 7.4 g
13. (a) −0.000125; **(b)** 78 percent   **15.** 7:10 A.M.

Exercises 9-3

1. $f^{-1} = (7, 1)(8, 2)(9, 3)$; $D_{f^{-1}} = 7, 8, 9$; $R_{f^{-1}} = 1, 2, 3$; $D_f = 1, 2, 3$; $R_f = 7, 8, 9$
3. $f^{-1} = (10, -2)(0, 0)(-1, 5)$; $D_{f^{-1}} = 10, 0, -1$; $R_{f^{-1}} = -2, 0, 5$; $D_f = -2, 0, 5$;
$R_f = 10, 0, -1$
5. $f^{-1} = (4, 4)(5, 5)(6, 6)$; $D_{f^{-1}} = R_{f^{-1}} = D_f = R_f = 4, 5, 6$
7. No   **9.** Yes   **11.** No   **13.** No   **15.** Yes   **17.** Yes
19. No   **21.** $f^{-1}(x) = x + 10$; 11   **23.** $g^{-1}(x) = \frac{1}{7}x + \frac{3}{7}$; $\frac{2}{7}$
25. $h^{-1}(x) = \sqrt[3]{x}$; -2 **27.** $f^{-1}(x) = 1/x$; $-\frac{1}{4}$ **29.** $x = 2^y$; 3
31. $f[f^{-1}(x)] = (x - 5) + 5 = x$; $f^{-1}[f(x)] = (x + 5) - 5 = x$
33. $f[f^{-1}(x)] = \sqrt[3]{x^3} = x$; $f^{-1}[f(x)] = (\sqrt[3]{x})^3 = x$

Exercises 9-4

1. $\log_3 9 = 2$ **3.** $\log_{1/2}(\frac{1}{4}) = 2$ **5.** $\log_4(\frac{1}{16}) = -2$ **7.** $\log_{25} 5 = \frac{1}{2}$
9. $\log_7 1 = 0$ **11.** $5^1 = 5$ **13.** $(\frac{1}{3})^2 = \frac{1}{9}$ **15.** $2^{-2} = \frac{1}{4}$
17. $49^{1/2} = 7$ **19.** $100^0 = 1$ **21.** 2   **23.** 1   **25.** -1 **27.** $\frac{1}{2}$
29. 0   **31.** 3   **33.** $\frac{1}{5}$ **35.** 5   **37.** 3   **39.** $b > 0, b \neq 1$
41. $3, -1, 0$ **43.** $4, 0, -7$ **45.** Figure K-62   **47.** Figure K-63
49. $x > 3$ or $x < -3$ **51.** All real numbers except 1

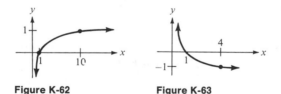

Figure K-62   **Figure K-63**

Exercises 9-5

1. $\log_{10} 7 + \log_{10} 5$ **3.** $\log_6 3 - \log_6 5$ **5.** $16 \log_5 11$ **7.** $\frac{1}{2} \log_2 3$
9. $\frac{1}{5} \log_{10} 16$ **11.** $2 + 3 \log_4 3$ **13.** $\frac{1}{2}(\log_b x + \log_b y)$ **15.** $\frac{5}{3} \log_b x$
17. $\frac{1}{4}(\log_b x + 2 \log_b y - 3 \log_b z)$ **19.** $\log_2 12$ **21.** $\log_4 4$ or 1
23. $\log_7 9$ **25.** $\log_{10} \frac{3}{8}$ **27.** $\log_b \sqrt[3]{xy^2}$ **29.** $\log_b(x - 1)$

31. $\log_b \sqrt{\dfrac{xz^3}{y^5}}$ **33.** 5   **35.** $3, -3$ **37.** 2   **39.** $x > 0$

41. **(a)** 0.7781;   **(b)** 0.6020;   **(c)** 0.2386;   **(d)** -0.1761;   **(e)** 1.4313
    (f) 1.8572;   **(g)** -0.2386;   **(h)** -0.0352;   **(i)** -0.3010;   **(j)** 0.6990
43. Let $x_1 = b^y$ and $x_2 = b^z$, then $y = \log_b x_1$ and $z = \log_b x_2$.

(a) $\dfrac{x_1}{x_2} = \dfrac{b^y}{b^z} = b^{y-z}$; thus, $\log_b \dfrac{x_1}{x_2} = y - z = \log_b x_1 - \log_b x_2$

(b) $(x_1)^k = (b^y)^k = b^{ky}$, thus, $\log_b(x_1)^k = ky = k \log_b x_1$

45. $\log_{10} 10^x = x \log_{10} 10 = x$, where x is any real number

Exercises 9–6

1. 0.3692   **3.** -0.6308 **5.** 3.3692   **7.** 1.9542   **9.** -0.9547
11. 6.59   **13.** 0.659   **15.** 36.0   **17.** 0.0387   **19.** 0.304   **21.** 1130
23. 63.5   **25.** 335   **27.** 9.85   **29.** 761,000   **31.** 6.34   **33.** 1.72
35. 55.8   **37.** 6.69   **39.** 3.85×10^{-6} **41.** 3.322   **43.** 1.044
45. 11.90   **47.** 5.83   **49.** 0.0159   **51.** 14.20 years   **53.** 7   **55.** 3.6
57. 1   **59.** 7.9×10^{-3}
61. **(a)** 0;   **(b)** 10;   **(c)** 20;   **(d)** 30; If the power ratio is multiplied by 10, add 10 dB to loudness.
63. 25 dB

Exercises 9–7

1. 3.170   **3.** 1.465   **5.** 0.8464   **7.** -3.190 **9.** -4.249
11. 1.608   **13.** 6.623   **15.** -0.7545 **17.** 1.9459   **19.** 3.7377
21. 8.7796   **23.** -0.5798 **25.** -7.7288 **27.** 1.2092   **29.** 9.9742
31. 0.7866   **33.** 0.0247

Exercises 9–8

1. Slope 2, intercept 1   **3.** Slope $\frac{1}{2}$, intercept 4   **7.** $y = 7.0x^{-0.35}$
9. $y = 3.0x^{0.65}$

SAMPLE TEST: CHAPTER 9

1. $(3, \frac{1}{64})(-1, 4)(0, 1)(\frac{3}{2}, \frac{1}{8})(-2, 16)$ **2.** $y = 500(1.06)^t$ **3.** \$3664.20
4. 35 days
5. $f^{-1} = (-1, 3)(-2, 4)$; $D_f = R_{f^{-1}} = 3, 4$; $D_{f^{-1}} = R_f = -1, -2$
6. **(a)** Yes;   **(b)** No   **7.** **(a)** $g^{-1}(x) = 2x + 5$;   **(b)** 3
8. D_f: all real numbers; D_g: $x > 0$
   R_f: $f(x) > 0$; R_g: all real numbers   Figure K-64
9. **(a)** $\log_4(\frac{1}{4}) = -1$;   **(b)** $2^3 = 8$ **10.** **(a)** 7;   **(b)** 0   **11.** $x > \frac{7}{3}$
12. **(a)** $\frac{1}{2}[3 \log_b x - \log_b y]$;   **(b)** $\log_b x^2 \sqrt[3]{y}$
13. -6 does not check; no solution   **14.** 0.2594
15. **(a)** False, $\log_{10} 10/\log_{10} 10 \neq \log_{10} 10 - \log_{10} 10$ since $\frac{1}{1} \neq 1 - 1$ **(b)** True;
   (c) False, $\log_{10} \frac{10}{10} \neq \log_{10} 10/\log_{10} 10$ since $0 \neq \frac{1}{1}$
16. **(a)** -0.9101;   **(b)** 0.00384   **17.** 8.86   **18.** 3.44×10^7

19. $\dfrac{0.9542}{0.6021}$ **20.** 2.5802

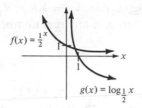

$f(x) = \frac{1}{2}^x$

$g(x) = \log_{\frac{1}{2}} x$

D_f: all real numbers; D_g: $x > 0$
R_f: $f(x) > 0$; R_g: all real numbers

Figure K-64

Exercises 10–1

1. 5   **3.** 8 yd   **5.** 9.6 meters   **7.** 5   **9.** $\dfrac{11}{2}, \dfrac{11}{4\pi}$

11. $\dfrac{25\pi}{2}$ in.^{2} **13.** 8 ft^2 **15.** $\dfrac{\pi}{6}$ **17.** $\dfrac{\pi}{3}$ **19.** $\dfrac{2\pi}{3}$ **21.** $\dfrac{5\pi}{6}$

23. $\dfrac{7\pi}{6}$ **25.** $\dfrac{4\pi}{3}$ **27.** $\dfrac{5\pi}{3}$ **29.** $\dfrac{11\pi}{6}$ **31.** $\dfrac{10\pi}{9}$ **33.** $\dfrac{\pi}{9}$ **35.** $\dfrac{61\pi}{36}$

37. 60°   **39.** 40°   **41.** 120°   **43.** 140°   **45.** 432°   **47.** 70°
49. 115°   **51.** 229°   **53.** 332°   **55.** 6 rad/second
57. (a) $(200\pi/3)$ rad/minute;   **(b)** 400π in./minute   **59.** 300π in./second
61. π in.   **63.** 860,000 mi

Exercises 10–2

1. 1   **3.** 0   **5.** 0   **7.** 0   **9.** Undefined   **11.** 1   **13.** -1

15. $\frac{1}{2}$ **17.** 1   **19.** $\frac{1}{2}$ **21.** $\dfrac{1}{\sqrt{3}}$ **23.** $\dfrac{\sqrt{3}}{2}$ **25.** $-\dfrac{2}{\sqrt{3}}$ **27.** $\dfrac{1}{\sqrt{3}}$

29. $-\dfrac{\sqrt{3}}{2}$ **31.** $d_1 = d_2$

$$\sqrt{x^2 + (y-1)^2} = \sqrt{(2y)^2}$$
$$x^2 + y^2 - 2y + 1 = 4y^2$$

since $x^2 + y^2 = 1$ we have

$$1 - 2y + 1 = 4y^2$$
$$0 = 4y^2 + 2y - 2$$
$$0 = 2(2y - 1)(y + 1)$$
$$y = \tfrac{1}{2}$$
$$x^2 + (\tfrac{1}{2})^2 = 1$$
$$x = \frac{\sqrt{3}}{2}$$

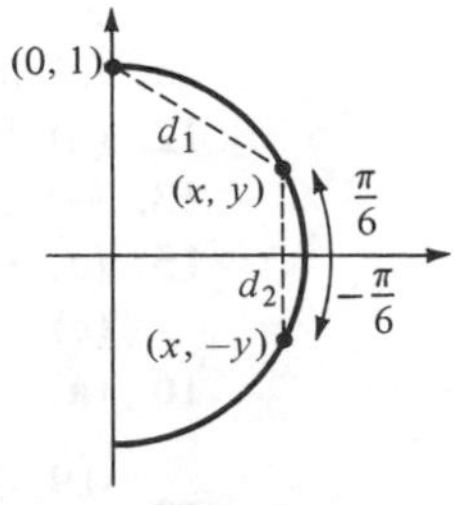

$(0, 1)$
d_1
(x, y)
$\dfrac{\pi}{6}$
d_2
$-\dfrac{\pi}{6}$
$(x, -y)$

Figure K-65

Exercises 10–3

1. $-\dfrac{\sqrt{3}}{2}$ **3.** $\sqrt{3}$ **5.** $-\dfrac{1}{\sqrt{2}}$ **7.** 2   **9.** $\frac{1}{2}$ **11.** $\frac{1}{2}$ **13.** -2

15. $-\dfrac{1}{\sqrt{2}}$ **17.** $-\sqrt{3}$ **19.** $-\dfrac{\sqrt{3}}{2}$ **21.** -0.4176 **23.** 1.162

25. -0.7643 **27.** 1.560   **29.** 0.2085   **31.** 0.6749   **33.** -1.041
35. 0.9521   **37.** 0.2190   **39.** -1.081
41. (a) 0.8417, 0.8415;   (b) 0.5417, 0.5403;   (c) 1, 1.000;   (d) 0.4794, 0.4794

Exercises 10–4

1. Figure K-66   **3.** Figure K-67   **5.** Figure K-68   **7.** Figure K-69
9. Figure K-70   **11.** Figure K-71   **13.** Figure K-72   **15.** Figure K-73
17. Figure K-74   **19.** Figure K-75   **21.** Figure K-76   **23.** Figure K-77
25. Figure K-78   **27.** Figure K-79   **29.** Figure K-80   **31.** $y = 3 \sin 2x$
33. $y = -4 \sin 3x$ **35.** $y = 1.5 \cos \frac{1}{2}x$ **37.** $y = -2 \cos 10x$
39. $y = 10 \sin \pi x$ **41.** d **43.** d **45.** c **47.** d **49.** d **51.** d
53. b **55.** d **57.** d **59.** a

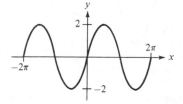

Amp.: 2; per.: 2π

Figure K-66

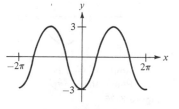

Amp.: 3; per.: 2π

Figure K-67

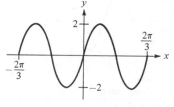

Amp.: 2; per.: $2\pi/3$

Figure K-68

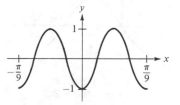

Amp.: 1; per.: $\pi/9$

Figure K-69

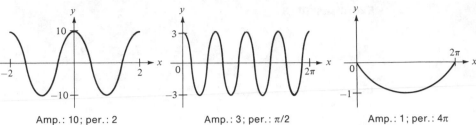

Amp.: 10; per.: 2

Figure K-70

Amp.: 3; per.: $\pi/2$

Figure K-71

Amp.: 1; per.: 4π

Figure K-72

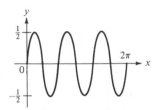

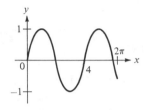

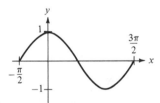

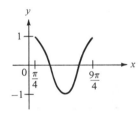

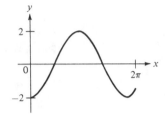

Amp.: $\frac{1}{2}$; per.: $2\pi/3$

Figure K-73

Amp.: 1; per.: 4

Figure K-74

Amp.: 2; per.: 6

Figure K-75

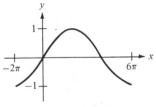

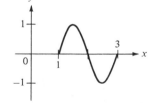

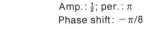

Amp.: 1; per.: 2π
Phase shift: $-\pi/2$

Figure K-76

Amp.: 1; per.: 2π
Phase shift: $\pi/4$

Figure K-77

Amp.: $\frac{1}{2}$; per.: π
Phase shift: $-\pi/8$

Figure K-78

Amp.: 1; per.: 8π
Phase shift: -2π

Figure K-79

Amp.: 1; per.: 2
Phase shift: 1

Figure K-80

Exercises 10–5

x	0	$\dfrac{\pi}{6}$	$\dfrac{\pi}{3}$	$\dfrac{\pi}{2}$	$\dfrac{2\pi}{3}$	$\dfrac{5\pi}{6}$	π	$\dfrac{7\pi}{6}$	$\dfrac{4\pi}{3}$	$\dfrac{3\pi}{2}$	$\dfrac{5\pi}{3}$	$\dfrac{11\pi}{6}$	2π
1. $\cot x$	und.	1.7	0.6	0	-0.6	-1.7	und.	1.7	0.6	0	-0.6	-1.7	und.
3. $\csc x$	und.	2	1.2	1	1.2	2	und.	-2	-1.2	-1	-1.2	-2	und.

5. D: all real numbers except $x = \dfrac{\pi}{2} + k\pi$ (k any integer); $R: y \le -1$ or $y \ge 1$; period; 2π

7. Inc., dec., dec., inc. **9.** Always increasing **11.** Inc., inc., dec., dec.,
13. The other trigonometric functions do not attain a maximum value.
15. Figure K-81 **17.** Figure K-82

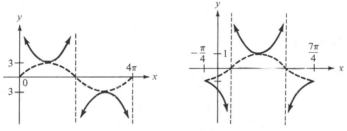

Figure K-81   Figure K-82

Exercises 10–6

1. $\dfrac{\pi}{3}, \dfrac{2\pi}{3}$ **3.** $\dfrac{2\pi}{3}, \dfrac{4\pi}{3}$ **5.** $\dfrac{3\pi}{4}, \dfrac{7\pi}{4}$ **7.** $0.12, 3.02$ **9.** $1.88, 5.02$

11. $\dfrac{\pi}{4}, \dfrac{7\pi}{4}$ **13.** No solution   **15.** $0.32, 3.46$

17. $\dfrac{\pi}{6} + k2\pi, \dfrac{5\pi}{6} + k2\pi$ (k any integer)

19. $3.60 + k2\pi, 5.82 + k2\pi$ (k any integer)

21. $\dfrac{\pi}{4}, \dfrac{3\pi}{4}, \dfrac{5\pi}{4}, \dfrac{7\pi}{4}$ **23.** $0, \dfrac{\pi}{2}, \dfrac{3\pi}{2}$ **25.** $\dfrac{\pi}{6}, \dfrac{\pi}{2}, \dfrac{5\pi}{6}, \dfrac{3\pi}{2}$

27. $1.25, 1.71, 4.39, 4.85$ **29.** $\dfrac{\pi}{6}, \dfrac{\pi}{2}, \dfrac{5\pi}{6}, \dfrac{7\pi}{6}, \dfrac{3\pi}{2}, \dfrac{11\pi}{6}$ **31.** $\dfrac{\pi}{12}, \dfrac{5\pi}{12}, \dfrac{13\pi}{12}, \dfrac{17\pi}{12}$

33. $\dfrac{3\pi}{4}$ **35.** $0, \dfrac{\pi}{4}, \dfrac{\pi}{2}, \pi, \dfrac{5\pi}{4}, \dfrac{3\pi}{2}$ **37.** $0.90, 5.38$ **39.** No solution

Exercises 10–7

1. $\dfrac{\pi}{3}$ **3.** $-\dfrac{\pi}{3}$ **5.** $\dfrac{\pi}{4}$ **7.** $-\dfrac{\pi}{2}$ **9.** $\dfrac{\pi}{4}$ **11.** $-\dfrac{\pi}{6}$ **13.** 0.32

15. 2.15   **17.** 1.05   **19.** 2.50   **21.** 1   **23.** $\frac{1}{2}$ **25.** 0   **27.** 1

29. -0.6552 **31.** $\frac{3}{5}$ **33.** $\frac{12}{5}$ **35.** $1/\sqrt{5}$

37. $D: -1 \le x \le 1; R: 0 \le y \le \pi$ **39.** $D:$ all real numbers; $R: 0 < y < \pi$

41. $D: x \le -1$ or $x \ge 1; R: -\dfrac{\pi}{2} \le y \le \dfrac{\pi}{2}, y \ne 0$ **43.** Figure K-83

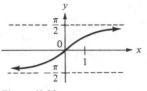

Figure K-83

Exercises 10–8

1. $\cos\left(\dfrac{\pi}{2} + x\right) = \cos\dfrac{\pi}{2}\cos x - \sin\dfrac{\pi}{2}\sin x$

$$= (0)\cos x - (1)\sin x$$
$$= -\sin x$$

3. $\cos(\pi - x) = \cos \pi \cos x + \sin \pi \sin x$
$$= (-1)\cos x + (0)\sin x$$
$$= -\cos x$$

5. $\cos\left(\dfrac{3\pi}{2} + x\right) = \cos\dfrac{3\pi}{2}\cos x - \sin\dfrac{3\pi}{2}\sin x$

$$= (0)\cos x - (-1)\sin x$$
$$= \sin x$$

7. $\sin(x_1 - x_2) = \sin[x_1 + (-x_2)]$
$$= \sin x_1 \cos(-x_2) + \cos x_1 \sin(-x_2)$$

Since $\cos(-x_2) = \cos x_2$ and $\sin(-x_2) = -\sin x_2$, we have
$$\sin(x_1 - x_2) = \sin x_1 \cos x_2 - \cos x_1 \sin x_2$$

9. $\sin\left(\dfrac{\pi}{2} - x\right) = \sin\dfrac{\pi}{2}\cos x - \cos\dfrac{\pi}{2}\sin x$

$$= (1)\cos x - (0)\sin x$$
$$= \cos x$$

11. $\sin(\pi + x) = \sin \pi \cos x + \cos \pi \sin x$

$$= (0)\cos x + (-1)\sin x$$

$$= -\sin x$$

13. $\sin\left(\dfrac{3\pi}{2} - x\right) = \sin \dfrac{3\pi}{2} \cos x - \cos \dfrac{3\pi}{2} \sin x$

$$= (-1)\cos x - (0)\sin x$$

$$= -\cos x$$

15. $\cos 2x = \cos^2 x - \sin^2 x$ (Example 5)

Since $\cos^2 x = 1 - \sin^2 x$, we have

$$\cos 2x = (1 - \sin^2 x) - \sin^2 x$$

$$= 1 - 2\sin^2 x$$

17. $\tan(x_1 + x_2) = \dfrac{\sin(x_1 + x_2)}{\cos(x_1 + x_2)}$

$$= \dfrac{\sin x_1 \cos x_2 + \cos x_1 \sin x_2}{\cos x_1 \cos x_2 - \sin x_1 \sin x_2}$$

$$= \dfrac{[(\sin x_1 \cos x_2)/(\cos x_1 \cos x_2)] + [(\cos x_1 \sin x_2)/(\cos x_1 \cos x_2)]}{[(\cos x_1 \cos x_2)/(\cos x_1 \cos x_2)] - [(\sin x_1 \sin x_2)/(\cos x_1 \cos x_2)]}$$

$$= \dfrac{\tan x_1 + \tan x_2}{1 - \tan x_1 \tan x_2}$$

19. $\tan(x_1 - x_2) = \tan[x_1 + (-x_2)]$

$$= \dfrac{\tan x_1 + \tan(-x_2)}{1 - \tan x_1 \tan(-x_2)}$$

Since $\tan(-x_2) = -\tan x_2$, we have

$$\tan(x_1 - x_2) = \dfrac{\tan x_1 - \tan x_2}{1 + \tan x_1 \tan x_2}$$

21. (a) $\dfrac{5}{13}$; **(b)** $\dfrac{5}{12}$; **(c)** $\dfrac{13}{12}$; **(d)** $-\dfrac{119}{169}$; **(e)** $\dfrac{120}{169}$; **(f)** $\dfrac{3\sqrt{13}}{13}$; **(g)** $\dfrac{2\sqrt{13}}{13}$

23. (a) $\dfrac{\sqrt{3}}{2}$; **(b)** $\dfrac{1}{2}$; **(c)** -1; **(d)** $\dfrac{-\sqrt{3}}{2}$

Exercises 10–9

(*Note:* Answers give suggested intermediate steps.)

1. $\sin x \dfrac{1}{\cos x} = \dfrac{\sin x}{\cos x} = \tan x$

3. $\sin^2 x \dfrac{1}{\sin x} = \sin x$

5. $\cos x \dfrac{\sin x}{\cos x} = \sin x = \dfrac{1}{\csc x}$

7. $\cos x \dfrac{\sin x}{\cos x}\dfrac{1}{\sin x} = 1$

9. $\sin^2 x \dfrac{\cos^2 x}{\sin^2 x} = \cos^2 x$

11. $\sin^2 x \dfrac{1}{\cos^2 x} = \tan^2 x$

13. $\dfrac{\cos^2 x}{\cos^2 x/\sin^2 x} = \sin^2 x$

15. $\dfrac{(\cos x/\sin x)(\sin x/\cos x)}{1/\cos x} = \cos x$

17. $1 - \cos^2 x = \sin^2 x = \dfrac{1}{\csc^2 x}$

19. $\dfrac{(1 + \sin x)(1 - \sin x)}{1 + \sin x} = 1 - \sin x$

21. $\dfrac{1 - \sin^2 x}{\sin^2 x} = \dfrac{\cos^2 x}{\sin^2 x} = \cot^2 x$

23. $\dfrac{\sec^2 x}{\csc^2 x} = \dfrac{1/\cos^2 x}{1/\sin^2 x} = \dfrac{\sin^2 x}{\cos^2 x} = \tan^2 x$

25. $\sin x \dfrac{\sin x}{\cos x} + \cos x = \dfrac{\sin^2 x + \cos^2 x}{\cos x} = \dfrac{1}{\cos x} = \sec x$

27. $1 - \dfrac{\sin^2 x}{\tan^2 x} = 1 - \dfrac{\sin^2 x}{\sin^2 x/\cos^2 x} = 1 - \cos^2 x = \sin^2 x$

29. $\dfrac{1/\sin x}{(\sin x/\cos x) + (\cos x/\sin x)} = \dfrac{\cos x}{\sin^2 x + \cos^2 x} = \cos x$

31. $(1 - \sin x)\left(\dfrac{1}{\cos x} + \dfrac{\sin x}{\cos x}\right) = \dfrac{1 - \sin^2 x}{\cos x} = \dfrac{\cos^2 x}{\cos x} = \cos x$

33. $\dfrac{1}{\sin x} 2 \sin x \cos x = 2 \cos x$

35. $\dfrac{2 \cos^2 x - 1 + 1}{2} = \cos^2 x$

37. $(1 - 2 \sin^2 x) + 2 \sin^2 x = 1$

39. $\dfrac{1 + 2 \cos^2 x - 1}{2 \sin x \cos x} = \dfrac{\cos x}{\sin x} = \cot x$

41. $\dfrac{\pi}{4}, \dfrac{5\pi}{4}$

43. $0, \pi$

45. $\dfrac{\pi}{4}, \dfrac{5\pi}{4}$

47. $\dfrac{\pi}{6}, \dfrac{5\pi}{6}$

49. $\dfrac{\pi}{2}, \dfrac{7\pi}{6}, \dfrac{11\pi}{6}$

51. $\dfrac{7\pi}{6}, \dfrac{3\pi}{2}, \dfrac{11\pi}{6}$

53. 1.84, 2.20, 4.98, 5.34

SAMPLE TEST: CHAPTER 10

1. $\dfrac{10\pi}{3}$ in. **2.** $\dfrac{50\pi}{3}$ in.^{2} **3.** 0 **4.** 0.4447 **5.** $-\sqrt{3}$ **6.** -1.010

7–8. Figure K-84 **9.** $y = 4 \sin \frac{3}{2}x$ **10.** Quadrant 2

11. All real numbers except $x = k\pi$ (k any integer) **12.** $\dfrac{7\pi}{6}, \dfrac{11\pi}{6}$

13. 1.00, 2.14, 3.77, 5.65 **14.** -0.46 **15.** 0

16. $\sin(3\pi - x) = \sin 3\pi \cos x - \cos 3\pi \sin x$
$= (0) \cos x - (-1) \sin x$
$= \sin x$

17. $\tan \dfrac{x}{2} = \dfrac{\sin(x/2)}{\cos(x/2)} = \dfrac{\pm\sqrt{(1 - \cos x)/2}}{\pm\sqrt{(1 + \cos x)/2}} = \pm\sqrt{\dfrac{1 - \cos x}{1 + \cos x}}$

18. $-\dfrac{120}{169}$

19. $\dfrac{\sin x + (\sin x/\cos x)}{1 + (1/\cos x)} = \dfrac{\sin x \cos x + \sin x}{\cos x + 1} = \dfrac{\sin x(\cos x + 1)}{\cos x + 1} = \sin x$

20. $(\sin x - \cos x)^2 = \sin^2 x - 2 \sin x \cos x + \cos^2 x$
$= 1 - 2 \sin x \cos x$
$= 1 - 2 \sin x$

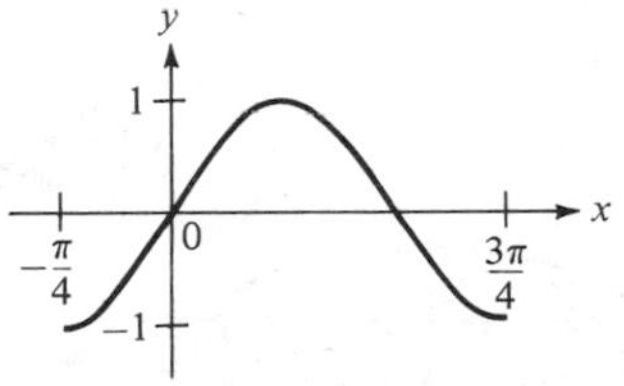

Amp.: 1; per.: π
Phase shift: $-\pi/4$

Figure K-84

Exercises 11–1

1. $0 = x - 3y + 11$ **3.** $0 = x^2 + y^2 + 4x + 6y - 3$
5. $0 = y^2 + 4y + 8x - 12$ **7.** $0 = 3x^2 + 4y^2 - 12$
9. $0 = 4x^2 - 12y^2 + 27$

Exercises 11–2

1. $(x + 3)^2 + (y - 4)^2 = 4$ **3.** $x^2 + y^2 = 1$ **5.** $(x + 4)^2 + (y + 1)^2 = 9$
7. $(x - 3)^2 + (y + 3)^2 = 18$ **9.** $(x + 1)^2 + (y + 1)^2 = 58$ **11.** (2, 5); 4

13. $(3, 0)$; $\sqrt{20}$ **15.** $(0, 0)$; 3   **17.** $(5, 3)$; 7   **19.** $(-4, 1)$; $\sqrt{18}$

21. $(-\frac{7}{2}, -\frac{3}{2})$; $\dfrac{\sqrt{42}}{2}$

Exercises 11–3

1. Figure K-85   **3.** Figure K-86   **5.** Figure K-87   **7.** Figure K-88

9. Figure K-89   **11.** Figure K-90   **13.** Figure K-91   **15.** $\dfrac{x^2}{16} + \dfrac{y^2}{25} = 1$

17. $\dfrac{x^2}{36} + \dfrac{y^2}{20} = 1$ **19.** $\dfrac{x^2}{9} + \dfrac{y^2}{4} = 1$ **21.** $\dfrac{(x-2)^2}{16} + \dfrac{(y-2)^2}{25} = 1$

23. $\dfrac{(x-2)^2}{25} + \dfrac{(y-3)^2}{16} = 1$ **25. (a)** 186,000,000 mi;   **(b)** 3,000,000 mi

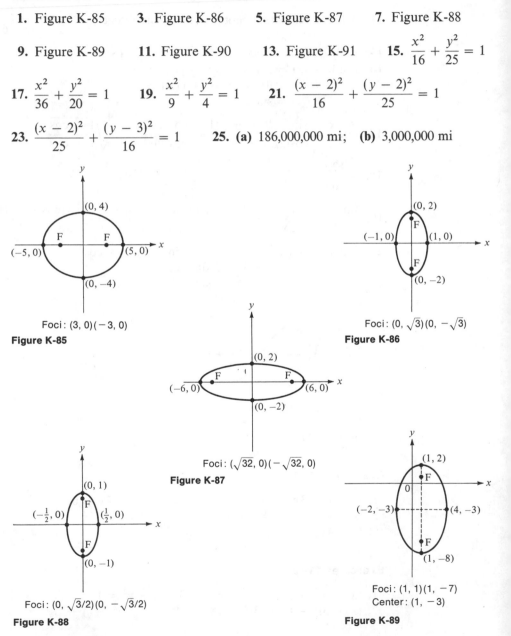

Foci: (3, 0)(−3, 0)
Figure K-85

Foci: (0, $\sqrt{3}$)(0, −$\sqrt{3}$)
Figure K-86

Foci: ($\sqrt{32}$, 0)(−$\sqrt{32}$, 0)
Figure K-87

Foci: (0, $\sqrt{3}/2$)(0, −$\sqrt{3}/2$)
Figure K-88

Foci: (1, 1)(1, −7)
Center: (1, −3)
Figure K-89

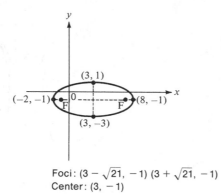

Foci: $(3 - \sqrt{21}, -1)$ $(3 + \sqrt{21}, -1)$
Center: $(3, -1)$

Figure K-90

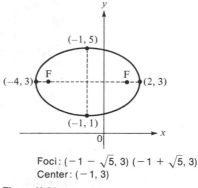

Foci: $(-1 - \sqrt{5}, 3)$ $(-1 + \sqrt{5}, 3)$
Center: $(-1, 3)$

Figure K-91

Exercises 11–4

1. Figure K-92   **3.** Figure K-93   **5.** Figure K-94   **7.** Figure K-95

9. Figure K-96   **11.** Figure K-97   **13.** Figure K-98   **15.** $\dfrac{y^2}{16} - \dfrac{x^2}{9} = 1$

17. $\dfrac{x^2}{9} - \dfrac{y^2}{7} = 1$ **19.** $\dfrac{x^2}{9} - \dfrac{y^2}{25} = 1$ **21.** $\dfrac{(x + 3)^2}{9} - \dfrac{(y + 4)^2}{16} = 1$

23. $\dfrac{(y - 1)^2}{4} - \dfrac{(x - 3)^2}{5} = 1$

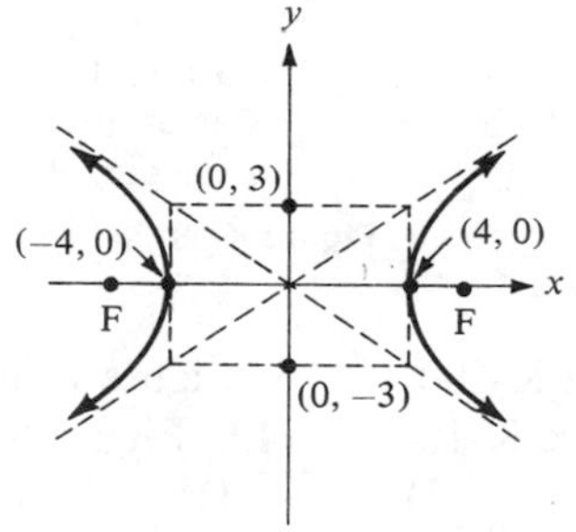

Foci: $(5, 0)(-5, 0)$
Asymp.: $y = \pm\frac{3}{4}x$

Figure K-92

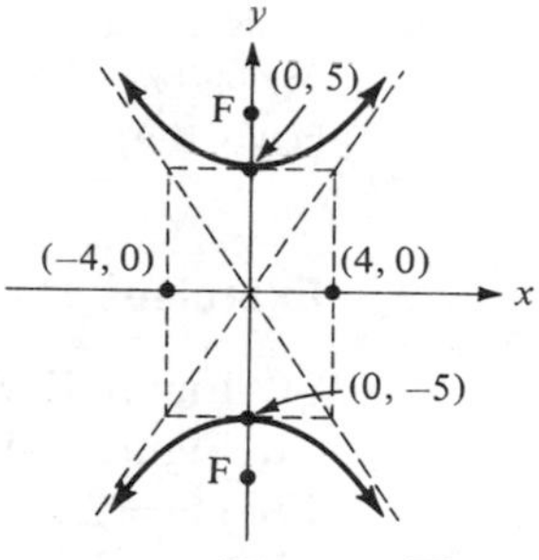

Foci: $(0, \sqrt{41})(0, -\sqrt{41})$
Asymp.: $y = \pm\frac{5}{4}x$

Figure K-93

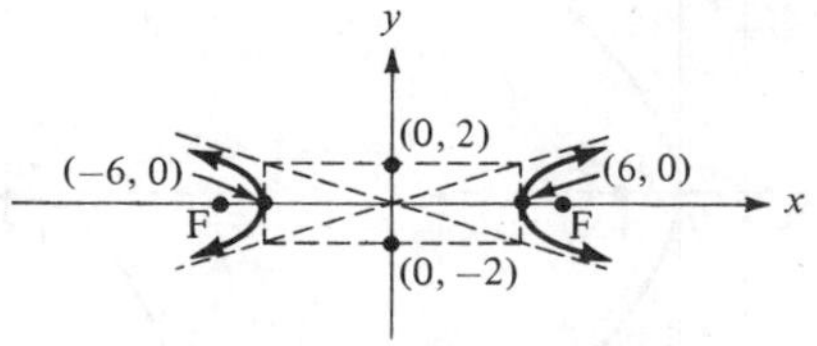

Foci: $(\sqrt{40}, 0)(-\sqrt{40}, 0)$
Asymp.: $y = \pm\frac{1}{3}x$

Figure K-94

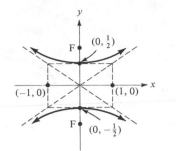

Foci: $(0, \sqrt{5}/2)(0, -\sqrt{5}/2)$
Asymp.: $y = \pm\frac{1}{2}x$

Figure K-95

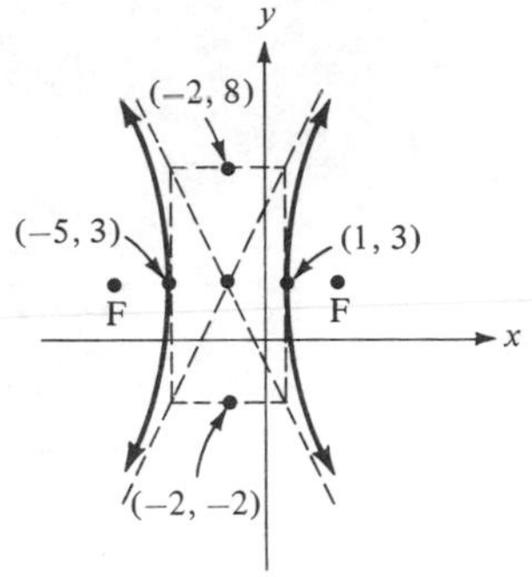

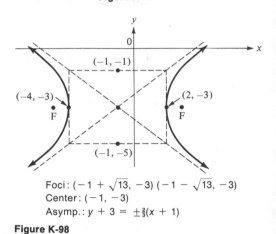

Foci: $(-2 + \sqrt{34}, 3)(-2 - \sqrt{34}, 3)$
Center: $(-2, 3)$
Asymp.: $y - 3 = \pm\frac{5}{3}(x + 2)$

Figure K-96

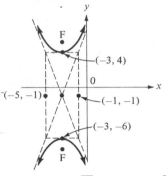

Foci: $(-3, -1 + \sqrt{29})(-3, -1 - \sqrt{29})$
Center: $(-3, -1)$
Asymp.: $y + 1 = \pm\frac{5}{2}(x + 3)$

Figure K-97

Foci: $(-1 + \sqrt{13}, -3)\ (-1 - \sqrt{13}, -3)$
Center: $(-1, -3)$
Asymp.: $y + 3 = \pm\frac{2}{3}(x + 1)$

Figure K-98

Exercises 11–5

1. Figure K-99 **3.** Figure K-100 **5.** Figure K-101 **7.** Figure K-102
9. Figure K-103 **11.** Figure K-104 **13.** Figure K-105 **15.** $x^2 = 12y$
17. $y^2 = 2x$ **19.** $x^2 = 16y$ **21.** $(y - 1)^2 = 12(x + 1)$
23. $(x - 3)^2 = 12y$ **25. (a)** $x^2 = 12y$; **(b)** 3 units from vertex **27.** 21 ft

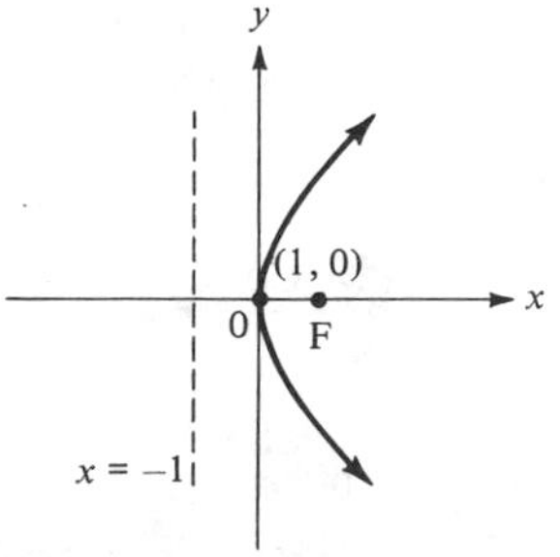

V: $(0, 0)$; F: $(1, 0)$; D: $x = -1$

Figure K-99

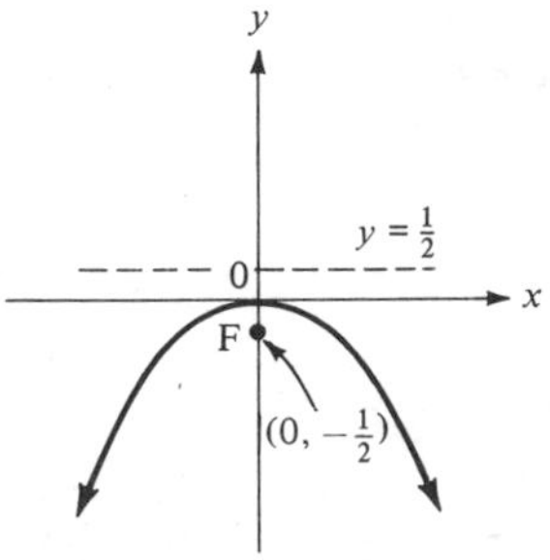

V: $(0, 0)$; F: $(0, -\frac{1}{2})$; D: $y = \frac{1}{2}$

Figure K-100

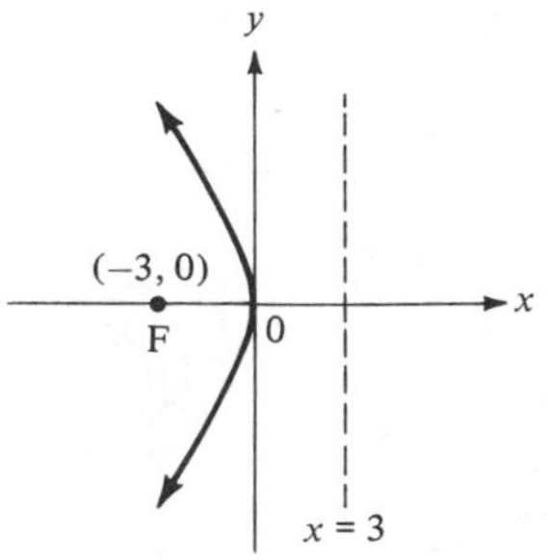

V: $(0, 0)$; F: $(-3, 0)$; D: $x = 3$

Figure K-101

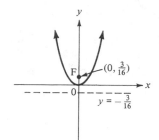

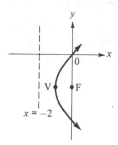

V: (0, 0); F: (0, $\frac{3}{16}$); D: $y = -\frac{3}{16}$

Figure K-102

V: ($-1, -2$); F: (0, -2); D: $x = -2$

Figure K-103

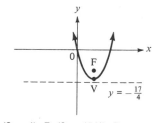

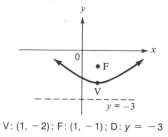

V: (2, -4); F: (2, $-15/4$); D: $y = -\frac{17}{4}$

Figure K-104

V: (1, -2); F: (1, -1); D: $y = -3$

Figure K-105

Exercises 11-6

1. Parabola **3.** Ellipse **5.** Hyperbola **7.** Parabola **9.** Circle

SAMPLE TEST: CHAPTER 11

1. $0 = 2x^2 + 2y^2 - 2x - 9y - 4$
2. (a) Hyperbola; **(b)** Parabola; **(c)** Line; **(d)** Ellipse; **(e)** Circle
3. $(-2, 3); 5$ **4.** $(x - 4)^2 + (y + 3)^2 = 20$ **5.** $(0, 3)(0, -3)$

6. $\dfrac{(x + 2)^2}{25} + \dfrac{(y - 2)^2}{9} = 1$ **7.** $y = \pm\frac{4}{5}x$ **8.** $\dfrac{x^2}{49} - \dfrac{y^2}{15} = 1$

9. $(0, -\frac{7}{4})$ **10.** $(y - 1)^2 = 12(x - 2)$

Appendix A-1

1. Five **3.** One **5.** Three **7.** Three **9.** Two **11.** Three
13. Two **15.** Three **17.** Four **19.** Two **21. (a)** 12.35; **(b)** 12
23. (a) 59,370; **(b)** 59,000 **25. (a)** 29,7$\overline{0}$0; **(b)** 3$\overline{0}$,000
27. (a) 84.10; **(b)** 84 **29. (a)** 0.06855; **(b)** 0.069
31. (a) 0.002001; **(b)** 0.0020 **33. (a)** 423; **(b)** 0.004
35. (a) Same accuracy; **(b)** 40.2 **37. (a)** 6.430; **(b)** same precision
39. (a) Same accuracy; **(b)** 0.0002 **41. (a)** Same accuracy; **(b)** 48.2
43. (a) 30$\overline{0}$; **(b)** 30$\overline{0}$

Appendix A-2

1. 14.1   **3.** −7.4   **5.** 1.9   **7.** 1.2   **9.** 32,800   **11.** 200,000
13. 200,000   **15.** 11.6   **17.** 90   **19.** 9.75   **21.** 51.00 ft
23. 274 mi   **25.** 143.94 ft   **27.** 18.71 cm^2 **29.** 2$\overline{0}$0 m^3

Index

Absolute value
  of complex numbers, 260; of real numbers, 12
Absolute value function, 70-71, 270
  graph of, 70, 273
Accuracy, of number, 490
Addition
  of algebraic expressions, 19; of complex numbers, 257; of fractions, 208; of functions, 281; of radicals, 243; of real numbers, 12; rules for signs, 14
Algebraic expression, 16
AM (amplitude modulation), 427
Amplitude
  of complex numbers, 260; of periodic functions, 431; of sine and cosine, 421
Angle
  coterminal. *See;* of depression, 106; of elevation, 106; of incidence, 34, 110, 111; measure of, 85; negative, 86; phase, 111; positive, 86; reference, 127; of reflection, 34, 111; of refraction, 100; standard position of, 86
Angular velocity, 400
Approximate numbers, 489
  rules for computation, 492, 494
Arc length, 397
Archimedes, 311
Area, of sector of circle, 399
Aristarchus, 156
Associative property
  of addition, 8; of multiplication, 9
Asymptotes
  horizontal, 350-353; of hyperbola, 471; vertical, 349
Axes, 64
Axis
  conjugate, 471; of ellipse, 464; real, 259; of symmetry, 319, 478; transverse, 471

Bar
  above approximate number, 490; above repeating decimal, 5
Binomial theorem, 172
  general term of, 173
Bounds of real zeros, 346
Break-even point, 305-306

Cancellation, 192
Carbon 14 dating, 368
Cartesian coordinate system, 63
Change of base of logarithm, 387
Circle
  area of sector of, 399; definition of, 460;

unit, 404
Circumference, of earth, 35
Coefficient, 19
Cofunctions, 122
Commutative property
  of addition, 9; of multiplication, 9
Complementary angles, 102
Complex fractions, 216
Complex numbers
  conjugate of, 256; definition of, 255; geometric form of, 259; imaginary part of, 255; powers and roots of, 258, 261-263; real part of, 255; summary of operational rules, 257; trigonometric form of, 259-260
Components of vector, 116
Composite function, 283
  domain of, 283
Compound interest, 362-364
Conic sections, 483
  characteristics of different, 483-484
Conjugate axis, 471
Conjugate of complex number, 256
Constant
  absolute, 19; arbitrary, 19; definition of, 19
Constant function, 68, 270
  graph of, 68
Constant of variation, 55
Coordinates, 64
Cosecant function
  definition of, 87, 406; graph of, 432; inverse, 441; right triangle definition, 102
Cosine function
  definition of, 87, 405; graph of, 422-423; inverse, 441; right triangle definition, 102
Cosines, law of, 149
Cotangent function
  definition of, 87, 406; graph of, 433; inverse, 441; right triangle definition, 102
Coterminal angles, 124
  rule to generate, 137
Counting numbers, 4
Cramer's rule, 313

Decibels (dB), 386-387
Decimals, repeating, 5
Degree
  angular measure, 94; of polynomial, 338; to radian measure, 398
DeMoivre's theorem, 261
Denominator, rationalizing, 248-250
Dependent
  system of equations, 304; variable, 48, 268

Descartes, Rene, 63, 100
rule of signs, 344
Determinant, 312
Digits, significant, 490
Diophantus, 226
Directrix, 477
Direct variation, 55
Discriminant, 331
Distance formula, 79
Distributive property, 9
Division
of algebraic expressions, 165; of complex numbers, 257; of fractions, 200; of functions, 281; long, 338; of polynomials, 338; of radicals, 247; of real numbers, 16; synthetic, 339; by zero, 5
Domain, 49, 52, 269, 272

e, definition of, 363
Ellipse
applications of, 468-469; axes of, 464; definition of, 463
Equal monthly payment formula, 386
Equation. *See also* System of equations
algebraic, 21; equivalent, 23; exponential, 359-360, 383; fractional, 219; logarithmic, 377; point-slope, 294; quadratic, 182, 327; radical, 252; rules to obtain equivalent, 23; solution of, 22; trigonometric, 134, 434
Equilibrium point, market, 308
Equilibrium price, 308
Eratosthenes, 35, 112
Evaluation, of algebraic expressions, 16
Even function, 279
Exact number, 494
Exponential functions
definition of, 357; graphed (on semilog paper), 393-394; graphs of, 358
Exponents
fractional, 237; laws of, 158-161, 229; negative, 230; positive integer, 15; zero, 230

Factor, 15
Factorability, test for, 180-181
Factoring, 173-181
Factor theorem, 344
Foci
of ellipse, 463; of hyperbola, 470
Focus, of parabola, 477
FOIL, multiplication method, 167-168
Fourier, Joseph, 426
Fractions, fundamental principle, 191
FM (frequency modulation), 427
Functional notation, 60, 277
Functions
composite, 283; composite, domain of, 283; composition of, 282-283; constant, 68, 270; difference quotient of, 278; even, 279; graph of, 65, 272; graphical test for, 71, 273; inverse, 369; linear. *See;* logarithmic. *See;* odd, 279; one-to-one, 370; operations with, 281; ordered pair, definition of, 52, 272; ordered pair, domain of, 52, 272; ordered pair, range of, 52, 272; periodic, 420; polynomial. *See;* quadratic. *See;* rate of change of, 288-289; rational, 348; rule definition, 49, 269; rule definition, domain of, 49, 269; rule definition, machine analogy, 49, 269; rule definition, range of, 49, 269; trigo-

nometric. *See;* value of, 60, 277; zero of, 343
Fundamental principle
of analytical geometry, 71, 457; of fractions, 191

General term, of binomial formula, 173
Geometric formulas, summary, 497-498
Good guy-bad guy scheme, 16
Graphs
definition of, 65, 272; on logarithmic paper, 389-392; misleading, 75-77; on semilogarithmic paper, 393-394
Greater than, 36
Greater than or equal to, 38
Greatest integer function, 280
Grouping, removing symbols of, 20

Hooke's law, 55
Horizontal
asymptote, 350-353; line test, 370
Hyperbola, definition of, 470

i, definition of, 255
Identities
guidelines for proving, 451; trigonometric, summary of, 499-500
Identity, definition of, 23, 408, 443
Imaginary axis, 259
Imaginary number, 255
Imaginary part of complex number, 255
Impedance triangle, 111
Inconsistent system of equations, 304
Independent variable, 48, 268
Indirect proof, 8
Inequalities
equivalent, 37; quadratic, 334; rules to obtain equivalent, 37
Initial ray, 85
Integers, 4
Interpolation, 95
Inverse function
definition of, 369; symbol for, 371; tests for, 370
Inverse trigonometric functions, 441
Inverse variation, 56
Irrational numbers, 6

j, definition of, 258-259
Joint variation, 57-58

Law of cosines, 149
Law of sines, 141
Laws of exponents, 158-161, 229
LCD (least common denominator), 210
LCM (least common multiple), 204
Less than, 36
Less than or equal to, 38
Line
point-slope form, 294; slope of, 289; slope-intercept form, 294
Linear function
definition of, 289, 293; graph of, 66, 289
Linear programming, 41
Linear velocity, 400
Lines. *See also* Line
parallel, 299; perpendicular, 300
Logarithmic function
definition of, 374; graph of, 374

Logarithms
    change of base, 387; characteristic of, 379;
    common (base 10), 379; computations using,
    382-383; definition of, 373; function. *See;*
    mantissa of, 379; natural (base *e*), 388; proper-
    ties of, 376; solving equations with, 383
Long division, 338

Magnitude, 12
Major axis of ellipse, 464
Mantissa, of logarithm, 379
Market equilibrium point, 308
Maximum-minimum problems, 323-325
Midpoint, formula for, 301
Minor axis of ellipse, 464
Minute (angular measure), 94
Multiplication
    of algebraic expressions, 164; of complex num-
    bers, 257; of fractions, 198; of functions, 281;
    of radicals, 244; of real numbers, 14; rules for
    signs, 15
Musical sounds, 426

Natural logarithm, 388
Negative, of a number, 10
Negative angle, 86
Negative exponent, 230
Negative number, 4
Newton's law
    of cooling, 366-367; of gravitation, 57
*n*th root, 236
Numbers
    complex, 255; exact, 494; imaginary, 255;
    imaginary, pure, 255; integers, 4; irrational, 6;
    negative, 4; negative of a, 10; positive, 4;
    precision of a, 490; prime, 204; rational, 4;
    real, 6; real number line, 12; reference, 413
Number tricks, 28

Odd function, 279
One-to-one function, 370
Operations, order of, 17
Ordered pair, 51, 271
Origin, 64
Oscilloscope, 426

Parabola, 69, 318, 477
    applications of, 323-325, 481-482; definition
    of, 477
Parallel lines, 299
Pascal's triangle, 172, 517
Period
    of any periodic function, 420; of sine and
    cosine, 420
Periodic function, 420
Perpendicular lines, 300
pH (hydrogen potential), 384-385
Phase angle, 111
Phase shift, 425
pi (π), 6
Point-slope equation, 294
Polynomial function
    definition of, 338; degree of, 338; graphs of,
    343-344; rational zeros of, 343
Population growth, 365-366
Positive angle, 86
Positive number, 4
Power, 15

Power function
    definition of, 390; graphs of, on log paper,
    390-392
Precision of number, 490
Prime number, 204
Principal root, 236
Proof, indirect, 8
Properties
    of logarithms, 376; of radicals, 240; of real
    numbers, 8
Proportion, 32
Pythagorean theorem, 498

Quadrant, 64
Quadrantal angle, 123
Quadratic equation
    definition of, 182, 327; nature of solutions,
    331; solution by factoring, 183; solution by
    formula, 327-329
Quadratic formula, 329
Quadratic function
    definition of, 317; graphs of, 69, 318-320;
    maximum or minimum of, 319; nature of *x*
    intercepts, 333
Quadratic inequalities, 334
Quadrefoil, 403

Radian
    definition of, 397; to degree measure, 398
Radical equation, 252
Radical sign, 6, 236
Radicals, properties of, 240
Radioactive decay, 365
Radio waves, 426-427
Range, 49, 52, 269, 272
Ratio, 31
Rational functions, 348
Rationalizing denominators, 248-250
Rational numbers, 4
Rational zero theorem, 344
Ray, 85
Real axis, 259
Real number line, 12
Real numbers, 6
Real part of complex number, 255
Reciprocal, 10
Reference angle, 127
Reference number, 413
Reflection, law of, 34, 111
Reflection property
    in ellipse, 468-469; in parabola, 481-482
Refraction, law of, 100-101
Regression analysis, 298-299
Remainder theorem, 341
Removing symbols of grouping, 20
Repeating decimals, 5
Resolving a vector, 116
Resultant, 113
Right triangles, 101
Root
    of a complex number, 262; of an equation,
    343; principal, 236
Rounding off, 489-490

Scaling, 75-76
Scientific notation, 233
Secant function
    definition of, 87, 406; graph of, 433;

Secant function (*Continued*)
    inverse, 441; right triangle definition, 102
Second (angular measure), 94
Sense of inequality, 37
Significant digits, 490
Similar terms, 20
Sine function
    definition of, 87, 405; graph of, 419; inverse, 439-441; right triangle definition, 102
Sines, law of, 141
Slope, 289
Slope
    of parallel lines, 299; of perpendicular lines, 300
Slope-intercept equation, 294
Solving triangles, 101
Standard form
    of circle, 461; of ellipse, 464, 465, 467; of hyperbola, 471, 473, 475; of parabola, 477, 478, 480
Standard position of angle, 86
Subscript, 17
Subtraction
    of algebraic expressions, 19; of complex numbers, 257; of fractions, 208; of functions, 281; of radicals, 243; of real numbers, 14
Supply and demand, 308
Symmetry, axis of, 319, 478
Synthetic division, 339
System of equations
    addition-subtraction method, 306; dependent, 304; determinant method, 311-316; determinant method, three variables, 314-316; determinant method, two variables, 312-313; inconsistent, 304; solution of, 303; substitution method, 305

Tangent function
    definition of, 87, 406;
    graph of, 431; inverse, 441; right triangle definition, 102

Terms, 19
Terminal ray, 85
Toffler, Alvin, 288
Transverse axis, 471
Trigonometric form of complex number, 259-260
Trigonometric functions
    of angles, definitions of, 87; as line segments, 412-413; of real numbers, definitions of, 405-406; right triangle definition, 102; signs of, 90, 414
Trigonometric identities, 408-409, 443-454
    definition of, 408, 443; for negative numbers, 408-409; summary of, 499-500
Trigonometric tables
    for $30°$, $45°$, $60°$, 121; for $0°$, $90°$, $180°$, $270°$, 124; for 0, $\pi/2$, π, $3\pi/2$, 407; for $\pi/6$, $\pi/4$, $\pi/3$, 411

Value of function, 60, 277
Variable
    definition of, 19; dependent, 48, 268; independent, 48, 268
Variation
    constant of, 55; direct, 55; inverse, 56; joint, 57-58
Vector
    components of, 116; definition of, 113; resolving, 116
Vertex
    of angle, 85; of parabola, 319, 478
Vertical asymptote, 349
Vertical line test, 71, 273
Vertices
    of ellipse, 465; of hyperbola, 471

Zero
    division by, 5; exponent, 230; of a function, 343; fundamental property of, 10; multiple, 346; real, bounds of, 346